Lecture Notes in Computer Science 1522

Edited by G. Goos, J. Hartmanis and J. van Leeuwen

T0216965

Springer

Berlin
Heidelberg
New York
Barcelona
Hong Kong
London
Milan
Paris
Singapore
Tokyo

Ganesh Gopalakrishnan   Phillip Windley (Eds.)

Formal Methods in Computer-Aided Design

Second International Conference, FMCAD '98
Palo Alto, CA, USA, November 4-6, 1998
Proceedings

Springer

Series Editors

Gerhard Goos, Karlsruhe University, Germany
Juris Hartmanis, Cornell University, NY, USA
Jan van Leeuwen, Utrecht University, The Netherlands

Volume Editors

Ganesh Gopalakrishnan
University of Utah, Department of Computer Science
50 S Central Campus, Salt Lake City, UT 841112-9205, USA
E-mail: ganesh@cs.utah.edu

Phillip Windley
Brigham Young University, Department of Computer Science
3361 TMCB, Provo, UT 84602-6576, USA
E-mail: windley@cs.byu.edu

Cataloging-in-Publication data applied for

Die Deutsche Bibliothek - CIP-Einheitsaufnahme

Formal methods in computer aided design : second international
conference ; proceedings / FMCAD '98, Palo Alto, CA, USA,
November 4 - 6, 1998. Ganesh Gopalakrishnan ; Phillip Windley
(ed.). - Berlin ; Heidelberg ; New York ; Barcelona ; Budapest ; Hong
Kong ; London ; Milan ; Paris ; Singapore ; Tokyo : Springer, 1998
    (Lecture notes in computer science ; Vol. 1522)
    ISBN 3-540-65191-8

CR Subject Classification (1998): B.1.2, B.1.4, B.2.2-3, B.6.2-3, B.7.2-3,
F.3.1, F.4.1, I.2.3, D.2.4, J.6

ISSN 0302-9743
ISBN 3-540-65191-8 Springer-Verlag Berlin Heidelberg New York

Typesetting: Camera-ready by author
SPIN 10692817   06/3142 – 5 4 3 2 1 0   Printed on acid-free paper

Preface

This volume contains the proceedings of the Second International Conference on *Formal Methods in Computer-Aided Design (FMCAD'98)*, organized November 4-6, in Palo Alto, California, USA. The first event of this series was organized by Mandayam Srivas and Albert Camilleri in 1996 in Palo Alto. FMCAD, which evolved from the series *Theorem Provers in Circuit Design (TPCD)*, strives to be a premier forum for disseminating research in Formal Verification (FV) methods for digital circuits and systems, including processors, custom VLSI circuits, microcode, and reactive software. In addition to significant case-studies and verification approaches, FMCAD also endeavors to represent advances in the driving technologies for verification, including binary decision diagrams, model checking, symbolic reasoning (theorem proving), symbolic simulation, and abstraction methods.

The conference included four invited lectures. The invited lectures were given by Kenneth McMillan (Cadence Berkeley Labs) on *Minimalist proof assistants: interactions of technology and methodology in formal system level verification*, by Carl-Johan Seger on *Formal methods in CAD from an industrial perspective*, by Randal E. Bryant and Bwolen Yang on *A performance study of BDD-based model checking*, and by Amir Pnueli on *Verification of data-insensitive circuits: an in-order-retirement case study*. Of the 55 regular paper submissions received, 27 were selected by the technical program committee for presentation at the conference. All four tools papers received were also selected.

We gratefully acknowledge the services of the technical program committee of FMCAD'98, which consisted of Adnan Aziz (Univ. of Texas at Austin, USA), Alan Hu (Univ. of British Columbia, Canada), Albert Camilleri (Hewlett-Packard, USA), Carl Pixley (Motorola, USA), Carlos Delgado Kloos (Univ. Carlos III de Madrid, Spain), Ching-Tsun Chou (Intel, USA), Eduard Cerny (Univ. of Montreal, Canada), Francisco Corella (Hewlett-Packard, USA), Jens (Stanford University, USA), Jerry Burch (Cadence Labs, USA), John van Tassel (Texas Instruments, USA), Limor Fix (Intel, Israel), Mandayam Srivas (SRI International, USA), Mark Aagaard (Intel, USA), Mary Sheeran (Chalmers University, Sweden), Masahiro Fujita (Fujitsu, USA), Ramin Hojati (HDAC, and UC Berkeley, USA), Randy Bryant (Carnegie-Mellon, USA), Ranga Vemuri (Univ. of Cincinnati, USA), Shiu-kai Chin (Syracuse Univ., USA), Steven German (IBM, USA), Steven Johnson (Indiana Univ., USA), Thomas Kropf (Univ. Karlsruhe, Germany), Tim Leonard (Compaq, USA), Tom Henzinger (UC Berkeley, USA), Tom Melham (Univ. of Glasgow, UK), Tom Shiple (Synopsys, USA), and Warren Hunt (IBM, USA).

The following researchers also helped in the evaluation of the submissions, and we are grateful for their efforts: Abdel Mokkedem, Mike Jones, and Rajnish Ghughal (University of Utah), Rob Shaw (Hewlett-Packard), Armin Biere, Bwolen Yang, and Yirng-An Chen (CMU), Andres Marin Lopez, Franz Josef

Stewing, and Peter T. Breuer (Univ. Carlos III, Madrid), Abdelkader Dekdouk, E. Mostapha Aboulhamid, and Otmane AIT MOHAMED (Univ. of Montreal, Canada), Chuck Yount, Marten van Hulst, and John Mark Bouler (Intel), Koichiro Takayama and Vamsi Boppana (Fujitsu), Orna Kupferman, Luca de Alfaro, Sriram K. Rajamani, and Shaz Qadeer (Berkeley), Jun Sawada (U. of Texas), Howard Wong-Toi (Cadence), Supratik Chakraborty, Clark Barrett, and Jeffrey Su (Stanford), Michaela Huhn, Ralf Reetz, Klaus Schneider, and Jürgen Ruf (Karlsruhe), Justin Chien and Jun Yuan (Compaq), Nazanin Mansouri, Naren Narasimhan, Elena Teica, and Rajesh Radhakrishnan (Univ. of Cincinnati). We also thank Ratan Nalumasu, PhD student at the Department of Computer Science, University of Utah, for helping us with the tool demo session in his capacity as the *Tools Chair* of FMCAD'98.

We thank Judith Burgess of SRI International, Menlo Park, CA, for her help and advice in organizing FMCAD'98. We gratefully acknowledge the services of Conferences and Institutes, University of Utah, notably of Jacqueline Brakey, Cathy Cunningham, and Linda Williams, for their work on registration, publicity, and conference facilities. We also gratefully acknowledge the services of the Springer-Verlag LNCS department, especially Alfred Hofmann and Anna Kramer, for their prompt help and communication. We thank the IFIP Working Group 10.5 for granting us the *in co-operation* status.

Last, *but not least*, FMCAD'98 has received financial support from Hewlett-Packard Company, Intel, Synopsys Inc., and Cadence Berkeley Labs. We thank all sponsors for their generosity.

Salt Lake City, UT　　　　　　　　　　　　　Ganesh C. Gopalakrishnan
Provo, UT　　　　　　　　　　　　　　　　　Phillip J. Windley

August 1998

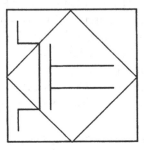

Table of Contents

Minimalist Proof Assistants: Interactions of Technology and Methodology in Formal System Level Verification

K. L. McMillan

Cadence Berkeley Labs
2001 Addison St., 3rd floor
Berkeley, CA 94704-1103
mcmillan@cadence.com

Abstract. The complexity of systems implemented in hardware continues to grow, and with it the the need for formal, automated, system level verification. Unfortunately, though automatic formal verification technologies continue to improve incrementally in terms of the size and complexity of systems they can handle, there is a widening gap between real designs and designs that can be verified automatically.

I will argue that proofs and proof assistants in some form, in combination with automated methods, are necessary to close this gap. However, the considerations that drive the design of a proof assistant for hardware verification and not necessarily those that have shaped existing general-purpose proof assistants. In particular, for a hardware proof assistant, the requirements in terms of logical expressiveness and the power of its deductive machinery are minimal. For example, the ability to reason about higher-order objects like sets and functions is probably superfluous in the hardware domain.

Rather, the primary consideration in constructing proofs of complex systems is that proofs be concise and maintainable. This means that a proof system must take maximum advantage of the strengths of model checking and automated decision procedures in order to minimize the need for manual decomposition of proofs. It is thus important to consider how inference rules and decision procedures (e.g., model checking) interact to allow concise proof decompositions in a particular domain of application. As an example, I will show how model checking combined with a few simple but domain-tailored inference rules allows surprisingly concise proofs about out-of-order instruction processors. This is chiefly because basing the proof on model checking eliminates the need to state and prove global invariants.

Along the way, I will also discuss some practical considerations for the design of large, formally verified hardware systems. In particular, the most concise proof decompositions for hardware systems are often non-hierarchical. Rather, proofs often decompose most naturally according to the paths followed by data and control through the system under various conditions, rather than according to structural hierarchy. Further, design for compositional verification differs strongly from the paradigm of design-by-debugging that is currently prevalent. The debugging approach leads to complex (and often unknown) interactions between design components, whereas the formal approach favors the design of "bullet-proof" components, that implement a given abstract model without any assumptions about environment behavior.

Reducing Manual Abstraction in Formal Verification of Out-of-Order Execution

Robert B. Jones[1,2], Jens U. Skakkebæk[1], and David L. Dill[1]

[1] Computer Systems Laboratory, Stanford University, Stanford, CA 94305, USA
{jus,dill}@cs.stanford.edu
[2] Strategic CAD Labs, Intel, JFT-104, 2111 NE 25th Ave., Hillsboro, OR 97124, USA
rjones@ichips.intel.com

Abstract. Several methods have recently been proposed for verifying processors with out-of-order execution. These methods use intermediate abstractions to decompose the verification process into smaller steps. Unfortunately, the process of manually creating intermediate abstractions is very laborious. We present an approach that dramatically reduces the need for an intermediate abstraction, so that only the scheduling logic of the implementation is abstracted. After the abstraction, we apply an enhanced *incremental-flushing* approach to verify the remaining circuitry by comparing the processor description against itself in a slightly simpler configuration. By induction, we demonstrate that any reachable configuration is equivalent to the simplest possible configuration. Finally, we prove correctness on the simplest configuration. The approach is illustrated with a simple example of an out-of-order execution core.

1 Introduction

Several techniques for formally verifying out-of-order microprocessor designs using theorem proving have recently been suggested [4, 10–12]. These techniques all use some form of intermediate abstraction to bridge the gap in abstraction level between the implementation and the specification, as defined by an *instruction-set architecture* (ISA).

Creating such intermediate abstractions manually and then showing the correspondence between the implementation and the intermediate abstraction is laborious, even for high-level models. Omitting the intermediate abstraction and manually developing the abstraction relation between the implementation and the ISA is even harder. First, the extended instruction parallelism in out-of-order architectures results in many complex interactions between executing instructions. This greater complexity makes it very difficult to devise an abstraction function. Second, large (≥ 40 element) buffers are used to record and maintain the program order of instructions.

Burch and Dill have devised an approach for pipelined microarchitectures that automatically generates the abstraction function by *flushing* the implementation state [3]. The technique has been extended to dual-issue and super-scalar architectures [2, 8, 13]. However, these techniques do not work for out-of-order architectures in practice because the number of cycles required to empty the buffer completely is so large. The logical formulas are too complex to manipulate in proofs and often too complex even to construct.

We have previously proposed *incremental flushing*, an extension to the Burch and Dill flushing approach that inductively empties the buffer in smaller proof steps [12]. We have applied it to automate part of the verification process of out-of-order designs. The approach requires, however, that the out-of-order core is abstracted into an in-order version. In this paper, we extend the incremental-flushing approach to directly reason about the out-of-order core also. This avoids the need for the in-order abstraction of our earlier approach. The implementation abstraction that is still required is comparatively minimal, and the automated incremental flushing approach can cover a much larger portion of the original design. This automates the generation of the abstraction function and significantly reduces the manual effort required.

The extended technique only requires that the internal scheduling logic of the processor be manually abstracted. An instruction is processed through a number of internal *steps*, which each may take several cycles. The scheduling logic affected determines which buffer entries, datapath resources, and busses different instructions and steps are assigned to. We apply induction to show that the implementation executing any number of instructions (up to the maximum allowed) is functionally equivalent with the same implementation executing only one instruction at a time. We finally complete the verification by checking the implementation with one instruction against the ISA. This proof is much simpler, since the bypass and buffering logic can be simplified away in the proofs. Note that to make the induction work, it must be possible to stall each stage of the out-of-order pipeline independently.

We use the same simple model of an out-of-order execution core to illustrate our approach that we used previously [12]. Although this example is not representative of industrial-scale designs, it captures essential features of out-of-order architectures: large queuing buffers, resource allocation within the buffers, and data-path scheduling of execution resources. We have discharged the proof obligations for the simple example using the Stanford Validity Checker (SVC).

2 Related Work

Sawada and Hunt's theorem-proving approach uses a table of history variables, called a *micro-architectural execution trace table* (MAETT) [10, 11]. The MAETT is an intermediate abstraction that contains selected parts of the implementation as well as extra history variables and variables holding abstracted values. It includes the ISA state and the ISA transition function. A predicate relating the implementation and MAETT is found by manual inspection and proven by induction to be an invariant on the execution of the implementation. In our approach, we do not need an intermediate abstraction of the circuit, only the scheduling logic is abstracted. We then use an incremental flushing technique to automatically generate the abstraction function, reducing the manual work required to relate the intermediate abstraction to the ISA.

Damm and Pnueli generalize an ISA specification to a non-deterministic abstraction [4]. They verify that the implementation satisfies the abstraction by manually establishing and proving the appropriate invariants. They have applied their technique to the Tomasulo algorithm [5], which has out-of-order instruction completion. In contrast, our out-of-order model features in-order retirement and the corresponding large buffers that are required. Damm and Pnueli's abstraction non-deterministically represents all possible instruction sequences which observe dataflow dependencies. Our non-deterministic scheduler abstraction also observes dataflow dependencies, but is additionally constrained by allowable resource allocations (e.g., buffer entries) in the implementation. Applying their method to architectures with in-order retirement would require manual proof by induction that the intermediate abstraction satisfies the ISA. We automate the proof obligations with incremental flushing.

Hosabettu et al. use a technique for decomposing the abstraction function and have applied it to the example of Sawada and Hunt with out-of-order retirement [7]. Although this aids in finding an appropriate abstraction function, manual intervention is needed in its construction.

Henzinger et al. use Tomasulo's algorithm to illustrate a method for manually decomposing the proof of correctness [6]. They manually provide abstract modules for parts of the implementation. These modules correspond to implementation internal steps. Similar to our approach, the abstractions are invariants on the implementation and are extended with auxiliary variables. Again, our new approach automates much of the abstraction process.

McMillan model checks the Tomasulo algorithm by manually decomposing the proof into smaller correctness proofs of the internal steps [9]. He also uses a reduction technique based on symmetry to extend the proof to a large number of execution units. Berezin et al. abstract the data path by introducing a data structure called a reference table. Each entry in the reference table corresponds to an uninterpreted term representing computation results of instructions [1]. They have applied their technique to Tomasulo's algorithm. However, the size of the state space grows exponentially with the number of concurrent instructions. Designs with in-order retirement contain a large reorder buffer and can contain many instructions executing simultaneously. In contrast to both automated model-checking approaches, our theorem-proving based method generalizes to arbitrary buffer sizes.

3 Preliminaries

The desired behavior of a processor is defined by an *instruction-set architecture* (ISA). The ISA represents the programmer-level view of a machine that executes instructions sequentially. The ISA for our example is shown in Figure 1a. The ISA state consists of a register file (RF), while the next-state function is computed with a generic execution unit (EU) that can execute any instruction. The ISA

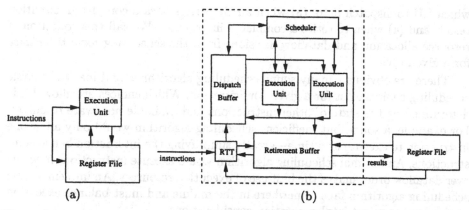

Fig. 1. (a) The simple ISA model. (b) Instruction flow in our out-of-order execution core IMPL.

also accepts a `bubble` input that leaves the state unchanged. Note that our ISA model does not include a program counter or memory state—as these are also omitted from our simplified out-of-order model.

Modern processors implement the ISA more aggressively. In out-of-order architectures, instructions are fetched, decoded, and sent to the execution core in program order. Internally, however, the core executes instructions out-of-order, as allowed by data dependencies. This allows independent instructions to execute concurrently. Finally, instruction results are written back to architecturally-visible state (the register file) in the order they were issued.

Consider our example out-of-order execution core (IMPL) shown in Figure 1b. The architectural register file (RF) contains the current state of the ISA-defined architectural registers. An instruction is processed in a number of *steps*, which may each last a number of cycles: When an instruction is *issued*, new entries are allocated in both the dispatch and retirement buffers, and the register translation table (RTT) entry for the logical register corresponding to the instruction destination is updated. The RTT is used to locate the instruction's source data. Instructions are *dispatched*, possibly out-of-order, from the dispatch buffer (DB) to individual execution units when their operands are ready and an execution unit is available. When an instruction finishes execution, the result is *written back* to the retirement buffer (RB). This data is also bypassed into the DB for instructions awaiting that particular result. Finally, the RB logic must ensure that instruction results are *retired* (committed to architectural state) in the original program order. When an RB entry is retired, the RTT is informed so that the logical register entry corresponding to the instruction's destination can be updated if necessary. IMPL also accepts a special `bubble` flushing input in place of an instruction. Intuitively, a `bubble` is similar to a `NOP` instruction but does not affect any state or consume any resources after being issued.

Figure 1b also shows the *scheduling logic*, which handles the allocation of hardware resources and instruction flow. Scheduling must determine (1) which slot in the DB to allocate at issue, (2) when to dispatch a ready instruction and which EU to dispatch it to, (3) when an EU writes back a completed execution result, and (4) when to retire a completed instruction. We call this collection of resource allocation and dataflow decisions from the scheduling logic the *choice* for a given cycle.

There are obviously many sound scheduling algorithms, and many allowable scheduling choices exist for a given configuration. Which choices are allowable is determined by the state of other instructions and available hardware resources. For example, a sound but inefficient scheduling algorithm would only allow one instruction to execute at a time—greatly simplifying the interaction between instructions. An optimal scheduling algorithm would execute instructions in whatever dataflow order makes the best use of execution resources. An implementable scheduling algorithm falls somewhere in the middle and must balance execution performance against implementation considerations.

We have made significant simplifying assumptions in our processor model: instructions have only one source operand, and only one issue and one retire can occur each cycle. We also omit a "front-end" with fetch, decode, and branch prediction logic. Omitting these features allowed our efforts to focus on the features which make the out-of-order verification problem difficult: the out-of-order execution and the large effective depth of the pipeline. The verification discussed in this paper uses a model with unbounded buffers.

4   The Approach

As in [12], the goal of our approach is to prove that the out-of-order implementation IMPL (as described by an HDL model) satisfies the ISA model. We define δ_i to be the implementation next-state function, which takes an initial state q_i and an input instruction i and returns a new state q_i', e.g., $q_i' = \delta_i(q_i, i)$. We extend δ_i in the obvious way to operate over input sequences $w = i_0 \ldots i_n$. We define δ_s similarly for ISA.

Let σ be a *size* function that returns the number of currently executing instructions, i.e., those that have been issued but not retired. We require that $\sigma(q_i^\circ) = 0$ for an initial implementation state q_i°. We define an instruction sequence w to be *completed* iff $\sigma(\delta_i(q_i^\circ, w)) = 0$, i.e., all instructions have been retired after executing w. We use the projection function $\pi_{\mathrm{RF}}(q_i)$ to denote the register file contents in state q_i – which we define as the specification state. For clarity in presentation, we define $q_{i1} \stackrel{RF}{=} q_{i2}$ to be $\pi_{\mathrm{RF}}(q_{i1}) = \pi_{\mathrm{RF}}(q_{i2})$, and we will sometimes use $\stackrel{RF}{=}$ when the projection π_{RF} is redundant on one side of the equality.

The overall correctness property for IMPL with respect to ISA is expressed formally as:

Correctness *For every completed instruction sequence w and initial state q_i°,*

$$\delta_i(q_i^\circ, w) \stackrel{RF}{=} \delta_s(\pi_{RF}(q_i^\circ), w).$$

That is, the architecturally visible state in IMPL and ISA is identical after executing any instruction sequence that retires all outstanding instructions in the implementation.

Our approach has three steps. First, we locate and abstract the IMPL scheduling logic and prove the abstraction correct. We refer to the abstracted implementation as SAI (scheduler-abstracted implementation). In the second step, we use incremental flushing to show that SAI with an abstracted scheduler calculates the same results as if the instructions were executed one at a time. Note that while the functional results should be identical, the timing of the results will of course be different. This proves the correctness of the reordering control logic. Finally, we show that SAI with an abstracted scheduler executing one instruction at a time satisfies the ISA.

5 First Step: Abstracting the Scheduling Logic

We first identify the scheduling logic in the design and its interface to the rest of the circuit. We wish to replace the original scheduling logic with the most general scheduling algorithm that still provides legal choices to the rest of the circuit. For example, the abstracted scheduling logic for our simple example will (1) issue an instruction to any empty slot in the DB, (2) dispatch an instruction to any available execution unit, (3) write back results from any execution unit that has finished executing, and (4) retire any instruction with result data. In a given state, the abstracted scheduling logic in SAI non-deterministically chooses an allocation based on the current state of the SAI. The non-determinism is implemented as an extra, unconstrained input.

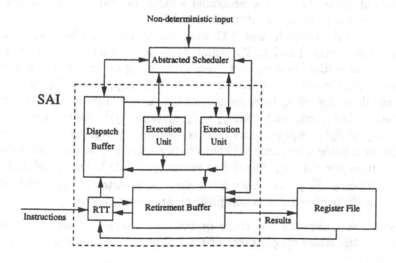

Fig. 2. Instruction flow in SAI with the abstracted resource allocator.

The SAI with an abstracted scheduler is illustrated in Figure 2. The abstract scheduler monitors the state of SAI and provides SAI with a scheduling choice for every instruction input. Naturally, we want the abstracted scheduler to make *legal* choices that only allocate free resources and advance only ready instructions from one stage to the next. For example, only instructions that have completed executing may be written back and retired. Identifying and abstracting the scheduling logic in a realistic design requires a detailed understanding of the circuit and may be error-prone. Fortunately, soundness of our approach is not compromised by a bad selection of abstracted scheduler. The later proof steps will fail if the abstracted scheduler in SAI is either incorrect or too general to verify its behavior against ISA. Note, however, that we do not require the scheduler to be centralized. The technique is equally applicable to a distributed scheduler, where each part of the scheduler is appropriately abstracted.

We first show that the abstract scheduler is sufficiently general to capture all the possible choice outputs that the implementation scheduler makes. We then extend this result with a composition argument to show that SAI with the abstracted scheduler is an appropriate abstraction of IMPL. Let S_i be the transition function of the implementation scheduler and let S_a be the transition function of the abstract scheduler. S_a takes an extra, non-deterministic input i_{nd}. We must show that for each step that S_i makes, there exists an S_a step such that the choice outputs are identical:

Proof Obligation 1 (Scheduler Abstraction Correctness) *For every reachable state q_i of IMPL and for every input i, there exists an input i_{nd} such that*

$$out(S_i(q_i, i)) = out(S_a(q_a, i, i_{nd})).$$

One way of instantiating the abstract scheduler for this proof is to use an oracle which observes the original scheduler's behavior and knows how the non-deterministic input affects the abstract scheduler.

Next, we must establish that SAI with the abstracted scheduler is an appropriate abstraction of IMPL. We define δ_a to be the SAI next-state function, which takes an initial state q_a and a pair consisting of an input instruction i and scheduler choice ch and returns a new state q'_a, e.g., $q'_a = \delta_a(q_a, \langle i, ch \rangle)$. We extend the definition of δ_a to sequences of instruction inputs w and choice sequence $w_{ch} = ch_0 \ldots ch_n$ such that $q'_a = \delta_a(q_a, \langle w, w_{ch} \rangle)$[1]. We say that a choice sequence w_{ch} is $S_a(q_a, w)$-generated, if it is obtained by stimulating the abstract scheduler to provide a sequence of choices corresponding to the instruction sequence w from the state q_a. We define states q_i of IMPL and q_a of SAI to be *consistent* when $q_i \stackrel{RF}{=} q_a$, i.e., they have identical architecturally visible states. Using Proof Obligation 1 and a composition argument, we can prove that:

IMPL-SAI Refinement *For every instruction sequence w and every pair of consistent initial states q_i°, q_a°, there exists a $S_i(q_a^\circ, w)$-generated choice sequence*

---

[1] The pair of sequences $\langle w, w_{ch} \rangle$ is easily derived from the corresponding sequence of pairs $\langle i_0, ch_0 \rangle, \ldots, \langle i_n, ch_n \rangle$.

w_{ch} such that

$$\delta_i(q_i^\circ, w) \stackrel{RF}{=} \delta_a(q_a^\circ, \langle w, w_{ch} \rangle).$$

We prove this by providing the following witness. By induction, we extend Proof Obligation 1 to work on sequences of inputs and obtain a $\mathcal{S}_a(q_a^\circ, w)$-generated sequence w_{ch} that is equal to the sequence that is output from the implementation scheduler. Since SAI was obtained from IMPL by abstracting only the resource allocation logic, the property follows trivially.

Note that this proof requires reachability invariants for IMPL and SAI. Finding the reachability invariant for IMPL is necessary for any inductive method, and is not unique to our approach. Finding the reachability invariant for SAI is straightforward, because of the minimal changes from IMPL.

6 Second Step: Functional Equivalence of SAI and ISA

The second step in the verification is to prove that SAI with the abstract scheduler satisfies ISA. Formally:

SAI-ISA Equivalence *For every completed instruction sequence w, initial SAI state q_a°, and $\mathcal{S}_a(q_a^\circ, w)$-generated sequence of choices w_{ch}:*

$$\delta_a(q_a^\circ, \langle w, w_{ch} \rangle) \stackrel{RF}{=} \delta_s(\pi_{RF}(q_a^\circ), w).$$

Recall that the Burch-Dill abstraction function *flushes* an implementation (by inserting bubbles) for the number of clock cycles necessary to completely expose the internal state. In the case of a simple five-stage pipeline, only five steps are required to complete the partially executed instructions. Following this approach with our model would compare a potentially full RB with the ISA model. The Burch-Dill flushing technique would unroll SAI to the depth of the RB, resulting in a logical expression too large for the decision procedure to check.

We extend the *incremental-flushing* approach presented in [12] to overcome this problem. Rather than flushing the entire pipeline directly, a set of smaller, inductive flushing steps is performed. Taken together, these proof obligations imply the full, monolithic flushing operation. To illustrate the approach, consider the graphical presentation of two different *executions* (state sequences) of SAI in Figure 3. We define the *execution* of a system as the sequence of states that the system passes through when executing a given input sequence. For instance, the execution indicated in Figure 3a is a result of executing the instruction sequence:

$$i_1, i_2, \texttt{bubble}, \texttt{bubble}, i_3, \texttt{bubble}, i_4, i_5, \texttt{bubble}, \texttt{bubble}, i_6, \texttt{bubble}, \texttt{bubble}.$$

with some choice sequence that appropriately allocates the resources so that all instructions have retired in the final state state. Apart from self-loops indicating internal execution, edges are only traversed when instructions are issued or retired.

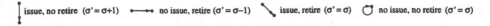

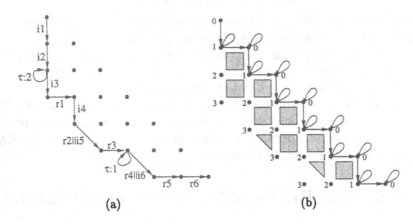

(a)   (b)

Fig. 3. (a) A *Max-n* execution ε_n. Labels i*n* and r*n* denote the issue and retirement of instruction number *n*. The label r*n*‖i*n* denotes simultaneous issue and retire. $\tau : n$ is a shorthand for *n* cycles where in each cycle, bubbles are issued and nothing is retired. (b) An equivalent *Max-1* execution ε_1. The squares indicate the distance between ε_n and ε_1.

We use $\varepsilon(q_a, \langle w, w_{ch} \rangle)$ to denote the execution resulting from the application of δ_a to a state q_a and the input sequence pair $\langle w, w_{ch} \rangle$. We define $last(\varepsilon(q_a, \langle w, w_{ch} \rangle))$ as the last state of the execution. Note that, by definition $last(\varepsilon(q_a, \langle w, w_{ch} \rangle)) = \delta_a(q_a, \langle w, w_{ch} \rangle)$. Each state in an execution is associated with the number of active instructions—defined earlier as the *size* function σ. This is illustrated in Figure 3b. We call an execution which contains states with most size *n* a *Max-n* execution (denoted ε_n). Accordingly, completely serialized executions with at most one outstanding element are *Max-1* executions (denoted ε_1). An example of a *Max-1* execution corresponding to the execution above could be

$$i_1, \mathtt{bubble}^4, i_2, \mathtt{bubble}^4, i_3, \mathtt{bubble}^4, i_4, \mathtt{bubble}^4, i_5, \mathtt{bubble}^4, i_6, \mathtt{bubble}^4.$$

where $\mathtt{bubble}^4 = \mathtt{bubble}, \mathtt{bubble}, \mathtt{bubble}, \mathtt{bubble}$. The execution is illustrated in Figure 3b.

The first step of the SAI-ISA verification establishes that:

Incremental-Flushing Induction Step *For every initial state q_a°, and for every Max-n execution $\varepsilon_n(q_a, \langle w, w_{ch} \rangle)$, there exists $\langle w^1, w_{ch}^1 \rangle$ derived from input pair $\varepsilon_n(q_a, \langle w, w_{ch} \rangle)$ and a corresponding Max-1 execution $\varepsilon_1(q_a, \langle w^1, w_{ch}^1 \rangle)$ such that:*

$$last(\varepsilon_n(q_a^\circ, \langle w, w_{ch} \rangle)) = last(\varepsilon_1(q_a^\circ, \langle w^1, w_{ch}^1 \rangle)).$$

A *Max-1* execution is derived from a *Max-n* execution by "stretching" the w and w_{ch} sequences with the appropriate $\mathtt{bubbles}$ and stalling choices, respectively, to stall the relevant parts of the out-of-order core. The intuition behind this approach is that the final results of *Max-n* and *Max-1* executions should be identical—because bubbles and stalling choices should not affect functional behavior. Clearly, if enough bubbles are inserted between subsequent instructions only one instruction will be in the pipeline at a time. In this situation it is computationally manageable to compare SAI with ISA, since the bypass control logic can be discarded in the proof. Section 6.1 details the proof obligations for this step and describes how we proved this property on our example.

The second SAI-ISA verification step shows that all *Max-1* executions produce the same result as the ISA model.

Incremental Flushing ISA Step *For every initial state q_a°, and every* Max-1 *execution ε_1 corresponding to an instruction sequence w^1 and every $S_i(q_a^\circ, w^1)$-generated choice sequence w_{ch}^1:*

$$last(\varepsilon_1(q_a^\circ, \langle w^1, w_{ch}^1 \rangle)) \overset{RF}{=} \delta_s(\pi_{RF}(q_a^\circ), w).$$

Proving this is much simpler than the original problem of directly proving SAI-ISA equivalence, since only one instruction is in the machine at any given time (because of the stretching bubbles and stalling choices). The proof is carried out by induction on the length of instruction sequences, as described in Section 6.2.

6.1 Inductive Step

The incremental flushing proof step can be split up into three proof obligations. First, we identify the maximum number of cycles required to symbolically simulate the implementation in order to ensure that at least one instruction is retired. This is used to prove termination of the induction proof. Let δ_a^n denote n cycles of symbolic execution. Formally, we must prove that:

Proof Obligation 2 (Retirement Upper-Bound) *There exists an upper bound u, such that for every reachable state q_a such that $\sigma(q_a) \geq 1$ and input sequence pair $\langle w, w_{ch} \rangle$, at least one active instruction from q_a will be retired between q_a and $\delta_a^u(q_a, \langle w, w_{ch} \rangle)$.*

That is, we make a progress assumption that the implementation retires an instruction within u cycles. We derive u by a *worst-case* analysis and determine the longest path that an issued instruction could potentially follow before being retired.

The upper bound u is assumed in the main induction. As we shall see, the induction case is used to inductively move the last issued instruction to the end of the execution sequence. In each application, independently executing instruction steps are reordered. This reordering is performed by moving the instruction till after the steps of the previously issued instructions.

In each application of the induction case, a subsequence is selected out of the execution such that an instruction i is issued in the first cycle of the subsequence. We denote the length of the subsequence by v, and will choose it to be $\geq u$. The length of the subsequence is doubled in the application of the induction case:

the v choices are split up in a way that the first v steps allow SAI to perform all steps that are not dependent on i. The steps related to i are then *replayed* in the remaining cycles. As a consequence, the freshly-issued instruction i and its steps are delayed by v cycles.

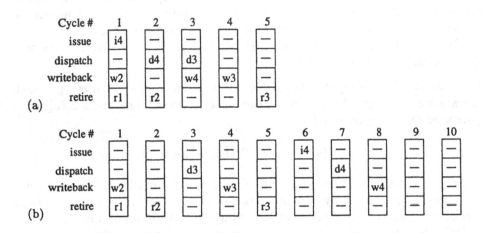

Fig. 4. (a) A choice sequence w_{ch}. (b) A stretched version w_{ch}' of the original choice sequence w_{ch}.

To illustrate, consider the scheduling sequence w_{ch} shown in Figure 4a. Each vertical box corresponds to a choice and the labels in, dn, wn, and rn respectively denote which dispatch buffer entry to store an issued instruction in, which dispatch buffer entry to dispatch, which completed instruction to write back, and whether or not to allow retirement of an instruction ready for retirement. Each number identifies a particular instruction n. For instance, the first choice retires instruction 1, writes back instruction 2, and issues instruction 4. A choice field which keeps a particular resource allocation unchanged is denoted with "—".

A scheduling sequence w_{ch}' is constructed by adding bubbles and stalling choices to w_{ch} (Figure 4b). Observe that the ordering of the issue, dispatch, writeback, and retirement choices for a given instruction are maintained. The only difference is the delayed issue of instruction 4 and its subsequent dispatch and writeback. On a per-instruction basis, the resources in w_{ch}' and w_{ch} must be the same and occur in the same order. This crucial requirement guarantees that the resulting partially-executed state is the same in both cases and facilitates an inductive proof over SAI state.

In the induction case, the length of the subsequence, v, must be chosen so that it is at least u cycles and long enough to make sure that the instruction can properly be moved passed the steps of other instructions. In our example, v must be at least double the maximum execution time in an execution unit, i.e., which in total is less than $2u$ (from Proof Obligation 2 we know that the time that any instruction spends in the execution unit is less than u). By doing this, we are able to delay the instruction sufficiently far to avoid resource contention when reordering.

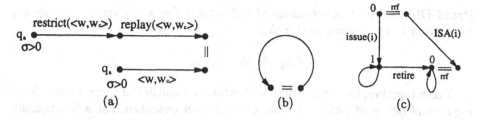

Fig. 5. (a) Illustration of Proof Obligation 3; the nodes are labelled with their sizes. (b) Illustration of Proof Obligation 4. We must prove that self loops return to the same state. (c) Illustration of Proof Obligation 5, the ISA induction step.

Given an input-sequence pair $\langle w, w_{ch} \rangle$, define $restrict_i(\langle w, w_{ch} \rangle)$ to be the projection of all elements of $\langle w, w_{ch} \rangle$ not depending on i. Similarly, we define $replay_i(\langle w, w_{ch} \rangle)$ to denote the projection of the elements of $\langle w, w_{ch} \rangle$ that depend on i. The proof obligation is then:

Proof Obligation 3 (Incremental Step) *For every reachable state q_a such that $\sigma(q_a) \geq 1$, and for every input-sequence pair $\langle w, w_{ch} \rangle$ such that the first element of w is a non-bubble instruction and w_{ch} is $S_i(q_a, w)$-generated:*

$$\delta_a^v(q_a, \langle w, w_{ch} \rangle) = \delta_a^{2v}(q_a, \langle w', w_{ch}' \rangle).$$

where $\langle w', w_{ch}' \rangle$ is the concatenation of $restrict_i(\langle w, w_{ch} \rangle)$ *and* $replay_i(\langle w, w_{ch} \rangle)$.

In other words, we must show that the stretched sequence results in the same state as the original sequence. The proof obligation is illustrated in Figure 5a.

As we shall see below, in the proof of Proof Obligation 3 it is sufficient to consider the cases where the necessary resource is available so that the instruction being moved can be scheduled appropriately and avoid resource contention. This weakening assumption can be added to the proof obligation.

Note that Proof Obligation 3 requires also that internal registers with auxiliary values to agree on the resulting states. To illustrate, the replayed instructions in our model may get their source operands from the RF rather than the RB. The fields in the dispatch buffer indicating the physical sources of the operands at issue may differ and should be set to some reset value after use.

Also observe that in each application of the induction step, more than one instruction may retire within the v steps. Naturally, the worst-case upper bound u (number of cycles before an instruction is guaranteed to retire) and therefore v may be quite large in some designs due to execution units with long latencies. This could result in symbolic expressions that are too large to check. In these cases, the execution units and associated arbitration logic must be abstracted separately.

The final proof obligation states that `bubble` inputs with stalling choices do not change SAI state (illustrated in Figure 5b):

Proof Obligation 4 (Correctness of Self-Loops) *For every reachable state q_a, instruction i, and stalling choice ch_{st}:*

$$\delta_a(q_a, \langle i, ch_{st} \rangle) = q_a.$$

Taken together, these three proof obligations establish the *Incremental Flushing* step of our verification, i.e. that every *Max-n* execution has a functionally equivalent *Max-1* execution. We next provide a brief sketch of the proof.

Proof Sketch:

We assume the three Proof Obligations shown above and must show that for every *Max-n* execution ε_n there exists a corresponding *Max-1* execution ε_1 such that

$$\varepsilon_n(q_a^\circ, \langle w, w_{ch} \rangle) \overset{\text{last}}{=} \varepsilon_1(q_a^\circ, \langle w^1, w_{ch}^1 \rangle).$$

We prove this by complete induction on the "distance" between the non-diagonal *Max-n* execution ε_n and the *Max-1* execution ε_1, where distance is the number of "squares" and "triangles" that separate the two executions. For example, eight squares and two triangles separate the executions in Figures 3a and 3b.

First, if all states in ε_n have $\sigma = 0$ in states where instructions are issued, then we have a *Max-1* sequence and are trivially done—no more than one instruction is ever executed at a time. This is the base case.

Otherwise, we reduce the distance by inductively moving the last instruction issued in a state of $\sigma \geq 1$ back until $\sigma = 0$. We repeat this until all instructions do not overlap in execution and thus obtain the base case.

In the induction, we repeatedly choose the last such instruction i and identify the choice subsequence of length v starting with i. If necessary, we can make the subsequence long enough, by extending ε_n with extra, trailing stalling choices, using Proof Obligation 4. We then apply Proof Obligation 3. If we have added the previously mentioned weakening assumption that resources are available at the end of v, we can satisfy this by locating the last place that the resource was freed and delay the following rescheduling till after the v cycles, using Proof Obligation 4[2].

We know that the number of internal steps between the instruction issue and the end of the execution sequence monotically decreases in each application, since we are moving the instruction passed at least one step of any kind in each application. We also know that we are able to move all the internal steps of the instruction, since the length v is greater than u. Furthermore, since the instruction sequence is completed, we know that we are also moving the instruction past

---

[2] In implementations where the freeing and scheduling of the resource overlap in time, we can prove a separate lemma that shows the correctness of the slight delay of the rescheduling after the freeing.

instruction retires, each time monotonically decreasing the distance as defined above and eventually reaching the base case. The induction is thus well-founded. **End Proof Sketch**

6.2 ISA Step

The final verification step is to show that all *Max*-1 executions of SAI are functionally equivalent with ISA. Because the instruction sequence w^1 completes all executions (i.e., leaves no outstanding instructions in the pipeline), we can divide it up into issue-retire fragments in the *Max*-1 execution. We can assume that each fragment has length u, since if one does not, we can apply Proof Obligation 4 to add or remove the necessary stalling cycles. The proof is an induction on the number of such fragments, comparing the execution and retirement of an arbitrary instruction from an arbitrary *Max*-1 initial state with the result that is retired by ISA. This is illustrated in Figure 5c. Formally:

Proof Obligation 5 (SAI-ISA Induction) *For every initial IA state* q_a°, *instruction* i, *and input sequence pair* $\langle w, w_{ch} \rangle$ *of length* u *containing only* i *as its first instruction:*

$$\delta_a^u(q_a^\circ, \langle w, w_{ch} \rangle) \stackrel{RF}{=} \delta_s(\pi_{RF}(q_a^\circ), i).$$

Because we have previously shown that a functionally equivalent *Max*-1 execution can be derived from an arbitrary *Max*-n execution, this step completes the proof of SAI-ISA equivalence.

7 Mechanical Verification

We have mechanically checked our simple SAI abstraction and Proof Obligations 3-5 for our example using the Stanford Validity Checker (SVC). The proofs finished in minutes. The three models (IMPL, SAI, and ISA) and the proof obligations are written in a Lisp-like HDL. The proof formulas are constructed by symbolically simulating the models in Lisp. SVC is invoked through a foreign-function interface to decide the validity of the formulas.

The mechanical verification of Proof Obligation 3 has exposed several bugs in the way the choice signals were introduced in the SAI. For instance, the original formulation did not stall retirement properly. This was detected in the verification with SVC when a stretched execution retired an instruction that the original execution did not. This illustrates the ability of the incremental-flushing step to detect possible bugs in the exposing of the scheduler interface.

We were able to locate the error using the counter example information that SVC produced when the error was reached. The counter example is a conjunction of predicates satisfied in the interpretation that falsifies the proof obligation. The user can apply this information in the context of the original system model to debug the error.

8 Discussion

This work addresses a recurring difficulty encountered in symbolic verification of out-of-order processor designs: the difficulty of creating an appropriate implementation abstraction. The extension of the incremental flushing technique enables significantly more automation than the basic technique alone and reduces the need for manual abstraction. On the down side, the computational complexity of the resulting proof obligations is higher, since more steps of symbolic simulations are performed in each proof step. This was not an issue in the verification of our very simple example. However, more research is needed to address the application of the approach to more realistic designs. Also, work is needed to establish if our techniques for avoiding resource contentions during reordering are sufficient for more complex architectures.

It has been argued that localizing the (possibly distributed) scheduling logic in a circuit will be difficult. Our assumption of practice is that optimal scheduling algorithms are determined empirically by simulation and that location and interfaces are clearly identifiable when plugging in different scheduler implementations. We expect that this knowledge can be exploited when locating the scheduling logic for verification purposes.

Acknowledgments

We would like to thank the anonymous reviewers for their comments to the paper. The first author is supported at Stanford by an NDSEG graduate fellowship. The other authors are partially supported by DARPA under contract number DABT63-96-C-0097-P00002. Insight about the difficulties associated with verifying pipelined processors was developed while the third author was a visiting professor at Intel's Strategic CAD Labs in the summer of 1995.

References

1. S. Berezin, A. Biere, E. Clarke, and Y. Zhu. Combining symbolic model checking with uninterpreted functions for out-of-order processor verification. Appears in this volume.
2. J. R. Burch. Techniques for verifying superscalar microprocessors. In *33rd ACM/IEEE Design Automation Conference*, pages 552–557, Las Vegas, Nevada, USA, June 1996. ACM Press.
3. J. R. Burch and D. L. Dill. Automatic verification of microprocessor control. In David L. Dill, editor, *Computer Aided Verification. 6th International Conference*, volume 818 of *LNCS*, pages 68–80, Stanford, California, USA, June 1994. Springer-Verlag.
4. Werner Damm and Amir Pnueli. Verifying out-of-order executions. In Hon F. li and David K. Probst, editors, *Advances in Hardware Design and Verification: IFIP WG10.5 Internation al Conference on Correct Hardware Design and Verification Methods (CHARME)*, pages 23–47, Montreal, Canada, October 1997. Chapman & Hall.

5. J. L. Hennessy and D. A. Patterson. *Computer Architecture: A Quantitative Approach.* Morgan Kaufmann, 1990.
6. T. A. Henzinger, S. Qadeer, and S. K. Rajamani. You assume, we guarantee: Methodology and case studies. Technical report, Electronics Research Lab, Univ. of California, Berkeley, CA 94720, 1998.
7. R. Hosabettu, M. Srivas, and G. Gopalakrishnan. Decomposing the proof of correctness of pipelined microprocessors. In A. J. Hu and M. Y. Vardi, editors, *Computer Aided Verification (CAV'98)*, volume 1427 of *Lecture Notes in Computer Science*, pages 122–134, Vancouver, Canada, June-July 1998. Springer-Verlag.
8. R. B. Jones, D. L. Dill, and J. R. Burch. Efficient validity checking for processor verification. In *Proceedings: IEEE International Conference on Computer-Aided Design (ICCAD)*, November 1995.
9. K. McMillan. Verification of an implementation of Tomasulo's algorithm by compositional model checking. In A. J. Hu and M. Y. Vardi, editors, *Computer Aided Verification (CAV'98)*, volume 1427 of *Lecture Notes in Computer Science*, pages 110–121, Vancouver, Canada, June-July 1998. Springer-Verlag.
10. J. Sawada and W. A. Hunt. Trace table based approach for pipelined microprocessor verification. In Orna Grumberg, editor, *Computer-Aided Verification, CAV '97*, volume 1254 of *Lecture Notes in Computer Science*, pages 364–375, Haifa, Israel, June 1997. Springer-Verlag.
11. J. Sawada and W. A. Hunt. Processor verification with precise exceptions and speculative execution. In A. J. Hu and M. Y. Vardi, editors, *Computer Aided Verification (CAV'98)*, volume 1427 of *Lecture Notes in Computer Science*, pages 135–146, Vancouver, Canada, June-July 1998. Springer-Verlag.
12. J. U. Skakkebæk, R. B. Jones, and D. L. Dill. Formal verification of out-of-order execution using incremental flushing. In A. J. Hu and M. Y. Vardi, editors, *Computer Aided Verification (CAV'98)*, volume 1427 of *Lecture Notes in Computer Science*, pages 98–109, Vancouver, Canada, June-July 1998. Springer-Verlag.
13. P. J. Windley and J. R. Burch. Mechanically checking a lemma used in an automatic verification tool. In *Proceedings: International Conference on Formal Methods in Computer-Aided Design*, pages 362–376, November 1996.

Bit-Level Abstraction in the Verification of Pipelined Microprocessors by Correspondence Checking[1]

Miroslav N. Velev[*]
mvelev@ece.cmu.edu
http://www.ece.cmu.edu/~mvelev

Randal E. Bryant[‡, *]
randy.bryant@cs.cmu.edu
http://www.cs.cmu.edu/~bryant

[*]Department of Electrical and Computer Engineering
[‡]School of Computer Science
Carnegie Mellon University, Pittsburgh, PA 15213, U.S.A.

Abstract. We present a way to abstract functional units in symbolic simulation of actual circuits, thus achieving the effect of uninterpreted functions at the bit-level. Additionally, we propose an efficient encoding technique that can be used to represent uninterpreted symbols with BDDs, while allowing these symbols to be propagated by simulation with a conventional bit-level symbolic simulator. Our abstraction and encoding techniques result in an automatic symmetry reduction and allow the control and forwarding logic of the actual circuit to be used unmodified. The abstraction method builds on the behavioral Efficient Memory Model [18][19] and its capability to dynamically introduce consistent initial state, which is identical for two simulation sequences. We apply the abstraction and encoding ideas on the verification of pipelined microprocessors by correspondence checking, where a pipelined microprocessor is compared against a non-pipelined specification.

1 Introduction

The increasing complexity of functional units in modern microprocessors and the need to begin the verification at the system level early in the design process, before the individual modules are implemented or even completely specified, requires the capability to abstract the details of functional blocks. The focus of this paper is how to achieve such abstraction in formal verification methods based on symbolic simulation, while keeping intact the control and forwarding logic, as well as the bit level connections in the actual circuit. We also present an efficient encoding technique, targeted to the logic of uninterpreted functions with equality [5], that can be used for representing uninterpreted symbols by means of BDDs [3]. This technique allows such uninterpreted symbols to be used while symbolically simulating the actual circuit at the bit-level, thus avoiding the need for the abstract model of the circuit required by previous methods based on uninterpreted functions [5][8][10]. The abstraction and encoding effectively achieve an automatic symmetry reduction of all data streams, while keeping the control and forwarding logic of the actual circuit intact.

Our abstraction method builds on the Efficient Memory Model (EMM) [18][19] and particularly on its capability to dynamically introduce new initial state (as required by a simulation sequence) which is consistent with previously introduced initial state. In this paper, we improve the efficiency of the EMM algorithms and data structures.

---

1. This research was supported in part by the SRC under contract 98-DC-068.

Furthermore, observing that every combinational block of logic can be implemented as a read-only memory with the logic block inputs serving as memory addresses, we abstract functional units at the bit level by replacing them with read-only EMMs. The definition of the EMM automatically enforces consistency of the output values for the present input pattern with output values returned for previous input patterns.

The presented abstraction and encoding techniques are combined with the *correspondence checking* method for verification of pipelined microprocessors by comparison to non-pipelined specifications. Correspondence checking was introduced by Burch and Dill [5], who used uninterpreted functions to abstract the details of functional units and memory arrays. However, their tool requires an abstract model of the circuit, leaving room for errors in its description and raising concerns about the correctness of the actual processor, given the correctness of its abstract model. Correspondence checking was extended to the bit-level and made applicable on actual circuits in [4]. However, preliminary results [19] showed that it did not scale well enough to be suitable for application to actual microprocessors. The major sources of complexity were the symbolic modeling of all the bits of data in the data path and the feedback loops, created by the forwarding logic. Hence, the need for abstracting complex functional units.

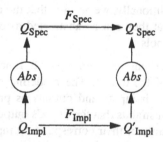

Fig. 1. Commutative diagram for the correctness criterion

The correctness criterion of correspondence checking, shown in Fig. 1, is due to Hoare [9] who used it for verifying computations on abstract data types in software. In a later work, Bose and Fisher [2] applied it to the verification of pipelined circuits. The implementation transformation F_{Impl} is verified by comparison against a specification transformation F_{Spec}. It is assumed that the two transformations start from a pair of matching initial states - Q_{Impl} and Q_{Spec}, respectively - where the match is determined according to some abstraction function *Abs*. The correctness criterion is that the two transformations should yield a pair of matching final states - Q'_{Impl} and Q'_{Spec}, respectively - where the match is determined by the same abstraction function. In other words, the abstraction function should make the diagram commute. Note that there are two paths from Q_{Impl} to Q'_{Spec}. We will refer to the one that involves F_{Impl} as the implementation side of the commutative diagram, while the one that involves F_{Spec} will be called the specification side.

Burch and Dill's contribution [5] is a conceptually elegant way to automatically compute the abstraction function *Abs* that maps the pipeline state of a processor to its user-visible state by symbolic simulation of the hardware design. Namely, starting from a general symbolic initial state Q_{Impl} they simulate a *flush* of the pipeline by stalling it for a sufficient number of cycles to allow all partially executed instructions to complete. Then, they consider the resulting state of the user-visible memories (e.g., the register file and the program counter) to be the matching state Q_{Spec}.

Burch [6] has extended the method to superscalar processor verification by proposing a new flushing mechanism (notice that the abstraction function can be arbitrary, as long as it makes the correctness criterion diagram commute) and by decomposing the commutative diagram into three more easily verifiable commutative diagrams. The correctness of this decomposition is proven in [21].

In bit-level correspondence checking, we use EMMs to represent both memories and uninterpreted functional units in the implementation and specification circuits. Essential to this is the EMM's property to dynamically introduce identical initial state to two simulation sequences [4]. In replacing these blocks, we assume that their actual implementations have been verified separately. For example, *symbolic trajectory evaluation* [16][11] has been combined with symmetry reductions [14] to enable the verification of very large memory arrays at the transistor level. An efficient representation of word-level functions has enabled the verification of complex functional units like floating-point multipliers [7]. Additionally, we assume that the data path connections have been verified to guarantee that they can be abstracted as only manipulating Boolean values and uninterpreted symbols.

Previous work on processing of uninterpreted functions with BDDs [8][10] required an abstract model of the circuit. The modeling of uninterpreted functional units was done by treating their inputs and outputs as primary outputs and inputs, respectively, and imposing constraints that the block's output values be consistent with previous ones, given the equality of their corresponding input patterns. The modeling of memory arrays was more complicated in that it also required these constraints to consider the effect of previous writes on the memory state. We achieve all these properties automatically by means of the EMM. While [8] and [10] generate a DAG-structured expression, that they call an IE netlist, which represents the correctness criterion and then process it off-line, our method works dynamically as part of a symbolic simulator. Finally, the techniques that these previous methods used for encoding uninterpreted symbols with BDDs are less efficient than ours. Symbolic model checking has also been combined with uninterpreted functions [1].

In the remainder of the paper, Sect. 2 defines the axioms of uninterpreted memories and functional units. Sect. 3 presents our technique for encoding of uninterpreted symbols with BDDs. Sect. 4 describes the EMM. Sect. 5 shows how to achieve bit-level abstraction of functional units by using the EMM. Dynamic generation of initial EMM state is presented in Sect. 6. The correspondence checking methodology is the focus of Sect. 7. Experimental results are presented in Sect. 8. Finally, conclusions are drawn and future work is outlined in Sect. 9.

2   Abstracting Memories and Functional Units

We will use the types address expression, **AExpr**, and data expression, **DExpr**, for denoting the kind of information that can be applied at the inputs or produced by the outputs of an abstract memory. Let m_0 : **AExpr** → **DExpr**, defined as a mapping from address expressions to data expressions, be the initial state of such a memory. Then, $m_0(a)$, where a is an address expression, will return the initial data of the memory at address a. The write operation for an abstract memory will be defined as $Write(m_i, a_1, d_1) \rightarrow m_{i+1}$ [13], i.e., taking as arguments the present state m_i of a memory, and address expression a_1 designating the location which is updated to contain data expression d_1, and producing the subsequent memory state m_{i+1}, such that $m_{i+1}(a_2) \rightarrow ITE(a_1 = a_2, d_1, m_i(a_2))$, where the ITE operator (for "If-Then-Else") selects d_1 when $a_1 = a_2$ is true, and $m_i(a_2)$ otherwise.

Based on the observation that any functional block can be represented as a read-only-memory (ROM), with the block's inputs serving as memory addresses, we will represent abstract functional units as abstract ROMs. According to the semantics of an abstract memory, an abstract ROM will always satisfy the property $a_1 = a_2 \Rightarrow f(a_1) = f(a_2)$, where $f()$ denotes the output function of the ROM-modeled abstract functional unit.

Motivated by application to actual circuits, we will represent address and data expressions by vectors of Boolean expressions having width n and w, respectively, for a memory with $N = 2^n$ locations, each holding a word of w bits. The type **BExpr** will denote Boolean expressions.

Address comparison is implemented as:

$$A1 = A2 \doteq \neg \bigvee_{i=1}^{n} A1_i \oplus A2_i, \tag{1}$$

while address selection $A1 \leftarrow ITE(b, A2, A3)$ is implemented by selecting the corresponding bits:

$$A1_i \leftarrow ITE(b, A2_i, A3_i), \quad i = 1, \ldots, n. \tag{2}$$

The definition of data operations is similar, but over vectors of width w.

An *uninterpreted symbol* is a compact representation of a word-level datum. Two uninterpreted symbols are compatible if they are compared for equality, stored in the same memory, or produced by the same memory in a given circuit. A *domain* is a set of compatible uninterpreted symbols. Typically, separate domains are introduced for instruction addresses, register identifiers, and register file data.

3   Encoding Uninterpreted Symbols

3.1 Background

Decision procedures based on the logic of uninterpreted functions with equality [13][17] use uninterpreted symbols to abstractly represent word-level values. Such symbols (e.g., $U1$, $U2$, and $U3$) are manipulated in two ways: 1) comparison for equality, $U1 = U2$, where the result is a Boolean expression, and 2) selection, $U3 = ITE(b, U1, U2)$, where b is a Boolean expression, meaning that $U3 = U1$ if b is true, and $U3 = U2$ otherwise. Boolean connectives - e.g., conjuction, disjuction, negation - can be applied on Boolean expressions and yield Boolean expressions. Although limited, this logic is sufficient for verification by correspondence checking. However, the initial decision procedures for correspondence checking [5][6] have not been based on BDDs, and thus have failed to exploit the simplification capabilities and manipulative power of BDD packages.

Previous research on adopting these decision procedures to manipulations with BDDs [10] has required a priori knowledge of the number of uninterpreted symbols in the same domain. Given that n uninterpreted symbols are required, Hojati *et al.* [10] encode each of them with $\lceil \log(n) \rceil$ Boolean variables. Thus, they require a total of $n \cdot \lceil \log(n) \rceil$ variables. Goel *et al.* [8] do not explicitly encode the symbols, but introduce a Boolean variable for every pair of symbols, indicating the conditions under which the two symbols are equal. This results in a total of $n \cdot (n - 1) / 2$ variables.

3.2 Our Encoding of Uninterpreted Symbols

Ideally, we would like to use the control and forwarding logic of the actual circuit intact in the simulations. Given that all this logic does with its input bit vectors is comparison for equality and selection, we would like to encode the input bit vectors with as few Boolean variables as possible and in a way that will allow the resulting expressions to be used for simulation of the actual circuit. Our technique to achieve this is illustrated in Table 1 for 4-bit vectors.

Uninterpreted Symbol	Encoding			
1	0	0	0	0
2	0	0	0	$a_{2,0}$
3	0	0	$a_{3,1}$	$a_{3,0}$
4	0	0	$a_{4,1}$	$a_{4,0}$
5	0	$a_{5,2}$	$a_{5,1}$	$a_{5,0}$
...		...		
8	0	$a_{8,2}$	$a_{8,1}$	$a_{8,0}$
9	$a_{9,3}$	$a_{9,2}$	$a_{9,1}$	$a_{9,0}$
...		...		
16	$a_{16,3}$	$a_{16,2}$	$a_{16,1}$	$a_{16,0}$
...		...		

Table 1. Encoding of 4-bit vectors that allows them to efficiently express the possibility that they be pairwise either equal or different, so that they can be treated as uninterpreted symbols

When there is a single bit vector generated in a given domain, then it does not need to be distinguished from other bit vectors, so that it can be represented with a vector of binary constants, e.g., 0s. When a second vector is generated, we need to express that it can be equal to or different from the first one. This can be done with a single Boolean variable in the least significant bit of the vector and the same binary constants in the other bit positions, as used in the first vector. When generating the nth vector, it could potentially have n possible values, so that we use $\lceil \log(n) \rceil$ new Boolean variables in the low order bits of the vector and the same binary constants in the remaining bit positions. If the vectors have a width of k bits, as determined by the circuit, then the number of variables generated for a new vector saturates at k. Note that the total number of Boolean variables that we need to encode n such vectors is:

$$\sum_{i=1}^{n} \min(\lceil \log(i) \rceil, k) \; .$$

In certain cases, we would like to allow distinguished constants in a given domain, e.g., registers 0 and 31 in the MIPS microprocessor [15] to be treated differently from the rest of the registers. The first one is hardwired to data value 0, while the second one is used to store the return address on a jump to subroutine instruction. We can incorporate such constant bit vectors in a given domain by introducing extra variables that will select one of the constant bit vectors or a new partially symbolic vector, generated according to our encoding. Additionally, we need to avoid exact replication of a constant vector in the encoding, so that when $\langle 0, 0, 0, 0, 0 \rangle$ is such a constant vector, the vector generator should start from $\langle 0, 0, 0, 0, a_{1,0} \rangle$. Then, the number of Boolean variables needed to encode the ith uninterpreted symbol will be $\min(\lceil \log(i+1) \rceil, k)$.

New bit vectors can be generated in each domain by function *GenDataExpr()*.

4 Efficient Modeling of Memory Arrays in Symbolic Simulation

4.1 Symbolic Decisions

We will use the term *context* to refer to an assignment of values to the symbolic variables. A Boolean expression can be viewed as defining a set of contexts, namely those for which the expression evaluates to **true**.

Note that predicate (1) in Sect. 2 is symbolic, i.e., it returns a symbolic Boolean expression and will be true in some contexts and false in others. Therefore, a symbolic predicate cannot be used as a control decision in algorithms. The function *Valid()*, when applied to a symbolic Boolean expression, will return **true** if the expression is valid or equal to **true** (i.e., true for all contexts), and will return **false** otherwise. We can make control decisions based on whether or not an expression is valid.

We have used BDDs [3] to represent the Boolean expressions in our implementation. However, any representation of Boolean expressions can be substituted, as long as function *Valid()* can be defined for it.

4.2 EMM Operation

The EMM models memory arrays with write and/or read ports, all of which have the same numbers of address and data bits - n and w, respectively - as shown in Figure 2.a.

An inward/outward triangle indicates an enable input of a write/read port. The assumption is that every memory system can be represented with such EMMs and possibly some extra logic. For example, a latch can be viewed as a memory array with a single address, so that it can be represented as an EMM with one write and one read port, both of which have the same number of data bits and only one address input, which is identically connected to the same Boolean constant (e.g., **true**) - see Fig. 2.b.

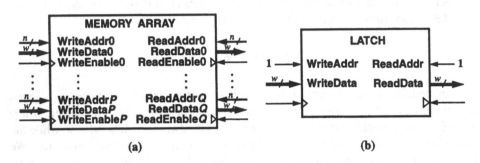

(a) (b)

Fig. 2. (a) A memory array that can be modeled by an EMM; (b) A latch modeled by an EMM

The interaction of the memory array with the rest of the circuit is assumed to take place when a port Enable signal is 1 (i.e., **true**). In case of multiple port Enables being 1 simultaneously, the resulting accesses to the memory array will be ordered according to the priority of the ports. It will be assumed that the memories respond instantaneously to any requests.

During symbolic simulation, the state of each EMM is represented with two lists - *init_list* and *write_list*. It should be pointed out that we used a single list for that purpose in our previous work [4][20]. The lists contain entries of the form $\langle c, a, d \rangle$, where c is a Boolean expression denoting the set of contexts for which the entry is defined, a is an address expression denoting a memory location, and d is a data expression denoting the contents of this location. The context information is included for modeling memory systems where the *Write* and *Read* operations may be performed conditionally depending on the value of a control signal. Initially the lists are empty.

In simulation, *write_list* is used for writes, and *init_list* is used for dynamic initialization of memory locations that have not been initialized or written before (as will be explained shortly), so that the state of the EMM is the concatenation of the two lists, with *init_list* having lower priority. An additional list, *previous_write_list*, will be used to store the contents of an EMM's *write_list* from a previous simulation sequence. The type **List** will be used to denote such memory lists, and **nil** will designate the end of a memory list. The list entries are kept in order from *head* (low priority) to *tail* (high priority). The entries towards the low priority end correspond to conceptually earlier memory state. Entries may be inserted at either end, using procedures *InsertHead*() and *InsertTail*().

The lists interacts with the rest of the circuit by means of a software interface developed as part of the symbolic simulation engine. The interface monitors the memory input lines. Should a memory input value change, given that its corresponding port

Enable value *c* is not 0, a *Write* or a *Read* operation will result, as determined by the type of the port. The Address and Data lines of the port will be scanned in order to form the address expression *a* and the data expression *d*, respectively. A *Write* operation takes as arguments both *a* and *d*, while a *Read* operation takes only *a*. These operations will be presented shortly.

A *Read* operation retrieves from the lists (see Sect. 4.4) a data expression *rd* that represents the data contents of address *a*. The software interface completes the read by scheduling the Data lines of the port to be updated with the data expression *ITE*(*c*, *rd*, *d*), i.e. to the retrieved data expression *rd* under the contexts *c* of the operation and to the old data expression *d* otherwise.

4.3 Implementation of the Memory *Write* Operation

procedure *Write*(**List** *write_list*, **BExpr** *c*, **AExpr** *a*, **DExpr** *d*)
/* Write data *d* to location *a* under contexts *c* */
    InsertTail(*write_list*, ⟨*c*, *a*, *d*⟩)

Fig. 3. Implementation of the *Write* operation

The *Write* operation, shown as a procedure in Fig. 3, is implemented by inserting an element into the *tail* (high priority) end of a memory write list, indicating that this entry should overwrite any other entries for this address.

4.4 Dynamically Introducing Consistent Initial States

In correspondence checking, we need a way to enforce the assumption that the two sequences, resulting from traversing the implementation and the specification sides of the commutative diagram in Fig. 1, start from the same initial state, Q_{Impl}, without explicitly initializing every memory location. We achieve this with function *Read*(), and by introducing a separate *write_list* for accumulating the effects of *Write* operations in each of the sequences, but a shared *init_list* for storing their common initial state. New initial state for locations that have never been accessed by either a read or a write, but are being read in the current execution sequence, will be introduced on-the-fly, as shown in Fig. 4.

Function *Read*() scans the concatenation of the *write_list* for the current execution sequence and the common *init_list*. It starts from the most recently written data in the *write_list* and proceeds backwards to the "conceptually oldest" initial state at the beginning of the *init_list*. Function *Tail*() takes a memory list and returns its tail entry. Function *Previous*() returns a memory list obtained from its argument memory list by removing the tail entry. The Boolean expression *found* is constructed to reflect the contexts under which the read location *ra* has been written or has been initialized. If *found* is not true for all contexts, a fresh data expression *g* is generated for the particular EMM by function *GenDataExpr*(). Then, the entry ⟨**true**, *ra*, *g*⟩ is inserted at the low priority end of the *init_list*, and *g* is reflected on the data expression *rd* that is returned. In this way, subsequent *Read* operations in either simulation sequence will encounter the same initial state for location *ra* in the given memory.

function *Read*(List *init_list*, List *write_list*, **AExpr** *ra*) : **DExpr**
/* read from location *ra* */
    found ← **false**
    l ← *ConcatenateLists*(*init_list*, *write_list*)
    if *l* ≠ **nil** **then**   /* scan backwards from most recently written data */
        ⟨*c*, *a*, *d*⟩ ← *Tail*(*l*)
        rd ← *d*
        found ← *c* ∧ (*a* = *ra*)
        l ← *Previous*(*l*)
        while (*l* ≠ **nil** ∧ ¬*Valid*(*found*)) **do**
            ⟨*c*, *a*, *d*⟩ ← *Tail*(*l*)
            match ← *c* ∧ (*a* = *ra*)
            rd ← *ITE*(*match*, *d*, *rd*)
            found ← *found* ∨ *match*
            l ← *Previous*(*l*)
    if ¬*Valid*(*found*) **then**
        g ← *GenDataExpr*(*init_list*) /* introduce new initial state for address *ra* */
        if *Valid*(¬*found*) **then**
            rd ← *g*   /* if *found* ≡ **false** */
        else
            rd ← *ITE*(*found*, *rd*, *g*)
        InsertHead(*init_list*, ⟨**true**, *ra*, *g*⟩)
    return *rd*

Fig. 4. Implementation of the *Read* operation

4.5 Comparing Final States

In order to verify the correctness criterion of correpondence checking, we need to check that the two simulation sequences modified the state of each memory array in the same way. This is done by function *Compare*(), presented in Fig. 5. It takes as arguments the two write lists and the *init_list* for a memory array and returns a Boolean expression representing the conditions under which the two sequences have equal effects on the memory state. Since the updates of memory state are reflected only on the *write_list* for the given execution sequence, while the initial state in *init_list* is the same for both sequences, we need examine only the memory locations in either *write_list*. We start from the heads of the write lists, as returned by function *Head*(). Function *Next*() returns a memory list obtained from its argument memory list by removing the head entry.

    The first while-loop skips pairs of identical writes made in both simulation sequences, since they would modify the common initial state identically and, hence, would preserve the equality of the memory state. When a pair of different updates is detected or one of the write lists terminates, we need to check if every memory location modified then or later has the same contents at the end of both simulation sequences. This is done in the subsequent while-loops. The memory list *substitution_list* stores new initial state, introduced during the comparison, which is used for common subexpression elimination of the identical state. It is assumed that applying function *Next()* to list *l* does not affect either of the write lists. As an optimization, which is not shown, one could keep a table of addresses that have been compared in order to avoid repetition of computations.

function *Compare*(List *write_list*$_1$, List *write_list*$_2$, List *init_list*): **BExpr**
    identical ← **true**
    while (*write_list*$_1$ ≠ **nil** ∧ *write_list*$_2$ ≠ **nil** ∧ *identical*) **do**
        /* eliminate identical entries at beginning of both write lists */
        ⟨c_1, a_1, d_1⟩ ← *Head*(*write_list*$_1$)
        ⟨c_2, a_2, d_2⟩ ← *Head*(*write_list*$_2$)
        identical ← $c_1 \equiv c_2 \wedge a_1 \equiv a_2 \wedge d_1 \equiv d_2$
        if *identical* **then**
            write_list$_1$ ← *Next*(*write_list*$_1$)
            write_list$_2$ ← *Next*(*write_list*$_2$)
    equal ← **true**
    substitution_list ← **nil**   /* for substitution of common subexpressions */
    l ← *write_list*$_1$
    while (*l* ≠ **nil** ∧ ¬*Valid*(¬*equal*)) **do**
        ⟨*c*, *a*, *d*⟩ ← *Head*(*l*)
        rd_1 ← *Read*(*substitution_list*, *write_list*$_1$, *a*)
        rd_2 ← *Read*(*substitution_list*, *write_list*$_2$, *a*)
        equal ← *equal* ∧ ($rd_1 = rd_2$)
        l ← *Next*(*l*)
    l ← *write_list*$_2$
    while (*l* ≠ **nil** ∧ ¬*Valid*(¬*equal*)) **do**
        ⟨*c*, *a*, *d*⟩ ← *Head*(*l*)
        rd_1 ← *Read*(*substitution_list*, *write_list*$_1$, *a*)
        rd_2 ← *Read*(*substitution_list*, *write_list*$_2$, *a*)
        equal ← *equal* ∧ ($rd_1 = rd_2$)
        l ← *Next*(*l*)
    return *equal*

Fig. 5. Comparing for equality the states of the two write lists for an EMM

The presented version of function *Compare()* would flag an error when a data expression has been read from a memory address and then written back unmodified to the same address during one of the simulation sequences, but has not been accessed during the other simulation sequence. Assuming that this situation would not occur allows us to use the optimization of skipping the identical initial state and identical sequence of writes, which results in reduced BDD sizes. In the case of a counterexample, the unoptimized version of function *Compare()* [4] can be used to ensure that the counterexample is not a false negative.

5 Bit-Level Uninterpreted Functions Modeled by the EMM

In order to abstract functional units, while keeping their bit-level inputs and outputs, we exploit the capability of the EMM to dynamically introduce initial state, as required by the given simulation sequence. Based on the observation in Sect. 2 and the definition of the EMM, we can model abstract functional units as read-only EMMs, which have a single read port that is constantly enabled, i.e., its Enable signal is connected to the Boolean constant **true**. Such EMM's address inputs are formed by concatenating the functional unit's data and control inputs (see Fig. 6), while the EMM's data outputs are connected with the functional unit's data outputs. In this way we use the EMM as a ROM whose contents are generated on-the-fly.

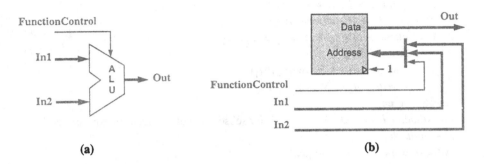

Fig. 6. (a) A functional unit; (b) Its abstraction by a read-only EMM

Since such an EMM is never written to, its two write lists will be empty and the *Read* operations will scan only the *init_list*. The implementation of function *Read()* guarantees the consistency of the output data expressions, i.e., that they be equal if their respective input patterns are equal. Thus, we avoid the need to explicitly impose such auxiliary constraints as done in [8][12].

6 Generating Initial Memory State

6.1 Motivation

Using BDDs requires a global ordering of the Boolean variables. However, when performing *Read* operations, an EMM's address expression will be compared for equality to other address expressions for the same EMM and the resulting Boolean expression will select data expressions. Hence, if the variables used in the address expressions are intermixed with or situated after the variables used in data expressions, then the data expression variables will need to be replicated for every possible outcome of the address expressions' comparisons for equality as these outcomes will become known at a lower level in the BDD shared DAG. Hence, the exponential complexity of the BDD representation of the Boolean expressions within that EMM's data expressions. This can be avoided by placing the address variables before the data variables in the global variable ordering.

Similarly, control variables (e.g., that will represent operation-codes and functional-codes) need to be placed before the data variables that they will affect. Again, the control variables will select one out of many operations to be performed on the data variables, so that the above argument still applies.

In order to account for the above two rules, we introduce the notion of ordered variable groups. Each variable group contains vectors of symbolic Boolean variables that are used for the same purpose (e.g., addresses, control, data). These variables can be either interleaved or placed sequentially in the variable order, but cannot be intermixed with variables from another variable group. Additional vectors from the same group can be generated dynamically, as required by the circuit for a given symbolic simulation sequence. Furthermore, when these vectors are used only for comparison for equality and selection, given the functional units are abstracted by EMMs, the vectors can be encoded as presented in Sect. 3. The variable groups are ordered based on their relation, according to the above two rules.

Therefore, the ideal global ordering of the variable groups, needed to verify the MIPS pipeline [15] for its register-register and register-immediate instructions (see Fig. 7), will be: instruction-addresses, operation-codes, functional-codes, register-identifiers, immediate-data, and data (when listed from first to last in the global ordering of the variables). The need to use two data groups is dictated by the format of the MIPS register-immediate instructions. Notice that the instruction-address variables represent the contents of the program counter (PC) and identify locations of the instruction memory (IMem), so that they need to be placed before the instructions (that will represent the contents of the IMem), i.e., before all other variable groups. The justification for the ordering of the other variable groups is similar.

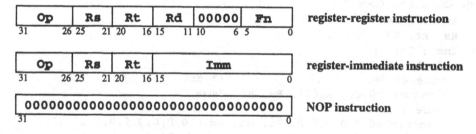

Fig. 7. The MIPS formats for register-register, register-immediate, and NOP instructions. The instructions are encoded with 32 bits, numbered 0 through 31 - see the digits below the rectangles. Op stands for the operation-code field of 6 bits, Fn represents the functional-code field of 6 bits, and Imm is the immediate field of 16 bits. Rs, Rt and Rd are 5-bit register identifiers. Rs and Rt are the source registers, while Rd is the destination register, for register-register instructions. Rs is a source register providing the second operand, and Rt is the destination register for register-immediate instructions

Furthermore, modern instruction set architectures consist of widely varying classes of instructions. As Fig. 7 shows, a given instruction bit may be used in different ways in the actual hardware under different conditions. For example, the lowest 6 bits of the MIPS instructions can represent the functional code in register-register instructions, or part of the immediate data in register-immediate instructions. Ideally, these 6 bits would be encoded with functional-code variables for the register-register instructions, and data variables for the register-immediate instructions. Furthermore, the functional-code variables need to be placed before the data variables, since the functional code determines what operations are to be performed on the data by the ALU. One drawback of BDDs [3], used as a representation of Boolean expressions in our implementation of the presented algorithms, is that they require a global ordering of the variables. Supplying a single vector of Boolean variables for the initial state of the instruction memory at a particular address location will mean that a dynamic-variable-reordering BDD package will need to find a different global variable order, conditional on the type of the instruction. This will be impossible for a symbolic instruction representing many instruction formats simultaneously. Similarly, the state bits in the pipeline latches may be used in different ways under different conditions.

Finally, not all possible binary states are reachable and we need the capability to restrict the set of initial states evaluated.

6.2 Indexing of Variable Groups

The state-encoding technique, called "variable-group indexing," allows the EMM to return a different data expression for each distinct control combination. The possible data expressions are selected by means of indexing variables, whose variable group has to be situated after the instruction-addresses and before the rest of the variable groups. The resulting data expression is generated according to a pattern, defined by the use for each EMM - see Fig. 8.

```
gendataexpr IMem (
   Ind : Index
   Rs, Rt, Rd : RegId
   Imm : ImmData
   switch <Ind> (
      case <0,0>:                         /* ori */
         return <0,0,1,1,0,1, Rs, Rt, Imm>
      case <0,1>:                         /* or  */
         return <0,0,0,0,0,0, Rs, Rt, Rd,0,0,0,0,0,1,0,0,1,0,1>
      default:                            /* nop */
         return <0,0,0,0,0,0,0,0,0,0,0,0,0,0,0,0,0,0,0,0,0,0,0,0,
                 0,0,0,0,0,0,0>
   )
)
```

Fig. 8. Definition of the variable-group indexing pattern for the MIPS instruction memory (IMem), when verifying the pipeline for only 3 instructions: ori, or, and nop. The operation and functional codes are represented with constants, as defined in the Instruction Set Architecture of the MIPS. For a complete verification of the pipelined processor, the above indexing pattern must be expanded to include every legal instruction. Such variable-group indexing patterns must be defined for all memories, pipeline latches, and functional units, abstracted with EMMs

Index, RegId, and ImmData are the declaration names for the indexing, register-identifier, and immediate-data variable groups, respectively. Ind, Rs, Rt, Rd, and Imm are vectors of fresh Boolean variables generated within the corresponding variable group by being interleaved with other such vectors in the same variable group.

The `switch` statement uses the vector of indexing variables in order to select one out of the three possible patterns for the initial state of the instruction memory, according to the `case` and `default` directives. A Boolean expression, *global_care_set*, can be used to accumulate (by means of conjunction) the effect of any conditions that the newly generated vectors of Boolean variables must satisfy. Notice that the variable-group indexing technique makes possible the verification of a microprocessor for a subset of its instruction set architecture.

As Fig. 8 shows, we might be imposing restrictions on the initial state that is generated on-the-fly by *GenDataExpr()*, therefore considering a subset of the system state space as a possible initial state. These restrictions are due to the sparse encoding of instructions and the sparse encoding of internal vectors of control signals. However, the correctness criterion expressed by the commutative diagram from Fig. 1 is a proof by induction. It proves that if the implementation starts from an arbitrary initial state Q_{Impl}, and is exercised with an arbitrary instruction, then the reachable state Q'_{Impl} will be correct, when compared (by means of an abstraction function) to the specification's behavior for an arbitrary instruction. Since the initial state Q_{Impl} is arbitrary, then it is a superset of the reachable state Q'_{Impl}, so that, by induction, the implementation will function correctly for an arbitrary instruction applied to Q'_{Impl}. However, if the initial state Q_{Impl} is restricted to a subset of the system state space, and Q'_{Impl} is not a subset of Q_{Impl}, then the inductive argument for the correctness criterion would not hold. Hence, we need to check that Q'_{Impl} is a subset of Q_{Impl}. This is equivalent to proving an invariant of the implementation - see [20] for details on how we do that.

7 Correspondence Checking Methodology

Step 1.   Load the pipelined implementation. Let *init_list* ← **nil**, *write_list* ← **nil** for every memory in the circuit. *global_care_set* ← **true**.

Step 2.   Simulate the implementation circuit for one clock cycle with a (legal) symbolic instruction. Verify the invariant of the pipeline initial state [20].

Step 3.   Simulate a flush of the implementation circuit.

Step 4.   Let *previous_write_list* ← *write_list*, *write_list* ← **nil** for each memory in the implementation.

Step 5.   Simulate a flush of the implementation circuit.

Step 6.   Swap the implementation and the non-pipelined specification circuits by keeping the contents of the memory lists for every user-visible memory.

Step 7.   Simulate the specification circuit for one clock cycle with the same symbolic instruction as used in Step 2.

Step 8.   Let *equality_i* ← *Compare(write_list_i, previous_write_list_i, init_list_i)* , $i = 1, ..., u$, where u is the number of user-visible memories.

Step 9.   Form the Boolean expression for the correctness criterion:

$$global\_care\_set \Rightarrow \bigwedge_{i=1}^{u} equality_i ,$$

where *global_care_set* is updated by function *GenDataExpr()* with conditions that constrain the initial memory state to be legal.

8 Experimental Results

We examined a 3-stage pipelined MIPS processor [20], which is comparable to the pipelined data paths used in [5][8]. It was compared to its non-pipelined version. The register file and pipeline latches, including the PC, were modeled as EMMs. Additionally, for the experiments studying abstraction of functional units by EMMs, we replaced the ALU in the data path and the Sign Extension logic with one EMM, and the adder that increments the PC with another EMM. The 3 stages in the pipeline are Fetch-and-Decode, Execute, and Write-Back. In order to investigate the potential of our abstraction and encoding techniques to scale for application to more complex pipelines, we performed experiments with pipelines where one, two, or three "Dummy" stages are inserted between the Execute and Write-Back stages. That increases the total number of pipeline stages to 4, 5, or 6, respectively. These Dummy stages do not perform any function, except that they temporarily store the result and each add another level of forwarding logic for the inputs of the ALU. Ten MIPS instructions [15] were supported by this processor and its non-pipelined version: five register-register instructions - or, and, add, slt, and sub; four register-immediate instructions - ori, andi, addi, and slti; and the nop.

The experiments were performed on an IBM RS/6000 43P-140 with a 233MHz PowerPC 604e microprocessor, having 256 MB of physical memory, and running AIX 4.1.5. Table 2 shows the results for CPU time and memory consumption, required for the verification. In the experiments labeled "Gate-Level," the ALU, the Sign Extension logic, and the adder that increments the PC were modeled at the gate level, while in the experiments labeled "Abstr.," these functional units were abstracted by read-only EMMs. In the cases of "Encoding," the uninterpreted symbols were generated as described in Sect. 3. When no encoding was used, complete vectors of Boolean variables were generated. Table 3 presents the total number of variables and the maximum number of BDD nodes, produced in the experiments studying abstraction.

Stages	Experiment	CPU Time [s]					Memory [MB]				
		Data Path Width					Data Path Width				
		4	8	16	32	64	4	8	16	32	64
3	Gate-Level	137	392	2555	---	---	25	97	193	---	---
	Abstr.	34	58	138	419	1569	13	23	43	96	241
	Abstr. & Encoding	2	2	2	2	3	2	2	2	2	2
4	Abstr.	582	1413	---	---	---	97	192	---	---	---
	Abstr. & Encoding	12	12	13	13	16	4	4	4	4	4
5	Abstr. & Encoding	115	115	117	118	124	25	25	25	25	25
6	Abstr. & Encoding	958	969	967	1018	1010	97	97	97	97	97

Table 2. Experimental results. The "Gate-Level" experiments ran out of memory for more than 3 pipeline stages, while the experiments with abstraction without encoding ran out of memory for more than 4 stages. Additionally, experiments designated with "---" ran out of memory too

The use of abstraction reduces the CPU time and memory consumption by an order of magnitude. However, the CPU time depends quadratically on the data path width, while the memory has a dependence that is between linear and quadratic. The combination of abstraction and encoding further reduces the CPU time and memory and makes them invariant with the data path width. As Table 3 shows, the max. number of BDD nodes is constant with the data path width for the cases of abstraction and encoding, while it grows linearly with the data path width for the cases of abstraction only. Hence, our abstraction and encoding techniques effectively achieve a symmetry reduction, while keeping intact the control and forwarding logic of the processor.

Extra levels of forwarding logic, due to dummy stages, increase the complexity of the Boolean expressions at the inputs of the abstracted ALU and result in an exponential complexity. This is due to the fact that the read-only EMM, that replaces the ALU, is addressed with Data uninterpreted symbols coming from the register file or from the subsequent pipeline latches by means of the forwarding logic. On the other hand, this EMM has to produce data expressions of Data uninterpreted symbols, since these will be written back to the register file. Hence, Data uninterpreted symbols are compared for equality as EMM addresses against other Data uninterpreted symbols, which are already used as addresses in the *init_list* for that EMM, and are used to select uninterpreted symbols of the same type. Clearly, this results in an exponential complexity.

Stages	Experiment	Total BDD Variables					Max. BDD Nodes ($\times 10^3$)				
		Data Path Width					Data Path Width				
		4	8	16	32	64	4	8	16	32	64
3	Abstr.	231	267	339	483	771	524	924	1,751	4,194	8,693
	Abstr. & Encoding	163	167	175	191	223	21	22	22	22	22
4	Abstr.	252	300	---	---	---	4,194	8,389	---	---	---
	Abstr. & Encoding	182	186	194	210	242	131	131	131	131	131
5	Abstr. & Encoding	202	206	214	230	262	1,049	1,049	1,049	1,049	1,049
6	Abstr. & Encoding	222	228	236	252	284	4,194	4,194	4,194	4,194	4,194

Table 3. Variables and max. BDD nodes generated for the experiments with abstraction. When no encoding was used, complete vectors of variables were generated. "---" means the experiment ran out of memory

For all of the experiments the variable groups were ordered in the following way: instruction addresses, indexing variables, register identifiers, immediate data, and data, when listed from first to last in the variable order. The optimal results for the experiments with abstraction and encoding for 3 or 4 pipeline stages were obtained by bit-wise interleaving the complete vectors of indexing variables, but placing sequentially in the variable order the variables for the uninterpreted symbols of type register identifier, immediate data, and data. For example, when generating a new uninterpreted symbol of type register identifier, its variables will be placed sequentially after the variables of previously generated uninterpreted symbols of the same type, but before the variables for uninterpreted symbols of type immediate data. Also, the optimal results were obtained by ordering the pipeline latches from last to first in the pipeline when generating their initial state, i.e., the latch before the Write-Back stage gets its initial state generated first, while the latch before the Execute stage gets its initial state generated last. For the experiments with pipelines of 5 or 6 stages, the optimal results were obtained by bit-wise interleaving the variables of the uninterpreted symbols of type register identifier and data.

Adding an extra pipeline stage after the Execute stage requires that initial state be produced for it. That leads to the generation of an extra uninterpreted symbol in the RegId group, in order to represent the destination register for the result stored in that stage, so that the number of RegId uninterpreted smbols needed is $4 + p$, where p is the number of pipeline stages. The number of Data uninterpreted symbols necessary for the verification is $6 + p$. The number of uninterpreted symbols generated is invariant with the data path width. The use of encoding reduces the total number of Boolean variables needed for the verification and makes it much less dependent on the data path width. That number still varies with the data path width because of the need to generate complete vectors of Boolean variables for checking the invariant of the pipeline initial state [20].

Introducing bugs in the forwarding logic or the instruction decoding PLA of the pipelined processor resulted in generation of counterexamples that increased the memory consumption up to 1.5 times but kept the CPU time almost the same, compared to the experiments with the correct circuit.

We also extended the initial 3 stage pipeline with a Memory stage between the Execute and Write-Back stages, and incorporated a load and a store instruction in the control of the processor. The verification experiments ran out of memory. The reason is that the Data Memory (we modeled it separately from the Instruction Memory in order to avoid the complexity of counterexamples due to self-modifying code) gets addressed with Data uninterpreted symbols produced by the ALU in the Execute stage. At the same time the Data Memory has to produce Data uninterpreted symbols to be written back to the register file. As discussed for the case of the ALU, that results in an exponential complexity.

9 Conclusions and Future Work

We proposed a way to abstract functional units at the bit level, by using the Efficient Memory Model in a read-only mode. We combined that with an encoding technique that allows uninterpreted symbols to be represented efficiently with Boolean variables, without a priori knowledge of the number of required symbols. An advantage of these ideas is that they keep the control and forwarding logic in the actual microprocessor intact, while achieving the effect of an automatic symmetry reduction.

A weakness of our methodology is that when using 0s for the high order bits in our encoding of uninterpreted symbols, we do not verify the correctness of the data transfer paths for these bits. This can be avoided by simulating the circuit entirely symbolically for one cycle and then analyzing the reached next state. Additionally, we need to check that the unabstracted logic of the processor does not use the uninterpreted symbols in a way that their encoding is not suitable for. These issues will be addressed in our future research. We will also work on automating the process of defining the initial state for pipeline latches and will explore techniques to overcome the exponential complexity of verifying pipelines with load/store instructions and many levels of forwarding logic.

References

1. S. Berezin, A. Biere, E.M. Clarke, and Y. Zhu, "Combining Symbolic Model Checking with Uninterpreted Functions for Out-of-Order Processor Verification," *FMCAD'98* (appears in this publication).
2. S. Bose, and A.L. Fisher, "Verifying Pipelined Hardware Using Symbolic Logic Simulation," *International Conference on Computer Design*, October 1989, pp. 217-221.
3. R.E. Bryant, "Symbolic Boolean Manipulation with Ordered Binary-Decision Diagrams," *ACM Computing Serveys*, Vol. 24, No. 3 (September 1992), pp. 293-318.
4. R.E. Bryant, and M.N. Velev, "Verification of Pipelined Microprocessors by Comparing Memory Execution Sequences in Symbolic Simulation,"[2] *Asian Computer Science Conference (ASIAN'97)*, R.K. Shyamasundar and K. Ueda, eds., LNCS 1345, Springer-Verlag, December 1997, pp. 18-31.
5. J.R. Burch, and D.L. Dill, "Automated Verification of Pipelined Microprocessor Control," *CAV'94*, D.L. Dill, ed., LNCS 818, Springer-Verlag, June 1994, pp. 68-80.
6. J.R. Burch, "Techniques for Verifying Superscalar Microprocessors," *DAC'96*, June 1996, pp. 552-557.
7. Y.-A. Chen, "Arithmetic Circuit Verification Based on Word-Level Decision Diagrams," Ph.D. thesis, School of Computer Science, Carnegie Mellon University, May 1998.
8. A. Goel, K. Sajid, H. Zhou, A. Aziz, and V. Singhal, "BDD Based Procedures for a Theory of Equality with Uninterpreted Functions," *CAV'98*, June, 1998.
9. C.A.R. Hoare, "Proof of Correctness of Data Representations," *Acta Informatica*, 1972, Vol.1, pp. 271-281.
10. R. Hojati, A. Kuehlmann, S. German, and R.K. Brayton, "Validity Checking in the Theory of Equality with Uninterpreted Functions Using Finite Instantiations," *International Workshop on Logic Synthesis*, May 1997.
11. A. Jain, "Formal Hardware Verification by Symbolic Trajectory Evaluation," Ph.D. thesis, Department of Electrical and Computer Engineering, Carnegie Mellon University, August 1997.
12. T.-H. Liu, K. Sajid, A. Aziz, and V. Singhal, "Optimizing Designs Containing Black Boxes," *34th Design Automation Conference*, June 1997, pp. 113-116.
13. G. Nelson, and D.C. Oppen, "Simplification by Cooperating Decision Procedures," *ACM Transactions on Programming Languages and Systems*, Vol. 1, No. 2, October 1979, pp. 245-257.
14. M. Pandey, "Formal Verification of Memory Arrays," Ph.D. thesis, School of Computer Science, Carnegie Mellon University, May 1997.
15. D.A. Patterson, and J.L. Hennessy, *Computer Organization and Design: The Hardware/Software Interface*, 2nd Edition, Morgan Kaufmann Publishers, San Francisco, CA, 1998.
16. C.-J.H. Seger, and R.E. Bryant, "Formal Verification by Symbolic Evaluation of Partially-Ordered Trajectories," *Formal Methods in System Design*, Vol. 6, No. 2, March 1995, pp. 147-190.
17. R.E. Shostak, "A Practical Decision Procedure for Arithmetic with Function Symbols," *J. ACM*, Vol. 26, No. 2, April 1979, pp. 351-360.
18. M.N. Velev, R.E. Bryant, and A. Jain, "Efficient Modeling of Memory Arrays in Symbolic Simulation,"[2] *CAV'97*, O. Grumberg, ed., LNCS 1254, Springer-Verlag, June 1997, pp. 388-399.
19. M.N. Velev, and R.E. Bryant, "Efficient Modeling of Memory Arrays in Symbolic Ternary Simulation,"[2] *International Conference on Tools and Algorithms for the Construction and Analysis of Systems (TACAS'98)*, B. Steffen, ed., LNCS 1384, Springer-Verlag, March-April 1998, pp. 136-150.
20. M.N. Velev, and R.E. Bryant, "Verification of Pipelined Microprocessors by Correspondence Checking in Symbolic Ternary Simulation,"[2] *International Conference on Application of Concurrency to System Design (CSD'98)*, IEEE Computer Society, March 1998, pp. 200-212.
21. P.J. Windley, and J.R. Burch, "Mechanically Checking a Lemma Used in an Automatic Verification Tool," *FMCAD'96*, M. Srivas and A. Camilleri, eds., LNCS 1166, Springer-Verlag, November 1996, pp. 362-376.

---

2. Available from: `http://www.ece.cmu.edu/~mvelev`

Solving Bit-Vector Equations*

M. Oliver Möller[1] and Harald Rueß[2]

[1] University of Århus
Department of Computer Science
Ny Munkegade, building 540
DK - 8000 Århus C, Denmark
omoeller@brics.dk

[2]SRI International
Computer Science Laboratory
333 Ravenswood Ave.
Menlo Park, CA, 94025, USA
ruess@csl.sri.com

Abstract. This paper is concerned with solving equations on fixed and non-fixed size bit-vector terms. We define an equational transformation system for solving equations on terms where all sizes of bit-vectors and extraction positions are known. This transformation system suggests a generalization for dealing with bit-vectors of unknown size and unknown extraction positions. Both solvers adhere to the principle of splitting bit-vectors only on demand, thereby making them quite effective in practice.

1 Introduction

Efficient automation of bit-vector reasoning is essential for the effective mechanization of many hardware verification proofs. It has been demonstrated, for example, in [SM95] that the lack of such a capability forms one of the main impediments to cost-effective verification of industrial-sized microprocessors.

Here we are specifically concerned with the problem of solving equational bit-vector constraints. Given an equation on bit-vector terms, a *solver* yields *true* if this equation is valid, *false* if it is unsatisfiable, and an equivalent system of solved equations otherwise. A solver can therefore not only be used to decide formulas but also to compute a compact representation of the set of all interpretations (witnesses) for a given formula. Consider, for example, the formula $x_{[6]} \otimes y_{[2]} = y_{[2]} \otimes x_{[6]}$, where $x_{[6]}, y_{[2]}$ are variables for bit-vectors of length 6 and 2, respectively, and $.\otimes.$ denotes concatenation of bit-vectors. A solved form for this formula is given by $x_{[6]} = y_{[2]} \otimes y_{[2]} \otimes y_{[2]}$ without further restrictions on $y_{[2]}$. Moreover, solving is the central concept for deciding formulas in the combination of theories, since Shostak's algorithm [Sho84] consists of a composition of solvers for the individual theories. Thus, solving bit-vector equations opens the door for deciding many hardware verification problems including operations from other fundamental theories like arithmetic, lists, or arrays.

It is our basic premise that the peculiarities of bit-vectors and the particular combination of bit-vector operations like concatenation, extraction of contiguous parts, bitwise Boolean operations, and finite arithmetic require specialized

* This research was supported in part by the Deutsche Forschungsgemeinschaft (DFG) project *Verifix* and by the Deutscher Akademischer Austauschdienst (DAAD). The work undertaken at SRI was partially supported by the National Science Foundation under grant No. CCR-9509931.

reasoners to effectively deal with bit-vector problems instead of simply encoding bit-vectors and operations there upon in, say, arithmetic. Such a specialized solver has been described, for example, in [CMR96, CMR97]. The key design feature of this algorithm is that it only splits bit-vectors on demand. In this way, run times of the solver are—in extreme cases—independent of the lengths of the data paths involved. The solver in [CMR97] includes a number of optimizations to improve usability. Unfortunately, these low-level details also tend to distract from the underlying, basically simple concepts of this algorithm.

In this paper we remedy the situation and reconstruct the solver in [CMR97] in terms of an equational transformation system. Besides the advantage of separating the conceptual ideas of the algorithm from low-level efficiency issues, the description of solving as an equational transformation system suggests a generalization to deal with equations on non-fixed size bit-vector terms, i.e., terms of unknown size or with integer variables as extraction positions.

We proceed as follows. In Section 2 we fix the notation used throughout this paper, review basic concepts of bit-vectors, and define the problem of solving. Hereby, we restrict ourselves to the theory of bit-vectors with the fundamental operations of concatenation and extraction only, since other bit-vector operations, like bitwise Boolean operations, can be added in a conceptually clean way using the notion of bit-vector OBDDs [CMR97, Möl98]. Section 3 forms the core of this paper and describes a rule-based algorithm for solving fixed size bit-vector equations. In Section 4 we extend the rule-based algorithm for solving equations on non-fixed size bit-vector terms. Section 5 concludes with some remarks.

Prototypical implementations of the fixed size and the non-fixed size bit-vector solvers are available from:

http://www.informatik.uni-ulm.de/ki/Bitvector/

2 Preliminaries

This section contains background material on Shostak's procedure for deciding combinations of quantifier-free theories and on the theory of bit-vectors.

Canonization. Shostak's procedure [Sho84] operates over a subclass of certain unquantified first-order theories called σ-theories. Informally, these theories have a computable canonizer function σ from terms (in the theory) to terms, such that an equation $t = u$ is valid in the theory if and only if $\sigma(t) \doteq \sigma(u)$, where $\doteq$ denotes syntactic equality. More precisely, the full set of requirements on canonizers—as stated in [CLS96]—are as follows.

Definition 1. *Let T be a set of terms, possibly containing subterms not in the theory. Then $\sigma : T \to T$ is called a* canonizer *if it fulfills the properties:*

1. *An equation $t = u$ in the theory is valid if and only if $\sigma(t) \doteq \sigma(u)$.*
2. *If $t \notin T$ then $\sigma(t) \doteq t$.*
3. *$\sigma(\sigma(t)) \doteq \sigma(t)$*
4. *If $\sigma(t) \doteq f(t_1, \ldots, t_n)$ for a term $t \in T$ then $\sigma(t_i) \doteq t_i$ for $1 \leq i \leq n$.*
5. *$vars(\sigma(t)) \subseteq vars(t)$*

Algebraic Solvability. To construct a decision procedure for equality in a combination of σ-theories, Shostak's method requires that the σ-theories have the additional property of *algebraic solvability*. A σ-theory is algebraically solvable if there exists a computable function *solve*, that takes an equation $s = t$ and returns *true* for a valid equation, *false* for an unsatisfiable equation, and, otherwise, an equivalent conjunction of equations of the form $x_i = t_i$, where the x_i's are distinct variables of $s = t$ that do not occur in any of the t_i's; notice that the t_i's may be constructed using some variables not occurring in the source equation $s = t$.

Definition 2. *Let $E = solve(e)$, t_i denote terms in some algebraic theory, and $x_i \in vars(e)$; then solve(.) is called a solver for this theory if the following requirements are fulfilled [CLS96].*

1. $\vdash e \quad \Leftrightarrow \quad solve(e) \quad$ is in the theory.
2. $E \in \{true, false\}$ or $E = \bigwedge_i x_i = t_i$.
3. If e contains no variables, then $E \in \{true, false\}$.
4. If $E = \bigwedge_i x_i = t_i$ then $x_i \in vars(e)$ and for all i, j: $x_i \notin vars(t_j)$, $x_i \neq x_j$, and $t_i \dot{=} \sigma(t_i)$.

Systems of equations E satisfying these requirements are also said to be in solved form *with respect to e.*

Solving an equation $t = u$ yields an explicit description of all satisfying models $\mathcal{I}$ of $t = u$, i.e., assignments of the variables to terms such that $\mathcal{I} \models t = u$ where semantic entailment $\models$ is defined in the usual way.

Example 1. A solved form for the equation $(x \lor y) = \neg z$ on Boolean terms is given by $\{x = a, y = b, z = \neg a \land \neg b\}$, where a, b are fresh variables.

Notice that solving is a rather powerful concept, since it can be used to decide fully-quantified equations by inspecting dependencies between the terms on the right-hand side of solved forms. The formula $\forall x, y. \exists z. (x \lor y) = \neg z$, for example, is valid since the solved form of the equation in the body (see Example 1) does not put any restrictions on the $\forall$-quantified variables. This argument holds in general.

Bit-vectors. Fixed size bit-vector terms have an associated (positive) length n, and the bits in a bit-vector of length n are indexed from 0 (leftmost) to $n - 1$ (rightmost). In the following, $s_{[n]}$, $t_{[m]}$, $u_{[k]}$ denote bit-vector terms of lengths n, m, k, respectively, and i, j are positions in bit-vectors; whenever possible we omit subscripts. Bit-vectors are either variables $x_{[n]}$, constants $\mathcal{C} := \{1_{[n]} | n \in \mathbb{N}^+\} \cup \{0_{[n]} | n \in \mathbb{N}^+\}$ (bit-vectors filled with 1's and 0's, respectively), subfield extractions $t_{[n]}[i : j]$ of $j - i + 1$ many bits i through j from bit-vector t, or concatenations $t \otimes u$; sometimes we use t^k, for $k \in \mathbb{N}^+$, to denote the k successive copies $t \otimes \ldots \otimes t$ of t. A bit-vector term $t_{[n]}$ is said to be *well-formed* if the

$$t_{[n]}[0:n-1] \rightarrow t_{[n]}$$
$$t[i:j][k:l] \rightarrow t[k+i:l+i]$$
$$(t_{[n]} \otimes u_{[m]})[i:j] \rightarrow t_{[n]}[i:j] \qquad \text{IF } j < n$$
$$\rightarrow u_{[m]}[i-n:j-n] \qquad \text{IF } n \leq i$$
$$\rightarrow t_{[n]}[i:n-1] \otimes u_{[m]}[0:j-n] \text{ IF } i < n \leq j$$
$$c_{[n]} \otimes c_{[m]} \rightarrow c_{[n+m]}$$
$$x[i:j] \otimes x[j+1:k] \rightarrow x[i:k]$$

Fig. 1. Canonizing Bitvecor Terms

corresponding set $wf(t_{[n]})$ of arithmetic constraints is satisfiable.

$$wf(p_{[n]}) := \{1 \leq n\}, \quad \text{if } p_{[n]} \text{ is a constant or a variable}$$
$$wf(t_{[n]} \otimes u_{[m]}) := wf(t_{[n]}) \cup wf(u_{[m]})$$
$$wf(t_{[n]}[i:j]) := wf(t_{[n]}) \cup \{0 \leq i, i \leq j, j < n\}$$

The set of well-formed terms of size n are collected in BV_n and BV denotes the union of all BV_n. Likewise, an equation $t_{[n]} = u_{[m]}$ is well-formed if

$$wf(t_{[n]} = u_{[m]}) := wf(t_{[n]}) \cup wf(u_{[m]}) \cup \{n = m\}$$

is satisfiable. Furthermore, $t \preceq u$ denotes the subterm relation on bit-vector terms, and the set of variables in term t (equation $t = u$) is denoted by $vars(t)$ ($vars(t = u)$).

Concatenation and extraction form a core set of operations that permits encoding other bit-vector operations like rotation or shift. Other extensions involve bitwise Boolean operations on bit-vectors (e.g. $t \, XOR \, u$) and finite arithmetic (e.g. addition modulo 2^n, $t +_{[n]} u$). Bit-vector terms of unknown size or terms with extraction at unknown positions are referred to as *non-fixed size* terms in the following.

Canonizing Bit-vector Terms. A bit-vector term t is called *atomic* if it is a variable or a constant, and *simple terms* are either atomic or of the form $x_{[n]}[i:j]$ where at least one of the inequalities $i \neq 0$, $j \neq n-1$ holds. Moreover, terms of the form $t_1 \otimes t_2 \otimes \ldots \otimes t_k$ (modulo associativity), where t_i are all *simple*, are referred to as being in *composition normal form*. If, in addition, none of the neighboring simple terms denote the same constant (modulo length) and a simple term of the form $t[i:j]$ is not followed by a simple term of the form $t[j+1:k]$, then a term in composition normal form is called *maximally connected*. Maximally connected terms are a *canonical form* for bit-vector terms. A canonical form for a term $t_{[n]}$, denoted by $\sigma(t_{[n]})$, is computed using the term rewrite system in Figure 1 (see [CMR97]).

Solvability. Fixed size bit-vector equations can readily be solved by comparing corresponding bits on the left-hand side and the right-hand side of the equation followed by propagating the resulting bit equations; such an algorithm is

$$(1) \quad p_{[n]} \otimes t = q_{[n]} \otimes u \quad \rightarrowtail \left\{ \begin{array}{l} p_{[n]} = q_{[n]} \\ t = u \end{array} \right\}$$

$$(1') \quad p_{[n]} \otimes t = q_{[m]} \otimes u, \; n < m \rightarrowtail \left\{ \begin{array}{l} p_{[n]} = \sigma(q_{[m]}[0 : n-1]) \\ \sigma(q_{[m]}[n : m-1]) \otimes u = t \\ q_{[m]} = \sigma(q_{[m]}[0 : n-1]) \otimes \sigma(q_{[m]}[n : m-1]) \end{array} \right\}$$

$$(1'') \quad p_{[n]} \otimes t = q_{[m]} \quad \rightarrowtail \left\{ \begin{array}{l} p_{[n]} = \sigma(q_{[m]}[0 : n-1]) \\ \sigma(q_{[m]}[n : m-1]) = t \\ q_{[m]} = \sigma(q_{[m]}[0 : n-1]) \otimes \sigma(q_{[m]}[n : m-1]) \end{array} \right\}$$

- -

$$(2) \quad c = d, \qquad c \neq d \rightarrowtail \textbf{FAIL}$$

$$(3) \quad t = t \qquad\qquad \rightarrowtail \{\}$$

$$(4) \quad \left\{ \begin{array}{l} p = t \\ q = u \end{array} \right\}, \quad q \preceq t \rightarrowtail \left\{ \begin{array}{l} p = t[q/u] \\ q = u \end{array} \right\}$$

$$(5) \quad \left\{ \begin{array}{l} p = q \\ q = r \end{array} \right\} \qquad \rightarrowtail \left\{ \begin{array}{l} p = r \\ q = r \end{array} \right\}$$

$$(6) \quad \left\{ \begin{array}{l} p = q \\ q = p \end{array} \right\} \qquad \rightarrowtail \{\, p = q\, \}$$

$$(7) \quad \left\{ \begin{array}{l} p = t \\ p = u \end{array} \right\} \qquad \rightarrowtail \left\{ \begin{array}{l} p = t \\ u = t \end{array} \right\}$$

$$(8) \quad c = t, \qquad t \notin C \rightarrowtail \{\, t = c\, \}$$

$$(9) \quad p = q \otimes t, \quad p \neq \sigma(q \otimes t) \rightarrowtail \{\, q \otimes t = p\, \}$$

Fig. 2. Equational Transformation System $\mathcal{C}_{\Re}$.

described, for example, in [CMR96]. This overly-eager bitwise splitting yields, not too surprisingly, exorbitant run times in many cases and the resulting solved forms are usually not succinct. The key feature of the solvers described below is to avoid case splits whenever possible.

3 Solving Fixed Size Bit-Vector Equations

In this section an equational transformation system for solving equations on fixed size bit-vector terms with concatenation and extraction is presented. Hereby, it is assumed that both the left-hand side and the right-hand side are canonized.

The rules in Figure 2 describe (conditional) transformations on sets of equations. They are of the form $E, \; \alpha \rightarrowtail E'$, where E, E' are sets of equations and α is a side condition. Such a rule is applicable if E matches a subset of the equations to be transformed and if the given side condition(s) are fulfilled. The formulation of the transformation rules in Figure 2 relies on the concept of chunks.

Definition 3. *A* chunk *is either a bit-vector variable $x_{[n]}$, an extraction from a bit-vector variable $x_{[n]}[i:j]$, or a constant $c_{[n]}$.*

In the sequel we use the convention that $c_{[n]}$, $d_{[m]}$ denote constants, $p_{[n]}$, $q_{[m]}$, $r_{[k]}$ denote chunks, and $s_{[n]}$, $t_{[m]}$, $u_{[k]}$ denote arbitrary bit-vector terms.

Rules (1)–(1") in Figure 2 split equations on terms with a chunk at the first position of a concatenation into several equations on subterms; recall from Section 2 that the canonizer σ computes maximally connected composition normal forms. Furthermore, rule (2) detects inconsistencies, rule (3) deletes trivial equations, and the remaining rules are mostly used to propagate equalities; $t[q/u]$ in rule (4) denotes replacement of all occurrences of q with u in t. Finally, rule (9) flips non-structural equations in order to make them applicable for further processing with rule (1"). Notice that none of the rules in Figure 2 introduces fresh variables.

Example 2. Let $x := x_{[16]}$, $y := y_{[8]}$, and $z := z_{[8]}$; then:

$$\{x[0:7]\otimes y = y\otimes z,\ x = x[0:7]\otimes x[8:15]\}$$
$$\overset{(1)}{\longmapsto}\quad \{x[0:7] = y,\ y = z,\ x = x[0:7]\otimes x[8:15]\}$$
$$\overset{(5)}{\longmapsto}\quad \{x[0:7] = z,\ y = z,\ x = x[0:7]\otimes x[8:15]\}$$
$$\overset{(4)}{\longmapsto}\quad \{x[0:7] = z,\ y = z,\ x = z\otimes x[8:15]\}$$

This derivation exemplifies the importance of so-called *structural* equations like $x = x[0:7]\otimes x[8:15]$. Intuitively, these equations are used to represent necessary splits of a variable. On the other hand, structural equations $p = s$ do not carry any semantic information, since $p \doteq \sigma(s)$.

Definition 4. *Given an equation $t = u$, the set of cuts for $x_{[n]} \in vars(t = u)$, denoted by $cuts(x_{[n]})$, is defined as follows:*

$$cuts(x_{[n]}) := \{-1, n-1\} \cup \{i-1, j \mid x_{[n]}[i:j] \preceq t = u\}$$

Now, the set of structural equations for $t = u$ is defined as:

$$\mathcal{SE}(t=u) := \bigcup_{x \in vars(t=u)} \{\ \sigma(x[i_0+1:i_k]) = x[i_0+1:i_1]\otimes \cdots \otimes x[i_{k-1}+1:i_k]\ \mid$$
$$i_j \in cuts(x),\ i_j < i_{j+1}$$
$$\text{and there is no } i' \in cuts(x),\ i' \neq i_j \text{ with } i_0 < i' < i_k\ \}$$

In the Example 2 above, $cuts(x_{[16]}) = \{-1, 7, 15\}$ and therefore

$$\mathcal{SE}(t=u) = \{\ x = x[0:7]\otimes x[8:15],\ y = y[0:7],\ z = z[0:7]\ \}.$$

The last two equations in this set of structural equations can be discarded, since they do not represent proper splits. Notice also that, in the worst case, the cardinality of $\mathcal{SE}(.)$ may grow quadratically with the size of the bit-vectors involved. Some simple implementation techniques for handling this kind of blow-up are listed in [Möl98].

Definition 5. *Let $t = u$ be a bit-vector equation. A set of equations $\Upsilon_\perp$ is called a solved set for $t = u$ if: 1. $\Upsilon_\perp$ is equivalent with $\{t = u\}$. 2. no rule of $\mathcal{C}_\mathfrak{R}$ is applicable for $\Upsilon_\perp$. 3. for each $x \in vars(t = u) \cap \Upsilon_\perp$, $\Upsilon_\perp$ contains an equation of the form $x = s$.*

A solved set does not constitute a solved form in the sense of Definition 2, since some variable might occur on both sides of an equation. Given a solved set, however, it is straightforward to construct an equivalent solved form by omitting equations not of the form $x_i = u$ and replacing occurrences of extractions right-hand sides with fresh variables. Given the solved set $\{x[0 : 7] = z, \; y = z, \; x = z \otimes x[8 : 15]\}$ of Example 2, the set $\{x = z \otimes a, \; y = z\}$, where a is a fresh variable, is an equivalent solved form. This can be done in general.

Lemma 1. *For each solved set $\Upsilon_\perp$ of equations there is an equivalent solved form $\Upsilon'_\perp$.*

Now we are in the position to state a soundness and completeness result for the equational transformation system in Figure 2. This theorem, together with the construction for proving Lemma 1, determines a solver in the sense of Definition 2.

Theorem 1. *Let $t = u$ be an equation on fixed size, canonized bit-vector terms. Starting with the initial set of equations $\{t = u\} \cup \mathcal{SE}(t = u)$, the equational transformation system $\mathcal{C}_\mathfrak{R}$ in Figure 2 terminates with an (equivalent) solved set for $t = u$.*

It can easily be checked that the transformation rules (1)–(9) in Figure 2 are equivalence-preserving. The transformation process terminates, since, first, rules (1)–(1") decrease the lengths of bit-vector terms, second, rules (2),(3),(6) decrease the number of equations in the respective target sets, and third, the rules (4),(5),(7),(8) do not enlarge target sets and, together with rule (6), they construct a unique representation of every chunk. This process of equality propagation is terminating. Finally, rule (9) flips an equation as a preprocessing step for applying rule (1"); it cannot be applied repeatedly. The form of the rules together with the initial set $\mathcal{SE}(t = u)$ guarantees that the terminal set is a solved set. In particular, the set $\mathcal{SE}(t = u)$ includes for each (non-arbitrary) variable x in $vars(t = u)$ an equation of the form $x = s$ in order to match the third requirement of Definition 5.

4 Solving Bit-Vector Equations of Non-Fixed Size

In this section we develop a conceptual generalization of the equational transformation system $\mathcal{C}_\mathfrak{R}$ that is capable of solving equations on non-fixed size bit-vector terms like the word problems below.

Example 3.

$$x_{[n]} \otimes 0_{[1]} \otimes y_{[m]} = z_{[2]} \otimes 1_{[1]} \otimes w_{[2]} \tag{1}$$

$$x_{[l]} \otimes 1_{[1]} \otimes 0_{[1]} = 1_{[1]} \otimes 0_{[1]} \otimes x_{[l]} \tag{2}$$

$$(1)\ \left(\{p_{[n]}\otimes t = q_{[m]}\otimes u\}\cup E,\ \Psi\right)\ \longmapsto\ \left[\begin{array}{l}\left[\left(\begin{array}{l}\left\{\begin{array}{l}p_{[n]} = \sigma(q_{[m]}[0:n-1]),\\ \sigma(q_{[m]}[n:m-1])\otimes u = t,\\ q_{[m]} = q_{[m]}[0:n-1]\otimes q_{[m]}[n:m-1]\end{array}\right\}\cup E,\\ \{n < m\}\cup\Psi\end{array}\right)\right]\\ \left(\begin{array}{l}\left\{\begin{array}{l}p_{[n]} = q_{[m]},\\ t = u\end{array}\right\}\cup E,\\ \{n = m\}\cup\Psi\end{array}\right)\\ \left[\left(\begin{array}{l}\left\{\begin{array}{l}q_{[m]} = \sigma(p_{[n]}[0:m-1]),\\ \sigma(p_{[n]}[m:n-1])\otimes t = u,\\ p_{[n]} = p_{[n]}[0:m-1]\otimes p_{[n]}[m:n-1]\end{array}\right\}\cup E,\\ \{n > m\}\cup\Psi\end{array}\right)\right]\end{array}\right]$$

$$(1)^{*}\ \left(\{x_{[n]}[i:j] = x_{[n]}[k:l]\}\cup E,\Psi\right)\ \longmapsto\ \left[\begin{array}{l}\left[\left(\begin{array}{l}\left\{x_{[n]}[i:l] = a_{[k-i]}{}^{\frac{l-i+1}{k-i}}\right\}\cup E,\\ \{(l-i+1)\mid (k-i)\}\cup\Psi\end{array}\right)\right]\\ \left(\begin{array}{l}\left\{x_{[n]}[i:l] = a_{[h]}\otimes(b_{[h']}\otimes a_{[h]})^{\frac{l-i-h+1}{k-i}}\right\}\cup E,\\ \{(l-i+1)\nmid (k-i)\}\cup\Psi\end{array}\right)\end{array}\right]$$

where $i < k$

where $h = (l-i+1)\mathrm{MOD}(k-i)$,
$h' = k-i-h$,
a, b fresh variables.

Fig. 3. Split Rules of $\mathcal{S}_{\Re}$.

Equation (1) is solvable if and only if $n = 1$, $m = 3$ or $n = 3, m = 1$, while Equation (2) is solvable if and only if l is even. Thus, both equations can not be solved in the strict sense of Definition 2. Instead, they motivate not only the need for representing all solutions as a disjunction of solved forms but also for integrating integer reasoning into the process of solving bit-vector equations.

Definition 6. *A frame is a pair (Υ, Ψ) consisting of a set of bit-vector equations Υ and a set Ψ of integer constraints of the form $n < m$, $n = m$, $n \mid m$ (divisibility), or $n \nmid m$ (non-divisibility) for $n, m \in \mathbb{N}^{+}$. Let $V_{BV}, V_{\mathbb{N}^{+}}$ be sets that include the bit-vector variables and the natural number variables of some frame (Υ, Ψ), respectively; then an interpretation of (Υ, Ψ) is a function $\mathcal{I} : V_{BV} \cup V_{\mathbb{N}^{+}} \to BV \cup \mathbb{N}^{+}$ such that $\mathcal{I}(x_{[n]}) \in BV_n$ and $\mathcal{I}(n) \in \mathbb{N}^{+}$. This determines notions like satisfiability or consistency of a frame in the usual way.*

A set of frames is called satisfiable if there is at least one satisfiable frame and two sets of frames are equivalent if their sets of satisfying interpretations coincide. Furthermore, a set Ξ of frames is called disjoint, if for each (Υ_i, Ψ_i), $(\Upsilon_j, \Psi_j) \in \Xi$, $i \neq j$, the conjunction of their integer constraints is inconsistent; i.e., if $\Psi_i \wedge \Psi_j \models \bot$.

$$(1'') \quad \left(\{p_{[n]} \otimes t = q_{[m]}\} \cup E, \ \Psi \right) \longmapsto \left(\left. \begin{cases} p_{[n]} = \sigma(q_{[m]}[0:n\text{-}1]), \\ \sigma(q_{[m]}[n:m\text{-}1]) = t, \\ q_{[m]} = q_{[m]}[0:n\text{-}1] \otimes q_{[m]}[n:m\text{-}1] \end{cases} \right\} \cup E, \ \Psi \right)$$

$$(2) \quad \left(\{c_{[n]} = d_{[n]}\} \cup E, \ \Psi \right), \ c \neq d \qquad \longmapsto \quad \textbf{FAIL}$$

$$(3) \quad \left(\{t = t\} \cup E, \ \Psi \right) \qquad\qquad\qquad \longmapsto \quad (E, \ \Psi)$$

$$(4) \quad \left(\begin{Bmatrix} p = t \\ q = u \end{Bmatrix} \cup E, \ \Psi \right), \ q \preceq t \qquad \longmapsto \quad \left(\begin{Bmatrix} p = t[q/u] \\ q = u \end{Bmatrix} \cup E, \ \Psi \right)$$

$$(5) \quad \left(\begin{Bmatrix} p = q \\ q = r \end{Bmatrix} \cup E, \ \Psi \right), \ a \text{ fresh} \quad \longmapsto \quad \left(\begin{Bmatrix} p = a \\ q = a \\ r = a \end{Bmatrix} \cup E, \ \Psi \right)$$

$$(6) \quad \left(\begin{Bmatrix} p = q \\ q = p \end{Bmatrix} \cup E, \ \Psi \right), \ a \text{ fresh} \quad \longmapsto \quad \left(\begin{Bmatrix} p = a \\ q = a \end{Bmatrix} \cup E, \ \Psi \right)$$

$$(7) \quad \left(\begin{Bmatrix} p = t \\ p = u \end{Bmatrix} \cup E, \ \Psi \right) \qquad\qquad \longmapsto \quad \left(\begin{Bmatrix} p = t \\ u = t \end{Bmatrix} \cup E, \ \Psi \right)$$

$$(8) \quad \left(\{c = t\} \cup E, \ \Psi \right), \qquad t \notin C \quad \longmapsto \quad \left(\{t = c\} \cup E, \ \Psi \right)$$

$$(9) \quad \left(\{p = q \otimes t\} \cup E, \ \Psi \right), \ p \neq \sigma(q \otimes t) \longmapsto \left(\{q \otimes t = p\} \cup E, \ \Psi \right)$$

$$(10) \quad (\Upsilon, \ \Psi), \qquad\qquad\qquad \Psi \models \bot \quad \longmapsto \quad \textbf{FAIL}$$

Fig. 4. Simple Rules of $\mathcal{S}_{\Re}$.

A frame transformation system called $\mathcal{S}_{\Re}$ is presented in Figures 3 and 4. The form of these rules is largely motivated by the equational transformation system in Figure 2 with the additional provision of case splits in rule (1) and (1)*. Application of rule (1), for example, replaces a source frame (Υ, Ψ) by three target frames (Υ_i, Ψ_i). It is easily seen that the property of disjointness is preserved by the frame transformation system $\mathcal{S}_{\Re}$. The rules (1")–(9) in Figure 4 only operate on sets of bit-vector equations and do not alter any integer constraints, while rule (10) is used to delete frames with inconsistent integer constraints. Hereby, the side condition $\Psi \models \bot$ of rule (10) is decidable. This can be shown by a simple reduction to the Diophantine problem for addition and divisibility that was proven decidable, for example, in [Lip78, Lip81].[1] Altogether, $\mathcal{S}_{\Re}$ yields $\{(\emptyset, \emptyset)\}$ for tautologies, whereas unsatisfiable formulae are eventually reduced to $\{\}$. Notice also that, in contrast to the rule system $\mathcal{C}_{\Re}$ in Figure 2, the frame transformation system $\mathcal{S}_{\Re}$ may introduce fresh variables a and b.

---

[1] The Diophantine problem for addition and divisibility is equivalent to the decidability of the class of formulas of the form $\exists x_1, \ldots, x_n. \bigwedge_{i=1}^{k} A_i$ in natural numbers, where the A_i have the form $x_m = x_j + x_k$, $x_m | x_j$, or $x_m = p$ and p is a natural number [DMV95].

Theorem 2 states a correctness result for this frame transformation system that can be proved similarly to Theorem 1; in contrast to Theorem 1, however, it does not imply termination of the transformation process.

Theorem 2. *Let $t = u$ be an equation on canonized and (possibly) non-fixed size bit-vector terms. Define the* initial frame *(Υ_0, Ψ_0) by*

$$(\{t = u\} \cup \mathcal{SE}(t = u), \; wf(t = u)).$$

If the process of applying rules $S_\mathfrak{R}$ to the singleton set $\Xi_0 := \{(\Upsilon_0, \Psi_0)\}$ terminates with a set of frames Ξ_f, then:

1. *Ξ_f contains only solved forms (for $t = u$) in the sense of Definition 2.*
2. *Ξ_f is disjoint.*
3. *Ξ_0 and Ξ_f are equivalent.*

The rules in Figures 3, 4 and the overall structure of the solver may be best explained by means of an example. Consider, for example, solving the word problem (2) in Example 3. In this example we freely simplify integer constraints by omitting weakest constraints.

Example 4. There are no structural equations for

$$x_{[l]} \otimes 1_{[1]} \otimes 0_{[1]} = 1_{[1]} \otimes 0_{[1]} \otimes x_{[l]}$$

and the corresponding well-formedness constraints are given by the singleton set $\{1 \leq l\}$. Thus, solving starts with the initial frame

$$\Xi_0 ::= \{((\{x_{[l]} \otimes 1_{[1]} \otimes 0_{[1]} = 1_{[1]} \otimes 0_{[1]} \otimes x_{[l]}\}, \{1 \leq l\}))\}.$$

Rule (1) in Figure 3 matches with the leftmost chunks $x_{[l]}$ and $1_{[1]}$ of the equation e, and application of this rule yields the following three frames:

$$\Xi_1 ::= \left\{ \begin{array}{ll} (\{\ldots\}, & \{1 \leq l, l < 1\}), \\ (\{x_{[l]} = 1_{[1]}, \; 1_{[1]} \otimes 0_{[1]} = 0_{[1]} \otimes x_{[l]}\}, & \{1 \leq l, l = 1\}), \\ (\{x_{[l]}[0:0] = 1_{[1]}, \; x_{[l]}[1:l\text{-}1] \otimes 1_{[1]} \otimes 0_{[1]} = 0_{[1]} \otimes x_{[l]}\}, & \{1 \leq l, l > 1\}) \end{array} \right\}$$

The first frame is inconsistent and can be deleted via rule (10). Likewise, the second frame in Ξ_1 vanishes, since rule (2) eventually triggers for the equation

$$1_{[1]} \otimes 0_{[1]} = 0_{[1]} \otimes x_{[1]}.$$

It remains to process

$$(\Upsilon_3, \Psi_3) ::= (\{x_{[l]}[0:0] = 1_{[1]}, \; x_{[l]}[1:l\text{-}1] \otimes 1_{[1]} \otimes 0_{[1]} = 0_{[1]} \otimes x_{[l]}\}, \{l > 1\}).$$

In an attempt to match the second equation of Υ_3 with rule (1), three new frames $(\Upsilon_{31}, \Psi_{31})$, $(\Upsilon_{32}, \Psi_{32})$ and $(\Upsilon_{33}, \Psi_{33})$ are generated with additional constraints

$l < 2$, $l = 2$ and $l > 2$, respectively. Now, Ψ_{31} yields *false* and $(\Upsilon_{32}, \Psi_{32})$ terminates, after simplifying the integer constraints, with the frame

$$(\{x_{[2]} = \mathbf{1}_{[1]} \otimes \mathbf{0}_{[1]}\}, \{l = 2\}).$$

In frame $(\Upsilon_{33}, \Psi_{33})$, the rule $(1)^*$ matches equation $x_{[l]}[0 : l - 3] = x_{[l]}[2 : l - 1]$. This yields two frames $(\Upsilon_{331}, \Psi_{331})$ and $(\Upsilon_{332}, \Psi_{332})$ such that l is required to be even in the first frame and odd in the second frame. In Υ_{331}, the equation $x_{[l]} = (a_{[2]})^{\frac{l}{2}}$ is added, thereby terminating with the frame

$$\left(x_{[l]} = (\mathbf{1}_{[1]} \otimes \mathbf{0}_{[1]})^{\frac{l}{2}}, \{l > 2,\ 2 \mid l\}\right).$$

Since Υ_{332} contains the equation $x_{[l]} = b_{[1]} \otimes (d_{[1]} \otimes b_{[1]})^{\frac{l-1}{2}}$, together with the equations $x_{[l]}[0 : 0] = \mathbf{1}_{[1]}$ and $x_{[l]}[l - 1 : l - 1] = \mathbf{0}_{[1]}$ this frame reduces to *false*. Altogether, solving terminates with the two frames:

$$\left\{ \begin{array}{l} (\{x_{[2]} = \mathbf{1}_{[1]} \otimes \mathbf{0}_{[1]}\},\quad \{l = 2\}) \\ (\{x_{[l]} = (\mathbf{1}_{[1]} \otimes \mathbf{0}_{[1]})^{\frac{l}{2}}\}, \{l > 2, 2 \mid l\}) \end{array} \right\}$$

Thus, the set of solutions is characterized by $x_{[l]} = (\mathbf{1}_{[1]} \otimes \mathbf{0}_{[1]})^{\frac{l}{2}}$ for all even l's greater or equal to 2. In special situations like the one above, a refinement of the solving algorithm could try to merge 'compatible' frames as soon as possible.

5   Conclusions

We have presented two specialized algorithms for solving equations on fixed size bit-vectors and for non-fixed size bit-vectors built up from concatenation and extraction operators. Both solvers adhere to the principle of splitting bit-vectors only on demand.

In the case of the fixed size bit-vector solver, this feature leads to moderate run times for large data paths. Even better, for some equations, solving time is independent of the size of data paths. However, the restriction to concatenation and extraction causes bit-wise splitting in situations where the regularity of solved forms cannot be expressed succinctly in terms of these operators only (for examples see [BDL98] or [BP98]). An extension of the fixed size solver with an iteration operator as proposed in [BP98] handles these cases well. More importantly, a useful bit-vector solver should support a rich set of operators including bitwise Boolean operators and finite arithmetic. We have demonstrated in [CMR97] that bitwise Boolean operators can be added in a conceptually clean way using the notion of *bit-vector OBDDs*. Again, splits are only performed on demand. OBDDs can also be used to encode finite arithmetic operations as Boolean functions. Using a ripple-carry adder for adding (modulo 2^n) two bit-vectors of length n in a naive way, however, yields overly-eager bitwise splitting and the resulting solver is, in our experience, not useful in practice. Barrett et al [BDL98] approach this problem by introducing a normal form geared to support arithmetic directly. In order to perform splits lazily, they introduce an overflow operator for ripple-carry additions. Their normalizer, however, does not possess the property of canonicity, and their algorithm does not seem to directly support bitwise Booleans operators.

The solver for non-fixed size bit-vector equations has been developed as a generalization of the algorithm for fixed size bit-vector equations. With a similar motivation in mind, Bjørner and Pichora [BP98]—independently—developed an algorithm for solving special cases of bit-vector equations of non-fixed size; their approach, however, is restricted to processing equations including one unknown size only, while our solver permits processing equations containing several unknowns. On the other hand, the solver in [BP98] is known to be terminating on the given fragment, while it is unknown if the algorithm described in Section 4 terminates for all input equations. If it is indeed terminating then it may be used to decide word equations [Mak92, PR98]. It has been shown, however, that any non-fixed size solver that supports a richer set of operators—as required for most hardware applications—is necessarily incomplete, since the halting problem can be reduced to solve non-fixed size equations on bit-vectors built up from concatenation, extraction, and bitwise Boolean operators only [Möl98].

Acknowledgements. We would like to thank Nikolaj Bjørner and David Cyrluk for their invaluable suggestions. In particular, Nikolaj pointed us to [Lip78]. Furthermore we thank the anonymous referees and Holger Pfeifer for their comments which helped to improve the presentation.

References

[BDL98] C.W. Barrett, D.L. Dill, and J.R. Levitt. A Decision Procedure for Bit-Vector Arithmetic. In *Proceedings of the 35th Design Automation Conference*, June 1998. San Francisco, CA.

[BP98] N.S. Bjørner and M.C. Pichora. Deciding Fixed and Non-Fixed Size Bit-Vectors. volume 1384 of *Lecture Notes in Computer Science LNCS*, pages 376–392, Heidelberg - New York - Berlin, April 1998. Springer.

[CLS96] D. Cyrluk, P. Lincoln, and N. Shankar. On Shostak's Decision Procedure for Combinations of Theories. In M. A. McRobbie and J. K. Slaney, editors, *Proceedings of the Thirteenth International Conference on Automated Deduction (CADE-96)*, volume 1104 of *LNAI*, pages 463–477, Berlin;Heidelberg;New York, 1996. Springer.

[CMR96] D. Cyrluk, M.O. Möller, and H. Rueß. An Efficient Decision Procedure for a Theory of Fixed-Size Bitvectors with Composition and Extraction. Ulmer Informatik-Berichte 96-08, Fakultät für Informatik, Universität Ulm, 1996.

[CMR97] D. Cyrluk, M.O. Möller, and H. Rueß. An Efficient Decision Procedure for the Theory of Fixed-Sized Bit-Vectors. In O. Grumberg, editor, *Computer Aided Verification. 9th International Conference (CAV97). Haifa, Israel, June 22-25, 1997: Proceedings*, volume 1254 of *Lecture Notes in Computer Science LNCS*, pages 60–71, Berlin - Heidelberg - New York, 1997. Springer.

[DMV95] A. Degtyarev, Yu. Matiyasevich, and A. Voronkov. Simultaneous Rigid E-Unification is not so Simple. UPMAIL Technical Report 104, Uppsala University, Computing Science Department, 1995.

[Lip78] L. Lipshitz. The Diophantine Problem for Addition and Divisibility. *Transactions of the American Mathematical Society*, 235:271–283, January 1978.

[Lip81] L. Lipshitz. Some Remarks on the Diophantine Problem for Addition and Divisibility. *Bull. Soc. Math. Belg. Sér. B*, 33(1):41–52, 1981.

[Mak92] G. S. Makanin. Investigations on equations in a free group. In K.U. Schulz, editor, *Proceedings of Word Equations and Related Topics (IWWERT '90)*, volume 572 of *LNCS*, pages 1–11, Berlin;Heidelberg;New York, 1992. Springer.

[Möl98] M.O. Möller. Solving Bit-Vector Equations—A Decision Procedure for Hardware Verification, 1998. Diploma Thesis, available from http://www.informatik.uni-ulm.de/ki/Bitvector/.

[PR98] W. Plandowski and W. Rytter. Application of Lempel-Ziv Encodings to the Solution of Words Equations. In *Proceedings of the 25th International Colloquium on Automata, Languages and Programming, ICALP'98*, pages 731–742, July 1998. Aalborg, Denmark.

[Sho84] R.E. Shostak. Deciding Combinations of Theories. *Journal of the ACM*, 31(1):1–12, January 1984.

[SM95] M.K. Srivas and S.P. Miller. Formal Verification of the AAMP5 Microprocessor. In M.G. Hinchey and J.P. Bowen, editors, *Applications of Formal Methods*, International Series in Computer Science, chapter 7, pages 125–180. Prentice Hall, Hemel Hempstead, UK, 1995.

The Formal Design of 1M-gate ASICs

Ásgeir Þór Eiríksson

Silicon Graphics Inc.
Mountain View, California
asgeir.eiriksson@computer.org

Abstract. We describe the refinement of a directory based cache coherence protocol specification, to a pipelined hardware implementation. The hardware that is analyzed is the most complex part of a 1M-gate ASIC. The design consists of 30000 lines of synthesizable register transfer-level verilog code. The design contains a pipeline that is 5 levels deep and approximately 150 bits wide. It has a 16 entry, 150 bit wide, context addressable memory (CAM), and has a 256x72 bit RAM. Refinement maps relate the high-level protocol model to the hardware implementation. We used the Cadence Berkeley Labs SMV model checker to create the maps and to prove their correctness. There are approximately 2000 proof obligations. The formal model has been used for three tasks. First, to formally diagnose, and then fix broken features in a legacy version of the design. Second, to integrate the legacy sub-system design with a new system design. Finally, it has been used to formally design additional sub-system features required for the new system design. The same hardware designer enhanced the design, created the refinement maps, and formally proved the correctness of the refinements.

1 Introduction

We have previously shown that it is possible to formally design and implement complex directory based cache coherence protocols [1, 2, 3]. The protocol that we refer to here is an integral part of the Origin 200, and Origin 2000 servers from Silicon Graphics Inc. [4, 5, 6]. These systems have now been shipping for 2 years, with up to 128 processors, and there have to date been no protocol bugs found in the lab or the field. We have however uncovered and had to fix a few sequencing and flow control logic problems.

The problems that were found after first silicon can all be characterized as subtle malfunctions involving extremely rare combinations of states in the interaction of 3-5 protocol message sequencing and flow control state machines. It is our experience that it is extremely difficult to hit these bugs in simulation, even after you know that they exist. We would therefore have had to formally analyze the sequencing and flow control of the protocol messages, in order to prevent these problems in first silicon.

The cache coherence protocol model has coherent processor read, and write-back requests, but in addition, it implements hardware coherent Direct Memory Accesses (DMA), and cache line size Block Transfer Engine (BTE) requests.

A majority of the bugs were in the DMA, and BTE block of the chip. This block implements a pipelined DMA/BTE protocol-processor. The block consists of 30.000 lines of synthesizable register-transfer-level (RTL) verilog code (approximately 400.000 gates after synthesis). It has1000 bits of pipeline register and control state, approximately 3000 bits of state in a Context Addressable Memory (CAM), and a merge cache RAM with approximately 18000 bits of state. The data consistency properties of the RAM control logic is ongoing work, and outside the scope of this paper.

Protocol message delivery to the pipeline is inherently asynchronous in nature, and we therefore require the full generality of model checking. Due to the asynchrony/non-determinism, we can for example not resort to the symbolic trajectory evaluation technique [7] that potentially are tractable for larger state spaces than model checking.

The capacity of symbolic model checking, in most cases, is limited to 150-250 bits of state. It is therefore evident that we have to resort to compositional methods in order to analyze the pipelined protocol-processor implementation. To achieve this we employ the refinement capabilities of the Cadence Berkeley Labs (CBL) SMV model checker [8, 9]. We base the analysis of the pipelined protocol implementation on refinement maps from an abstract protocol model to the detailed hardware implementation.

The paper presents the results of using the refinement capabilities of the CBL SMV symbolic model checker [8, 9] to refine the abstract protocol model, to the above DMA/BTE pipelined protocol-processor implementation. Deciding on an overall refinement strategy, and creating the refinement maps require deep insights into the design. Therefore the hardware engineer responsible for the design devised the refinement strategy, wrote the refinement maps, and proved their correctness.

The major contribution of this paper is to demonstrate how compositional model checking techniques can be used to derive tractable hierarchical formal models, which are then used to analyze large and complex hardware designs. The paper also demonstrates that compositional model checking techniques have reached a level of maturity where it's possible for a hardware designer to formally design complex hardware within the strict time schedule of a design project.

The rest of the paper is organized as follows. We first give an overview of design by refinement with the CBL SMV tool. Then we go on to describe the system that we want to analyze. We then describe the refinement strategy, and refinement maps for the design. We finally present our conclusions and possible future extensions.

2 Compositional Model Checking

The refinement framework within CBL SMV is based on signal layers and trace containment [8,9]. When a particular signal appears at multiple layers, this automatically generates a refinement proof obligation from the high level version of the signal to the lower level version. The default order of signal layers, from higher, e.g. `layer1`, to lower, e.g. `layer2`, is based on textual order, or it can be specified explicitly using the 'refines' construct, e.g. `layer2 refines layer1`. The following CBL SMV code fragment shows a signal stack, where `s1` is used at `layer1` and `layer2`:

```
stack {
  layer1 : {
   s1 := f_layer1;
  }
  layer2 : {
   s1 := f_layer2;
  }
}
```

When proving a particular refinement of signal `s1`, at `layer1`, CBL SMV by default uses the highest level version of all the signals in the transitive fan-in of signal `s1//layer1`, and `s1//layer2`. This default behavior is changed with the 'using-prove' construct. For example if we need signal `s2` from `layer2`, to prove the refinement of `s1//layer1`, then we would write 'using `s2//layer2` prove `s1//layer1`'. CBL SMV then uses the default versions, i.e. highest available versions, of all signals in the transitive fan-in of `s1//layer1`, and `s1//layer2`, except it will use the `layer2` version of `s2`.

The refinement proof amounts to CBL SMV proving the invariant temporal logic assertion G (`s1//layer2` ⊂ `s1//layer1`), where ⊂ is set inclusion. An informal interpretation of this assertion is that `s1//layer2` is a refinement of `s1//layer1`, if for all reachable states of the model, the values of `s1//layer2` are contained in the set of possible values of `s1//layer1`.

The CBL SMV tool manages the list of refinements that need to be proven. It is possible to instruct CBL SMV to automatically prove all the out of date refinement proofs, or to print the list of all refinements that need to be proven. The latter option is useful for large designs. In this case we take the list of properties, and use a batch-scheduling tool to issue multiple CBL SMV jobs, where each job proves one refinement.

The most crucial task, in our experience, when refining a complex large design, is the decomposition of the design into partitions that are tractable for model checking, i.e. partitions with no more than 200 bits of state.

3 System Overview

The Origin 2000 directory-based distributed shared memory machine, shown in Figure 1, consists of nodes of one or two processors, physical memory, directory memory, a node controller Hub, XIO crossbar interconnect, IO devices, and a scalable interconnect with 2-512 different nodes [4, 5, 6]. The Hub chip, shown in Figure 1, is composed of a crossbar, processor interface, network interface, IO Interface, and a Memory/Directory controller. The coherence protocol is implemented in the Hub chip. The sub-system that we analyze implements the hardware coherent DMA and cache line sized BTE requests.

A programmer of the Origin machine is presented with a single, linear, virtual address space that is shared among all the processes running on the different nodes, that have distributed physical memory.

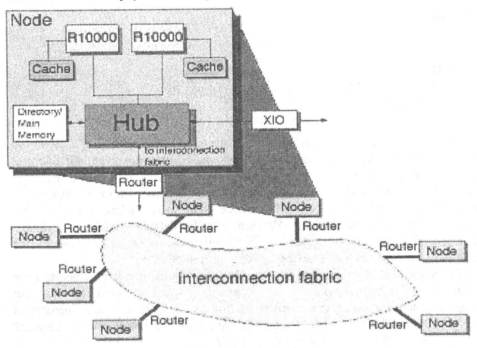

Fig. 1. Origin2000 System Overview

The cache coherence protocol is the set of rules that ensure that the different processor and I/O caches contain coherent values for the same memory location, and that the order of writes to different locations, as seen by the programmer, are consistent with the memory consistency model. The coherence protocol is invalidation/intervention based, and together with the processor, supports the sequentially consistent memory model. The directory memory stores information about the state of a particular cache line. The protocol is non-blocking i.e. it never buffers requests at the directory memory. If the memory does not have ownership, the directory-state is modified to signal that the request was handled. The requests are forwarded as an intervention or invalidate to the processor or I/O merge cache that does have ownership. An invalidate-message is forwarded if a processor has a shared copy, an intervention is sent if a processor or I/O merge cache has an exclusive copy. In order to be independent of the network topology, the protocol does not rely on network ordering. For further detail on the cache coherence protocol the reader is referred to [1,5].

4 Design Refinement

In this section we begin by describing the protocol model in more detail and how we map it onto the pipelined hardware implementation. We then describe the refinement strategy, and finally the refinement maps.

4.1 Cache Coherence Protocol Model

The cache coherence protocol specification consists of a collection of multiple input, multiple output state machine tables, that determine the response to incoming messages in terms of state changes and outgoing messages. The tables serve as input to CBL SMV, and to verilog RTL output and next-state function generation programs.

The protocol model is shown in Figure 2. The model consists of the following three types of clusters: directory memory cluster M, processor cluster P, and an I/O cluster. In addition there is a model for the interconnection network. For details we refer the reader to [1] which describes the model in more detail.

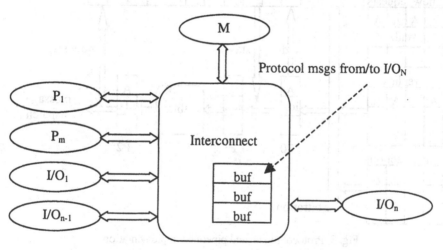

Fig. 2. Cache coherence protocol model

The M, P, and I/O models consist of translated version of the protocol tables, and logic to respond to messages, and in the case of the P and I/O models, logic to generate the possible types of requests. The M cluster receives requests from the P and I/O clusters, and also receives directory cache line ownership transfer responses.

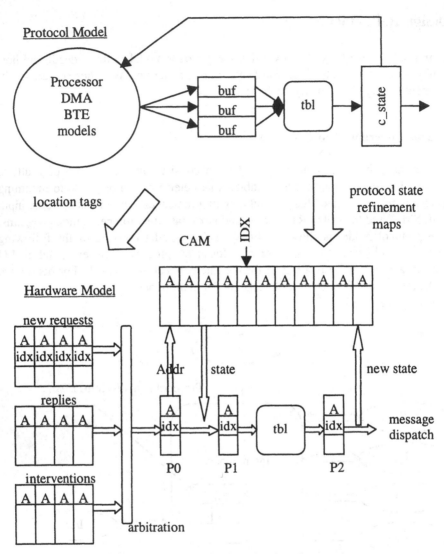

Fig. 3. Protocol model and pipelined implementation

The P requests consist of the following coherence requests: read shared, read exclusive, an upgrade of a shared line to exclusive state, and a write back to memory of a dirty exclusive line. The I/O requests include hardware coherent DMA read and write requests for partial and a full cache line; write back of exclusive lines from the merge cache. It also includes BTE block transfer, pull requests, and block transfer push requests. To simplify the description of the refinement strategy and refinement maps, and avoid repetition, we will only describe the DMA requests in the following. The refinement model we have developed also includes the BTE requests. Wherever we say DMA in the following, the actual refinement model also contains cases for each of the two BTE streams.

The network interconnect is modeled as a collection of in-flight message buffers. The decision as to which messages (if any) to deliver at any given time is non-deterministic. Thus, the time from sending a message to its delivery is completely arbitrary. This is important, since the protocol does not rely on any ordering properties of the network. If no message is delivered at a given time, a P or I/O cluster is non-deterministically chosen to issue a request; it either selects one of the possible requests or stays idle. This ensures that interconnect can drain all messages in flight to their destination without producing new messages.

4.2 Pipeline Description

The protocol message-processing pipeline, shown in the bottom half of Figure 3, consists of the following hardware units: an arbiter, a CAM, a protocol table, pipeline registers, and control logic to sequence information between the different units. The CAM stores the state information associated with a particular DMA request. It is read by presenting an address A and performing a context addressable lookup. It is written using a CAM entry index Idx. The index Idx is contained in new requests, but is derived from the CAM lookup for other protocol messages. The protocol table is combinational logic that determines the next protocol state and possibly new protocol messages, for a particular CAM entry, based on the arriving message, and the current state in the CAM entry. The registers are used to implement a pipelined protocol message processor. The control logic is complex and subtle. It sequences message headers to the protocol table, prevents hazards in the pipeline, and ensures data consistency when sequencing data into and out of the merge cache.

Initially when a new request arrives from the I/O bus, it is issued a CAM index from the pool of free CAM entries, then stored in an I/O pending request register. Once in the pending register, it requests a grant from the arbiter controlling access to the p0 stage of the pipeline. Before getting the arbitration grant the new request needs to contend with other new requests, and interconnect intervention, acknowledge, and reply messages. When a new request eventually advances to the p0 stage of the pipeline it does an address lookup in the CAM to determine if the new request is in conflict with another outstanding request to the same address. Next the new request proceeds to the p1 stage of the pipeline. The p1 stage of the pipeline contains the state contained in the new request, e.g. allocated CAM index Idx, request type, address of request, and source of the request. When the new request, address, hits in the CAM, the p1 stage of the pipeline also contains state information fetched from the CAM. The p1 state, and various error control, and presence/absence of congestion, is used to look up the next protocol state, and protocol messages generated by the new request, which then proceeds to the p2 stage. The CAM entry with index Idx is updated from the p2 stage, and if a protocol message is generated, state is sent to the new message dispatch engine, which is stage p3.

Protocol intervention, acknowledge, and reply messages from interconnect proceed through the pipeline in a similar fashion. The main difference is that they initially don't have any CAM index information, rather they use an address to access the CAM state, and to determine the CAM index Idx that in turn is used at stage p2 when updating the CAM state. Once they win arbitration, they proceed to the p0 stage, and use a 40-bit address to look up the protocol state for a particular transaction in the CAM. The protocol state, and the CAM state then proceed to the p1 stage, look up the next protocol action, and proceed to stage p2. The CAM state at entry Idx is again updated from stage p2, and if necessary state is sent to the new message dispatch unit at stage p3. If a particular message from interconnect concludes the request, the CAM entry used by the request is returned to the pool of free CAM entries.

When we refine the protocol model to the pipelined implementation, we break the delivery of a protocol message to its destination into hardware sequencing steps. We use the hardware pipeline control logic to sequence the message through the pipeline and we tag the message with its location within the pipeline. Instead of immediately producing a new protocol request, a new protocol request is required to sequence through the hardware. In turn it raises an arbitration request, receives a grant, does a CAM lookup, and proceeds to lookup the protocol message that is finally sent by the dispatch unit.

4.3 Outline of Refinement Strategy

Our strategy is to refine the behavior of the protocol for a single non-deterministically chosen CAM entry IDX, that we associate with a particular address A. We associate all the other CAM indices nIDX with non-deterministic addresses ~A. The ~A addresses differ from A in one bit position, but are otherwise unconstrained.

This strategy is motivated by the following two key observations. First, that in the p0 stage of the pipeline, the address of a particular request/reply is used to fetch the state of the correct CAM entry. It is therefore sufficient, for our purposes, to be able to distinguish our address A from all other addresses ~A. Specifically we do not need to be able to distinguish all the different ~A addresses. Finally, in the p2 stage of the pipeline, the CAM index is used to update the correct CAM entry. Again we only need to determine if the CAM index at stage p2 is IDX or nIDX.

We therefore refine a single CAM index IDX protocol model, and keep track of enough pipeline state information for requests and replies to other indices nIDX to ensure no state corruption for the chosen index IDX. In order to achieve this it is sufficient to remember if the state of a particular pipeline stage is state for our chosen CAM index IDX.

The above strategy hinges on a CAM correctness assumption that we do not prove. We are assuming that the CAM operates correctly for all 40-bit addresses not just the chosen address A, and the derived addresses ~A that we use in the abstract model. We will conclude, based on this assumption, that all our abstract protocol refinement results are valid for all possible 40-bit addresses.

A new request proceeds through the abstract pipeline in the following manner: First, when a request is being stored into a pending request register the model non-deterministically decides to associate it with (A,IDX) or (~A,nIDX). If it decides the former, the pipeline sequencing logic sets the valid bit for the pending register, and a refinement map also sets a tag variable dma_pnd, that indicates that a request we are refining from the protocol model is currently in a pending request register. Based on this initial non-deterministic decision, a refinement map tracks the progress of the (A,IDX) request through the pipeline, e.g. tag variable dma_p0 is set when it proceeds to stage p0, and a tag variable dma_p1 when it reaches stage p1. No tag variables are set for the (~A,nIDX) requests.

When address A is looked up in the CAM, the state is fetched for entry IDX, modeled by the protocol state c_state (see Figure 3). When any other address ~A is looked up in the CAM, the fetched state is undefined. When a request has progressed to the p1 stage, the model looks up an actual protocol action if the state at p1 is associated with (A,IDX), otherwise the protocol action is undefined. It is still possible to prove refinement of the high-level protocol model under these circumstances, because we keep track of the sequencing of the IDX throughout the pipeline, and the model therefore prevents corruption of the (A,IDX) CAM state information by the other address index pairs (~A,nIDX).

4.4 Sequencing Protocol Message through the Pipeline

We refine DMA node N of the protocol model shown in Figure 2. The upper half of Figure 3 shows the protocol model with everything abstracted except the state associated with the node N, c_state, and the protocol message buffers carrying protocol messages to/from node. The DMA protocol message request tag variables are named dma_pnd, dma_p0, dma_p1, dma_p2, for each of the possible locations where the DMA request can reside within the pipeline. The protocol message reply (request) tag variables are named rp_pnd (rq_pnd), rp_p0 (rq_p0), rp_p1 (rq_p1), and rp_p2 (rq_p2).

Initially dma_pnd, dma_p0, dma_p1, and dma_p2 are all zero. When a DMA request arrives it asserts the input signal to the pnd stage, and once there is no request occupying the pnd stage, it moves to the pending stage. If the new request non-deterministically chooses the address CAM index pair (A,IDX), the new request also sets dma_pnd. The pending stage requests an arbitration grant to proceed to the p0 stage of the pipeline, and once the grant is given the request moves to p0 stage, clears dma_pnd, and sets dma_p0. At this point the pending stage is free for new requests to pipeline behind the first request, i.e. new requests can set dma_pnd. The p0 stage presents the address to the CAM, and moves to the p1 stage once no pipeline stall conditions are present, setting dma_p1, and clearing dma_p0, unless there is another requests moving into the p0 stage. The p1 stage presents the request type, and CAM-state to the protocol table, to determine the next protocol action. If there was a CAM address match in the p0 stage, the request stream is stalled otherwise a new protocol message is generated by the protocol. Once the p2 stage is free, the state that will be stored back into the CAM, and the new message information moves to the p2 stage. The message in the p2 stage moves to the dispatch engine once the required resources become available. The dispatch engine generates the new message, which will eventually result in the arrival of protocol reply and acknowledge messages.

When the protocol model selects a particular reply/acknowledge (intervention) message the valid bit in the pending register is set. If the non-deterministic choice variable chooses to deliver an abstract protocol message, i.e. it chooses (A,IDX), and it's selecting the message for the first time, the deliver variable is set, and the rp_pnd (rq_pnd) variable is also set. The reply (reply) message then proceeds through the pipeline in the same manner as the new request message. Once it has the arbitration grant, it moves to the p0 stage setting rp_p0 (rq_p0), once there are no pipeline stall conditions, it moves to the p1, stage setting the rp_p1 (rq_p1) variable. It then moves to the p2 stage setting the rp_p2 (rq_p2) variable and finally updates the state of the CAM, and if necessary generates further protocol messages. If at the pending register the choice is (~A,nIDX) none of tag variables are set for that reply (request), but the pipeline flow control logic will still set the valid bits for those requests.

4.5 Details of Protocol Message Delivery

The abstract protocol model uses a two-phase non-deterministic message delivery mechanism [1]. First, it non-deterministically selects a buffer for delivery (sel_buf), and then it non-deterministically selects to deliver this buffer (deliver_buf). In the abstract model this facilitates the delivery of the protocol messages in all possible orders, and it also allows each of the requesting modules, at each time step, to generate new requests. One of the reasons this message delivery mechanism was chosen is that it refines in a straightforward manner to the hardware arbitration and sequencer logic.

This refinement is accomplished in several steps. First, the pipeline is stalled unless the sel_buf variable selects module N. Module N is chosen, cur_dst=N, when either sel_buf selects a protocol message, or an empty buffer, destined for N. Second, we add a state variable (deliver) to each buffer, to indicate if a particular message buffer is being sequenced through the hardware pipeline. Finally, when module N is chosen, and a message is being delivered through the pipeline, we constrain the deliver_buf as shown in the following CBL SMV pseudo-code:

```
default {
  deliver_buf := cur_dst=N ? (buf=EMPTY ? 1 : 0) : {0,1};
  } in {
  case {
    dma_p1          : deliver_buf := 0;
    rp_p1 | rq_p1 : deliver_buf := 1;
}
```

When no (A,IDX) messages are at state p1, the deliver_buf variable is 0 if a non-empty buffer destined for N is selected, 1 if an empty buffer destined for N is seleced, or it's non-deterministically 0 or 1. When an (A,IDX) message reaches stage p1, in case of a new request there is non-deterministic choice of the possible DMA requests, or delivery of a reply, acknowledge, or intervention message.

4.6 Details of CAM Index Abstraction

An example of the CAM index abstraction is shown in the following fragment of CBL SMV pseudo-code:

```
next(IDX) := IDX;
nIDX_p0   := choose_IDX_p0=IDX ? ~IDX : choose_IDX_p0;
```

The nIDX_p0 variable is the name we use for nIDX at stage p0, and choose_IDX_p0 is a non-deterministic choice of a CAM-index value not equal to IDX. The IDX variable is not initialized, but retains it's non-deterministically chosen initial value, so we prove refinement properties for all possible IDX values in one CBL SMV run. This same IDX abstraction method is used for the pending request register, and each of the pipeline registers.

4.7 Details of Address Abstraction

An example of message address abstraction is shown in the following CBL SMV pseudo-code:

```
if(~RST & p1_vld) {
    p1_address := (dma_p1 | rp_p1 | rq_p1) ? A : ~A_p1;
}
```

The example shows the address abstraction at the p1 stage in the pipeline. When the p1 stage state is valid p1_vld, then if an (A,IDX) message is present the p1 stage address maps to A, otherwise it maps to ~A_p1. This same abstraction method is used for each entry in the CAM, and for each pipeline register.

4.8 Details of CAM Abstraction

A CAM entry stores an address, a valid bit, and protocol state information. The CAM address abstraction is shown above. The valid bit abstraction is shown in the following CBL SMV pseudo code fragment:

```
if (~RST & p0_vld & WrP1) {
 forall(I in CAM_INDEX)
  CAM[I].vld = I=IDX ? c_state.V: {0,1};
}
```

The valid bit is derived from the protocol valid bit c_state.V if the index of the CAM is the same as the chosen index IDX, otherwise the valid bit is unconstrained.

The CAM abstraction used for the protocol state information is shown in the following CBL SMV code:

```
if (~RST & (rp_p0 | rq_p0 | dma_p0) & WrP1) {
  forall(I in CAM_INDEX)
   if(I = IDX)
     CAM output = c_state;
}
```

The CAM protocol state information is undefined except when the CAM entry with index IDX is accessed and written to stage p1, in which case it's equal to the protocol state c_state.

The state of the p1 and p2 stages are abstracted in a similar fashion, i.e. the state of each of these registers is the same as the protocol state, when a protocol message is present at the stage, otherwise the state of these registers is undefined.

4.9 Overview of Different Abstraction Layers

This concludes the overview of the most critical abstractions that were used to derive a tractable model of the hardware pipeline implementation. For completeness we summarize, in the following pseudo-code all the abstraction layers that were used in the refinement maps:

```
stack {

abstract_protocol_model : {
 abstract protocol model consisting of an M node, and a
 module modeling a collection of P and I/O modules
 -   this model has approximately 60 bits of state
}

configuration_registers : {
 set various hardware configuration registers
 disable timeout triggers
 -   this layer is not refined
}

refine_protocol_model : {
 non-deterministically select a CAM (A,IDX) pair
 sequence abstract protocol messages through pipeline
 sequence (A,IDX) information using presence bits
 lookup next protocol state and action

eliminate_micro_pkt_counters : {
 replace micro-packet counters with end-bit logic
 }
```

```
eliminate_redundancy : {
 replicated state eliminated
 register state undefined when vld bit clear
}

rtl_layer : {
30.000 lines of synthesizable verilog RTL
}

}
```

We've described all the layers except the following: `abstract_protocol_model`, `configuration_registers`, `eliminate_micro_pkt_counters`, and `eliminate_redundancy`.

The `abstract_protocol_model` layer contains a protocol model consisting of an M node and an abstraction of a collection of P and II nodes. Various ACTL properties have been proven for the abstract protocol model. See [1] for details. This model is refined in a different refinement hierarchy, to the protocol model as described in the protocol specification document.

Hardware typically contains a large collection of programmable configuration registers, and timeout counters. The configuration registers are used to set various mode bits. The timeout counters are used to detect error conditions where particular requests do not complete. The `configuration_registers` signal layer sets the mode bits to typical values, i.e. eliminates the registers used to store the mode bit values. This layer also ties all timeout triggers to 0, i.e. the model will never have any timeout events, and thereby eliminates all state associated with the timeout counters. This layer is not refined.

The `eliminate_micro_pkt_counters` layer abstracts all the end of message state machines, which use micro-packet counters, with micro-packet end bit detectors. The hardware has various message sequencers that count message micro-packets until the final micro-packet is reached. The messages have an end bit set in the last micro-packet to signal the last packet, but for hardware signal propagation delay reasons it is necessary to derive the same information directly from register outputs. To eliminate the state associated with the micro-packet counter, the end-bit detection logic is rewritten to look directly at the end bit. Refining the rewritten function back to the original hardware function then proves the equivalence of the rewritten end-bit detection functions.

The `eliminate_redundancy` layer eliminates two sources in the hardware of model checking inefficiencies. First, multiple copies of the same state are rewritten such that all except one state variable is eliminated. The eliminated state variables are then rewritten as combinational functions of that remaining state variable. Hardware timing considerations are the source of the multiple copies of the same state. Finally, all control registers with valid bits are abstracted such that all state is undefined when the valid bit is clear.

5 Size of Models and run-time Requirements

The refinement hierarchy has approximately 1500 refinement obligations. We utilize a farm of 200 CPUs to run the regressions. A full regression completes overnight; it takes approximately 7 hours. The address and CRB index abstractions complete in 1-2 min. and require 50M of memory, but the protocol state refinements complete in 30-60 min requiring up to 500M of memory. We use no pre-derived BDD variable order files, but we enable sifting of the BDD variables at run-time.

6 Conclusions

Compositional model checking techniques make the analysis of the most complex hardware design possible. We have successfully refined a complex directory based cache coherence protocol to an equally complex pipelined hardware implementation.

A person can only devise the compositional model checking refinement strategy if they have deep insight into the design. One person with the necessary insight is the designer of the hardware, and in our case, this person devised the refinement strategy, wrote the refinement maps, and proved their correctness.

Compositional model checking techniques have reached a level of maturity where it is possible for a hardware designer to formally design complex hardware within the strict time schedule of a design project.

7 Future Research

We are working on several extensions to this work. First, we are developing extensions that involve data consistency properties; e.g. ensuring that data is not read from the merge cache RAM before the required data has been written into the merge cache.

We are also working on extensions that prove fault tolerance properties of the hardware. These include proving that runt messages are padded to the correct length, that long messages are truncated to the correct length, and that both runt and long messages cause the setting of the correct error bits.

Finally we are working on proving that phantom protocol messages, are rejected by the logic, and that they have no side effects, i.e. they can cause no state changes in the CAM.

8 Acknowledgements

We are indebted to Ken McMillan of Cadence Berkeley Labs, who provided us with the newest version of CBL SMV, and graciously shared his deep insight into formal methods. We would also like to thank Kianoosh Naghshineh, formerly of Silicon Graphics Inc., for sharing his deep insight into hardware design.

References

[1] Ásgeir Th. Eiriksson, and Ken L. McMillan, "Using Formal Verification/Analysis Methods on the Critical Path in System Design", In P. Wolper, editor, Computer-Aided Verification Conference: 7th International Conference, CAV'95, pp. 367-380. Springer Verlag, 1995. Lecture Notes in Computer Science Number 939.

[2] Ásgeir Th. Eiríksson, "Integrating Formal Verification Methods with A Conventional Project Design Flow", 33rd Design Automation Conference, Las Vegas NV, 1996, pp. 666-671.

[3] Ásgeir Th. Eiríksson, John Keen, Alex Silbey, Swami Venkataramam, Michael Woodacre, "Origin System Design Methodology and Experience: 1M-gate ASICs and Beyond", Proceedings of the Compcon Conference, San Jose, 1997.

[4] Jim Laudon, and Dan Lenoski, "System Overview of the Origin 200/2000 Product Line", Proceedings of the Compcon Conference, San Jose, 1997.

[5] Jim Laudon, and Dan Lenoski, "The SGI Origin: A ccNUMA Highly Scalable Server", Proceedings from the International Symposium on Computer Architecture (ISCA), Denver, Colorado, 1997

[6] Silicon Graphics Incorporated, 2011 N. Shoreline Blvd., Mountain View, CA, "Origin Technology", http://www.sgi.com/origin/technology.html

[7] Carl-Johan H. Seger and Randal E. Bryant, "Formal verification by symbolic evaluation of partially-ordered trajectories", Formal Methods in System Design, Kluewer Academic Press, New York, Volume 6, pp. 147-189, 1995.

[8] Ken L. McMillan, "A Compositional Rule for Hardware Design Refinement", In Orna Grumberg, editor, Computer-Aided Verification Conference: 9th International Conference, CAV '97, pp. 24-35, Haifa, Israel, June 1997, Lecture Notes in Computer Science Number 1254.

[9] Ken L. McMillan, "Verification of an Implementation of Tomasulo's Algorithm by Compositional Model Checking", In Alan J. Hu & Moshe Y. Vardi, editors, Computer-Aided Verification Conference: 10th Conference, CAV '98, pp. 110-121, Vancouver, Canada, July 1998, Lecture Notes in Computer Science Number 1427.

Design of Experiments for Evaluation of BDD Packages Using Controlled Circuit Mutations

Justin E. Harlow III[1] and Franc Brglez[2]

[1] Duke University, ECE Department
Durham, NC 27705 USA
jharlow@ee.duke.edu
[2] N.C. State University, Collaborative Benchmarking Laboratory
Department of Computer Science
Raleigh, NC 27695 USA
brglez@cbl.ncsu.edu

Abstract. Despite more than a decade of experience with the use of standardized benchmark circuits, meaningful comparisons of EDA algorithms remain elusive. In this paper, we introduce a new methodology for characterizing the performance of Binary Decision Diagram (BDD) software. Our method involves the synthesis of large equivalence classes of functionally perturbed circuits, based on a known reference circuit. We demonstrate that such classes induce controllable distributions of BDD algorithm performance, which provide the foundation for statistically significant comparison of different algorithms.

1 Introduction

We introduce methods rooted in the *Design of Experiments*, first formalized in [1], to evaluate the properties of programs which implement Reduced, Ordered Binary Decision Diagrams [2] (hereafter referred to as BDDs). The choice of BDD variable order has a profound impact on the size of the BDD data structure. Determining an optimal variable ordering is an NP-hard problem upon which much research has focused, *e. g.* [3–5].

Our approach to evaluating the performance of BDD variable ordering algorithms is radically differerent from the ones reported in the past, including [4, 6]. Rather than evaluating the algorithms on basis of *relatively few single* unrelated instances of known benchmark circuits, such as available in [7–9], we evaluate the algorithms on the basis of *a relatively large number* of related instances of circuits, each belonging to an *equivalence class*. Any circuit from [7–9] can serve

---

* This research has been supported by contracts from the Semiconductor Research Corporation (94–DJ–553), SEMATECH (94–DJ–800), DARPA/ARO (P–3316–EL/DAAH04–94–G–2080), and (DAAG55-97-1-0345), and a grant from Semiconductor Research Corporation.

as a reference circuit to form an equivalence class in the proposed experiments. In this paper, we introduce *Class F mutants*, based on controlled perturbations of circuit functional characterics. This class complements the *Class W mutants*, based on controlled perturbations of circuit wiring characterics, introduced recently [10]. The background and a broader context for this approach is given in [11–13]. A special case of *Class W mutants* is the graph-isomorphic and logically equivalent class, *Class WD*, whose properties include:

P1: the order of all I/Os and interior nodes in the netlist is random relative to all other members in the class,

P2: the names of all I/Os and interior nodes in the netlist are assigned randomly relative to all other members in the class[1].

Properties **P1** and **P2** are essential to good experimental design and must be maintained universally for all mutant classes, not only for *Class WD* mutants. Without such properties, we could not have exposed the true performance of BDD variable algorithms for the *Class WD* in [13] and the families of *Class F mutants* introduced in this paper.

The paper is organized into sections that include: principles of experimental design to support evaluation of BDD algorithms, synthesis of circuit mutant equivalence classes under controlled functional perturbations, profiles of representative mutant classes, comparative experiments with BDD packages using the mutant classes, and conclusions.

2   Design of Experiments: Treatments for BDDs

NP-hard problems, such as optimal variable ordering for BDDs, are solved by devising a polynomial-time heuristic, often with no guarantee whatsoever of the quality of the solution. Changing the starting point for the problem instance can induce unpredictable variability of results when experiments are repeated.

Two of the fundamental principles of experimental design are *randomization* and *replication*. We adopt these principles for the experimental evaluation of algorithms by (1) creating a *circuit equivalence class* and (2) repeating the experiments for each member in the class. The basic abstractions for such experiments include [12]:

1. an equivalence class of experimental subjects, eligible for a treatment;
2. application of a specific treatment;
3. statistical evaluation of treatment effectiveness.

Here, a *treatment* is synonymous with an *algorithm* and an equivalence class of experimental subjects is synonymous with a circuit mutant class. Figure 1 illustrates these abstractions in a generic flow. The cost index, minimized in the context of variable ordering for BDDs, is the BDD size.

---

[1] The node renaming treats I/Os in a manner that preserves the original I/O names as post-fixes, so the original I/O names can be restored to facilitate pairwise logic verification.

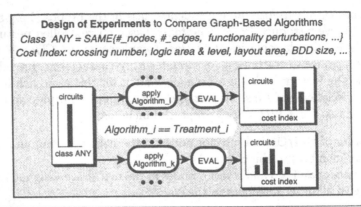

Treat- ment	Initial Ordering	Dynamic Ordering (in VIS [14])
0	Natural*	None
6	Natural	Sift enabled during construction
7	Natural	Sift enabled during construction; Sift forced once after construction
8	Natural	Sift enabled during construction; Sift forced twice after construction
9	Natural	Sift-converge enabled during construction; Sift-converge forced after construction

Natural implies a variable ordering given by the circuit netlist.

Fig. 1. Design of experiments with circuit equivalence classes to compare the sizes of ROBDDs under distinctive variable ordering algorithms (treatments).

Classification of BDD experiments. The table in Figure 1 itemizes a total of 5 distinctive treatments (BDD variable ordering algorithms) that we apply to a number of circuit equivalence classes in this paper.

- *Treatment 0* is based on the 'natural' order of variables in the circuit netlist only; no algorithm is applied to change the variable order when constructing the BDD. This treatment parallels the 'placebo treatment' in biomedicine.
- *Treatments 6–9* start with a 'natural' order of variables in the circuit netlist, followed by a number of 'sift' operations in VIS [14].

These are identical treatments to those applied to the *Class WD* mutants and reported in [13]. We are omitting *Treatments 1–5*, described in [13], since we encountered problems with the static variable ordering.

An Experiment with ALU4r-WD. Here, we consider Treatment 8, available from the three BDD packages in VIS [14]: CU from [15], CAL from [14], and CMU from [16]. Treatment 8 from each package is applied to *one hundred* instances of the graph-isomorphic and logically equivalent circuits from the the

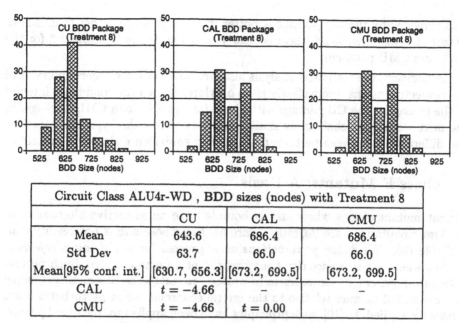

$t_{0.975} = \pm 1.972$ (two-tailed test at the $\alpha = 0.05$ significance level)

Fig. 2. Comparative performance of three BDD packages, observed for *one hundred* instances of an isomorphic and logically equivalent circuit class.

equivalence class ALU4r-WD, based on the reference circuit ALU4r[2]. The results of 300 experiments are summarized in Figure 2. The near-normal distributions of BBD sizes for *each* of the three variable ordering algorithms should not be a surprise: they merely point out that most solutions based on heuristics are highly sensitive to the initial order of variables. Since there is a significant overlap in the distributions, we have to formulate and put to test the null hypothesis H_0 [18,19], now paraphrased for treatments in place of algorithms:

Null Hypothesis 1 (H_0) *There is no difference between treatments: any observed differences of population means are due to random fluctuations in sampling from the same population.*

For the distributions shown in Figure 2, we apply the t-test [18] to accept or reject a decision about the hypothesis H_0. Using a two-tailed test at the $\alpha = 0.05$ significance level, we would reject H_0 if t were outside the range $-t_{0.975}$ to $+t_{0.975}$. In the case shown, $t_{0.975} = \pm 1.972$ and $t = -4.66$ when comparing CAL and CMU to CU, hence we reject H_0 at the significance level of 0.05. In other words, we are 95% confident that for the equivalence class ALU4r-WD, Treatment

---

[2] This version is based on a 4-bit ALU circuit textbook schematic, independently entered by J. Calhoun [17], re-entered by D. Ghosh in 1998 and verified against each other. The "benchmark alu4" version in [9] differs from this version in three minterms, verified by J. Harlow in 1998.

8 as provided by the CU package is better due to its design rather than due to chance. On the other hand, there is clearly no difference in Treatment 8 for the CAL and CMU packages.

In addition, the statistical analysis also provided 95% confidence intervals of the respective means: there clearly is no overlap between the confidence interval of the means for the CU package with those for the CAL and CMU package. In the next section, we shall apply similar analysis to test the hypothesis that we can differentiate between perturbation-induced equivalence classes of mutants.

3 Class F Mutants: A Basis

Circuit mutant classes whose size is bounded are an attractive alternative to random circuits that are logically invariant, but whose size is not readily controllable [20]. We define *perturbations* as a process for modifying a *reference circuit*, where a a reference ciruit may represent an actual design. Such perturbations can generate vast numbers of *mutant circuits*, each of which is changed in a controlled manner relative to the reference circuit, when perturbations are randomly applied. In [10], such principles have been applied to controlled pertur-

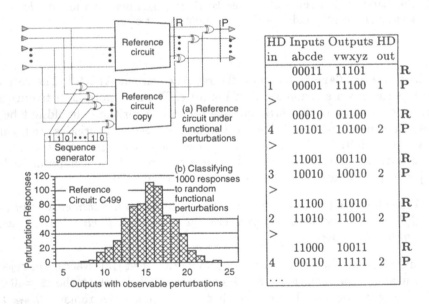

Fig. 3. Classification of randomly generated functional perturbations of the reference circuit. The illustrative response measures the Boolean difference in terms of *Hamming distance* (HD) between the inputs/outputs of the reference circuit **R** and the copy of the perturbed reference circuit **P**.

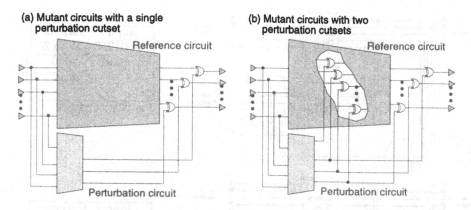

Fig. 4. A reference circuit with candidate cutsets to control the placement of functional perturbations, giving rise to mutants in the equivalence class F circuit families. Mutant classifications are defined in terms such perturbations.

bation of wires. Here, we apply *functional* perturbations that modify the circuit functional characteristics, giving rise to the *Class F* mutants.

Classification of random functional perturbations. Consider the conceptual experiment of Figure 3(a), in which a reference circuit and a copy of itself are connected as shown. Each input of the copy is perturbed with outputs from the sequence generator, which may generate any sequence except {0000...00}. This condition ensures that at least one input of the reference circuit copy is always perturbed relative to the inputs applied to the reference circuit. The outputs of both circuits are XOR-ed pairwise, so that the circuit outputs are perturbed under certain input patterns and random input perturbations.

We (1) apply random patterns to the circuit primary inputs, (2) apply random sequences to perturb the inputs of the reference circuit copy, and (3) classify the XOR-ed outputs of both circuits by the number of outputs with observable perturbations. The results of such classification characterize the circuit behavior under randomly applied input perturbations: the near-normal distribution that peaks at 16 outputs in Figure 3(b) is characteristic for the circuit C499 [9] already after the simulation of some 100 patterns. There are 111 instances out of 1000 trials in which perturbations are observable on exactly 16 outputs. Subsequently, we can interpret the table of simulation results in Figure 3 in two ways: (1) as results of a nominal simulation applied to inputs of the reference circuit only and measured at the primary outputs of the reference circuit, or (2) as responses **R** and **P** on the wire cutsets **R** and **P** that probe the outputs of the reference circuit alone and the Boolean difference of the two circuits. We apply (1) to get the circuit responses to random patterns, and we apply the interpretation in (2) as the basis to form the most important *Class F* mutant circuits.

Structures implementing functional perturbations. We apply the concept of p-in-q perturbation introduced in [21] as the basic primitive of our functional

70

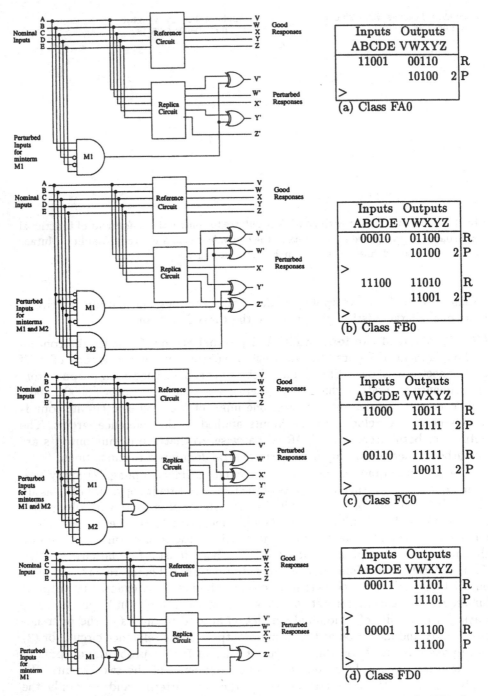

Fig. 5. Generation of mutant classes FA0 (one minterm perturbation), FB0 (two minterm perturbation), FC0 (two minterm perturbation, entropy–invariant), and FD0 (one–minterm perturbation on input and output, logic invariant).

perturbation: decoding signal values on p wires to generate a control signal c as an input to q XOR-gates that are inserted as perturbations onto a cutset of q wires. We illustrate a generalization of this concept in Figure 4 for two cases: (a) a single perturbation cutset P_1, and (b) two perturbation cutsets P_1 and P_2. When the cutsets stricly include only primary outputs (POs), or primary inputs and primary outputs (PIs and POs), respectively, we can classify the simulation responses such as shown in Figure 3 to readily generate a number of well-defined *Class F* mutants. The implementation details for four mutant classes are shown in Figure 5:

Class FA0: 1-minterm decoder, perturbing a fixed number of wires in the PO cutset, chosen randomly. The example in Figure 5(a) is a special case of 1 minterm perturbing two outputs, inducing a Hamming distance of 2 between the two responses shown.

Class FB0: 2-minterm decoder, *each* perturbing the same fixed number of wires in the PO cutset, *each* chosen randomly. The example in Figure 5(b) is a special case of 2 minterms, each perturbing two outputs, inducing a Hamming distance of 2 between the two responses shown.

Class FC0: 2-minterm decoder, combined with an OR-gate perturbing a fixed number of wires in the PO cutset – such that the combined effects of two perturbations maintains the *entropy invariance* of the circuit functionality. The example in Figure 5(c) is a special case of 2 minterms, combined with an OR-gate perturbing two outputs, inducing a Hamming distance of 2 between the two responses shown such that (1) the perturbed response to first pattern is identical to the nominal response to the second pattern, and (2) the perturbed response to second pattern is identical to the nominal response to the first pattern.

Class FD0: 1-minterm decoder, perturbing a fixed number of wires in the PO cutset *as well as* another fixed number of wires in the PI cutset – such that the combined effects of two perturbations maintain the *logic invariance* of the circuit functionality. The example in Figure 5(d) is a special case of 1-minterm perturbing one output and one input, inducing a Hamming distance of 1 between the two responses shown such that (1) the perturbed response to first pattern is identical to the nominal response to the first pattern, and (2) the perturbed response to second pattern is identical to the nominal response to the second pattern.

Contrasting Class W and Class F Mutants. *Class W* mutants are induced by wiring perturbations [10], and in contrast to *Class F* mutants, there is no control on the changes of circuit functionality. It is therefore instructive to analyze the performance of BDD algorithms using the two classes. A statistical summary of of a small-scale experiment with *Class W* and *Class F* mutants of *ALU4r* is shown in Figures 6 and 7. The histograms in Figure 6(a-c) refer to *Class W* mutants and the histograms in Figure 6(d-f) refer to *Class F* mutants. It is clear that the two classes induce two very different BDD behaviors when subjected to the same Treatment 8, using the CU package in VIS [14].

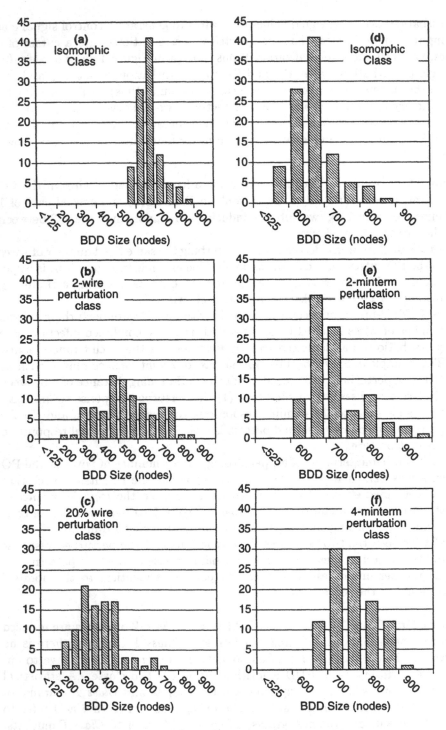

Fig. 6. Distribution of BDD sizes for Class W and Class F mutants.

Class W histograms are not well-defined in this context since each circuit may represent a function very different from all others in a class. Also, in this case, the functions appear simpler than the one represented by the reference circuit isomorphic and logically equivalent class. This in general is not true. For some cases of *Class F* mutants of circuit c499, we have not succeeded in finding an order for which a BDD could be constructed [10]. In contrast, *Class F* histograms in Figure 6(d-f) are clearly related, with distributions pointing to progressively more difficult functions, in this case, as the size of the functional perturbation increases.

The t-scores in Figure 7 bring out several statistically significant observations about the two equivalence classes. *Class W* is being evaluated relative to the isomorphic class in terms of 1-wire (WG), 2-wire (WA), 20% (WE), 40% (WC) and 100% (WB) wiring perturbations. All circuits in *Class W* have the same number of wires and gates. *Class F* is being evaluated relative to the isomorphic class in terms of 4-minterm logic-invariant perturbations (FI0), 2-minterm entropy-invariant perturbations (FC0), 4-minterm entropy-invariant perturbations (FG0), 1-minterm random perturbations (FA0), 2-minterm random perturbations (FB0), 3-minterm random perturbations (FE0), and 4-minterm random perturbations (FF0). Details about the sizes of these circuits are given in Table 1. Briefly, we conclude:

- Treatment 6 has too much variability to be of any use in evaluating the sigificance of t-scores. We settle on Treatment 8 for this task.

(a) Table of t scores for wiring mutant classes.

Treatment	Iso Class		1-wire	2-wire	20%	40%	100%
	Mean	Std Dev	WG	WA	WE	WC	WB
6	1,236.0	168.9	5.35	11.07	20.92	34.31	46.79
7	710.3	74.3	2.462	7.77	21.65	35.71	54.39
8	643.6	63.7	2.64	8.099	22.26	37.18	59.32
9	631.3	62.9	2.07	7.796	22.51	36.27	55.20

(b) Table of t scores for functional mutant classes.

Treatment	Iso Class		FI0	FC0	FG0	FA0	FB0	FE0	FF0
	Mean	Std Dev							
6	1,236.0	168.9	-2.14	-1.49	-4.42	-2.225	-0.582	.1099	-2.110
7	710.3	74.3	-1.68	-5.05	-10.5	-2.507	-1.909	-3.019	-3.011
8	643.6	63.7	-0.967	-6.29	-12.6	-1.166	-2.107	-3.617	-4.219
9	631.3	62.9	0.102	-6.05	-10.9	0.101	-1.297	-2.898	-3.818

Fig. 7. The t test is used to determine whether circuit equivalence classes are significantly different. A t value such that $|t| \geq t_{crit}$ indicates a statistically significant difference in the populations being compared. The two–tailed t test for the 95% confidence level implies a t_{crit} value of 1.972.

– Given the t-score for the 1-wire class (WG) in Figure 7(a), $t = 2.64$, we are at least 95% confident that this mutant class is different from the isomorphic class. This difference increases rapidly for all other mutants in the *Class W*.

– Given the t-score for the 4-minterm logic-invariant perturbation class (FIO) in Figure 7(b), $t = -0.967$, we are at least 95% confident that this mutant class is *not different* from the isomorphic class, i.e. the perturbations did not confound the algorithms with this circuit size.

– Given the t-score for the 2-minterm entropy-invariant perturbation class (FCO) in Figure 7(b), $t = -6.29$, we are at least 95% confident that this mutant class is *is significantly different* from the isomorphic class.

– Given the t-score for the 4-minterm entropy-invariant perturbation class (FG0) in Figure 7(b), $t = -12.6$, we are at least 95% confident that this mutant class is *is significantly different* from the isomorphic class, much more so than the class (FC0).

– Given the t-scores for the 1-, 2-, 3-, 4-minterm random perturbation classes (FA0, FB0, FE0, FF0) in Figure 7(b), we are at least 95% confident that 1-minterm mutant class is *not different* from the isomorphic class, while others increasingly are, although the increase is significantly below the rate induced by entropy-invariant perturbations.

Additional *Class F* mutants, generalizations of the ones introduce here, are discussed in Section 4. Random pattern simulation, such as illustrated in Figure 3 is required to generate simulation responses for the generation of entropy-invariant and logic-invariant classes as follows:

– For entropy-invariant classes, find the largest subclass of pattern pairs whose responses induce the same number of perturbations in the PO cutset. For circuit c499, 1000 random pattern simulations, there are 111 instances of patterns that induce observable perturbations on exactly 16 outputs. Each pair of patterns that induces 16 perturbations on the *same* primary outputs is a candidate pair to induce an entropy-invariant perturbation.

– For logic-invariant classes, find the largest subclass of pattern pairs whose responses induce the same number of perturbations in the PO cutset, say 16 as is the case for c499. Since the two input patterns are different, perturbations can be induced on the primary inputs *and* primary outputs such that the perturbations on the PIs are canceled at the POs. Pairs of patterns are selected such that the number of perturbations on the PIs and POs is the same for all mutants.

4   Class F Mutants: Representative Implementations

Similarly to *Class W* mutants whose names are pre-fixed with W, we assign short names to each *Class F* mutant, and pre-fix with F for a functional perturbation class. Figure 5 illustrates the specific structure of mutant classes FA0, FB0, FC0, FD0. The additional classes in Table 1 are straightforward extensions of

Table 1. Summary of gate–equivalent sizes of mutants in several functional-perturbation equivalence classes.

Class Name	Equivalence Properties	2-input gate equivalent size			
		ALU4r	C432	C499	C1355
<circuit>-WD	Isomorphic	91	179	206	518
... -FA0	1 minterm perturbation in PO cut	105	217	262	574
... -FB0	2 minterm perturbation in PO cut	125	255	318	630
... -FC0	2 minterm perturbation in PO cut Entropy–invariant	122	253	303	615
... -FD0	1 minterm perturbation in PI cut 1 minterm perturbation in PO cut Logic–invariant	115	235	282	594
... -FE0	3 minterm perturbation in PO cut	142	293	374	686
... -FF0	4 minterm perturbation in PO cut	159	331	430	742
... -FG0	4 minterm perturbation in PO cut Entropy–invariant	153	327	400	712
... -FH0	2 minterm perturbation in PI cut 2 minterm perturbation in PO cut Logic–invariant	139	291	358	670
... -FI0	4 minterm perturbation in PI cut 4 minterm perturbation in PO cut Logic–invariant	187	403	510	822

the classes in Figure 5. Table 1 summarizes the complete set of *functional perturbation* classes we have studied to date and posted on the Web.

Each mutant of a given class is of a constant size in terms of 2–input gate equivalents. The size of mutants in a given class is determined by the size of the reference circuit and the specific additional circuitry used to realize the perturbations. All additional circuitry is directly implemented in terms of 2–input gates, so as to preserve identical class size and the most simple form of the logic. The size of this circuitry grows linearly in terms of the number of decoding minterms, primary inputs, and the total size of perturbations placed onto the primary input and primary output wires. For example:

– The reference circuit (or any member of the isomorphic class ALU4r-WD) is realized as 91 2–input gate equivalents, with 14 inputs and 8 outputs.
– We perturb a randomly chosen set of 4 primary outputs with a single minterm to synthesize class ALU4r-FA0.
– The size of the 1-minterm decoding gate is 13 2–input gates.
– The total size of any mutant circuit in Class ALU4r-FA0 is thus $91 + (4 + 13) = 108$ 2–input gates.
– To generate the entropy-invariant class ALU4r-FC0 we require a total of $91 + (4 + 2 \times 13 + 2) = 123$ gates, etc.

We have used *Class F* mutants, synthesized as described above, to design a number of experiments that characterize the performance of the static and dynamic variable ordering algorithms in VIS [14]. The results of representative experiments are discussed next.

5 Comparative Experiments

This section puts to final test the principles we introduced earlier: (1) statistical validation of *Class F* mutants for two representative reference circuits; (2) pairwise comparisons of three BDD packages with a number of *Class F* mutants based on two reference circuits; (3) status of the experiments in progress.

(a) Sample statistics for 100 mutants of C432 in four classes, under four treatments.

Treatment	C432-WD Mean/Std Dev [Conf. Int.]	C432-FIO Mean/Std Dev [Conf. Int.]	C432-FCO Mean/Std Dev [Conf. Int.]	C432-FGO Mean/Std Dev [Conf. Int.]
6	1,528.8/120.3 [1,504.7 1,552.9]	1,811.4/389.1 [1,719.5 1,903.3]	1,668.7/125.5 [1,643.6 1,693.8]	1,826.7/148.1 [1,797.1 1,856.3]
7	1,515.2/114.8 [1,492.2 1,538.2]	1,538.2/168.9 [1,498.3 1,578.1]	1,657.7/120.8 [1,633.6 1,681.8]	1,806.1/118.2 [1,782.5 1,829.7]
8	1,512.9/116.7 [1,489.6 1,536.2]	1,491.5/118.2 [1,463.6 1,5194]	1,656.1/120.3 [1,632.0 1,680.2]	1,805.8/118.1 [1,782.2 1,829.4]
9	1,505.1/118.6 [1,481.4 1,528.8]	1,502.7/124.1 [1,473.4 1,532.0]	1,674.7/123.4 [1,650.0 1,699.4]	1,813.5/112.6 [1,791.0 1,836.0]

(b) *t* tests for differences of sample means.

Treatment	C432 classes Mean—Std WD		t-stat FIO	FCO	FGO	ALU4r classes Mean—Std WD		t-stat FIO	FCO	FGO
6	1,528.8	120.3	-6.83	-8.05	-15.5	1,228	156.0	-2.14	-1.49	-4.42
7	1,515.2	114.8	-1.06	-8.55	-17.7	707	76.6	-1.68	-5.05	-10.5
8	1,512.9	116.7	1.18	-8.54	-17.6	640	57.9	-0.97	-6.29	-12.6
9	1,505.1	118.6	0.13	-9.91	-18.9	629	65.8	0.10	-6.05	-10.9

Fig. 8. The *t* test is used to distinguish significant differences between the mean of a given class and the corresponding isomorphic class.

Validating classes of c432 and ALU4r. Having generated the *Class F* mutants of c432 and ALU4r, the question arises whether a number of treatments applied to each class results in distributions that are distinguishable. Results of such experiments are summarized in Figure 8.

The table in Figure 8(a) gives the sample mean, sample standard deviation, and the 95% confidence interval of the mean for the isomorphic class C432-WD and three functional perturbation classes, under four treatments (defined in Figure

1). These figures represent 100 samples (mutant circuits) per treatment. The following observations can be made from these data:

- The isomorphic class C432-WD consists of identical circuits, differing only in their ordering and variable naming, yet it is evident that the algorithms do not achieve optimal variable ordering. The variability of the mean of this class represents the statistical noise which will be present in experiments with other classes. We use statistical tests of significance to distinguish other classes from the isomorphic class, on the premise that any significant difference in the class means represents the effect of the perturbations in those classes.
- Treatments 6–9 represent increasingly 'strong' attempts to optimize the variable order, and thus to reduce the BDD size, of the circuits. Note here, however, that Treatment 9 performs slightly worse than Treatment 8 in most cases, indicating that the additional cost of sifting to convergence may not be justified, at least for this circuit.
- A logic–invariant class such as C432-FI0 should be indistinguishable from an isomorphic class such as C432-WD if the optimization is effective. We will illustrate below how the t-test can be used to determine whether the difference between this class and the isomorphic class is significant.
- Entropy–invariant classes such as C432-FC0 and C432-FG0 do not add or subtract minterms relative to the reference circuit function, and are expected to be of nearly the same complexity as the reference circuit. Again, a t-test is used to expose differences in class mean between this class and the isomorphic class.

The table in Figure 8(b) presents the results of a test of significance between the mean BDD size of several classes against the isomorphic class, for several treatments. The sample mean and standard deviation of the isomorphic class for each circuit is given for reference, and the value of the t statistic for each test is listed. We use a 2–tailed, uncorrelated t-test to evaluate the class mean differences. This test assumes that the two classes have approximately equal means and variances, and that the samples in each class are uncorrelated with the samples of the other class. For this test, if $|t| \geq 1.98$, the mean of the subject class is significantly different from the mean of the isomorphic class at the 95% confidence level. The t statistics clearly show that after optimization, logic–invariant classes C432-FG0 and ALU4r-FG0 are indistinguishable from their respective isomorphic classes. However, the entropy–invariant classes remain distinct from the isomorphic classes despite optimization.

Pairwise comparisons of BDD packages. Unlike the illustrative experiment with ALU4r-WD only in Section 2, we apply Treatment 8 from the three BDD packages, CU from [15], CAL from [14], and CMU from [16], to three equivalence classes (4-minterm logic-invariant, 2- and 4-minterm entropy-invariant, and two reference circuits (ALU4r and C432) – a total of $3 \times 100 \times 2 = 600$ experiments.

The results are summarized in Figure 9. For each class and each packages we tabulate the sample mean and standard deviation of the BDD size, along with

(a) Package comparisons for ALU4r classes under Treatment 8.

Class	CU Package Mean—Std Dev [Conf. int.]		CAL Package Mean—Std Dev [Conf. int.]		CMU Package Mean—Std Dev [Conf. int.]	
ALU4r-FIO	640.1	59.6	692.1	70.4	692.1	70.4
	[678.0 706.1]		[678.0 706.1]		[628.2 652.0]	
CAL	$t = 6.48$		–		–	
CMU	$t = 6.48$		$t = 0.00$		–	
ALU4r-FCO	712.7	73.5	749.6	74.2	749.6	74.2
	[698.0 727.4]		[734.8 764.4]		[734.8 764.4]	
CAL	$t = 4.28$		–		–	
CMU	$t = 4.28$		$t = 0.00$		–	
ALU4r-FGO	755.2	71.9	791.2	70.9	791.2	70.9
	[740.8 769.6]		[777.0 805.3]		[777.0 805.3]	
CAL	$t = 4.48$		–		–	
CMU	$t = 4.48$		$t = 0.00$		–	

(b) Package comparisons for C432 classes under Treatment 8.

Class	CU Package Mean—Std Dev [Conf. int.]		CAL Package Mean—Std Dev [Conf. int.]		CMU Package Mean—Std Dev [Conf. int.]	
C432-FIO	1,491.6	118.2	1,720.5	354.7	1,702.5	343.6
	[1,463.6 1,519.4]		[1,636.7 1,804.3]		[1,621.4 1,783.6]	
CAL	$t = 5.62$		–		–	
CMU	$t = 5.33$		$t = 0.84$		–	
C432-FCO	1,656.1	120.3	1,917.7	347.3	1,928.7	360.3
	[1,632.0 1,680.2]		[1,848.2 1,987.2]		[1,856.6 2,000.8]	
CAL	$t = 7.39$		–		–	
CMU	$t = 7.43$		$t = -0.84$		–	
C432-FGO	1,805.8	118.1	2,034.2	353.9	2,045.3	369.1
	[1,782.2 1,829.4]		[1,963.4 2,105.0]		[1,971.5 2,119.1]	
CAL	$t = 6.54$		–		–	
CMU	$t = 6.61$		$t = -0.93$		–	

Fig. 9. Differences in performance between two BDD packages can be evaluated using pairwise t-tests. Several classes of ALU4r and C432 are compared under Treatment 8.

the 95% confidence interval for the mean. In addition, the t statistic is given, comparing each distribution pairwise with the other two. These tests indicate the relative performance of one package against another under the same experimental conditions, but cannot be used to say which of the three packages performs best; more sophisticated multiple comparison tests [19] are available for this kind of study, but are not discussed here. The statistics in Figure 9 strongly suggest that the CU package performs better, on the average than the other two packages in terms of its ability to optimize variable ordering. However, other factors such as maximum BDD size before sifting, run time, memory usage, etc. are also important, but are not included in the studies reported here. A comprehensive evaluation of one package against another would include all of these factors, and would require evaluation of results against a wide variety of circuits to be considered conclusive.

Status of experiments in progress. We encountered a number of factors that required repetitions of experiments. The major factor is nontrivial, unexpected, and often unexplainable variability of results under slightly different input conditions. Cases of such points include:

- dependence of results not only on the variable order in the netlist but also on naming of the input variables. This matter was resolved by consistent introduction of mutant classes with randomly re-named variables for each instance of the class. Extensive results on these experiments, including on the ineffectivness of static order in VIS, are reported in [13].
- grossly inconsistent results with slight changes in scripts that invoke 'sifting' in VIS. The current Treatment 6 as described in Figure 1 becomes equivalent to Treatment 0 (i.e. random variable order assignment) for any class of C432 if we change the treatment from 'Sift enabled during construction' to 'Sift forced during construction'.
- On the other hand, Treatment 6 as described in Figure 1 is miracoulusly effective for c432, but not subcircuits of c432 nor the simple ALU4r circuit. For these circuits, Treatment 6 and Treatment 0 are statistically equivalent.
- Circuits such as c499 and c1355 that execute in a reasonable time for the original instances of the circuits take hours to execute under random reordering of nodes and names – hence the hesitation and lack of time to proceed with the rigor of experiments as presented in this paper.

6 Conclusions

The lessons learned about the experiments with BDD variable ordering provide a new beginning for experimental design, in evaluating variable ordering algorithms for BDDs as well as when evaluating the performance of any polynomial-time heuristics applied to NP-hard problems.

The home page http://www.cbl.ncsu.edu/experiments/ has pointers to circuit mutant equivalence classes, tables and statistical summaries of experiments reported in this paper. Utilities such as renaming I/O and node names in a netlist can be found under ../software/. Executable demonstrations of mutant generation and workflows that execute domain-specific designs of experiments can be viewed under ../demos/.

ACKNOWLEDGMENTS. Most of the isomorphic and logically equivalent classes of circuits used in our experiments were already generated by Debabrata Ghosh. He also provided generous assistance in generating additional instances of isomorphic classes, including the software to generate random netlist orders and random re-naming of all circuit nodes.

References

1. R. A. Fisher. *Statistical Methods, Experimental Design, and Scientific Inference.* Oxford University Press, 1993. Reprinted, with corrections, from earlier versions, 1925-1973.
2. R. E. Bryant. Graph-based algorithms for Boolean function manipulation. *IEEE Trans. Computers*, C–35(8):677–691, August 1986.
3. R. Rudell. Dynamic variable ordering for ordered binary decision diagrams. In *Proceedings of the International Conference on Computer Aided Design*, pages 42–47, November 1993.
4. M. R. Mercer, R. Kapur, and D. E. Ross. Functional approaches to generating orderings for efficient symbolic representations. In *Proceedings of the 29th ACM/IEEE Design Automation Conference*, pages 624–627, June 1992.
5. S. Panda, F. Somenzi, and B. F. Plessier. Symmetry detection and dynamic variable ordering of decision diagrams. In *Proceedings of the International Conference on Computer Aided Design*, pages 628–631, November 1994.
6. E. M. Sentovich. A brief study of BDD package performance. In *Proceedings of FMCAD 96*, number 1166 in LNCS, pages 389–403, 1996.
7. F. Brglez and H. Fujiwara. Special Session on ATPG (Also introducing 'A Neutral Netlist of 10 Combinational Benchmark Circuits'). In *Int. Symp. On Circuits and Systems*, 1985. Now a benchmark directory ISCAS85 at http://www.cbl.ncsu.edu/benchmarks/Benchmarks-upto-1996.html.
8. F. Brglez, D. Bryan, and K. Kozminski. Combinational profiles of sequential benchmark circuits. In *IEEE 1989 International Symposium on Circuits and Systems – ISCAS89*, pages 1924–1934, May 1989. A basis for a benchmark directory ISCAS89, now archived at http://www.cbl.ncsu.edu/benchmarks/.
9. S. Yang. Logic Synthesis and Optimization Benchmarks User Guide. Technical Report 1991-IWLS-UG-Saeyang, MCNC, Research Triangle Park, NC, January 1991. Now available from http://www.cbl.ncsu.edu/publications/-#1991-IWLS-UG-Saeyang and benchmarks from http://www.cbl.ncsu.edu/-benchmarks/Benchmarks-upto-1996.html.
10. D. Ghosh, N. Kapur, J. E. Harlow, and F. Brglez. Synthesis of Wiring Signature-Invariant Equivalence Class Circuit Mutants and Applications to Benchmarking. In *Proceedings, Design Automation and Test in Europe*, pages 656–663, Feb 1998. Also available at http://www.cbl.ncsu.edu/publications/#1998-DATE-Ghosh.
11. N. Kapur, D. Ghosh, and F. Brglez. Towards A New Benchmarking Paradigm in EDA: Analysis of Equivalence Class Mutant Circuit Distributions. In *ACM International Symposium on Physical Design*, April 1997. Also available from http://www.cbl.ncsu.edu/publications/#1997-ISPD-Kapur.
12. F. Brglez. Design of Experiments to Evaluate CAD Algorithms: Which Improvements Are Due to Improved Heuristic and Which Are Merely Due to Chance? Technical Report 1998-TR@CBL-04-Brglez, CBL, CS Dept., NCSU, Box 7550, Raleigh, NC 27695, April 1998. Also available at http://www.cbl.ncsu.edu/-publications/#1998-TR@CBL-04-Brglez.
13. J. E. Harlow and F. Brglez. Design of Experiments in BDD Variable Ordering: Lessons Learned. In *Proceedings of the International Conference on Computer Aided Design.* ACM, November 1998. Also available from http://www.cbl.ncsu.edu/publications/#1998-ICCAD-Harlow.
14. The VIS Group. VIS: A system for verification and synthesis. In R. Alur and T. Henzinger, editors, *Proceedings of the 8th International Conference*

on Computer Aided Verification, number 1102 in Lecture Notes in Computer Science, pages 428–432, New Brunswick, NJ, July 1996. Springer. Version 1.2 is available from UC Berkeley Design Technology Warehouse at `http://www-cad.eecs.berkeley.edu/Software/software.html`.

15. F. Somenzi *et al.* Colorado University Decision Diagram package (CUDD), release 2.1.2, 1997. Available from `ftp://vlsi.colorado.edu/pub/cudd-2.1.2.tar.gz`.

16. D. E. Long. CMU BDD package, 1993. Available from `http://emc.cmu.edu/pub/bdd/bddlib.tar.Z`.

17. J. Calhoun and F. Brglez. A Framework and Method for Hierarchical Test Generation. *IEEE Transactions on Computer-Aided Design*, 11(1):45–67, January 1992.

18. K. A. Brownlee. *Statistical Theory and Methodology In Science and Engineering.* Krieger Publishing, 1984. Reprinted, with revisons, from second edition, 1965.

19. J. C. Hsu. *Multiple Comparisons: Theory and Methods.* Chapman & Hall, 1996.

20. K. Iwama and K. Hino. Random generation of test instances for logic optimizers. In *Proceedings of the 31st ACM IEEE Design Automation Conference*, pages 430–434, 1994.

21. A. Žemva and F. Brglez. Detectable Perturbations: A Paradigm for Technology Specific Multi-Fault Test Generation. In *VLSI Test Symposium*, pages 350–357, April 1995. Also available from `http://www.cbl.ncsu.edu/publications/-#1995-VTS-Zemva-p350`.

A Tutorial on Stålmarck's Proof Procedure for Propositional Logic

Mary Sheeran and Gunnar Stålmarck

Prover Technology AB and Chalmers University of Technology, Sweden

Abstract. We explain Stålmarck's proof procedure for classical propositional logic. The method is implemented in a commercial tool that has been used successfully in real industrial verification projects. Here, we present the proof system underlying the method, and motivate the various design decisions that have resulted in a system that copes well with the large formulas encountered in industrial-scale verification. We also discuss possible applications in Computer Aided Design of electronic circuits.

1 Introduction

In the computer aided design of electronic circuits, a key function is tautology checking, that is testing whether a Boolean expression is true for all truth assignments of its variables. Tautology checking is used not only in hardware verification, but also in synthesis and optimisation. All known methods of tautology checking take time exponential in the size of the input formula, in the worst case. Since the problem is known to be co-NP complete [6], it seems unlikely that we can do any better than this in the worst case. But what about the formulas that actually arise in practice? In many cases, Binary Decision Diagrams (BDDs) and their variants work well, both for tautology checking and for other applications, such as model checking. A glance through the proceedings of the first conference on Formal Methods for CAD confirms that BDDs have become ubiquitous in hardware verification [23].

In this tutorial, we explain Stålmarck's method of tautology checking that may well rival BDDs [5] in some applications, but that is relatively unknown in the hardware verification community [12]. This patented method is implemented in a commercial tool that has been used in many industrial system verification projects [3]. Complex devices, such as engine management units or railway interlocking systems, are modelled in propositional logic, either directly or by translation from industry-standard formats. The required properties of the system are also expressed in propositional logic, and to verify the system is to check that the formula *system* → *properties* is a tautology. Often, the verification problem can be expressed as an inductive proof, and the base case and the step checked using Stålmarck's method. Many of these real-world verifications give rise to enormous formulas that could not be handled by current BDD packages. Groote has found Stålmarck's method to be very efficient compared to BDD-based methods and the Otter prover in the verification of the safety guaranteeing system at a particular Dutch railway station [11]. The largest formula encountered so far arose in railway interlocking; it had 350,000 connectives and the log recording the proof (for later independent checking) was 780 megabytes long.

The exciting thing about Stålmarck's method is that it copes with such formulas with aplomb, provided that they are *easy* according to a proof-theoretic measure that we will discuss later. And what is even more surprising is that real-world problems do indeed give rise to large but easy formulas.

In this tutorial, we first briefly present two standard proof systems for propositional logic: Gentzen's cut-free sequent calculus and semantic tableaux. Next, we show how cut-free proofs are intrinsically redundant, and motivate a rather different proof method, in which we use relations on formulas, rather than just sets of formulas that are known to be true or false. The system KE and the Davis-Putnam procedure can be seen as special cases of this approach. Next, we add a new kind of rule, to give the Dilemma proof system. This is the system that underlies Stålmarck's method. Finally, we outline the Dilemma proof procedure, give some important complexity results, and discuss applications.

2 Necessary Background: Proof Systems

In order to understand why a proof procedure is efficient from a practical point of view, we *must* study the underlying proof system. A proof procedure consists of two parts: an inductive definition of the classical consequence relation (that is rules about what it means to be a tautology) and a related algorithm for generating proofs. The algorithm can really only be understood by showing how it relates to a particular way of defining the consequence relation. And the way in which one defines the consequence relation – the choice of underlying proof system – has a surprisingly large effect on the performance of the resulting algorithm. So the first step in designing an efficient proof procedure is to choose a suitable proof system. Let us review some standard proof systems for classical propositional logic, and their properties.

2.1 Gentzen's Sequent Calculus with Cut

Gentzen introduced the sequent calculus during the thirties in order to prove his *Hauptsatz*. It states that all proofs can be brought into a form without roundabouts, obeying the subformula property, so that a proof uses only concepts (and subformulas) from the formula to be proved. Proofs in this calculus contain expressions of the form $\{A_1, \ldots, A_n\} \vdash \{B_1, \ldots, B_m\}$, so called *sequents*, informally read as "if all of the A_is are true, then one of the B_js is true".

Proofs start from obviously valid sequents of the form $A \vdash A$, the axioms. Complex formulas are then built up in the sequents by applications of operational rules. The calculus also includes a rule for introducing new formulas, the thinning rule, and one rule that removes formulas, the cut rule. Thinning can be used to minimize the number of different sequents in proofs and hence, to reduce proof complexity when proofs are viewed as directed acyclic graphs, rather than trees. When presenting the rules, we write Γ, A for the set $\Gamma \cup \{A\}$.

Axiom

$$A \vdash A$$

Structural Rules

(*Thinning*) $\dfrac{\Gamma \vdash \Delta}{\Gamma, \Theta \vdash \Delta, \Lambda}$ (*Cut*) $\dfrac{\Gamma, A \vdash \Delta \quad \Gamma \vdash \Delta, A}{\Gamma \vdash \Delta}$

Operational Rules

(*Or-left*) $\dfrac{\Gamma, A \vdash \Delta \quad \Gamma, B \vdash \Delta}{\Gamma, A \vee B \vdash \Delta}$ (*Or-right*) $\dfrac{\Gamma \vdash \Delta, A, B}{\Gamma \vdash \Delta, A \vee B}$

(*And-left*) $\dfrac{\Gamma, A, B \vdash \Delta}{\Gamma, A \wedge B \vdash \Delta}$ (*And-right*) $\dfrac{\Gamma \vdash \Delta, A \quad \Gamma \vdash \Delta, B}{\Gamma \vdash \Delta, A \wedge B}$

(*Imp-left*) $\dfrac{\Gamma \vdash \Delta, A \quad \Gamma, B \vdash \Delta}{\Gamma, A \to B \vdash \Delta}$ (*Imp-right*) $\dfrac{\Gamma, A \vdash \Delta, B}{\Gamma \vdash \Delta, A \to B}$

(*Neg-left*) $\dfrac{\Gamma \vdash \Delta, A}{\Gamma, \neg A \vdash \Delta}$ (*Neg-right*) $\dfrac{\Gamma, A \vdash \Delta}{\Gamma \vdash \Delta, \neg A}$

2.2 Removing Cut, and Gaining the Subformula Principle

The sequent calculus is complete, even if we remove the Cut rule. That is what Gentzen's Haupsatz says. All of the rules of the cut-free sequent calculus are elimination rules, so the system trivially obeys the *subformula principle*. Every proof uses only subformulas of the formula to be proved. No extraneous formulas or definitions are used, so that only concepts that were already there in the formula to be proved are used. The proof is direct, or as Gentzen put it, it is not roundabout [10]. As we shall see later, having the subformula principle allows us to place bounds on proof size, and this is very important in practice.

2.3 Removing Thinning

If, in addition, we take as axioms sequents of the form $\Gamma, A \vdash A, \Delta$, then we can add extra formulas using axioms, and the Thinning Rule becomes redundant and can be removed. The resulting system is essentially the same as Kleene's system G4 [15]. This system has been particularly important in automated deduction because it lends itself to a goal oriented proof search. One can start with the sequent that is to be proved and use the rules backwards, aiming to reach axioms or obviously unprovable sequents. This proof procedure works because all of the rules in G4 are *invertible*: the provability of the sequent below the line implies the provability of the sequents above the line. Note that the thinning rule is not invertible.

2.4 The Semantic Tableau Method

Smullyan's system of 'analytic tableaux' [22] is another classic proof method. Any valuation of a formula (an assignment of $\top$ (true) or $\bot$ (false)) to the propositional variables) must make the formula either true or false. So, for each connective, we examine the possible cases. If $A \wedge B$ is true, then A must be true and B must be true (the And rule). If $A \wedge B$ is false, then A is false or B is false and we explore both possibilities (the Not-And rule). This combination of the law of the excluded middle with the usual semantic interpretation of the connectives gives the following tableau rules for propositional logic:

$$(And) \quad \frac{A \wedge B}{\begin{array}{c} A \\ B \end{array}} \qquad\qquad (Not\text{-}Or) \quad \frac{\neg(A \vee B)}{\begin{array}{c} \neg A \\ \neg B \end{array}}$$

$$(Or) \quad \frac{A \vee B}{A \mid B} \qquad\qquad (Not\text{-}And) \quad \frac{\neg(A \wedge B)}{\neg A \mid \neg B}$$

$$(Impl) \quad \frac{A \rightarrow B}{\neg A \mid B} \qquad\qquad (Not\text{-}Impl) \quad \frac{\neg(A \rightarrow B)}{\begin{array}{c} A \\ \neg B \end{array}}$$

$$(Not\text{-}Not) \quad \frac{\neg\neg A}{A}$$

This method analyses a formula by progressively breaking it into its component parts, using these rules. The branching rules (Or, Impl and Not-And) are such that the disjunction of the formulas in the branches is a consequence of the formula above the line. Rules such as Not-Or that have two formulas below the line are just shorthand for a pair of rules, each giving a single formula. We start with the formula to be proved at the root of a tree and repeatedly apply the above rules. Along any path through the tree, we build up information about one possible valuation of the subformulas, by gathering a set that contains all of the formulas along that path. Think of this as being all of the formulas that we know to be true in a particular valuation. If we end up with an explicitly contradictory set, one containing both A and $\neg A$ for some formula A, then the exploration of that branch has failed to find a model (a setting of the propositional variables the makes the formula true). If all of the branches of the tree are contradictory in this way, then we know that there is no model of the formula, so it is contradictory and its negation is valid.

Semantic Tableaux systems were first introduced during the fifties, by Beth, Kanger, Hintikka and Shütte almost simultaneously [1, 14, 13, 18]. Because the system has only elimination rules, it trivially obeys the subformula principle. This has the important effect of placing a limit on the size of proofs in relation to the size of the formula to be proved. We write the size of formula A as $|A|$. It is the number of variable occurrences plus the number of connectives. The number of subformulas of a formula is the same as the size of the formula, and is also the maximum length of any path in a semantic tableau, since each step along the path adds one formula to a set of formulas.

2.5 The Intrinsic Redundancy of Cut-free Proofs

No matter what procedure is used to search for proofs in G4 or the tableau system, the search tree can grow explosively, even for simple commonly occuring examples. Such growth happens even for the smallest possible proof, so the problem is not in the procedure but in the cut-free nature of the proof system. D'Agostino, in his thesis [7], presents a small and enlightening example: the minimal tableau refutation (without Cut or Thinning) of the formula

$$(A \vee B) \wedge (A \vee \neg B) \wedge (\neg A \vee C) \wedge (\neg A \vee \neg C)$$

Each of the seven paths explored results in a contradiction. Note, however, that in the righthand side of the tree, we explore a part of the search space that has already been explored in the left subtree. After building the left subtree, we know that assuming that A is true leads to a contradiction, yet we repeat this search even after assuming that B is true. (Don't be misled by the picture into thinking that using graphs instead of trees might help! Our trees are really decorated at each node with the set of formulas along the path from the root, and each node in the above tree corresponds to a different set.) This kind of redundant pattern can be repeated inside the redundant subtrees, so that a combinatorial explosion results. The semantic tableaux rules given above don't really match the search space that we are trying to explore. We would like our refutation trees to be a better match with the search space. The solution is to put back in a form of cut, while keeping the subformula principle.

To understand this step, we must study the space that we are searching, and what rules and proofs look like.

2.6 Rules

Rules in tableau systems correspond to clauses in the defintion of true in a valuation. So, for example, if $A \vee B$ is true and A is false, then B is true. This and the other two elimination rules for $\vee$ are written

$$\frac{\begin{array}{c}A \vee B \equiv \top\\ A \equiv \bot\end{array}}{B \equiv \top} \qquad \frac{\begin{array}{c}A \vee B \equiv \top\\ B \equiv \bot\end{array}}{A \equiv \top} \qquad \frac{A \vee B \equiv \bot}{\begin{array}{c}A \equiv \bot\\ B \equiv \bot\end{array}}$$

The introduction rules for $\vee$ are

$$\frac{A \equiv \top}{A \vee B \equiv \top} \qquad \frac{B \equiv \top}{A \vee B \equiv \top} \qquad \frac{\begin{array}{c}A \equiv \bot\\ B \equiv \bot\end{array}}{A \vee B \equiv \bot}$$

Here, when analysing a connective $\circ$, we consider not only $A \circ B$ but also its immediate subformulas and their complements. This gives a different set of rules from the classic tableau rules that we have already seen.

We can extend this idea by considering not just truth values of subformulas, but also whether two formulas must have the same truth value. For example, if A and B have the same value, then $A \vee B$ also has that value, and if A and B have different values, then $A \vee B$ must be true.

$$\frac{A \equiv B}{\begin{array}{c}A \equiv A \vee B\\ B \equiv A \vee B\end{array}} \qquad \frac{A \equiv \neg B}{A \vee B \equiv \top}$$

By examining each connective in turn, we can generate a large set of propagation rules. We include only the *proper* rules. In general, the rules for a connective $\circ$ look like

$$\frac{F_1 \equiv G_1, \dots F_n \equiv G_n}{F \equiv G}$$

where each $F_i, G_i \in \{A, B, A \circ B, \neg A, \neg B, \top, \bot\}$. A rule is *proper* if and only if

$$\{F_1 \equiv G_1, \dots F_n \equiv G_n\} \models F \equiv G$$
$$\{F_1 \equiv G_1, \dots F_n \equiv G_n\} \not\models \bot$$
$$\{F_1 \equiv G_1, \dots F_n \equiv G_n\} - \{F_i \equiv G_i\} \not\models F \equiv G$$

2.7 The Systems KE/I and KE

The subset of the proper rules for which $G, G_i \in \{\top, \bot\}$ corresponds to the introduction and elimination rules of the propositional fragment of the system KE/I. This proof system was introduced by Mondadori, and has been further studied by D'Agostino [17, 7]. Using just the introduction and elimination rules does not give a system that is complete for propositional logic. However, adding a single branching rule, the principle of bivalence,

$$(PB) \qquad \frac{}{A \equiv \top | A \equiv \bot}$$

gives a proof system that is complete for propositional logic and that does not suffer from the kind of redundancy that we illustrated earlier. We have put back the cut rule! A is restriced to be a subformula of the formula to be proved, so we again have the subformula principle (with the associated bound on proof size) and because of this the PB rule is known as an *analytic* form of cut. Indeed, just the elimination rules plus the PB rule form a complete system, KE, which is extensively studied in D'Agostino's thesis [7].

The formula $(A \vee B) \wedge (A \vee \neg B) \wedge (\neg A \vee C) \wedge (\neg A \vee \neg C)$, for which we earlier showed the minimal tableau refutation, gives the following KE-refutation.

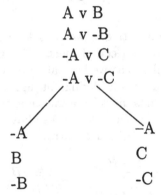

Let us start from KE and extend the language of formulas to include generalised conjunction and disjunction. If we now add two simplification rules (affirmative-negative and subsumption), we get a proof procedure that is equivalent to the well-known Davis-Putnam procedure [9] (in the version of [8]) for formulas in conjunctive normal form (CNF). So, KE can be seen as a generalisation of Davis-Putnam that does not require reduction to CNF.

The proof system underlying Stålmarck's method uses the larger set of propagation rules in which the G and G_i are no longer constrained to be in $\{\top, \bot\}$. It is no longer sufficient to maintain sets of formulas that are known to be true or false; we must also maintain information about sets of formulas that are known to have the same value. To do this, we introduce formula relations.

2.8 Formula Relations

The complement of a formula A, written A', is B if $A = \neg B$ and is $\neg A$ otherwise. Let $S(X)$ be the set containing all the subformulas of X (including $\top$) and their complements.

A *formula relation* $\sim$ on X is an equivalence relation with domain $S(X)$, with the constraint that if $A \sim B$ then $A' \sim B'$. If $A \sim B$, that means that A and B are in the same equivalence class and must have the same truth value. Working with $S(X)$, which includes the complements of subformulas of X, allows us to encode both equalities and inequalities between subformulas. $A \not\sim B$ is encoded as $A' \sim B$.

We write $R(A \equiv B)$ for the least formula relation containing R and relating A and B. The smallest formula relation is the identity relation on $S(X)$, written X^+. It simply places each element of $S(X)$ in its own equivalence class. Rather more interesting is $X^+(X \equiv \top)$, which we abbreviate to $X^\top$. This *partial valuation* will later be the starting point when we attempt to refute X.

Example Let C and D be propositional variables, and $X = C \wedge D$. Then,

$$X^\top = \{[C \wedge D, \top], [C], [D], [\neg(C \wedge D), \bot], [\neg C], [\neg D]\}$$

where equivalence classes are shown using square brackets.

2.9 Applying Rules to Formula Relations

Each schematic rule

$$\frac{F_1 \equiv G_1, \dots F_n \equiv G_n}{F \equiv G}$$

corresponds to a partial function on formula relations. It takes a formula relation R in which each $F_i \equiv G_i$ for i in $\{1..n\}$, and returns the larger $R(F \equiv G)$.

Continuing the previous example: applying the two $\wedge$-elimination rules

$$\frac{A \wedge B \equiv \top}{A \equiv \top} \qquad \frac{A \wedge B \equiv \top}{B \equiv \top}$$

in sequence to $X^\top = \{[C \wedge D, \top], [C], [D], [\neg(C \wedge D), \bot], [\neg C], [\neg D]\}$ gives R_1 and then R_2.

$$R_1 = \{[C \wedge D, C, \top], [D], [\neg(C \wedge D), \neg C, \bot], [\neg D]\}$$
$$R_2 = \{[C \wedge D, C, D, \top], [\neg(C \wedge D), \neg C, \neg D, \bot]\}$$

R_2 contains exactly two equivalence classes, and no further simple rules are applicable to it. It gives a *model* of $C \wedge D$, which therefore cannot be refuted.

If, instead, we take $Y = C \wedge \neg C$ and apply the two $\wedge$-elimination rules to $Y^\top$, we get R_3 and R_4.

$$R_3 = \{[C \wedge \neg C, C, \top], [\neg(C \wedge \neg C), \neg C, \bot]\}$$
$$R_4 = \{[C \wedge \neg C, C, \top, \neg(C \wedge \neg C), \neg C, \bot]\}$$

R_4 groups all of the formulas in its domain into a single equivalence class, and so is the largest formula relation on Y. It is explicitly contradictory since it (several times) places a formula and its complement in the same equivalence class. So, this sequence of rule applications constitues a refutation of $C \wedge \neg C$. From the assumption that $C \wedge \neg C$ is true, we have derived a contradiction.

3 The Dilemma Proof System

We are now ready to present the Dilemma proof system that underlies Stålmarck's method. The large set of *proper* rules that we have introduced allows us to reach conclusions that would require branching in systems with a smaller set of simple rules. Since the lengths of paths in refutation graph proofs are bounded by the size of the formula to be proved, it is branching that is the critical factor for complexity speed up.

Of course branching cannot be avoided altogether, since the simple rules are not complete for propositional logic. We introduce a special 'branch and merge' rule called the Dilemma rule. The Dilemma proof system is just this rule plus the simple rules.

3.1 The Dilemma Rule

The Dilemma rule is pictured as

$$
\frac{
\begin{array}{cc}
\dfrac{\quad\quad\quad\quad R \quad\quad\quad\quad}{} \\
\begin{array}{cc}
R(A \equiv B) & R(A \equiv \neg B) \\
(derivation) & (derivation) \\
R_1 & R_2
\end{array}
\end{array}
}{R_1 \sqcap R_2}
$$

Given a formula relation R, we apply the Dilemma rule by choosing A and B from different (and non-complementary) equivalence classes in R. We make two new Dilemma derivations starting from $R(A \equiv B)$ and $R(A \equiv \neg B)$, to give R_1 and R_2 respectively. Finally, we intersect R_1 and R_2, to extract the conclusions that are common to both branches. If an explicitly contradictory formula relation is always extended to the relation with a single equivalence class, then the intersection operation is simply set intersection ($\cap$) of the relations viewed as sets of pairs. In practice, we stop a derivation as soon as two formulas A and $\neg A$ are placed in the same equivalence class. Then $R_1 \sqcap R_2$ is defined to be R_2 if R_1 is explicitly contradictory, R_1 if R_2 is explicitly contradictory, and $R_1 \cap R_2$ otherwise.

Note that R is a subset of both R_1 and R_2, since a derivation only adds to the relation. This means that R is a subset of $R_1 \sqcap R_2$, and the rule is sound. The rule can be seen as the combination of a cut (the branch) with two (backwards) applications of thinning, from $R_1 \sqcap R_2$ to R_1, and from $R_1 \sqcap R_2$ to R_2. In the context in which these thinning applications appear, they are invertible. If we can refute $R_1 \sqcap R_2$, then we have refuted R, so it must be possible to refute both R_1 and R_2. If we omit the use of thinning, and simply refute R_1 and R_2 separately, then the two proofs are likely to have much in common (since R_1 and R_2 have R in common). Using thinning avoids this repetition, and so has an effect akin to that of having lemmas.

3.2 Dilemma Derivations

The following three clauses define what it means to be a Dilemma derivation, and also the related notion of proof depth.

1. *Simple Rules.* If the application of one of the simple rules to R_1 gives R_2, then $\Pi = R_1 R_2$ (R_1 followed by R_2) is a Dilemma derivation of R_2 from R_1. We write that assertion as $\Pi : R_1 \Rightarrow R_2$. If none of the simple rules applies to R, we say that R itself is a derivation of R from R. In both cases, the proof depth, written $depth(\Pi)$, is zero.

2. *Composition.* If $\Pi_1 : R_1 \Rightarrow R_2$ and $\Pi_2 : R_2 \Rightarrow R_3$, then we can compose the proofs: $\Pi_1 \Pi_2 : R_1 \Rightarrow R_3$. The proof depth of the composition is defined to be $max(depth(\Pi_1), depth(\Pi_2))$.

3. *Dilemma Rule.* If $\Pi_1 : R(A \equiv B) \Rightarrow R_1$ and $\Pi_2 : R(A \equiv \neg B) \Rightarrow R_2$, then

$$\frac{R}{\dfrac{\Pi_1 \qquad \Pi_2}{R_1 \sqcap R_2}}$$

is a derivation from R to $R_1 \sqcap R_2$ and its depth is $max(depth(\Pi_1), depth(\Pi_2)) + 1$

Proofs built using these rules have a series-parallel shape. The depth of a proof is the same as the maximum number of simultaneously open branches.

Example

$$
\begin{array}{c}
R_1 \\
R_2 \\
R_3 \\
\end{array}
$$

$$
\begin{array}{cc}
R_3(A \equiv B) & R_3(A \equiv \neg B) \\
R_4 & \\
R_5 & R_{10} \\
\end{array}
$$

$$
\begin{array}{cc}
R_6(C \equiv D) \quad R_6(C \equiv \neg D) & \\
R_7 & \\
R_8 \qquad R_9 & \\
R_8 \sqcap R_9 & R_{11} \\
(R_8 \sqcap R_9) \sqcap R_{11} & \\
\end{array}
$$

This proof has depth 2. The *size* of a Dilemma proof Π, denoted $|\Pi|$, is the number of occurrences of formula relations in Π. This proof has size 16.

To make the link back from derivations involving formula relations to proofs about formulas, note that the derivation of a contradictory formula relation from $X^\top$ constitutes a refutation of the formula X. Similarly, we can check whether or not X is a tautology by attempting to refute $X^+(X \equiv \bot)$.

3.3 Proof hardness

A formula relation R is k-easy if and only if there is a derivation $\Pi : R \Rightarrow S$ such that S is explicitly contradictory and $depth(\Pi) \leq k$. A relation R is k-hard if and only if there is no derivation $\Pi : R \Rightarrow S$ such that S is explicitly contradictory and $depth(\Pi) < k$. A relation R has *hardness degree* k, $h(R) = k$, if and only if R is k-easy and k-hard.

The formula $(A \wedge (B \vee C)) \to ((A \wedge B) \vee (A \wedge C))$ has hardness degree 0, while reversing the direction of the implication gives a formula of hardness degree 1. Harrison's paper on Stålmarck's algorithm as a HOL derived rule lists many 1- and 2-hard problems with performance statistics for Harrison's implementation of the method [12]. Many of the examples are circuits taken from the set of examples presented at the IFIP International Workshop on Applied Formal Methods for Correct VLSI Design, held at IMEC in 1989.

In practice, it turns out that real industrial verification problems often produce formulas with hardness degree 0 or 1. Why is this? One reason must be that the Dilemma system has a large set of propagation rules, reducing the need for branching. The only other explanation that we can offer is the observation that systems that have been designed by one person, or in which one person understands and can argue informally for the correctness of the system tend to result in easy formulas. Systems in which components are coupled together in a relatively uncontrolled way, so that behaviour is hard to predict by informal analysis, are also hard to analyse formally. These empirical observations indicate that it is a good idea to introduce design rules that improve verifiability. Indeed, the main supplier of railway interlocking software in Sweden has made an important move in this direction, by introducing software coding rules that guarantee the verifiability of the resulting interlocking control software, see section 6.1. Figuring out how to place constraints on the designer so that it is easy to reason about the correctness of his products seems likely to be an important area of research in formal methods.

4 From proof system to proof procedure

We have now studied the Dilemma proof system. How do we turn it into an efficient proof procedure? The first step is to find a good data structure for representing formulas. The second is to provide an efficient algorithm that searches exhaustively for shallow Dilemma proofs.

4.1 Triplets

For ease of manipulation, compound subformulas are represented by triplets, having the form $x : y \circ z$. The triplet variable x represents the compound formula obtained by applying a binary operator $\circ$ to the triplet operands y and z. x represents a subformula, and y and z are literals, that is either variables (real or triplet variables) or negated variables.

Example The formula $(A \wedge (B \vee C)) \to ((A \wedge B) \vee (A \wedge C))$ is reduced to triplets as follows:

$$t : \top$$
$$a : A$$
$$b : B$$

$$c : C$$
$$d : b \lor c$$
$$e : a \land d$$
$$f : a \land b$$
$$g : a \land c$$
$$h : f \lor g$$
$$i : e \rightarrow h$$

The saturation algorithm

A relation R is k-saturated if and only if for every Dilemma derivation $\Pi : R \Rightarrow S$ with $depth(\Pi) \leq k$, it holds that $R = S$. In other words, proofs of depth k or less add no new equivalences between subformulas.

The k-saturation procedure exhaustively searches for a proof of depth k. If a relation R has hardness degree k, then $saturate(R, k)$ must be explicitly contradictory, and k-saturation finds a disproof of R. The procedure is defined recursively. 0-saturation applies the propagation rules to a relation until no more rules are applicable. It chooses a compound subformula, applies a related simple rule and then continues to apply simple rules on those triplets whose variables were affected by the result of the first rule. The process continues until no further simple rules can be applied.

The following pseudo-code fragment presents 0-saturation.

```
saturate(R,0) =

Q := Compound(R)
while non-empty(Q)
 do
   remove some q from the set Q
   if contradictory(R(q))
      then return R(q)
      else Q := Q union parents(R(q)-R)
         R := R(q)
 od
return R
```

The set $Compound(R)$ is the set of all the compound subformulas in the domain of R, and this is the initial value of the pool of subformulas to be processed. We chose one element q from the pool and apply to R a simple rule related to q. $R(q)$ is $R(F \equiv G)$ if there is a simple rule whose premises all involve formulas from the set containing q and its immediate subformulas and their complements, and whose conclusion is $F \equiv G$. If no such simple rule applies, then $R(q)$ is just R. The application of a rule related to q in this way leads to the discovery of a set of new equivalences, $R(q) - R$. We add back to the pool those subformulas that contain as an immediate subformula one (or more) of the formulas involved in these equivalences - the 'parents' of the subformulas about which we have just discovered something new. These are the triplets that might be affected by the new information gained in applying the rule. Having thus (possibly) augmented the pool, we update R with the new equivalences and repeat, until the pool is empty. At that point, no more simple rules apply to R and 0-saturation is complete.

$(k + 1)$-saturation is defined in terms of branching and k-saturation. The version presented here branches on the truth or falsity of a subformula, rather than on equivalence between two arbitrary subformulas. This is the version that is currently implemented. A better strategy might perhaps be to try to merge the two largest equivalence classes.

```
saturate(R,k+1) =

repeat
  L : = Sub(R)
  R' := R
  while non-empty(L)
   do
    remove some l from L
    R1 := saturate(R'(l equiv FALSE),k)
    R2 := saturate(R'(l equiv TRUE) ,k)
    if contradictory(R1) and contradictory(R2)
       then return R1 union R2
       else if contradictory(R1)
              then R' := R2
              else if contradictory(R2)
                     then R' := R1
                     else R' := R1 union R2
   od
until R' = R
return R
```

The set $Sub(R)$ contains the subformulas of R, both variables and compound subformulas. We repeatedly branch on each subformula in turn until there are no further consequences in the form of new equivalences. Of course, one does not use $(k + 1)$-saturation until after k-saturation has been performed, since branching should be minimised. Information gained during k-saturation is available during the subsequent $(k + 1)$ saturation. It is the continuous gathering of information (in the form of equivalences) that distinguishes the algorithm from both breadth first search and interative deepening.

Note how the saturation algorithm propagates information both upwards and downwards in the syntax tree of the formula. In this, it is rather relational.

After the development of the algorithm described here, Kunz et al independently developed *recursive learning*, a method of solving the Boolean satisfiability problem, with applications in digital test generation [16]. The method has much in common with that presented here, and also discovers logical relationships between subformulas of a formula (or nodes in a circuit).

5   Complexity Results

An upper bound on proof lengths for Dilemma (and obviously for KE/I that is a subsystem of Dilemma) in relation to hardness degree is $n^{k+1} + n^k$, where n is the size of the formula of hardness degree k to be proved. This bound assumes that the Dilemma rule branches on the truth or falsity of a subformula

A lower bound is 2^k for analytic KE/I and $2^{k/2}$ for Dilemma. The difference is not due to the extension to formula relations but to the extension to series parallel graphs. It should be noted that the upper bound is "under exponential" over the lower bound.

Is Dilemma superior to analytic KE/I? There are infinite formula sequences for which the hardness of each F_i is less than a constant k in Dilemma, but grows logarithmically in analytic KE/I. It is interesting to note that the hardness degree for analytic KE/I and Dilemma are instead related linearly if either the formula relations are replaced by sets or if the Dilemma rule is replaced by analytic cut, but in the relational version. The relational version seems to open opportunities to remove repetitions by the invertible thinning used in the Dilemma rule.

Using the saturation procedure, the time required to search exhaustively for a proof of depth k of a formula A is bounded by $O(n^{2k+1})$, where n is the size of A. Note that the upper bound on proof search is not more than the square of the upper bound on proof length, and so it is still "under exponential" related to the shortest possible proof.

6   Industrial applications

In Sweden, Adtranz is the main supplier of railway interlocking software. Tools based on Stålmarck's method have been used to find potential runtime errors, so called double values, in railway interlocking systems since 1990. The checking is done entirely automatically, and the tools are incorporated into Adtranz' development environment. Adtranz report a 90% reduction in time spent on testing, and a reduction in overall development cost of about 15%. Here, we describe in a little more detail a different kind of verification project, also from the railway field [4].

6.1   Safety Analysis in Railway Interlocking Software

The railway interlocking software developed at Adtranz is written in a domain-specific synchronous declarative language called Sternol. That this is (rather surprisingly) the case has undoubtedly aided verification efforts in the area! It is also one of the reasons why we think that it is fair to say that many of the systems verified using tools based on Stålmarck's method have been *hardware-like*. From our point of view, there is very little difference between a Sternol program and a circuit.

A railyard implementation consists of a set of specialised instances of generic railyard objects such as points and signals, one for each physical object in the yard. Specialisation involves setting individual variables, for example the number of lights on a signal object. A control program containing code for each type of generic object then 'runs' the railyard implementation. On each cycle, every object receives some inputs and sets some of its variables depending on the values of boolean guards. Some of the variables are outputs of the object. Inputs and outputs are either booleans or integers. (The double value checking referred to earlier checks that this process is actually deterministic and does not try to set a variable to two different values.) The *same* control program runs *every* Sternol-based yard in Sweden.

A simple way to check safety properties for a railyard is to analyse for each object with safety critical outputs all the objects that follow it in the direction of travel. When a particular safety critical output is emitted by the first object, a certain number of the following objects must be unoccupied, and if they are points, they must point in the direction of travel. Such requirements are easily translated into propositional logic with arithmetic (PA). (The commercial implementations of Stålmarck's method incorporate a weak form of integer arithmetic. Simple rules for the arithmetic operators are added to the propositional rules described here.) The Sternol program for the yard itself encodes a synchronous finite state machine with boolean and arithmetic variables. It can be automatically translated to a PA formula expressing initialisation and the state transition relation. (This kind of translation is routinely used in hardware verification.) Proving (or disproving) the requirements is then straightforward.

This simple but effective verification method has been used in earnest in several projects. In the validation of the interlocking software for the Madrid Subway Station Lago, it was found that some points could be in a neutral position, rather than being locked in the correct position. Such an error might cause minor damage to train and track and is not considered very safety critical. However, the error was corrected. A similar verification was carried out in 1997 for a larger railyard in Finland. There, a serious safety critical error was found, which might have led to a derailment. The error is currently being corrected.

The small Madrid verification produced formulas containing around 36,000 triplets and took 40 man-hours of work in total. The largest yard analysed so far resulted in a formula containing about half a million triplets.

Based on their experience in performing this kind of formal verification, Adtranz have introduced coding rules for Sternol that guarantee the verifiability of the resulting interlocking control software. These rules constrain the programmer, forcing him to construct hierarchical state machines in a particular style. For example, if resources are claimed in a particular order, then they must be released in reverse order. The result of following these guidelines has been that

all formulas generated from Sternol programs since their introduction have been
of hardness degree 1 or less, and so quickly proved.

Using Stålmarck's method in CAD

Stålmarck's method provides an efficient way to do tautology checking on large
formulas. We expect it to be applicable in CAD. Some of the industrial verification problems that have been tackled using Stålmarck's method are very similar
to post hoc hardware verification of systems built using boolean operations and
arithmetic. Typically, systems are modelled as synchronous boolean automata.
For example, the reverse flushing control in a nuclear power station's cooling
system has been verified [24]. There seems to be no reason why circuits should
not be verified in the same style, and indeed the Swedish consulting company
LUTAB has had some projects in this area.

In an assignment from the Swedish Defence Material Administration, LUTAB
verified the hardware and software controlling the landing gear of a Saab mili-
tary aircraft. Both the hardware and software were modelled manually in propositional logic. Only the electronic part of the hardware (operating in parallel
with the software) was modelled, so valves and relays were not covered. The
hardware model was based on the functional schematics of the actual implementation. The combined system controlling the opening and closing of the doors to
the compartment in which the landing gear is stowed, as well as the lowering of
the landing gear, was modelled as a hierarchy of modules. Using the manually
constructed system model, it was possible to ask questions like "can the landing
gear be lowered at the same time as the doors are closed?". (It is important to
check such properties; there have been cases where functional test on a physical test rig has revealed this kind of error and the damage that it can cause.)
The analysis checked a set of safety requirements on the system. It also covered
failure modes, and so could analyse the behaviour of the system under a single
fault, two simultaneous faults, and so on. In those cases, countermodels to the
safety requirements were generated and analysed.

The modelling took LUTAB one man-month to do. A lot of that time was
spent modelling integer values (which at that time were not supported in the
implementation of Stålmarck's method used) by means of boolean variables.
Similar projects have since been undertaken for civil aircraft. More recent work
in the area is described in [25].

Just as with BDDs, some classes of circuits have proved difficult to verify
using Stålmarck's method. Multipliers, in particular, give rise to hard formulas,
and the hardness grows with circuit size. We have been experimenting with ways
to systematically reduce proof hardness for regular circuits by adding definitions
of fresh variables to the system formula – something akin to lemmas. The trick
is to know exactly what new definitions to add. Here, we benefit from the fact
that we describe circuits not directly in propositional logic but in the functional
programming language Haskell, in a combinator-based style that has its roots
in μFP and Ruby [2]. We generate propositional logic formulas for instances of

generic circuits by symbolic evaluation. The combinator style makes it easier to discover where new definitions should be added. We are guided by the fact that the proof of a circuit instance is closely related to the shape of the circuit. Using this approach, large combinational multipliers have been verified [20]. We are developing novel methods to verify sequential circuits.

7 Summary

We have presented Stålmarck's patented proof procedure for propositional logic. The underlying Dilemma proof system is efficient for two reasons.

1. *Efficient propagation.* It provides efficient propagation of information about subformulas of the formula to be refuted. This efficiency of propagation comes both from the large set of introduction and elimination rules and from the fact that relations are used to maintain equivalences between subformulas.
2. *Series-parallel graphs instead of trees.* The Dilemma rule recombines the results of the two sides of a branch. It is a combination of cut and an invertible form of thinning. This avoids repetition in proofs.

The proof procedure itself is efficient because of the careful design of the k-saturation algorithm, which guarantees to find short proofs if they exist.

The method has been used to perform industrial-strength formal verification of many hardware-like systems. The question of whether or not it can be applied in practice in CAD remains to be answered. We hope that this paper will stimulate others to join the quest for an answer.

Acknowledgements

. Thanks to Koen Claessen for his careful reading of earlier drafts and to the members of the Formal Methods group at Chalmers, with whom we have had many enjoyable discussions. Sheeran's work at Prover Technology is partly funded by the EU project SYRF (Synchronous Reactive Formalisms).

References

1. E.W. Beth: Semantic entailment and formal derivability. Mededelingen der Kon. Nederlandes Akademie van Wetenschappen. Afd. letterkunde, n.s., 18, 309-342, Amsterdam, 1955.
2. P. Bjesse, K. Claessen, M. Sheeran and S. Singh: Lava: Hardware Design in Haskell. Proc. Int. Conf. on Functional Programming, ACM Press, 1998.
3. A. Borälv: The industrial success of verification tools based on Stålmarck's method. Proc. 9th Int. Conf. on Computer Aided Verification, Springer-Verlag LNCS vol. 1254, 1997.
4. A. Borälv and G. Stålmarck. Prover Technology in Railways, In Industrial-Strength Formal Methods, Academic Press, 1998.

5. R. Bryant: Graph-Based Algorithms for Boolean Function Manipulation. IEEE Trans. Comp., vol. c-35, no.8, 1986.

6. S.A. Cook: The complexity of theorem-proving procedures. In Proc. 3rd ACM Symp. on the Theory of Computing, 1971.

7. M. D'Agostino: Investigation into the complexity of some propositional calculi. D. Phil. Dissertation, Programming Research Group, Oxford University, 1990.

8. M. Davis, G. Logemann and D. Loveland: A machine program for theorem proving. Communications of the ACM, 5:394-397, 1962. Reprinted in [21].

9. M. Davis and H. Putnam: A computing procedure for quantification theory. Journal of the ACM, 7:201-215, 1960. Reprinted in [21].

10. G. Gentzen: Untersuchungen über das logische Schliessen. Mathematische Zeitschrift, 39, 176-210, 1935. English translation in The Collected Papers of Gerhard Gentzen, Szabo (ed.), North-Holland, Amsterdam, 1969.

11. J.F. Groote, J.W.C. Koorn and S.F.M. van Vlijmen: The Safety Guaranteeing System at Station Hoorn-Kersenboogerd. Technical Report 121, Logic Group Preprint Series, Utrecht Univ., 1994.

12. J. Harrison: The Stålmarck Method as a HOL Derived Rule. Theorem Proving in Higher Order Logics, Springer-Verlag LNCS vol. 1125, 1996.

13. J.K.J. Hintikka: Form and content in quantification theory. Acta Philosophica Fennica, VII, 1955.

14. S. Kanger: Provability in Logic. Acta Universitatis Stockholmiensis, Stockholm Studies in Philosopy, 1, 1957.

15. S. C. Kleene: *Mathematical Logic*. John Wiley and Sons Inc., New York, 1967.

16. W. Kunz and D.K. Pradhan: Recursive Learning: A New Implication Technique for Efficient Solutions to CAD-problems: Test, Verification and Optimization. IEEE Trans. CAD, vol. 13, no. 9, 1994.

17. M. Mondadori: An improvement of Jeffrey's deductive trees. Annali dell'Universita di Ferrara; Sez III; Discussion paper 7, Universita di Ferrara, 1989.

18. K. Schütte: *Proof Theory*, Springer-Verlag, Berlin, 1977.

19. G. Stålmarck: A system for determining propositional logic theorems by applying values and rules to triplets that are generated from a formula, 1989. Swedish Patent No. 467 076 (approved 1992), U.S. Patent No. 5 276 897 (approved 1994), European Patent No. 0403 454 (approved 1995).

20. M.Sheeran and A. Borälv: How to prove properties of recursively defined circuits using Stålmarck's method. Proc. Workshop on Formal Methods for Hardware and Hardware-like systems, Marstrand, June 1998.

21. J. Siekman and G. Wrightson (editors): *Automation of Reasoning*. Springer-Verlag, New York, 1983.

22. R.M. Smullyan: *First Order Logic*. Springer, Berlin, 1969.

23. M. Srivas and A. Camilleri (editors): Proc. Int. Conf. on Formal Methods in Computer-Aided Design. Springer-Verlag LNCS vol. 1146, 1996.

24. M. Säflund: Modelling and formally verifying systems and software in industrial applications. Proc. second Int. Conf. on Reliability, Maintainability and Safety (ICRMS '94), Xu Ferong (ed.), 1994.

25. O. Åkerlund, G. Stålmarck and M. Helander: Formal Safety and Reliability Analysis of Embedded Aerospace Systems at Saab. Proc. 7th IEEE Int. Symp. on Software Reliability Engineering (Industrial Track), IEEE Computer Society Press, 1996.

Almana: A BDD Minimization Tool Integrating Heuristic and Rewriting Methods

Macha Nikolskaïa, Antoine Rauzy, and David James Sherman
{macha,rauzy,david}@labri.u-bordeaux.fr

LaBRI, Université Bordeaux-1, France

Abstract. Constructing a small BDD from a given boolean formula depends on finding a good variable ordering. Finding a good order is NP-complete. In the past, methods based on *heuristic analysis* and *formula (circuit) rewriting* have shown to be useful for specific problem domains. We show that these methods need to be integrated: heuristic analysis can drive rewriting, which in turn simplifies analysis. We support this claim with experimental results, and describe *Almana*, an integrated tool for exploring the combination of analysis and rewriting on large boolean formulae.

1 Introduction

Finding a good variable order is a central issue in the practical use of binary decision diagrams (BDDs) for industrial problems. Unfortunately, finding the order that minimizes BDD size for a given boolean formula is an NP-complete problem[Bry86,THY93]. This is particularly unfortunate as the wrong order can give exponentially-worse results, as Bryant shows in [Bry92], and moreover some common formulae do not admit small (polynomial) BDDs at all.

Nonetheless BDDs are astonishingly useful for a wide variety of practical industrial problems, and are indeed the state-of-the-art representation for boolean formulae representing combinational circuits, fault trees, and other complex systems. We can only suppose that BDDs exploit a hidden and not well understood internal structure in these domains that permits a compact representation. Capturing some version of this hidden structure in the variable order seems to be the key to minimizing BDD size. On this capturing depends the success of BDDs in their application domains. For example, in circuit analysis, heuristic methods manage to use the knowledge of the formula structure rather well[FFK88,FOH93].

Our particular interest is the use of BDDs for risk analysis of complex systems, where the biggest hurdle is *constructing* the BDD in memory. Formal methods used to analyze risk in industrial systems do not fundamentally differ from those used to verify circuits or protocols. Since the high-level notations used to design a model of the system cannot be assessed directly, these descriptions are compiled into combinational circuits (plus some additional information for probabalistic studies), and the resulting circuit is assessed by low-level tools. As with circuits and protocols, the compilation of these models is a difficult process, not for syntactic reasons but because the way a model is written dramatically influences the performance of the low-level tool.

Typical formulae for risk assessment problems have 100 to several thousand variables and consequently from 10^{10^2} to 10^{10^4} different possible variable orders. Typical BDDs for these problems generate 10^6 nodes in memory even when a good variable order is found. Preprocessing of the low-level representation is thus required for practical use of BDD-based tools. The tool we present in this paper, *Almana*, is a generic preprocessing tool for combinational circuits.

Due to the NP-complete nature of finding a good variable order, different approximation methods are found in the literature. *Heuristic methods* based on analysis of the original boolean formula abound, and can be subdivided into *static* techniques, that inspect the formula off-line[Bry86,MWBSV88,FFM93,FOH93], and *dynamic* techniques, that adjust the variable order on-line ([Rud93], see also [FMK91,ISY91,FYBSV93]). Though very popular, dynamic methods present many problems. The first is that they require that we have already constructed the BDD or some part of it in memory, which is impossible for large systems. A more troublesome problem is that existing techniques are based on *sifting*, which exchanges adjacent variables. Unfortunately, in real systems variables come in *blocks* of related variables, that need to be kept together in the final order or the size explodes. A promising approach is that of *group sifting* [PS95]. As far as we have seen dynamic techniques are not yet appropriate for large complex systems. We restrict our discussion to static methods in this paper.

Rewriting methods perform semantics-preserving transformations of the original boolean formula to obtain an equivalent formula that admits a good natural variable ordering[IH93,Nie94,Bou96]. These formulae are in fact *circuits*, since they are maximally shared; formula rewriting is in fact DAG rewriting of boolean circuits. Formula rewriting is a new and very promising technique, especially since the boolean formulae produced by automatic tools are generally *badly written* in terms of topological variable orderings[NR98].

The central hypothesis of this paper is that these two approximation methods complement each other nicely, and need to be integrated into a single tool for analyzing and rewriting boolean formulae. Such a tool would be used by engineers and other specialists to *program* rewriting strategies for converting formulae produced by modeling tools into something more easily handled by BDD-based analysis tools. Almana, developed at the LaBRI (Université Bordeaux-1), is just such a tool.

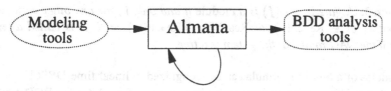

In what follows we present a brief overview of the BDD size minimization problem and the best existing methods for finding good variable orders (section 2), followed by a description of the Almana toolbox and the way that it integrates heuristic- and rewriting-based methods (section 3). Section 4 presents experimental results, notably on large formulae provided by industrial partners in the Club Aralia,[1] where the complemen-

---

[1] A group of industrial and academic partners concerned with the Aralia BDD package (Commission d'Énergie Atomique, Dassault Aviation, Électricité de France, ELF Aquitaine, LaBRI

tary nature of these methods becomes evident. Section 5 generalizes this approach and describes the next generation tool we feel is necessary, one version of which is under construction at our laboratory. Section 6 concludes.

2 BDD Size Minimization

In order to assess a formula using a BDD technique, it is obviously necessary first to construct its BDD. As the size of a BDD is extremely sensitive to variable order, and the problem of finding the best ordering is NP-complete, it is critical to have appropriate heuristic methods. The examples from the industrial problems we deal with have on the order of 100 to several thousand variables. There is no way to assess such formulae other than to use some specific heuristics, or simplify them, or both.

The best known techniques for minimizing the size of the BDD induced by a boolean formula are heuristic analyses, that try to find a good variable order; and rewriting techniques, that try to convert the formula into a form that is easier to process.

Definition 1. *A* boolean formula *is a term constructed from constants* 0 *and* 1, *a set of variables V and usual logical connectives* $\land$, $\lor$, $\neg$ *etc.*

We consider boolean formulae presented as a set of equations $g = f$ where g is a variable denoting a logic gate and f is a boolean formula. Such representation implies that formulae are at once circuits and directed acyclic graphs (DAG) labeled by variables, from, respectively, engineering and graph-theoretic standpoints.
Example: $f = g \land h, g = (b \lor a), h = (a \lor c)$.

Let $\mathcal{T}(f)$ be the set of *terminal variables* on which f depends, and $\mathcal{G}(f)$ the set of *gate variables* used in its definition; $V(f) = \mathcal{G}(f) \cup \mathcal{T}(f)$. For the previous example $\mathcal{G}(f) = \{g, h\}$ and $\mathcal{T}(f) = \{a, b, c\}$.

Definition 2. *For a variable* $v \in V(f)$, Reach(v) *is the set of descendant variables in the DAG defined by f reachable from v.*

Definition 3. *A gate* $v \in V(f)$ *is a* module *if and only if, for any other variable* $w \in V(f)$, *either* $w \in$ Reach(v), *or* Reach(v) $\cap$ Reach(w) $= \emptyset$. *It is called a* terminal module *if the DAG rooted at the gate v is a tree.*

The modules of a boolean formula can be recognized in linear time[DR96].
The following theorem expresses the importance of modules for BDD size minimization.

Notations: For a formula f we will note by $\sigma(f)$ an ordering of the variables of $\mathcal{T}(f)$, and $\mathcal{S}(f)$ the size of the corresponding BDD.

(Univ. Bordeaux-1), LADS (Univ. Bordeaux-1), Renault, Schneider Electric, SGN, Technicatome). Aralia[Rau95b] was originally produced at the LaBRI, and is commercialized by IXI, Inc. (contact tony.hutinet@ixi.fr).

Theorem 1. *Let f and g be two formulae with disjoint sets of terminal variables $T(f)$ and $T(g)$. Then the following is true for both $\sigma(f), \sigma(g)$ and $\sigma(g), \sigma(f)$ orderings.*

$$S(f * g) = S(f) + S(g)$$

*where * denotes either $\vee$ or $\wedge$.*

For proof see for example [Ber89].

The important corollary of this theorem is the following. If for a formula f it is possible to find a set of modules such that the union of their terminal variables is equal to the set of boolean variables of f, then the size of the BDD for f is equal to the sum of sizes of BDDs of the modules. And this whatever the ordering of modules.

Proposition 1. *The size of a BDD corresponding to a terminal module is equal to the number of variables.*

This comes from the fact that a formula f is a terminal module if and only if $\forall v \in T(f)$, v appears only once in f.

2.1 Heuristic Methods

The best known algorithm for finding an optimal order is of complexity $O(n^2 3^n)$ where n is the size of the BDD in nodes ([ISY91], improving [FS90]), and hence is useless for practical applications. The nature of the variable ordering problem requires heuristic techniques.

The main problem with existing heuristic methods for variable ordering is that the same heuristic, applied to different rewritings of the same formula, can give wildly different results. [BBR97] shows standard deviations for industrial examples so large that order-of-magnitude differences in BDD size are typical. It is consequently difficult to apply these methods with any confidence.

One very important common feature for an ordering heuristic is to respect modules. This is essential because there always exists an optimal ordering that respects modules [Ber89].

The basic technique used in static ordering heuristics is the top-down left-most traversal of the formula. Used as such it gives the *topological order* to the variables. Different improvements have been made such as in [FFK88,FFM93,Rau95a] and others. Some of them try to attach more semantic sense to the position of the variable in the ordering, such as *shuffle* [NR98] based on [FOH93]; and that in [TI94], which works well for formulae encoding perfect matchings of bipartite graphs.

The point is that most of the known heuristics have been studied for combinational or sequential circuits. As stated in [BBR97] it is not clear whether thay are also efficient for other boolean formulae. Some research has to be done in the direction of the fine-tuning of heuristic methods according to the domain of application.

For some comparative studies see for example [BRKM91] or [BBR97]. In the following we summarize the popular static variable ordering heuristics, all of which are implemented in Almana (described in section 3). They all respect modules.

Heuristic topology: This is the most frequently used and the simplest heuristic. It consists of a depth-first left-most traversal of the formula.

Heuristic weight: This heuristic from [MIY90] propagates weights bottom-up in the formula. Every terminal variable is attributed weight 1 and the weight of each intermediate variable is equal to the sum of the weights of its children. For each variable its children are reordered in order of increasing weight. The resulting order is obtained by the depth-first left-most traversal.

$$W(v) = \begin{cases} 1, \forall v \in \mathcal{T} \\ \sum_i W(v_i), \forall v \in \mathcal{G}, v_i \in Ch(v) \end{cases}$$

where the $Ch(v)$ are the children of variable v and and W stands for weight.

Heuristic flow: This heuristic from [Rau93] propagates flows. The roots are attributed flow 1. The flow of a variable v is divided by the number of its children. The flow for each variable is the sum of flows coming from its parents. For each variable its children are reordered in order of decreasing flow. The final order is obtained by a depth-first left-most traversal.

$$F(v) = \begin{cases} 1, \forall v \in Roots \\ \frac{1}{n} \sum_i F(v_i), v_i \in Pa(v), n = \#Pa(v) \end{cases}$$

where the $Pa(v)$ are the parents of variable v and F stands for flow.

Heuristic parents: This heuristic from [FFK88] and [FFM93] counts the circuit fan-out, that is, the number of parents, of each variable. For each variable its children are reordered in decreasing order of number of parents. The final order is obtained by a depth-first left-most traversal.

$$P(v) = \begin{cases} 0, \forall v \in Roots \\ \#Pa(v), \forall v \in \mathcal{T} \bigcup \mathcal{G} \end{cases}$$

where $Roots$ is the set of roots of the formula and P stands for parents.

Heuristic level: This heuristic from [MWBSV88] classifies variables based on their distance from the root.

$$L(v) = \begin{cases} 0, \forall v \in Roots \\ \max_{v_i \in Pa(v)}(L(v_i)) + 1, \forall v \in \mathcal{T} \bigcup \mathcal{G} \end{cases}$$

where L stands for level.

Heuristic shuffle: This is an interleaving heuristic based on [FOH93] that we proposed in [NR98]. It was originally designed to exploit the topology of combinational circuits and we adapted it to the case of multi-level redunduncies of components. Shuffle tries to assure that variables that are semantically close end up near each other in the variable ordering. Let us note $\sim$ an operation defined over lists of variables in the following way (where [ ] denotes the empty list and $\cdot$ denotes concatenation).

$$[\,] \sim L = L$$
$$L \sim [\,] = L$$
$$x \cdot L_1 \sim x \cdot L_2 = x \cdot (L_1 \sim L_2)$$
$$x \cdot L_1 \sim y \cdot L_2 = x \cdot (L_1 \sim y \cdot L_2) \text{ if } x \notin L_2$$
$$x \cdot L_1 \sim y \cdot L = (x \cdot L_1 \sim y \cdot L_2) \sim L_3 \text{ if } L = L_2 \cdot x \cdot L_3$$

Note that $\sim$ is a nonassociative operator,[2] for which the empty list $[\,]$ is an identity element. The variable ordering corresponding to the variable $v = v_1 * v_2 * \dots * v_n$ is defined as $L = L_1 \sim L_2 \sim \dots \sim L_n$ where the v_i are children of v, $*$ is any boolean operator, and the L_i are the results of the $\sim$ operation for v_i.

Heuristic HMA: This is a hybrid heuristic that we proposed in [NR98], specifically tailored to sums-of-products forms.

2.2 Rewriting Methods

In [BBR97] it is reported that a criterion that may strongly influence the size of the BDD is the way the formula is written and in particular the order of its arguments. The use of rewriting techniques for improving the tractability of boolean formulae is described in [IH93] and [Nie94]. In all of these cases the motivating observation is that the original formulae resulting from system modeling seem highly redundant and can be rewritten in a way that improves their tractability while preserving their meaning. The use of rewriting methods does not change the inherent difficulty of finding a good variable order, but does prove a useful adjunct to heuristic analysis and can have some effect on the size of intermediate BDDs.

The profound difficulty is that the relations between formula structure and BDDs are not well understood. We lack a theoretical basis that lets us know when one version of a formula is better than another when it comes to constructing the corresponding BDDs. Consequently, there will be for the forseeable future a strong heuristic component to rewriting methods as well. Indeed, as we describe in section 4.2, heuristic analyses can aid formula rewriting by identifying candidate nodes.

One observation that has practical interest is that the more the maximally-shared formula resembles a tree, the smaller the corresponding BDD. That is, some *factorization* of the formula can lead to good improvements. We say "some" because there is experimental evidence that too much factorization leads to larger BDDs. The intuition is the following. We want, on the one hand, variables that are related in the formula to be near each other in the variable order. On the other hand, we want variables that are important for the entire formula to appear early in the variable order. Judicious factorization (to depth four or five) permits us to maximize the latter need while preserving as much as possible the locality property of the former.

The theoretical reason why it is interesting to make the formula tree-like is expressed by theorem 1 (page 4) and its corollary.

---

[2] The nonassociativity of $\sim$, by the way, explains its sensitivity to random rewriting.

3 The Almana Toolkit

Despite the theoretical and practical difficulty of minimizing BDD sizes for complex systems, we must nonetheless do our best to handle real problems. Since a general solution is intractable, we need tools that can be adapted to specific application domains, to do the best that we can.

Almana is a collection of software components that provides heuristic analyses, rewriting transformations, and utilities in an integrated setting. All of these components operate on a shared DAG-based representation of a boolean formula, built from the standard boolean operators provided by the Aralia BDD package. The nodes of this shared DAG are labeled with boolean operators, variables, and constants. We can further attach arbitrary *properties* to nodes for use in analysis.

The Almana software components, shown schematically in figure 1, fall into the following five categories.

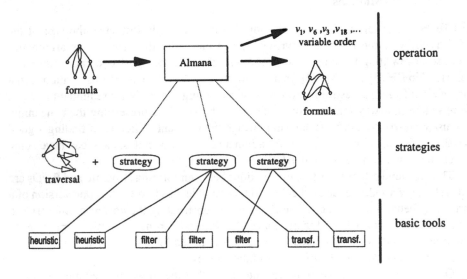

Fig. 1. Almana components and their relations

Heuristics. A heuristic analysis in Almana traverses the formula, maintaining partial results in node properties. Each heuristic is decomposed into preprocessing, midprocessing, and postprocessing phases. These phases can be purely analytic, that is, based solely on inspection of the formula; or they can locally rewrite the formula to find a better configuration. For example, the weight heuristic works as follows.

preprocessing Traverse the formula bottom-up. For each variable encountered, 1) mark whether it takes part in a trivial module or not; 2) mark it with its *weight*, that is, the number of variables below it; and 3) sort its child edges by weight.

midprocessing Traverse the formula **again**. For every node and for every interval of child edges with the same weight, *randomly permute* these edges.

postprocessing Clear all of the marks and return the best order found.

If we skip the middle step of random rewriting, we obtain the analytic heuristic proposed by [Bou96]. However, as we reported in [NR98], the combination of analysis and random rewriting gives much better results.

Figure 2 shows the heuristics currently available in Almana.

Heuristic	Principle
topology	topological (depth-first, left-right) order
flow	top-down propagation of flow [Rau95a]
weight	bottom-up propagation of weights [MIY90]
weight_op	same with distinctions for boolean operators
weight_cat	same with categories of weights
weight_as_flow	combination of weight and flow
parents	preference to variables often referenced [FFM93]
level	preference to variables high in the formula [MWBSV88]
shuffle	[NR98] based on variable interleaving [FOH93]
HMA	group strongly related variables [NR98]

Fig. 2. Predefined variable-order heuristics in Almana

Filters. A filter determines whether a given node responds to certain criteria, determined by inspecting it and neighboring nodes. A filter is the imperative equivalent of the left-hand side of a rewriting rule. An example is the *repeated_children* filter, which returns `true` if at least two children of the given associative and commutative operator are the same.

Figure 3 summarizes the filters (and transformations) currently available in Almana, described as rewrite rules. In the notation used in that figure, $v/\mathbf{op}[\vec{f_i}]$ denotes a gate v with operator **op** and subterms $f_1, \ldots, f_i, \ldots, f_n$. Informally, the *one_child* rule shortens trivial definitions; the *flatten* rule removes nested appliations of the same operator; the *repeated_children* rule removes duplicate inputs to conjunctions and disjunctions; the *regroup* rule regroups a term into k subterms; the *factorize* rule factorizes a sum of products (or a product of sums) with respect to a variable u; and the *multiple_definition* rule globally shares equivalent (congruent) definitions.

Transformations. A transformation rewrites a given node, by modifying it and its neighboring nodes. Nodes and edges can be created and deleted by transformations. A transformation is the imperative equivalent of the right-hand side of a rewriting rule. An example is the *one_child* transformation, which replaces a node by its (unique) child while maintaining the DAG structure. Figure 3 shows the transformations currently available in Almana.

Filter	$\Rightarrow$ Transformation

one_child
$\quad v/\mathbf{op}\,[f]$ $\qquad\qquad\qquad\qquad\Rightarrow\quad f$

$\qquad$ where $\mathbf{op} \neq \mathbf{not}$

flatten $\qquad\qquad\qquad\qquad\qquad\Rightarrow\quad v/\mathbf{op}\,[\ldots f_{i-1}, f_{i+1}, \ldots, \vec{g_i}]$
$\quad v/\mathbf{op}\,[\vec{f_i}]$

$\qquad$ where $\mathbf{op} \in \{\mathbf{and}, \mathbf{or}\}$
$\qquad$ and $f_i = v_i/\mathbf{op}[g_1, \ldots, g_n]$

repeated_children $\qquad\qquad\qquad\Rightarrow\quad v/\mathbf{op}\,[\ldots f_i \ldots f_{j-1}, f_{j+1} \ldots]$
$\quad v/\mathbf{op}\,[\ldots f_i \ldots f_j \ldots]$

$\qquad$ where $\mathbf{op} \in \{\mathbf{and}, \mathbf{or}\}$
$\qquad$ and $f_i = f_j$

regroup k $\qquad\qquad\qquad\qquad\Rightarrow\quad v/\mathbf{op}\,[g_1, \ldots, g_k]$
$\quad v/\mathbf{op}\,[f_1, \ldots, f_n]$ $\qquad\qquad$ where each $g_i =$

$\qquad$ where $s = \lceil \frac{n}{k} \rceil \geq 2$ $\qquad\qquad\quad v_i/\mathbf{op}\,[f_{s(i-1)+1}, \ldots, f_q]$
$\qquad$ and $\mathbf{op}$ is associative $\qquad\qquad$ and $q = s{\cdot}i$ if $s{\cdot}i < n$, n otherwise

factorize u $\qquad\qquad\qquad\qquad\Rightarrow\quad v/\mathbf{op}\,[v'/\overline{\mathbf{op}}\,[u, \vec{f_i'}], \ldots]$
$\quad v/\mathbf{op}\,[\ldots \vec{f_i} \ldots]$ $\qquad\qquad$ where each $f_i' = v_i'/\mathbf{op}\,[\vec{g_i}]$

$\qquad$ where $\overline{\mathbf{and}} = \mathbf{or}$, $\overline{\mathbf{or}} = \mathbf{and}$
$\qquad$ and $f_i = v_i/\overline{\mathbf{op}}\,[u, \vec{g_i}]$

multiple_definition $\qquad\qquad\quad\Rightarrow\quad x/\mathbf{op}\,[\vec{f_i}]$
$\quad v/\mathbf{op}\,[\vec{f_i}]$

$\qquad$ where $\exists x/\mathbf{op}\,[\vec{f_i}]$

Fig. 3. Predefined filters and transformations in Almana

Strategies. A strategy determines in what order to apply filters and transformations. Basic strategies apply corresponding filters and transformations globally to the entire formula. More elaborate strategies, such as that presented in section 4.2, use the results of analysis and other filters to traverse the formula and rewrite it. An important element of the component-based design of Almana is that filters, transformations, and heuristics can be freely combined in one strategy, should the application domain so require.

Utilities. Various utility components are provided in Almana, for I/O in Aralia format, node property manipulation, and random number generation.

4 Experimental Results

Different application domains provide boolean formulae with different structures, and one of the goals of Almana is to provide scope for domain-specific improvements. It seems for the moment unrealistic to support that general-purpose variable ordering heuristics will work well in all cases. By allowing a specialist to freely combine different analytic and rewriting techniques, Almana lets one tailor improvement strategies to specific application domains.

To support this claim, we present two kinds of experimental result. The first illustrates the influence of argument permutation on BDD size. The second is a concrete example where the interaction between heuristic analysis and rewriting permitted improvements that those methods were separately unable to provide.

4.1 Effect of Fan-in Permutation on Size

In order to illustrate the influence of permuting the arguments of gates we tested boolean formulae from the Aralia benchmark[3] following the experimental procedure proposed by [BBR97]. By simply changing gate fan-in orders we observe BDD size variations of up to a factor of 100. The results we present were obtained by applying all of the Almana heuristics on 500 random rewritings (permutations of fan-ins) of each circuit. Figure 4 shows formula characteristics and some numeric results. The columns give respectively the name of the formula; the number of terminal variables; the number of gate variables; and the minimum, maximum, mean, and standard deviation of the number of nodes in the computed BDDs. Further comparisons and similar results can be found in [BBR97].

4.2 Combining Analysis and Rewriting

All of the examples in section 4.1 could be handled by existing heuristic methods, even though they show large differences in BDD size depending on different rewritings of the formula. However, there are a lot of practical examples that cannot be treated in this way. Such an example was provided by a member of the Club Aralia, and is presented in this section.

The problem formula is given in the form of a root **OR** of some 500 **AND** expressions, that describes the different possible configurations of an automobile. These formulae belong to the large class of 2-level sums-of-products formulae, which are of great interest for quantitative reliability analysis since the failure scenarii of a system are in this form.

The example we consider has 156 terminal variables and 515 gates. Despite its seemingly small size, it cannot be handled by any known heuristic methods even on our Sparc Ultra-1 processor with 512 Mb of core memory. In regarding the formulae, we note that they are, like many other automatically-generated formulae, "badly written" and highly redundant. This suggests that we might be able to develop appro-

---

[3] Available from www.labri.u-bordeaux.fr.

Name	#$\mathcal{T}$	#$\mathcal{G}$	min	max	mean	st.dev
baboab1	61	81	1817	30121	5750.7	2212.37
baboab2	32	40	145	1943	457.952	142.404
baboab3	80	107	3305	27349	9926.87	3229.01
das9204	53	30	53	153	79.4012	15.9311
das9208	103	145	2614	20614	4833.58	1332.81
edf9r01	458	434	551	289421	9437.65	22412.5
edfpa02	278	251	7562	128847	53581.8	25209.2
edfpa09	196	142	255	4727	1094.16	774.067
edfpr01	548	484	8705	691766	185871.	118286.
edfrbd1	120	178	145	34557	1325.65	2517.62
isp9602	116	122	708	2993	1629.68	334.158
isp9603	91	95	703	10950	42037.71	1783.94
isp9607	71	65	149	805	409.923	64.4269

Fig. 4. Influence of fan-in permutations on BDD size

priate rewriting techniques for this problem. Indeed, we were able to use Almana to quickly develop formula factorizations driven by our HMA heuristic, which is used in the rewriting strategy to choose the next variable with respect to which the factorization is performed. Using this strategy, we obtain a formula for which the resulting BDD is quite small and poses no problem.

Some observations are interesting. First, HMA turned out to be the only heuristic that could efficiently guide the rewriting process. This is not a standard heuristic, but one that we previously programmed for the specific case of sums of products. Second, if the rewriting is done too much, the formula becomes once more intractable. Consequently, it is necessary to be able to program different termination cases in rewriting strategies. Third, this is a typical example of a formula for which a good variable order exists, but for which it is difficult to compute the corresponding BDD.

Can we apply these results to other kinds of formulae? The quick answer is no: heuristic methods try to capture domain knowledge, implicitly or explicitly, and methods that work are too intimately linked to their domain to be easily applied to others. Heuristics that work for fly-by-wire systems may work for automobiles but not for nuclear power plants, and almost certainly fail for pipelined microprocessors. What *does* work is to look methodically for stable and robust methods, by *adapting* existing methods that work for similar problems—or for similar modelling tools. In formulae investigated in collaboration with Dassault Aviation, for example, we have seen heavily-nested sums of products that can be more compactly represented as k-out-of-n formulae. That these conjunctions are consistently generated by the same tool for the same kinds of system permits us to fine-tune heuristics for Dassault. Once you know the tool and the kind of system, it is usually not too hard to adapt existing heuristics. So what we *can* apply is the approach and the basic tools, and insist on the need for flexibility in combining them.

5 Towards an Exploratory Tool

One of the key advantages of the component-based approach in Almana for the specialist is the ease with which new strategies for specific application domains can be explored. To integrate this exploration into the engineering process, however, a more elaborate user interface is required. In this section we briefly describe the requirements for a next-generation interactive tool for exploring boolean formula analysis and rewriting. One version of such a tool is currently under development at Bordeaux.

The first unavoidable observation is that some kind of programming interface seems necessary. Unlike well-understood domains like signal processing and circuit design, it is not clear what basis set of components is necessary and sufficient for expressing the desired analyses and rewritings. It must be possible to define new components and add them to the system.

The second unavoidable observation is that it needs to be an *imperative* programming interface. While traditional declarative methods are conceptually very clean and seemingly applicable—boolean expressions with maximal sharing are circuits, and circuits are labeled, ordered directed acyclic graphs—handling circuits with existing graph-rewriting techniques poses some problems. First, unlike usual term rewriting, transformation rules in this case may need to take sharing into account, rendering effectively useless standard fast rewriting techniques. Second, the particular syntactic restrictions imposed by declarative approaches are too constraining for their use to be natural in the practical case of boolean expressions.

What seems more useful is an *imperative* language for describing concrete rewriting rules and strategies. While less satisfying than a declarative solution based on automatic handling of classes of rules, such a language permits hands-on experimentation and immediate solutions to recognizably-solvable problems.

One valuable aspect of declarative rules is that they provide visually recognizable patterns—simply looking at the rule gives a clear idea of which elements of the local neighborhood of a candidate node in memory contribute to the pattern and take part in the replacement. An imperative approach to formula rewriting needs an equivalent idea of *locality*, as an aid to understanding the imperative code. Such an idea is provided by the notion of *cursors* pointing into the subject formula. Cursors are placed on edges in the graph: a cursor is therefore always either between two nodes, before the root, or after a leaf in the DAG. Technical report [NRS98] provides technical details on an imperative approach to formula manipulation based on local operations defined over a set of cursors into the graph. These operations were consciously designed to be added as primitive operations to an existing *extension language* (scripting language), such as Guile, Tcl, or others, that provides a concrete syntax, function definitions, and appropriate control structures.

In such an interactive programming tool, boolean formulae become a predefined data types and existing Almana components become library routines that manipulate these objects. Basic data structures, such as arrays for representing variable orders, are provided by the extension language.

For example, consider the repeated-children filter, which returns *true* if at least two children of the given associative and commutative operator are the same. This filter is easily coded in the following manner:

```
fun repeated_children_filter (Cursor c)
{
  if (c.get_below()->has_name());
    c.down(0);
  if (!is_associative_commutative(c))
    return false;
  table = [];
  for (c.down(0); !c.is_rightmost(); c.right())
    table.append(c.get_below());
  sort(table);
  return table == uniq(table);
}
```

The idea is that the cursor inspects the current gate; if it is not an associative or commutative operator, return *false*. Otherwise, accumulate the fan-in variables in an array, sort the array, and remove any repeated elements. If the resulting array contains the same elements as the sorted array, then there were no repeated elements; otherwise, there were. What is clear in any event is that the ability to program different filters and transformations greatly simplifies the task of adapting Almana to a given application domain.

In the version under development at Bordeaux, this programming interface is rounded out by an on-screen interactive interface. During rewriting, for example, subformulae recognized by a filter are identified one-by-one on the screen, giving the user the opportunity to apply different transformations or not.

The reader is referred to [NRS98] for further information and examples.

6 Conclusion

Minimizing the size of the BDD associated with a boolean formula is a hard problem. Existing techniques based on heuristic analysis and formula rewriting can give good results in specific cases, but do not represent a general solution. Given the theoretical problems with BDD minimization and the fact that most formulae do not admit polynomial BDDs, it seems unlikely that a general-purpose solution can be developed. Of course this does not reduce the need for real tools to handle real problems.

We have seen that an integrated tool that permits one to freely combine heuristic and rewriting techniques can be very useful. Such a tool lets us develop solutions that take advantage of domain knowledge and application-specific formula structure. Indeed we saw experimental evidence that combining these techniques permits significant improvement in BDD size that was impossible when the techniques were applied separately.

The tool we have developed, Almana, has also proved useful for exploration of different optimization strategies and rewriting techniques. Almana proposes a collection of freely-combinable tools for applying known and newly-developed techniques.

Currently, Almana is used by computer scientists working in concert with reliability engineers. An important goal of the Almana project is to develop an interactive interface that reduces the notational burden for specialists and clearly presents the combination of tools used for a particular application domain. Such an interface is currently under development.

Many application domains produce similar formulae and have similar needs for BDD minimization. It would be interesting to see whether this domain knowledge can be codified in a way that stresses their similarities and differences. A taxonomy of application domains producing boolean formulae would be an important step towards understanding the meaning of variable orders and the precise relations between boolean formulae and the corresponding BDDs. The search for a *semantics of variable order* is a theoretical question at the heart of the BDD minimization problem.

References

[BBR97]   M. Bouissou, F. Bruyère, and A. Rauzy. BDD based fault-tree processing: A comparison of variable ordering heuristics. In *Proceedings of European Safety and Reliability Association Conference, ESREL'97*, 1997.

[Ber89]   C.L. Berman. Ordered Binary Decision Diagrams and Circuit Structure. In *Proceedings of the IEEE International Conference on Computer Aided Design, IC-CAD'89*, September 1989. Cambridge MA, USA.

[Bou96]   M. Bouissou. An Ordering Heuristics for Building Binary Decision Diagrams from Fault-Trees. In *Proceedings of the Annual Reliability and Maintenability Symposium, ARMS'96*, 1996.

[BRKM91]   K.M. Butler, D.E. Ross, R. Kapur, and M.R. Mercer. Heuristics to Compute Variable Orderings for Efficient Manipulation of Ordered BDDs. In *Proceedings of the 28th Design Automation Conference, DAC'91*, June 1991. San Francisco, California.

[Bry86]   R. Bryant. Graph Based Algorithms for Boolean Fonction Manipulation. *IEEE Transactions on Computers*, 35(8):677–691, August 1986.

[Bry92]   R. Bryant. Symbolic Boolean Manipulation with Ordered Binary Decision Diagrams. *ACM Computing Surveys*, 24:293–318, September 1992.

[DR96]   Y. Dutuit and A. Rauzy. A Linear-Time Algorithm to Find Modules of Fault Trees. *IEEE Transactions on Reliability*, 45:422–425, 1996.

[FFK88]   M. Fujita, H. Fujisawa, and N. Kawato. Evaluation and Improvements of Boolean Comparison Method Based on Binary Decision Diagrams. In *Proceedings of IEEE International Conference on Computer Aided Design, ICCAD'88*, pages 2–5, 1988.

[FFM93]   M. Fujita, H. Fujisawa, and Y. Matsunaga. Variable Ordering Algorithm for Ordered Binary Decision Diagrams and Their Evalutation. *IEEE Transactions on Computer-Aided Design of Integrated Circuits and Systems*, 12(1):6–12, January 1993.

[FMK91]   M. Fujita, Y. Matsunaga, and N. Kakuda. On the Variable Ordering of Binary Decision Diagrams for the Application of Multilevel Logic Synthesis. In *Proceedings of European Conference on Design Automation, EDAC'91*, pages 50–54, 1991.

[FOH93]   H. Fujii, G. Ootomo, and C. Hori. Interleaved Based Variables Ordering Methods for Ordered Binary Decision Diagrams. In *Proceedings of the IEEE International Conference on Computer Aided Design*, pages 38–41, 1993.

[FS90]   S.J. Friedman and K.J. Supowit. Finding the Optimal Variable Ordering for Binary Decision Diagrams. *IEEE Transactions on Computers*, 39(5):710–713, May 1990.

[FYBSV93] E. Felt, G. York, R. Brayton, and A. Sangiovanni-Vincentelli. Dynamic Variable Reordering for BDD Minimization. In *Proceedings of the European Design Automation Conference EURO-DAC'93/EURO-VHDL'93*, pages 130–135. IEEE Computer Society Press, 1993.

[IH93]   K. Iwama and K. Hino. Random Generation of Test Instances for Logic Optimizers. In *Proceedings* 31*th* *ACM/IEEE Design Automation Conference, DAC'93*, pages 430–434, San Diego, 1993.

[ISY91]   N. Ishuira, H. Sawada, and S. Yajima. Minimization of Binary Decision Diagrams Based on Exchanges of Variables. In *Proceedings of the IEEE International Conference on Computer Aided Design, ICCAD'91*, pages 472–475, November 1991. Santa Clara CA, USA.

[MIY90]   S. Minato, N. Ishiura, and S. Yajima. Shared Binary Decision Diagrams with Attributed Edges for Efficient Boolean Function Manipulation. In L.J.M Claesen, editor, *Proceedings of the 27th ACM/IEEE Design Automation Conference, DAC'90*, pages 52–57, June 1990.

[MWBSV88]   S. Malik, A.R. Wang, R.K Brayton, and A. Sangiovanni-Vincentelli. Logic Verification using Binary Decision Diagrams in Logic Synthesis Environment. In *Proceedings of the IEEE International Conference on Computer Aided Design, ICCAD'88*, pages 6–9, November 1988. Santa Clara CA, USA.

[Nie94]   I. Niemel. On simplification of large fault trees. In Elsevier Science Limited, editor, *Reliability Engeneering and System Safety*, pages 135–138, 1994.

[NR98]   M. Nikolskaia and A. Rauzy. Heuristics for BDD handling of sum-of-products formulae. In Balkema, editor, *Proceedings of the European Safety and Reliability Association Conference, ESREL'98*, June 1998.

[NRS98]   M. Nikolskaïa, A. Rauzy, and D. J. Sherman. Editable DAG (eDAG) specification. Technical Report 1198-98, LaBRI, Université Bordeaux-1, April 1998.

[PS95]   S. Panda and F. Somenzi. Who Are the Variables in Your Neighborhood. In *Proceedings of IEEE International Conference on Computer Aided Design, ICCAD'95*, pages 74–77, 1995.

[Rau93]   A. Rauzy. New Algorithms for Fault Trees Analysis. *Reliability Engineering & System Safety*, 05(59):203–211, 1993.

[Rau95a]   A. Rauzy. Aralia Version 1.0 : Developer's Guide. Technical report, LaBRI – URA CNRS 1304 – Université Bordeaux I, 1995.

[Rau95b]   A. Rauzy. Aralia version 1.0 : the Toolbox Manual. Technical report 1093-95, LaBRI – URA CNRS 1304 – Université Bordeaux-I, 1995.

[Rud93]   R. Rudell. Dynamic Variable Ordering for Ordered Binary Decision Diagrams. In *Proceedings of IEEE International Conference on Computer Aided Design, ICCAD'93*, pages 42–47, November 1993.

[THY93]   S. Tani, K. Hamaguchi, and S. Yajima. The complexity of the optimal variable ordering problems of shared binary decision diagrams. In *Proceedings of the 4th International Symposium on Algorithms and Computations, ISAAC'93*, volume 762 of *LNCS*, pages 389–398. Springer Verlag, 1993.

[TI94]   S. Tani and H. Imai. A reordering operation for an ordered binary decision diagram and an extended framework for combinatorics of graphs. In *Proceedings of the 5th International Symposium on Algorithms and Computations, ISAAC'94*, volume 834 of *LNCS*, pages 575–583. Springer Verlag, 1994.

[Weg88]   I. Wegener. On the Complexity of Branching Programs and Decision Trees for Clique Functions. *J. ACM*, 35(2):461–471, 1988.

Bisimulation Minimization in an Automata-Theoretic Verification Framework

Kathi Fisler and Moshe Y. Vardi*

Department of Computer Science
Rice University
6100 S. Main, MS 132
Houston, TX 77005-1892
{kfisler, vardi}@cs.rice.edu
http://www.cs.rice.edu/{~kfisler, ~vardi}

Abstract. Bisimulation is a seemingly attractive state-space minimization technique because it can be computed automatically and yields the smallest model preserving all μ-calculus formulas. It is considered impractical for symbolic model checking, however, because the required BDDs are prohibitively large for most designs. We revisit bisimulation minimization, this time in an automata-theoretic framework. Bisimulation has potential in this framework because after intersecting the design with the negation of the property, minimization can ignore most of the atomic propositions. We compute bisimulation using an algorithm due to Lee and Yannakakis that represents bisimulation relations by their equivalence classes and only explores reachable classes. This greatly improves on the time and memory usage of naïve algorithms. We demonstrate that bisimulation is practical for many designs within the automata-theoretic framework. In most cases, however, the cost of performing this reduction still outweighs that of conventional model checking.

1 Introduction

Symbolic model checking has had a dramatic impact on formal verification [4, 11]. Binary decision diagrams (BDDs) provide an efficient and canonical symbolic representation for digital systems, which enables a complete analysis of systems with exceedingly large state-spaces. BDD-based model checkers have enjoyed substantial and growing industrial use over the last few years, discovering very subtle flaws in designs of considerable complexity. Nevertheless, the growing size of current and future semiconductor designs seriously challenges current model checking technology. Despite technological advances, the unremitting increase in design complexity, known as the *state-explosion* problem, remains a serious obstacle to industrial-scale verification.

Research into combating state explosion takes many forms, including alternate representations of transition systems [3, 32], compositional reasoning [20,

---
* Supported in part by NSF grants CDA-9625898, CCR-9628400 and CCR-9700061, and by a grant from the Intel Corporation.

27], and various state-space reduction techniques [10, 14, 25]. Most of the latter techniques are a form of *abstraction*. An abstraction suppresses information from a concrete state-space by mapping it into a smaller, abstract state-space. Each set of concrete states mapping to a single abstract state forms an *equivalence class* under the abstraction. Hopefully, the abstract model is more amenable to model checking. The abstraction must, however, be *safe*: verification in the abstract model must "carry over" to the concrete model. More formally, the abstraction must preserve the truth or falsehood of the properties of interest. Abstractions preserving both truth and falsehood provide *strong preservation*, while those preserving only truth provide *weak preservation*.

Ideally, we want abstractions to meet two conditions. First, they should yield the smallest safe abstract models. Second, constructing them should require minimal manual involvement. The second is crucial for abstraction to work well in an industrial environment. Current techniques generally fail to satisfy this criterion, which significantly limits their applicability. For the first, we need a means to compare the size of abstract models. We say that abstraction A_1 is *coarser* than abstraction A_2 if A_1 collapses some subsets of A_2's equivalence classes together. The coarser the abstraction, the smaller the abstract state space. The properties to be preserved generally dictate the maximal coarseness of safe abstractions.

Bisimulation [31] is a general abstraction satisfying both these conditions. Bisimulation collapses states that have both the same local properties and bisimilar successor states. It is the coarsest abstraction that strongly preserves all properties expressible in the propositional μ-calculus, which expressively subsumes CTL* (an extension of both the linear temporal logic LTL and the branching temporal logic CTL) [23]. Furthermore, it can be computed in a completely automated fashion, requiring no understanding of the design under consideration. It is straightforward to compute symbolically, and therefore suggests a good approach to state-space reduction in the context of symbolic model checking. Unfortunately, as reported by Clarke *et al.* [9], the BDDs used to compute the bisimulation relation get overly large in practice; accordingly, they concluded that computing bisimulations is not a feasible approach to algorithmic state-space reduction.

This paper revisits bisimulation reductions, this time in an automata-theoretic framework [28, 33]. The change of framework is significant. In an automata-based framework, model checking involves intersecting the design automaton with an automaton representing the negation of the property, then checking whether a fair path exists. Restricted to this single (and simple) property, the bisimulation abstraction is even coarser than bisimulation on the design alone because the atomic propositions of the design need not be considered.

Performed naïvely, this approach still suffers from the problem of "exploding" BDDs because the bisimulation relation is a binary relation on the state space (thus, if the states are described in terms of n Boolean variables, the bisimulation relation is over $2n$ Boolean variables, resulting in very large BDDs). We overcome this problem by using an algorithm due to Lee and Yannakakis [29]. This algorithm has two key ideas: first, it computes bisimulations only on the

reachable state space; second, it computes the equivalence classes of the bisimulation relation rather than the bisimulation relation itself. By combining the
automata-theoretic approach with the Lee-Yannakakis algorithm, we are able to
abstract much larger designs using bisimulation than was previously possible.
Unfortunately, while computing our abstraction can result in significant reduction in the size of the state space, the gain obtained by the reduction is often
offset by the cost of computing it. Thus, state-space reduction via bisimulation
abstractions is a useful technique, but its applicability is not universal.

The outline of the paper is as follows. Section 2 discusses our experimental
framework. Section 3 reviews bisimulation and the algorithms used for computing
it. Section 4 describes our experiences using the Lee-Yannakakis algorithm in
the automata-theoretic framework. We discuss a variant of the algorithm in
Section 5 and comparative results with traditional model checking in Section 6.
Section 7 presents related work. Concluding remarks and future directions appear
in Section 8.

2 Experimental Framework

We conducted our experiments in a framework built upon the VIS model checker
[19]. We chose VIS for several reasons: first, it supports automata-theoretic model
checking; second, as a tool, it is engineered to make extensions relatively easy;
third, it uses Verilog as an input language, for which examples are easier to find
than for other model checker input languages.

Our experiment suite contains nine designs, most taken from the VIS distribution. They include a 3-bit counter, a traffic-light controller, an arbiter, two
mutual exclusion protocols, a cache coherence protocol, a railway controller, an
elevator, and a tree-structured arbiter. The tested properties are taken from
the properties distributed with the examples. Whenever possible, we test both
a safety property and a liveness property of each design. Our experiments also
include properties that do not hold of their respective designs. We manually
translate the negated properties into Verilog state machines. For each experiment described, we use VIS as a front-end to get a symbolic representation
of the corresponding transition system. Our algorithms work on these symbolic
representations, using operations provided by VIS for reachability computations.
In an attempt to control the intermediate BDD sizes, we use VIS's partitioned
transition relation representation for the original transition system. The minimized transition system has a monolithic transition relation. All experiments
were run on an UltraSparc 140 with 128 megabytes of memory running Solaris.

One important note regarding memory statistics is in order. Within VIS,
we used Somenzi's CUDD package [13]. Depending upon the setting of the
LooseUpTo parameter, CUDD may use substantially more memory than is actually needed for a computation (as much as a factor of two in our experience)
in an attempt to improve time efficiency. We have set this parameter to a low
value (100) for our runs to make our memory statistics as tight as possible. As
these figures may still be imprecise, we also report the maximum number of live

BDD nodes during each computation. CUDD does not provide this figure, but we have approximated it manually by inserting checks throughout our code.

3 Computing Bisimulation

Bisimulation is an equivalence relation on the states of a transition system [31]. It distinguishes states based on two requirements: whether they agree on the values of a given subset of atomic propositions, and whether transitions from each state lead to bisimilar states. The subset of atomic propositions on which states must agree depends on the application. A general bisimulation-based minimization would consider all variables (both state variables and atomic propositions) that could be referenced in properties. When verifying a particular property, however, only those state variables and atomic propositions affecting the property need to be considered [31]. Our experiments heed this observation.

Formally, we define transitions systems and bisimulation as follows. A *transition system* is a tuple $\langle S, R, AP, L, F \rangle$, where S is the set of states, $R \subseteq S \times S$ is the transition relation, AP is the set of atomic propositions, $L \subseteq S \times AP$ indicates which atomic propositions are true in each state, and $F \subseteq 2^{AP}$ captures the fairness condition. Each element of F indicates a set of states in which a single fairness constraint holds. We use Büchi fairness constraints, so fair paths pass through a state of each element of F infinitely often. The sets of states comprising F may be specified at the user level using formulas. For purposes of discussion, however, we prefer the state view. We assume that there is a unique atomic proposition ap_{fi} corresponding to each set F_i in F such that for all states s in S, (s, ap_{fi}) is in L iff s is contained in F_i. These atomic propositions are used to ensure that the fairness condition is preserved under bisimulation.

Given a transition relation $\langle S, R, AP, L, F \rangle$ and a subset AP' of AP, a binary relation $B \subseteq S \times S$ is a *bisimulation relation relative to* AP' iff the following conditions hold for every pair of states (s_1, s_2) in B:

1. For every atomic proposition ap in AP', (s_1, ap) is in L iff (s_2, ap) is in L.
2. For all s_1' such that (s_1, s_1') is in R there exists s_2' such that (s_2, s_2') is in R and (s_1', s_2') is in B.
3. For all s_2' such that (s_2, s_2') is in R there exists s_1' such that (s_1, s_1') is in R and (s_1', s_2') is in B.

We can compute bisimulation symbolically by computing its complement, *i.e.*, those pairs of states that fail to meet one of the above three conditions. We are interested in the maximum bisimulation relation: that which distinguishes as few states as possible. Computing the least fixpoint of the following equation yields the complemented maximum relation:

$$
\begin{aligned}
&\exists\, ap \in AP' \ \neg(L(s_1, ap) \Leftrightarrow L(s_2, ap)) \ \vee \\
&\exists\, s_1' \ R(s_1, s_1') \wedge \neg(\exists s_2'(R(s_2, s_2') \wedge B(s_1', s_2'))) \ \vee \\
&\exists\, s_2' \ R(s_2, s_2') \wedge \neg(\exists s_1'(R(s_1, s_1') \wedge B(s_1', s_2')))
\end{aligned}
$$

Table 1. Computing bisimulation with the symbolic algorithm. The numbers after the design names distinguish the various properties under consideration. The types of each property appear in Table 3. The "Atomic Propositions" column indicates how many variables appear in the respective property.

	State Variables	Atomic Propositions	Equivalence Classes	Max Live Nodes	Memory (MB)
counter	3	1	8	127	2.68
tlc-1	7	2	26	991	2.86
tlc-2	7	3	38	1184	2.80
k-elev	10	3	246	15576	3.4
eisenberg-1	17	2	9804	422642	34
eisenberg-2	17	2	10100	424757	20

In our experiments, this bisimulation computation completed only for the smaller examples, as shown in Table 1.[1] These results strongly indicate that plain symbolic bisimulation computation is an infeasible state-space reduction technique for large designs and support the similar conclusion drawn by Clarke *et al.* [9]. Section 4 shows that bisimulation becomes more feasible for large designs under three conditions: (1) when we represent bisimulation by its equivalence classes rather than the binary relation, (2) when we consider only the reachable equivalence classes while performing bisimulation, and (3) when we minimize after intersecting the automaton for the negation of the desired property.

4 Making Bisimulation More Tractable

Bisimulation as presented in the previous section becomes intractable largely because the computed relation is defined over $2n$ variables, where n is the number of boolean state variables in the original design. This doubling yields extremely large BDDs even for moderate-size designs. In addition, the relation may distinguish large numbers of states by considering many atomic propositions in the original design. Although in general large relations do not necessarily imply large BDDs, our evidence shows that the bisimulation relations are large in practice. Reducing the size of the relation itself is a reasonable approach to try to make the resulting BDDs more tractable.

Our experiments address both problems. For the former, we use an algorithm due to Lee and Yannakakis [29] that represents the bisimulation relation by its set of equivalence classes (each defined over n variables), rather than as a monolithic binary relation (over $2n$ variables). Section 4.1 provides an overview of the algorithm. For the latter, we intersect the negation of the property with the design before minimization, thereby reducing the number of atomic propositions

---

[1] These runs used Cudd's default `LooseUpTo` value of 800000 in order to save computation time. The eisenberg computations each took over a full day to complete.

that need to be considered. The remaining parts of this section show that running the algorithm on the intersected automaton does render bisimulation more tractable as a minimization technique.

4.1 The Lee-Yannakakis Algorithm

The Lee-Yannakakis bisimulation algorithm has two desirable features. First, it represents the relation by its equivalence classes, instead of by its pairs of related states. Second, it only computes bisimulation on the reachable state space. Model checkers have long ignored unreachable states for sake of efficiency. The traditional formalization of bisimulation, however, does not distinguish between equivalence classes that contain reachable states and those that do not. As our experiments show, the number of equivalence classes containing no reachable states can be substantial. This parallels the large number of unreachable states in most systems.

The combination of minimization and reachability offers two further advantages. First, the algorithm can minimize on-line. By augmenting the algorithm with an on-the-fly model checker, it is conceivable to verify designs which are too large for straightforward model checking, or for traditional minimization techniques. Second, the combination allows representatives to be selected such that if there is an edge between two blocks in the minimized system, then there is an edge between their representatives in the original system. This is useful for reporting error traces during model checking. Any path through the minimized system corresponds to a legitimate path through the original design by following the representatives.

The algorithm starts with a partition of the states into equivalence classes based only on the atomic propositions. Each reachable equivalence class also stores a single reachable state acting as the representative of the class. Our implementation represents each of these with a BDD. Two structures, a stack and a queue, store equivalence classes that need further processing. Classes on the stack must be processed to construct their outgoing edges. Classes in the queue must be split into multiple equivalence classes in order to preserve the bisimulation relation. The algorithm terminates when both the stack and the queue are empty. As the classes are represented explicitly (even though their elements are represented symbolically), this algorithm can be viewed as converting a symbolically-represented transition system into an explicitly-represented, minimized transition system. A summary appears in Figure 1.

4.2 Minimized Bisimulation on Designs

Our first experiment used the Lee-Yannakakis algorithm to compute bisimulation on the designs alone (without intersecting the negated property). Each minimization considered only the atomic propositions appearing in the respective property and those for the fairness constraints. As shown in Table 2, representing the equivalence classes and accounting for reachable classes does enable us to compute bisimulation more efficiently and on a larger collection of designs.

```
begin
    stack := {⟨B, i⟩}, where i is the initial state and B is the block containing i;
    queue := ∅ ;
    partition contains one block per combination of atomic propositions ;

search:
    while stack ≠ ∅ do begin
        ⟨B, p⟩ := pop(stack) ;
        D := R(B) ;
        foreach block ⟨C, q⟩ containing a state in R(p) do begin
            if B contains a state with no successor in C then enqueue ⟨B, p⟩ ;
            if q is not defined then select p_c in C ∩ R(p) and set q := p_c ;
            add edge ⟨B, p⟩ → ⟨C, q⟩ ;
            D := D − C ;
        end ;
        if D ≠ ∅ then enqueue ⟨B, p⟩ ;
    end

split:
    while queue ≠ ∅ do begin
        ⟨B, p⟩ := delete (queue) ;
        B' := {q ∈ B : blocks(R(q)) = blocks(R(p))} ;
        B'' := B − B' ;
        B := B' ;
        add a block with stateset B'' to the partition ;
        foreach edge ⟨C, q⟩ → ⟨B, p⟩ in the minimized system do begin
            if some elements of C reach B and some do not then enqueue ⟨C, q⟩ ;
            if some elements of C reach B'' and some do not then enqueue ⟨C, q⟩ ;
            if no successor or q is in B then delete edge ⟨C, q⟩ → ⟨B, p⟩ ;
            if some successor of q is in B'' then
                if the block for B'' is not marked then
                    select p''_b in R(q) ∩ B'', mark, and push block onto stack ;
                add edge ⟨C, q⟩ → ⟨B'', p''_b⟩ ;
        end
        if stack ≠ ∅ then goto search ;
    end
```

Fig. 1. The Lee-Yannakakis combined bisimulation and reachability algorithm [29]. R is the transition relation of the system being minimized. The term "block" refers to an equivalence class and its representative. In this pseudocode, the *enqueue* operation checks whether a block is already in the queue before enqueuing it.

Table 2. Computing bisimulation plus reachability with the Lee-Yannakakis algorithm. The properties for the new designs (bakery, arbiter, and treearbiter) each referred to two atomic propositions. The figures for bakery-1 also hold for bakery-2 as both properties use the same atomic propositions. The "% Reach. Saved" column reflects the ratio of reachable classes to reachable states.

	State Variables	Reachable States	Reachable Classes	% Reach. Saved	Max Live Nodes	Memory (MB)
counter	3	8	8	0%	45	2.7
tlc-1	7	20	13	35%	392	2.87
tlc-2	7	20	20	0%	439	2.87
k-elev	10	262	132	50%	2436	2.99
eisenberg-1	17	1611	1214	25%	28810	4.56
eisenberg-2	17	1611	1076	33%	28859	4.58
bakery-1	20	2886	2546	18%	68543	5.47
arbiter	23	73	64	12%	8477	3.29
treearbiter4-1	23	5568	288	95%	24129	4.7
treearbiter4-2	23	5568	1344	76%	58805	5.82

Comparing the number of reachable classes to the total number of equivalence classes shown in Table 1, we see that the percentage of classes that are reachable can be rather small on larger examples. Based on these figures, representation and reachability appear to be significant factors in bisimulation's tractability.

4.3 Minimizing Relative to Properties

The bisimulation computations in the previous section considered only those atomic propositions appearing in properties or required to support fairness constraints. However, it is possible to further reduce the number of atomic propositions under consideration. As we are working within an automata-theoretic framework, once the negation of the property has been intersected with the original design automaton, the property to be checked reduces to "does there exist a fair path" (EG*true* in CTL). Algorithmically, we customize bisimulation minimization to this property at the level of the atomic propositions. Rather than consider the atomic propositions in the property, we now use only those corresponding to the fairness constraints. Specializing to the property therefore does not require any change to the Lee-Yannakakis algorithm or any additional machinery.

We refer to the intersection of the design and the negated property as the *DesignProp automaton*. Table 3 shows the reductions achieved by computing bisimulation on these automata. As the table shows, we are now able to minimize certain designs (coherence and dcnew) that were not amenable to previous computations. The numbers of reachable states and reachable classes are now relative to the DesignProp automaton, rather than to the original design as in Table 2. The transition system for DesignProp has additional state variables as

Table 3. Computing bisimulation on the DesignProp automaton. Variables p and q in the properties may refer to boolean combinations of atomic propositions; they are independent across properties. Asterisks indicate properties that fail to hold of their respective designs. The newly handled designs, dcnew and coherence, have 24 and 37 state variables (before intersection with their properties), respectively. The coherence property refers to four atomic propositions; each of the dcnew properties refers to two.

	Property	Reach. States	Reach. Classes	% Reach. Saved	Max Live Nodes	Memory (MB)
counter	GFp	23	16	30%	139	2.77
tlc-1	Gp	20	13	35%	472	2.91
tlc-2	$G(p \rightarrow Fq)$	46	30	35%	649	2.95
k-elev	$G(p \rightarrow Fq)$	648	266	59%	5035	3.26
eisenberg-1*	Gp	3198	2074	35%	50447	5.39
eisenberg-2	$G(p \rightarrow Fq)$	4155	2713	35%	53580	5.72
bakery-1*	$G(p \rightarrow Fq)$	8079	7258	10%	118871	8.04
bakery-2	$G(p \rightarrow Fq)$	5822	5302	9%	103102	7.4
arbiter	$G(p \rightarrow Fq)$	185	158	15%	21518	3.86
treearbiter4-1	Gp	5568	1	100%	19938	6.54
treearbiter4-2	$G(p \rightarrow Fq)$	13392	3320	75%	162212	12
dcnew-1	Fp	186876	1	100%	2286	4.03
dcnew-2*	GFp	445404	242	100%	15579	4.5
coherence	Gp	94738	1	100%	18099	7.73

compared to the design. The number of additional variables depends on the type of property involved. A safety property contributes one state variable, while the liveness properties considered here contribute two.

These extra variables account for the additional states and any increased numbers of live nodes needed for the computation. The numbers of live nodes are generally much higher for computing bisimulation on the DesignProp automaton, often roughly 50% higher. For treearbiter4-1, however, the maximum number of live nodes actually decreases. We attribute this to the small number of reachable classes (in this case, only one) in the DesignProp automaton.

Given that we can now handle more designs, minimizing the DesignProp automaton seems promising. From the perspective of the Lee-Yannakakis algorithm, we did nothing more than reduce the number of atomic propositions considered in forming the initial partition. Given the definition of bisimulation presented in Section 3, it is not surprising that reducing the number of atomic propositions used can greatly impact the degree of minimization achieved, because the atomic propositions dictate the initial partitioning into equivalence classes. Accordingly, further reducing the number of required atomic propositions might prove even more promising. The next section discusses an experiment that requires only one atomic proposition per design, regardless of the number of fairness constraints.

4.4 Minimizing with One Atomic Proposition

The experiments reported in the previous section considered one atomic proposition per fairness constraint. Further reducing the number of atomic propositions therefore requires reducing the number of fairness constraints. The fairness constraints indicate sets of states that must be passed through infinitely often. Any method that satisfies the constraints is sufficient, as long as it requires each set to be visited infinitely often. One such method checks the fairness constraints in some linear order [8], as shown in the following diagram.

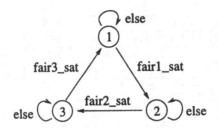

This diagram shows the linearization of three fairness constraints using an automaton. The automaton has one state for each fairness constraint. Each state has two transitions: one going to the next state in the ordering when the associated fairness constraint is satisfied, and one looping back to the same state for when the constraint is not satisfied. If the fairness constraint for this automaton is "state=1 and fair1_sat", then each fairness constraint is satisfied infinitely often along the fair paths. Such a linearized automaton therefore enforces the fairness requirements while requiring only a single fairness constraint.

We used this linearization technique on each design containing fairness constraints. For each such design, we added an automaton of the above form to the original Verilog description of DesignProp. This automaton had one state for each fairness constraint in the design, plus one state for the fairness constraint of the property (each property contributed exactly one fairness constraint). This new automaton was intersected with DesignProp before minimization; we call this the *OneFair automaton*. The minimization considered only the fairness constraint of this new automaton. As a result, each minimization began with an initial partition containing exactly two equivalence classes (states that satisfied the fairness constraint, and those that did not).

Table 4 summarizes the results of this experiment. The order in which we linearized the fairness constraints had a non-trivial impact on the amount of reduction achieved. For the arbiter, which has three fairness constraints, the number of blocks ranged from 95 to 106 (over six permutations). For bakery-2, which has four fairness constraints, the number of blocks ranged from 1331 to 1417 (over six permutations). For eisenberg-1, which has six fairness constraints, the number of blocks ranged from 3590 to 4130 (over six permutations). We leave the question of how to choose a good ordering to future research.

We can now compare the reductions achieved on each of our three automata: the design (Table 2), the DesignProp (Table 3), and the OneFair (Table 4). With

Table 4. Single fairness statistics. The "Num Fair" column indicates the number of constraints of the original design; it does not account for the fairness constraint added by the property automaton. The "% Classes Saved" column reflects the ratio of reachable classes in the OneFair automaton to the reachable classes in the DesignProp automaton (Table 3). Designs not covered in this table had no fairness constraints. We did not produce a OneFair automaton for treearbiter4-1 because its DesignProp automaton already yields only one reachable block.

	Num Fair	Reach. States	Reach. Classes	% Reach. Saved	% Classes Saved	Max Nodes	Memory (MB)
tlc-1	2	28	1	96%	92%	554	2.98
tlc-2	2	56	10	82%	67%	717	3.03
k-elev	4	3664	128	97%	52%	12349	3.96
eisenberg-1*	6	10039	3590	64%	-73%	113354	11
eisenberg-2	6	16228	2096	87%	23%	81224	7.1
bakery-1*	4	25081	5297	79%	27%	146466	9.26
bakery-2	4	21815	1417	94%	73%	78743	6.19
arbiter	3	353	97	73%	39%	55937	5.55
treearbiter4-2	4	26712	2365	91%	29%	461598	38

the exception of the bakery example, which achieves slightly better reductions on the original design than on the DesignProp automaton for either property, each example yields more substantial reductions the fewer atomic propositions are considered during minimization.

Conversely, these tables also show an increase in live nodes and memory to minimize the OneFair as opposed to the DesignProp automaton. As in the comparison between Tables 2 and 3, we attribute the increase in the number of live nodes to the additional state variables required by the new automaton. The increase from Table 3 to Table 4, however, is more substantial than the increase noted between Tables 2 and 3. Here, only one experiment (bakery-2) uses fewer live nodes for minimizing OneFair as opposed to DesignProp. The remaining experiments show an average increase of 114% in the number of live nodes.

Based on these contradictory figures, we are unable to say whether the final reduction to a single fairness constraint is worthwhile. Ideally, we would like to find some examples for which the minimization is only tractable under this reduction. Such examples, however, would likely require large numbers of fairness constraints. Our current examples do not meet this requirement. Until we find such evidence, the utility of this extra minimization is questionable.

5 Tuning the Original Algorithm

Attempts to run the previous experiments on even larger designs failed because the runs ran out of memory before finishing. This section discusses an attempt to reduce the algorithm's memory requirements.

The Lee-Yannakakis algorithm improves on the original symbolic computation described in Section 3 by processing only those equivalence classes containing reachable states. However, the algorithm still maintains the unreachable classes because they may become reachable as the minimized graph is constructed. We compared the number of reachable classes to the number of classes being maintained at the end of each run of the algorithm. On average, only 50% of the maintained classes are actually reachable.

This raises a possible approach to improving memory usage. We represent classes as BDDs. In practice, BDDs representing large sets of states may be much smaller than BDDs representing smaller sets of states. This suggests that we could try maintaining only a single class containing all those states not yet assigned to a reachable class.[2] Such a merger would create a coarser partition than the Lee-Yannakakis algorithm with respect to the unreachable classes. Hopefully, it would also save on the BDD nodes needed to represent those classes.

Unfortunately, this approach is likely to be computationally expensive. Recall that an unreachable class becomes reachable if it is found to contain a reachable state. In the original algorithm, the previously unreachable class is converted to a reachable block and placed on the stack for processing. The algorithm requires all states in this newly reachable class to agree on the values of the atomic propositions of interest. Maintaining only a single unreachable class violates this requirement. A more complicated procedure would therefore be required to extract a new reachable class that satisfied at least this requirement.

Fortunately, this requirement is sufficient as well as necessary. We could therefore maintain one unreachable class for each combination of atomic propositions. If the number of combinations is smaller than the number of unreachable classes, we might still save on BDD nodes without incurring a high computational overhead. Our experiments indicate that the number of combinations is indeed substantially smaller than the number of unreachable classes for almost all runs.

Accordingly, we created a modified version of the Lee-Yannakakis algorithm. For each subset of atomic propositions satisfied by some state, the new version maintains a *holding set* of states not currently assigned to any reachable equivalence class. The algorithm views the holding set as a single (large) equivalence class. When a state in a holding set is found to be reachable, the holding set is converted into a reachable block as in the original algorithm. The main change to the pseudocode lies in the "split" section (Figure 1). If the new equivalence class B'' is unreachable at the end of the **foreach** loop, the newly created block is destroyed and the elements of B'' are returned to the holding set.

Comparing the live node and memory statistics obtained from the original algorithm and our new version yields some interesting results. For the smaller examples, the two algorithms have similar live node and memory figures. For the eisenberg and larger examples, the new algorithm performs substantially worse than the original on the OneFair automata in terms of memory usage (44% worse on average). In contrast, the new algorithm performs substantially

---

[2] Note that this single class does not necessarily contain all of the unreachable states. Reachable classes often contain unreachable states.

better on the DesignProp automata in terms of live nodes (12% better on average, as high as 25% on several examples). The results are mixed for the design minimizations. The actual figures and other details on the new algorithm are available in a technical report version of this paper [17] and on our web site (http://www.cs.rice.edu/CS/Verification/).

6   A Comparison to Model Checking

We evaluate the utility of our minimizations by comparing model checking on minimized systems to model checking on the original designs. We used the VIS CTL model checker to test properties on the original designs (minimized and unminimized) and the VIS language emptiness checker to test for fair paths on the minimized DesignProp and OneFair systems. Our implementation of the Lee-Yannakakis algorithm produces monolithic (rather than partitioned) transition relations. Generating partitioned minimized transition relations is an interesting research challenge. In order to make the comparisons meaningful, we also performed the VIS CTL model checking runs using monolithic transition relations. We were able to run VIS on our minimized designs by replacing the transition relation from the original design with the minimized transition relation in VIS's internal data structures. This works because our minimization does not change how states are represented. The minimized transition relation is merely restricted to relate only the representatives of the reachable equivalence classes.

Table 5 provides four sets of figures. The first two columns give the memory and time needed to model check the CTL properties against the original designs (Table 3 indicates the type of each property). The third column shows the time required for model checking after the design has been minimized with respect to the atomic propositions in the property and in the fairness constraints. The next two columns give the minimization time (under the original Lee-Yannakakis algorithm) and language emptiness test time for the corresponding DesignProp automaton. The last two columns give similar figures for the OneFair automaton. These last two columns are empty for any design that did not have fairness constraints (and thus had no OneFair automaton).

Looking only at the model checking and language emptiness times, the results are mixed. For eisenberg-2, model checking in the original design is actually faster than checking language emptiness in the minimized DesignProp automaton. The eisenberg design has 1611 reachable states (Table 2), while the corresponding DesignProp automaton has 2713 reachable classes (Table 3). This increase in reachable states explains the increased model checking time. The savings in the remaining cases vary from little to substantial, with the best savings on the coherence example. The original coherence design has 94738 reachable states; after minimization, it has a single reachable block (Table 3). The savings in this case are therefore not surprising.

Performing the language emptiness test on the OneFair automata as compared to the DesignProp automata offers no benefits. In several cases, such as both eisenberg tests and bakery-1, the results are decidedly worse for the OneFair

Table 5. Model checking statistics from VIS. "Min Design MC Time" shows the time taken by model checking after minimizing the design with respect to the atomic propositions in the respective property. DesProp abbreviates DesignProp. Memory figures for performing the minimizations appear in Tables 2 through 4. The language emptiness checks did not significantly increase memory usage over that required for minimization. Maximum live node statistics were unavailable for the model checking runs.

	Design MC Mem	Design MC Time	Min Design MC Time	DesProp LY Time	DesProp LE Time	OneFair LY Time	OneFair LE Time
counter	2.6	0.1	0	0.3	0		
tlc-1	2.76	0.2	0	0.4	0	0.3	0
tlc-2	2.76	0.2	0	0.7	0.1	0.6	0
k-elev	2.92	0.5	0.2	14.9	0.2	39.6	0.2
eisenberg-1*	3.87	3.1	1.4	287	2.3	1508.9	13.1
eisenberg-2	3.85	3.1	5.9	438.4	14.7	667.9	36.6
bakery-1*	3.64	8.4	7.8	1729.3	15.6	1462.8	55.9
bakery-2	3.62	8.4	7.0	971.5	3.7	455.4	3.6
arbiter	2.98	0.6	0.2	56	0.5	256.6	0.4
treearbiter4-1	3.7	2.1	0.2	67.9	0		
treearbiter4-2	4.03	2.1	1.5	2578.2	7.2	31084.2	14.3
dcnew-1	4.12	1.8		1.4	0		
dcnew-2*	4.12	1.8		91.7	2.2		
coherence	6.24	45.5		28.2	5.5		

automaton. In all three cases, the number of reachable states in the original design is smaller than the number of reachable classes in the corresponding OneFair automaton (Tables 2 and 4).

Accounting for minimization time, performing model checking on the original design is faster than the combination of minimization and language emptiness checking on all but the dcnew1 and coherence examples. Both of these examples have large numbers of reachable states (186876 and 94738, respectively) and reduce to a single reachable block. The total cost of using the OneFair automaton is significantly worse than that for the DesignProp automaton for all but the bakery-2 example.

7 Related Work

Several researchers have explored symbolic minimization techniques. None of these efforts, however, computed reachability simultaneously; they are therefore difficult to use for on-line minimization. Bouali and de Simone provided a symbolic bisimulation algorithm over partitioned transition relations [7]. A variant of this algorithm is implemented in the XEVE verification environment, which also supports VIS's input format [6]. Aziz *et al.* take a compositional approach to bisimulation minimization in a symbolic setting [2]. This approach also performs minimization relative to a given property. However, it represents equivalence relations (rather than classes). Kick uses a symbolic technique to reduce models relative to properties [26]. This technique is not guaranteed to produce transition systems as coarse as those obtained through bisimulation; furthermore, it has not been tested on realistic examples.

Bouajjani *et al.* minimize state graphs from programs using a combination of bisimulation and reachability [5]. Lee and Yannakakis show that the Bouajjani *et al.* algorithm is less efficient than theirs in theory due to the order in which they split classes [29]. In practice, Alur *et al.* found the Bouajjani *et al.* algorithm faster for checking language emptiness of timed automata [1]. Dams proposes a minimization technique for strong preservation of CTL* properties [14]. His technique resembles bisimulation, but produces a potentially coarser relation that preserves only a given set of properties. Rather than split classes based on transitions, he splits a class when its members disagree on subproperties relevant to verifying the given properties. While his approach is fully automatic, it requires many intermediate model-checking analyses, and thus appears computationally prohibitive. Dams, Gerth, and Grumberg present an earlier version of this work for universal CTL formulas [15]. Neither version has been implemented.

Dams, Gerth and Grumberg suggest using abstract interpretations to automatically generate abstractions from program-level descriptions of transition systems [16]. While they aim for strong preservation of formulas, they do not consider minimization relative to particular properties. Furthermore, their generation technique does not necessarily produce the coarsest possible abstraction. This work extends earlier work by Clarke, Grumberg, and Long that considers only weak preservation of universal CTL (∀CTL) properties [10]. Other abstraction frameworks, such as that supported in COSPAN, require the user to manually construct the abstract models [28]. Proposed approaches to symmetry reductions also require various degrees of manual intervention [9, 21, 25]. Furthermore, only Gyuris and Sistla's framework supports fairness constraints.

8 Conclusion and Future Work

We have explored bisimulation minimization as a state-space reduction technique in an automata-theoretic verification framework. Bisimulation is theoretically attractive because it requires no user-intervention and produces the coarsest abstract model that strongly preserves all μ-calculus properties. Practically, however, it has been considered unsuitable for use in symbolic model checking because the BDDs needed to compute the bisimulation relation grow too large.

This paper shows that bisimulation can be viable in a symbolic framework. Using an algorithm due to Lee and Yannakakis, which combines reachability analysis with bisimulation on symbolically-represented transition systems, we have minimized designs far larger than those amenable to conventional bisimulation computations. Much of our success results from being in the automata-theoretic framework. By intersecting the negated property with the design before minimization, we only need to consider the atomic propositions corresponding to fairness constraints during minimization. Reducing the number of atomic propositions used in minimization increases the amount of reduction achieved and the size of design amenable to algorithmic minimization.

As with many proposed techniques to advance the state of model checking [22], our approach works well on some designs and not as well on others. In particular, it appears useful for verifying safety properties of designs with no fairness constraints. Such designs should yield only one reachable equivalence class if the safety property holds. This small number of reachable classes keeps the memory requirements down and reduces the time of convergence for the algorithm. Our approach fits in with a general strategy of providing several model-checking algorithms within a single framework, allowing a user to choose algorithms with a good chance of succeeding based on the nature of the design.

The time and memory requirements of our current approach motivate much of our future work. First, as Table 5 shows, minimization time can be substantial. Alur *et al.* experienced similar time requirements for the Lee and Yannakakis algorithm [1]. Minimizing relative to the DesignProp automata prohibits reuse of results because each property yields a separate machine to be minimized. We therefore plan to experiment with more reusable reductions. For example, if a design has several properties relying on the same few atomic propositions, the design minimization could be performed relative to only those propositions. This reduces the number of atomic propositions used in the minimization as we do here, but yields a more reusable minimized transition system.

Second, the memory requirements for performing our reductions are still significant for larger designs. In particular, the costs of minimization usually outweigh those for performing model checking on the original design. BDD size seems to be the bottleneck to our handling larger examples. Based on analyses not discussed here, the problem appears to lie more in intermediate explosion during reachability computations than in the sizes of the BDDs needed to represent the equivalence classes themselves. We therefore intend to explore alternate symbolic representations in conjunction with the Lee-Yannakakis algorithm.

We have done some preliminary experiments with other forms of decision diagrams, such as multi-terminal BDDs (MTBDDs, ADDs) [3] and zero-suppressed BDDs (ZDDs) [32]. ZDDs offer little to no advantage for this application, but ADDs appear promising. ADDs can represent the equivalence relation as a single structure by relating each state to a constant identifying its equivalence class. While this structure uses more than the number n of state variables, in practice it will not use the $2n$ variables needed to represent the bisimulation relation as a single BDD. Preliminary experiments indicate that the number of ADD nodes in the single structure is often substantially smaller than the sum total of the nodes used to represent the classes separately. Furthermore, a monolithic representation allows us to explore purely symbolic versions of the Lee-Yannakakis algorithm. We have already begun investigations in this area.

Another obvious extension to this work involves integrating the minimization with on-the-fly model checking, as mentioned earlier. On-the-fly model checking is often employed in an automata-theoretic framework [12,30]. Such integration might improve the utility of this minimization technique with respect to failed properties. We also plan to explore coarser abstractions for the automata-theoretic framework. Bisimulation preserves details about the lengths of paths that are irrelevant for checking whether a fair path exists. Coarser abstractions, such as those based on two-sided simulation [15, 24] or stuttering equivalence [18], may be even better suited to algorithmic state-space reduction in the automata-theoretic framework. Finally, we plan to explore applying our method to abstract part of the state space. For example, it could be used for data path abstraction by collapsing the space of data values to a small set of relevant values [34].

References

1. R. Alur, C. Courcoubetis, D. Dill, N. Halbwachs, and H. Wong-Toi. An implementation of three algorithms for timing verification based on automata emptiness. In *Proc. of the IEEE Real-Time Systems Symposium*, pages 157–166, 1992.
2. A. Aziz, V. Singhal, G.M. Swamy, and R.K. Brayton. Minimizing interacting finite state machines: A compositional approach to language containment. In *Proc. of the International Conference on Computer Design (ICCD)*, 1994.
3. R. I. Bahar et al. Algebraic decision diagrams and their applications. In *Proc. of the International Conference on Computer-Aided Design*, pages 188–191, 1993.
4. I. Beer, S. Ben-David, D. Geist, R. Gewirtzman, and M. Yoeli. Methodology and system for practical formal verification of reactive hardware. In *Proc. 6th Conference on Computer Aided Verification*, volume 818 of *Lecture Notes in Computer Science*, pages 182–193, Stanford, June 1994.
5. A. Bouajjani, J.-C. Fernandez, N. Halbwachs, P. Raymond, and C. Ratel. Minimal state graph generation. *Science of Computer Programming*, 18:247–269, 1992.
6. A. Bouali. XEVE, an ESTEREL verification environment. In *Proc. of the International Conference on Computer-Aided Verification (CAV)*, pages 500–504, 1998.
7. A. Bouali and R. de Simone. Symbolic bisimulation minimization. In *Proc. of the International Conference on Computer-Aided Verification (CAV)*, pages 96–108, 1992.
8. Y. Choueka. Theories of automata on ω-tapes: A simplified approach. *Journal of Computer and System Sciences*, 8:117–141, 1974.
9. E.M. Clarke, R. Enders, T. Filkorn, and S. Jha. Exploiting symmetry in temporal logic model checking. *Formal Methods in System Design*, 9(1/2):77–104, August 1996.
10. E.M. Clarke, O. Grumberg, and D.E. Long. Model-checking and abstraction. In *Proc. of the 19th ACM Symposium on Principles of Programming Languages*, pages 343–354, 1992.
11. E.M. Clarke and R.P. Kurshan. Computer aided verification. *IEEE Spectrum*, 33:61–67, 1986.
12. C. Courcoubetis, M.Y. Vardi, P. Wolper, and M. Yannakakis. Memory efficient algorithms for the verification of temporal properties. *Formal Methods in System Design*, 1:275–288, 1992.
13. The CUDD package. Available from http://vlsi.colorado.edu/~fabio/.
14. D. Dams. *Abstract Interpretation and Partition Refinement for Model Checking*. PhD thesis, Technische Universiteit Eindhoven, 1996.
15. D. Dams, O. Grumberg, and R. Gerth. Generation of reduced models for checking fragments of CTL. In *Proc. 5th Int.l Conference on Computer-Aided Verification*, pages 479–490, 1993.

16. D. Dams, O. Grumberg, and R. Gerth. Abstract interpretation of reactive systems. *ACM Transactions on Programming Languages and Systems (TOPLAS)*, 19(2), March 1997.

17. K. Fisler and M.Y. Vardi. Bisimulation minimization in an automata-theoretic verification framework (extended version). Technical report, Rice University, Department of Computer Science, 1998.

18. J.F. Groote and F. Vaandrager. An efficient algorithm for branching bisimulation and stuttering equivalence. In *Proc. of the International Conference on Automata, Languages, and Programming*, pages 626–638, 1990.

19. The VIS Group. VIS: A system for verification and synthesis. In R. Alur and T. Henzinger, editors, *In the Proc. of the 8th International Conference on Computer Aided Verification*, pages 428–432. Springer Verlag, July 1996.

20. O. Grumberg and D.E. Long. Model checking and modular verification. *ACM Trans. on Programming Languages and Systems*, 16(3):843–871, 1994.

21. V. Gyuris and A. P. Sistla. On-the-fly model checking under fairness that exploits symmetry. In *Proc. of International Conference on Computer-Aided Verification (CAV)*, pages 232–243, 1997.

22. R. Hardin, R.P. Kurshan, S. Shukla, and M.Y. Vardi. A new heuristic for bad cycle detection using BDDs. In *Proc. of the International Conference on Computer-Aided Verification*, 1997.

23. M. Hennessy and R. Milner. Algebraic laws for nondeterminism and concurrency. *Journal of ACM*, 32:137–161, 1985.

24. T.A. Henzinger, O. Kupferman, and S. Rajamani. Fair simulation. In *Proc. 8th Conference on Concurrency Theory*, volume 1243 of *Lecture Notes in Computer Science*, pages 273–287, Warsaw, July 1997. Springer-Verlag.

25. C. N. Ip and D. L. Dill. Better verification through symmetry. *Formal Methods in System Design*, 9(1/2):41–76, August 1996.

26. A. Kick. Formula dependent model reduction through elimination of invisible transitions for checking fragments of CTL. Technical Report 1995-27, Universität Karlsruhe, 1995.

27. O. Kupferman and M.Y. Vardi. On the complexity of branching modular model checking. In *Proc. 6th Conference on Concurrency Theory*, volume 962 of *Lecture Notes in Computer Science*, pages 408–422, August 1995. Springer-Verlag.

28. R.P. Kurshan. *Computer Aided Verification of Coordinating Processes*. Princeton Univ. Press, 1994.

29. D. Lee and M. Yannakakis. Online minimization of transition systems. In *Proc. 24th ACM Symposium on Theory of Computing*, pages 264–274, May 1992.

30. David E. Long. *Model-checking, Abstraction, and Compositional Verification*. PhD thesis, Carnegie-Mellon University, 1993.

31. R. Milner. *A Calculus of Communicating Systems*, volume 92 of *Lecture Notes in Computer Science*. Springer Verlag, Berlin, 1980.

32. S. Minato. *Binary decision diagrams and applications for VLSI CAD*. Kluwer Academic Publishers, 1996.

33. M.Y. Vardi and P. Wolper. An automata-theoretic approach to automatic program verification. In *Proc. First Symposium on Logic in Computer Science*, pages 322–331, Cambridge, June 1986.

34. P. Wolper. Expressing interesting properties of programs in propositional temporal logic. In *Proc. 13th ACM Symp. on Principles of Programming*, pages 184–192, St. Petersburgh, January 1986.

Automatic Verification of Mixed-Level Logic Circuits

Keith Hanna

Computing Laboratory
University of Kent
Canterbury, Kent, UK
f.k.hanna@ukc.ac.uk
http://www.cs.ukc.ac.uk/people/staff/fkh

Abstract. An approach is described to the specification and verification of digital systems implemented wholly or partly at the analog level of abstraction. The approach relies upon specifying the behaviours of analog components (such as transistors) by piecewise-linear predicates on voltages and currents. A decision procedure is described that can, for a wide class of specifications, automatically establish the correctness of an implementation.

1 Introduction

The specification and formal verification of digital systems described entirely at the *digital* level of abstraction (that is, described in terms of ideal digital devices) is a well-understood art. Assuming that behavioural specifications take the form of predicates over tuples of digital signals and assuming that a behavioural specification is available for each component in an implementation, then it is a trivial matter to write down a predicate (the *derived specification*) that specifies the behaviour of the implementation. The implementation is *correct* with respect to a given *requirements specification* if every tuple of digital waveforms that satisfies the derived specification also satisfies the requirements specification.

In practice, however, many digital systems are not implemented entirely at the digital level of abstraction. In order to optimise performance, power consumption or chip area, parts of the system are often implemented at the *analog* level. For instance, it is relatively common to find carry propagation chains (in carry lookahead adders or priority networks) implemented in terms of single transistors. By eliminating the logically-redundant buffering stages that would be present were ideal gates used instead, the propagation delay along the critical path may be significantly reduced. In applications where power consumption must be minimised, the use of *pass-transistor logic* (in which input signals are routed to their destinations along paths formed by pass transistors) for realising combinational logic (as an alternative to networks of conventional gates) is commonplace. Further, in distributed or modularised systems, the use of *wired-or* configurations or *tristate* drivers in bus circuits is almost universal.

The use of such *mixed-level* implementation techniques renders the simple approach to specification and verification outline above invalid for a number of reasons:

Lack of compositionality. Viewed at the digital level of abstraction, *compositionality* no longer holds. That is, the behaviour of a circuit cannot, in general, be described in terms of the behaviours of its component parts.

Adirectionality. Whereas ideal digital components have a well-defined sense of directionality, analog-level components (resistors, transistors, etc) do not. The notion of "inputs" and "outputs" is no longer relevant.

Analog signal levels. In general, the signals may not be at well-defined digital levels. For instance, a signal transmitted through a pass-transistor network will suffer a degree of degradation depending upon the particular route it takes through the network and the loading imposed by following stages.

Malfunctioning. With some circuit configurations (such as those involving tristate drivers), inappropriate patterns of excitations can result in short circuits of the power rails. Depending on the duration, this may lead either to transient malfunctioning of the overall circuit (due to glitches) or irreversible damage to the drivers (due to excessive power dissipation).

In order to be able to apply the techniques of specification and verification to mixed-level implementations, it is necessary to be able to work at both the digital and the analog levels of abstraction. The latter entails working in terms of the voltages *and currents* at the terminals of the individual analog electronic devices.

1.1 Approach

A variety of approaches have been advocated to the task of specification and verification at the analog level based upon *idealised* approximations to the characteristics of analog components (for example, treating transistors as ideal switches) and/or simplifying assumptions (for example, assuming that current flows are negligible). In some circumstances such approaches can provide a useful degree of insight and can allow the identification of many kinds of design errors. In general, however, conclusions drawn from reasoning about the properties of systems defined in terms of idealised components (or arbitrary approximations to them) cannot be relied upon to predict the properties of actual systems.

The approach advocated in this paper (which refines and extends the approach described in [3]) is based not on specifications of idealised components but rather on *conservative approximations* to the behaviour of actual components. That is, the specifications place bounds on the possible *ranges* of behaviours that actual components may exhibit rather than attempt to describe the actual behaviours the components exhibit. The approach is thus compatible with the limited knowledge a designer has of the characteristics of analog components (due to production spreads, ageing, environmental factors, etc.).

1.2 Overview

In §2 we consider the formulation of analog behavioural specifications for typical electronic components and the formulation of derived specifications for circuits. We base the examples given on bipolar transistors (and TTL technology) rather than on FETs since the former present a much greater technical challenge to specify and to reason about. In §3 we consider the formulation of a requirements specification (defined at the analog level and taking account of loading and drive capabilities) for a typical TTL gate. In §4 we define the verification condition for the correctness of an implementation, we show that (for a wide class of specifications) such verification conditions are formally decidable and we describe an implementation of a decision procedure.

2 Specification of Analog Devices and Circuits

We define a *behavioural specification* at the analog level of abstraction as a predicate on the voltages and currents at the terminals of a device. There are two ways such a predicate can be formulated. Either it can defined over the voltages and currents at all terminals, or one terminal can be adopted as a reference and the predicate can be defined over the voltages (relative to this terminal) and currents at the remaining terminals. The first approach has the advantage of treating all terminals symmetrically but the disadvantage of involving a redundant voltage and current. The second approach is particularly convenient for two terminal devices (such as resistors or diodes) since it allows one to talk about *the* current through the device and *the* voltage across it.

A behavioural specification can be either *static* (involving steady-state signals) or *dynamic* (involving time-varying signals). In this paper we deal only with the former but we note that, since the principles of the approach are independent of temporal considerations, it extends naturally to dynamic specifications.

Notation. We work within the context of a typed, higher-order predicate logic and assume the existence of a primitive type *prop* (containing the two propositional truth values T and F) and the type $\mathbb{R}$ (the reals). For clarity, we introduce type synonyms of V (for voltages) and I (for currents), defining both as equivalent to $\mathbb{R}$.

Examples. The behavioural specification, *res*, for an *ideal resistor*, is a predicate *res*, parametrized by the device's resistance, that asserts that the voltage across the device and current through it are related by Ohm's law:

$$res: \mathbb{R} \to (V \times I) \to prop$$
$$res\ r\ (v,i) \triangleq (v = i \times r)$$

The probability of a real resistor obeying (exactly) an idealised specification like this is zero. A more useful specification is for a resistor whose resistance is not defined exactly but only to within a specified tolerance:

$$Res: \mathbb{R} \to \mathbb{R} \to (V \times I) \to prop$$
$$res \; r_1 \; tol \; (v, i) \triangleq \exists r: \mathbb{R}.(|r - r_1| \leq tol \times r_1) \wedge (res \; r \; (v, i))$$

The specification, *diode*, for an *ideal junction diode*, is given by the well-known *ideal diode equation*, cast in relational form

$$diode: V \to I \to (V \times I) \to prop$$
$$diode \; v_{th} \; i_{sat} \; (v, i) \triangleq (i = i_{sat} \times exp(v/v_{th} - 1))$$

where the parameters v_{th} and i_{sat} define the device's *threshold voltage* and *saturation current*. Again, a looser version of the same specification is more useful. The following specification is for a diode whose parameters are only known to within bounds:

$$Diode \; V^2 \to I^2 \to (V \times I) \to prop$$
$$Diode \; (v_{th1}, v_{th2}) \; (i_{sat1}, i_{sat2}) \; (v, i) \triangleq$$
$$\exists v_{th}, i_{sat}.$$
$$(v_{th1} \leq v_{th} \leq v_{th2}) \wedge (i_{sat1} \leq i_{sat} \leq i_{sat2}) \wedge$$
$$diode \; v_{th} \; i_{sat} \; (v, i)$$

The above specifications are for 2-terminal devices. Specifications in this form for typical 3-terminal devices (such as transistors) tend to be too complex to be of practical utility.

2.1 Linear and Rectilinear Specifications

In general, a behavioural specification of an analog component is a arbitrary predicate on the voltages and currents at its ports. Such a degree of generality is not, however, either necessary or desirable.

If one examines the (informal) justification that a designer of an analog implementation of a digital system might typically give as to why the proposed design is correct, one can observe that almost never is there any reliance on the detailed behavioural characteristics of the components. Rather, the justification is invariably couched in terms of extremal or limiting characteristics. For instance, a diode might be referred to as being "reverse biassed" or a transistor as being "saturated" or "hard on", it being understood that the magnitude of the leakage current though the diode or the saturation voltage across the transistor can safely be assumed to be negligible.

This suggests that, for the purposes of digital design, it may be possible to work in terms of relatively coarse behavioural specifications of the analog components with no loss in the range of implementations that can be shown to be correct. Provided that these specifications are *conservative approximations* to the range of behaviours the actual components may exhibit then the soundness of the overall approach will not be prejudiced. We can exploit this fact by limiting the set of behavioural specifications we consider to a syntactically defined subset whose members are amenable to automatic decision procedures.

We have investigated two such classes of specification; these we term *piecewise-linear* specifications and *rectilinear* specifications. The former we define as being predicates whose atomic predicates consist only of (in)equalities of linear combinations of the parameters and the latter as being ones whose atomic predicates consist only of (in)equalities on single parameters. A pwl specification can be visualised as consisting of a set of regions in parameter space bounded by hyperplanes and a rl specification as one consisting of regions bounded by hyperplanes parallel to the coordinate axes.

Even restricted to one or other of these forms, a specification can still have a relatively complex syntactic form. It is therefore worth adopting (without loss of generality) a standard syntactic form in which to present specifications. A good choice turns out to be to structure a specification as the conjunction of a set of implications. Each implication can be thought of as forming a *partial* specification of the behaviour; the hypothesis of the implication delimits a region of the parameter space and the conclusion specifies the behaviour within that region. (Note that the regions are not necessarily disjoint nor all inclusive.) This form has the advantage that omitting (by accident or intent) one or more of the implications does not invalidate the specification; it merely weakens it.

The following subsections illustrate pwl and rl specifications for typical components.

Resistor. A pwl specification for a resistor of resistance $r \pm \delta r$ is (see Fig. 1(a)):

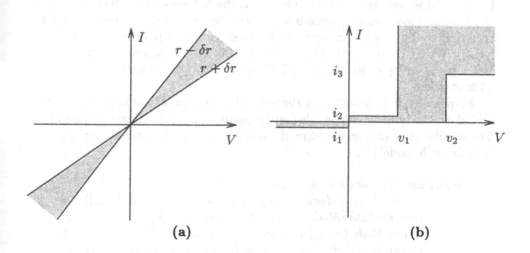

(a)   (b)

Fig. 1. (a) Graph of a pwl specification for a resistor of value $r \pm \delta r$;   (b) graph of a rl specification for a diode

$$res \ r \ \delta r \ (v,i) \ \stackrel{\triangle}{=}$$
$$((i \geq 0) \Rightarrow (i \times r_1 \leq v \leq i \times r_2)) \qquad \wedge$$
$$((i \leq 0) \Rightarrow (i \times r_1 \geq v \geq i \times r_2))$$
$$\textbf{where } r_1 = r - \delta r \textbf{ and } r_2 = r + \delta r$$

Bipolar junction diode. A rl specification for a bipolar junction diode is formed from the conjunction of four partial specifications, one for each of the following regions of its parameter space:

1. When it is reverse biassed (or *hard off*);
2. When it is forward biassed, but not yet conducting significantly;
3. When it is forward biassed and conducting an indeterminate amount; and
4. When it is forward biassed and conducting strongly (or *hard on*).

This leads (see Fig. 1(b)) to a rectilinear specification of the form:

$$diode \ v_1 \ v_2 \ i_1 \ i_2 \ i_3 \ (v,i) \ \stackrel{\triangle}{=}$$
$$((v < 0) \Rightarrow (i_1 \leq i \leq 0)) \qquad \wedge$$
$$((0 \leq v \leq v_1) \Rightarrow (0 \leq i \leq i_2)) \qquad \wedge$$
$$((v_1 \leq v \leq v_2) \Rightarrow (0 \leq i)) \qquad \wedge$$
$$((v_2 \leq v) \Rightarrow (i_3 \leq i))$$

The specification of a typical silicon signal junction diode can be obtained by specialisation:

$$signal\_diode \ \stackrel{\triangle}{=} \ diode \quad 0.3 \quad 0.6 \quad -10^{-6} \quad 10^{-6} \quad 0.01$$

Bipolar junction transistor. Informally, the behaviour of a BJT [4] is usually split into four modes (*forward active, saturation, cutoff* and *reverse active*), according to the relative polarities of its three electrodes (emitter, base and collector). Whilst in linear circuits, operation is usually restricted to the forward active mode, in logic circuits (as in TTL technology) transistors may operate in all four modes.

Empirically, it turns out that the properties of a transistor of interest to the digital circuit designer can be adequately captured by a rectilinear specification. The specification, cast in the form of a conjunction of four partial specifications (one for each mode), is:

$$transistor \ < parameters > \ (v_b, i_b, v_c, i_c) \ \stackrel{\triangle}{=}$$
$$(forwardActiveMode \ (v_b, v_c) \Rightarrow forwardSpec \ (v_b, i_b, v_c, i_c)) \quad \wedge$$
$$(reverseActiveMode \ (v_b, v_c) \Rightarrow reverseSpec \ (v_b, i_b, v_c, i_c)) \quad \wedge$$
$$(cutoffMode \ (v_b, v_c) \Rightarrow cutoffSpec \ (v_b, i_b, v_c, i_c)) \qquad \wedge$$
$$(saturationMode \quad (v_b, v_c) \Rightarrow saturationSpec \ (v_b, i_b, v_c, i_c))$$

The hypotheses of the implications describe the extent of the modes and the conclusions of the implications specify the behaviour within each mode. For example, for a typical silicon NPN transistor, the extent of its cutoff mode is defined by

$$cutoffMode\ (v_b, v_c) \triangleq (v_b \leq 0) \wedge (v_b \leq v_c)$$

and its behaviour in that mode is defined by

$$cutoffSpec\ (v_b, i_b, v_c, i_c) \triangleq (-10^{-9} \leq i_b \leq 0) \wedge (-10^{-9} \leq i_c \leq 0)$$

The behavioural properties of multi-emitter transistors (as used in TTL gates) are more complex but, again, they can be adequately approximated by rectilinear specifications.

TTL characteristics In addition to defining conservative approximations to the behaviour of primitive components (as above) it is also possible to define specifications for the analog behavioural interfaces associated with particular technologies. As an example, the load and drive characteristics of the input and output terminals for standard TTL can be can be concisely and unambiguously defined by rectilinear specifications:

- The *load* that a standard TTL input terminal imposes on its environment (often referred to as a *TTL unit load*) is defined by the predicate *load*: $V \times I \rightarrow$ *prop* whose graph is shown in Fig. 2(a).
- The *loading* that the environment can impose on a standard TTL output terminal without limiting its capability of reaching well-defined high and low voltage output levels is defined by the predicate *loading*: $V \times I \rightarrow$ *prop* whose graph is shown in Fig. 2(b).

2.2 Derived Specifications

Given a circuit, the *derived specification* of the circuit is defined to be the strongest relation on the voltages and currents at the terminals of that circuit that can be inferred from the constraints imposed by the behavioural specifications of the individual components, the conservation laws of analog electronics and the circuit topology.

There are a variety of standard ways that circuit designers use to represent circuits. We will view a *circuit* (see Fig. 3) as consisting of;

- a set of *nodes*;
- a set of *components* whose *terminals* (called the *internal* terminals of the circuit) are connected to the nodes;
- a set of *external terminals* also connected to the nodes.

We assume that a voltage and a current are associated with each terminal of each component and with each of the external terminals.

The *derived specification* for a circuit is a predicate on the voltages and currents associated with the external terminals. It is obtained by taking the conjunction of the following assertions:

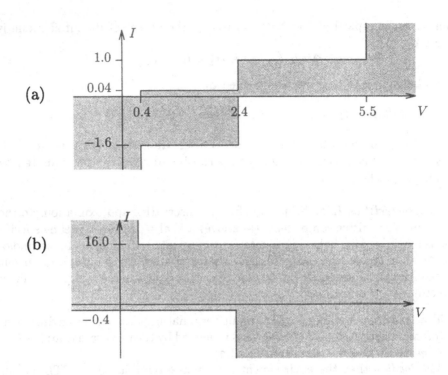

Fig. 2. Graphs for (a) the predicate *load* and (b) the predicate *loading* (currents are in mA).

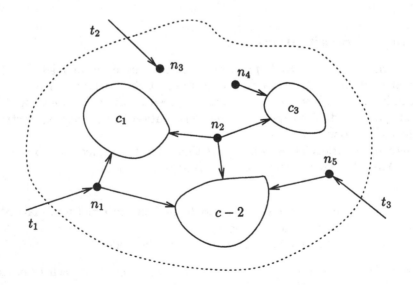

Fig. 3. A circuit consisting of a set of nodes $\{n_1, \ldots, n_5\}$, a set of components $\{c_1, c_2, c_3\}$ and a set of terminals $\{t_1, t_2, t_3\}$.

- the voltages and currents at the terminals of each component obey the behavioural specification associated with that component;
- the currents on the terminals (internal and external) incident on each node sum to zero and the voltages on the terminals are equal to each other;

and existentially quantifying over the voltages and currents associated with the internal terminals.

An important distinction between derived specifications defined at the digital and at the analog levels of abstraction is that whilst the former are valid only for well-formed circuits (ie, circuits that obey the *design rules* specific to a particular technology) the latter are unconditionally valid.

Example. A typical TTL implementation for a Nand gate is shown in Fig. 4. Assuming that $Q1$ is the specification for transistor Q1, $R1$ the specification for

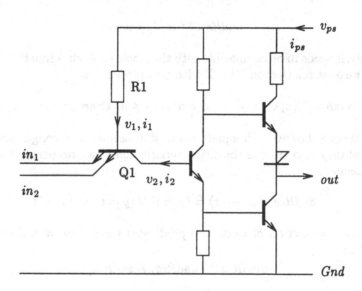

Fig. 4. A typical implementation, in TTL technology, of a Nand gate.

resistor R1, and so on, the derived specification for this circuit takes the form:

$$NANDimp: (V \times I) \to (V \times I)^2 \to (V \times I) \to prop$$
$$NANDimp\ (v_{ps}, i_{ps})\ ((v_{in_1}, i_{in_1}), (v_{in_2}, i_{in_2}))\ (v_{out}, i_{out}) \triangleq$$
$$\exists v_1, v_2, \ldots, v_n : V.$$
$$\exists i_1, i_2, \ldots, i_m : I.$$
$$\quad Q1\ (v_{in_1}, i_{in_1}, v_{in_2}, i_{in_2}, v_1, i_1, v_2, i_2) \wedge$$
$$\quad R1\ (v_{ps} - v_1, i_1) \wedge$$
$$\quad \ldots$$

3 Specification of Digital Devices

Our aim in this section is to formulate specifications, at the digital level of abstraction, for the steady-state behaviour of ordinary gates. This apparently trivial task needs to be handled with care in order to avoid inconsistency arising when these specifications are related to the corresponding ones at the analog level of abstraction. We illustrate the approach by considering the specification for a 2-input And gate.

Our starting point is the set $\mathbb{B} = \{t, f\}$ of *ideal* digital signal levels to which we add a third element, n, to allow *non-digital* signal levels to be represented, giving the set $\mathbb{T} \triangleq \{t, n, f\}$ of *non-ideal* signal levels.

Since the output of an And gate with signals of t and n at its inputs could be any of t, n or f, a *relational* description is required; a functional one would involve arbitrary overspecification. The specification, a predicate on the tuple of signals at the inputs and output of the gate, is of type

$$andReq: \mathbb{T}^2 \times \mathbb{T} \to prop$$

Evidently, it needs to be compatible with the Boolean algebra function $and: \mathbb{B}^2 \to \mathbb{B}$. The weakest relation on $\mathbb{T}^2 \times \mathbb{T}$ with this property is

$$andReq_1 \ ((x, y), z) \triangleq \textbf{if } (x \neq \mathsf{n}) \wedge (y \neq \mathsf{n}) \textbf{ then } z = and \ (x, y)$$

In practice, however, designers invariably assume a stronger specification than this; they also assume the characteristic non-strict property that f acts as a zero element:

$$andReq_2 \ ((x, y), z) \triangleq (x = \mathsf{f}) \vee (y = \mathsf{f}) \Rightarrow (z = \mathsf{f})$$

Taking the conjunction of these two predicates yields the overall specification for an And gate:

$$andReq \triangleq andReq_1 \wedge andReq_2$$

This can equivalently be expressed as

$$\begin{aligned} andReq \ ((x, y), z) \triangleq \\ ((x = \mathsf{t}) \wedge (y = \mathsf{t}) &\Rightarrow (z = \mathsf{t})) \qquad \wedge \\ ((x = \mathsf{f}) \vee (y = \mathsf{f}) &\Rightarrow (z = \mathsf{f})) \end{aligned}$$

Relational specifications for other types of gate can be derived in a similar way. For example, the specification for a Nand gate is

$$\begin{aligned} nandReq: \mathbb{T}^2 &\times \mathbb{T} \to prop \\ nandReq \ ((x, y), z) &\triangleq \\ ((x = \mathsf{t}) \wedge (y = \mathsf{t}) &\Rightarrow (z = \mathsf{f})) \qquad \wedge \\ ((x = \mathsf{f}) \vee (y = \mathsf{f}) &\Rightarrow (z = \mathsf{t})) \end{aligned}$$

3.1 Specification at the Analog Level

We now describe how specifications of digital devices can be formulated at the analog level (where both voltages *and* currents are involved). We illustrate the discussion with examples based on TTL (transistor-transistor logic), a technology based on bipolar junction transistors (and therefore a challenging one to handle formally).

We begin by describing how device specifications expressed at the digital level of abstraction (such as the Nand gate specification just defined) can be mapped down to the analog level of abstraction. There are two aspects that have to be considered: the *abstraction function* that defines the relation between signals at each of these levels and the *loading predicates* that define the loading a gate is allowed to impose on its environment and the amount of drive it can supply to its environment.

Voltage abstraction function. The voltage abstraction function relates analog voltage levels to digital signal levels. For a standard TTL technology (and assuming the *true-as-high* convention) voltages below 0.4V represent false and ones above 2.4 represent true. Thus, the abstraction function for TTL is:

$$a\colon V \to \mathbb{T}$$
$$a\,v \stackrel{\wedge}{=} \quad \text{if } v < 0.4 \text{ then f else}$$
$$\text{if } v > 2.4 \text{ then t else n}$$

Using the abstraction function, a device specification defined at the digital level can be mapped down to a voltage specification at the analog level (notice the contravariance: whilst the abstraction function maps analog signals up to digital ones, it maps digital specifications *down to* analog ones).

As an example, the digital-level Nand gate specification, *nandReq*, defined earlier, maps to the analog voltage specification, *NandReq*, defined by

$$NandReq\colon V^2 \to V \to prop$$
$$NandReq\,(v_1, v_2)\,v \stackrel{\wedge}{=} nandReq\,((a\,v_1), (a\,v_2))\,(a\,v)$$

This predicate, however, both over and under specifies the behaviour required of a real Nand gate:

- It overspecifies it since it demands *bidirectionality* — because it implies that, if the output terminal were to be forced low (for example, by shorting it to ground), then the input terminals must be driven to high.
- It underspecifies it since it is not *compositional* — because it would be satisfied by an implementation that permanently shorted both its inputs to ground and its output to high.

In order to avoid these problems, it is necessary for the specification to take account of currents as well as voltages.

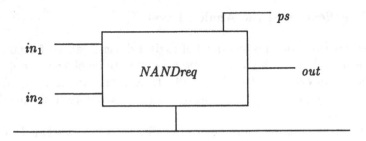

Fig. 5. Interface for analog-level specification of a Nand gate

As an example, here is a specification that captures the properties that a designer would normally require and expect of a nand gate (with terminals as shown in Fig. 5):

$$NANDreq: (V \times I) \to (V \times I)^2 \to (V \times I) \to prop$$
$$NANDreq\ (v_{ps}, i_{ps})\ ((v_{in_1}, i_{in_1}), (v_{in_2}, i_{in_2}))\ (v_{out}, i_{out}) \triangleq$$
$$(4.5 \leq v_{ps} \leq 5.5) \Rightarrow$$
$$(i_{ps} \leq 0.1)\ \wedge$$
$$load\ (v_{in_1}, i_{in_1})\ \wedge\ load\ (v_{in_2}, i_{in_2})\ \wedge$$
$$(loading\ (v_{out}, i_{out}) \Rightarrow NandReq\ (v_{in_1}, v_{in_2})\ v_{out})$$

This specification states that, provided the power supply voltage, v_{ps}, is within correct limits, the maximum current the gate will draw must also be within limits, the load the gate imposes on each of its input terminals must satisfy the *load* specification and that, provided the load the environment imposes on the output terminal satisfies the *loading* specification, the voltages on the gate's input and output terminals must satisfy the voltage-only specification for a Nand gate. (The requirement that the current i_{ps} drawn from the supply be within bounds is a means of excluding improper implementations, such as one that attempts to have its output terminal simultaneously assume the values high and low by the simple expedient of connecting it directly to both the power rail and ground.)

4 Verification

Given an analog-level requirements specification *req* and an implementation having a derived specification *imp*, the implementation is said to be *correct* if every tuple x of voltage-current pairs that satisfies the derived specification also satisfies the requirements specification. That is, if

$$\forall x.\ imp\ x \Rightarrow req\ x$$

or, in higher-order form, $imp \sqsupseteq req$ (that is, if *imp* is a stronger predicate than *req*).

For example, the verification condition asserting that the behaviour of the TTL nand-gate implementation satisfies the requirements specification for a nand gate is $NANDimp \sqsupseteq NANDreq$, or, in first-order form:

$$\forall v_{ps}, v_{in_1}, v_{in_2}, v_{out} : V.$$
$$\forall i_{ps}, i_{in_1}, i_{in_2}, i_{out} : I.$$
$$NANDimp \; (v_{ps}, i_{ps}) \; ((v_{in_1}, i_{in_1}), (v_{in_2}, i_{in_2})) \; (v_{out}, i_{out}) \;\Rightarrow$$
$$NANDreq \; (v_{ps}, i_{ps}) \; ((v_{in_1}, i_{in_1}), (v_{in_2}, i_{in_2})) \; (v_{out}, i_{out})$$

Although, as already noted, this approach can be applied to *any* implementation (there are no restrictions to "well-formed" analog circuits), there is one caveat that must be born in mind. The approach implicitly *assumes* that the voltages and currents in the circuit are in steady state conditions. However, a circuit with active components may potentially exhibit oscillatory behaviour and the approach used does not exclude the possibility of such behaviour. This is, however, a well-known problem in analog design and well-established methods (based on feedback analysis) exist to demonstrate that a given circuit is stable.

Establishing correctness. There are two approaches that can be used to establish the truth of analog-level verification conditions. One (a syntactic approach) is to use conventional theorem-proving techniques, the other (a semantic approach) is to use model-checking techniques.

An account of a theorem-proving approach was presented in [3]. Whilst this approach is certainly possible, it turns out to be highly labour-intensive. Essentially it involves attaching assertions to the output terminals of an implementation and then using goal-directed techniques to push them back step by step through the circuit (using the device specifications as predicate transformers) to the input terminals.

The model-checking approach involves using *constraint satisfaction* techniques. In principle, a general purpose constraint satisfaction program (such as CLP(R) [2]) could be used. In practice, such an approach turns out to be hopelessly inefficient. A specialised decision procedure (similar to those described in [1]) has therefore been tailored to the characteristics of this particular kind of problem.

4.1 Decision Procedure

The verification condition we wish to establish is of the form

$$\forall e.\; imp[e] \Rightarrow req[e]$$

where:

- e is the tuple of *external* variables (for instance, in the verification condition for the Nand gate (above), the tuple would comprise the variables $v_{ps}, \ldots, i_{out}$);

- $imp[e]$ is the derived specification of the implementation;
- $req[e]$ is the requirements specification.

In general, a derived specification, will be of the form $\exists i.\ implem[e, i]$ where i is a tuple of *internal* variables in the implementation (for instance, the variables $v_1, \ldots, v_n$ and $i_1, \ldots, i_m$ in the *NANDimp* specification). Thus, the verification condition can be cast in the form

$$\forall e.\ (\exists i.\ implem[e, i]) \Rightarrow req[e]$$

By moving the existential quantifier to the outer level (where it becomes a universal), and then by replacing the universal quantifier by a negated existential formula, the verification condition can be cast in the form

$$\neg \exists e, i.\ implem[e, i] \wedge \neg req[e]$$

Assuming that the requirements specification, *req*, does not contain any embedded quantifiers (in practice, this is almost invariably the case), the body of this formula can then (by using simple propositional reasoning) be expressed in disjunctive normal form:

$$\neg \exists e, i.\ \phi_1[e, i]) \vee \ldots \vee \phi_n[e, i]$$

where each subformula ϕ_j is a conjunction of literals, and each literal is a simple linear (in)equality in the external and internal variables.

Finally, by distributing the existential quantifier over the disjunction, the verification condition is converted to the form

$$\neg((\exists e, i.\ \phi_1[e, i]) \vee \ldots \vee (\exists e, i.\ \phi_n[e, i]))$$

The verification condition can now be established simply by showing that each of the n formulae, $\phi_1[e, i], \ldots, \phi_n[e, i]$, has no feasible solution.

4.2 Practical Implementation

An algorithm to implement the above decision procedure has been written in Haskell (with a Fortran back-end). The algorithm takes as its input a set of definitions of the behaviours of the primitive components (resistors, diodes, transistors) and a statement of the verification condition. As an example, the definition for a diode is

```
diode v i =
    let i1 = -1.0E-6;
        v2 = 0.3;
        i2 =  1.0E-6;
        v3 = 0.6;
        i3 = 10.0E-3
    in
    (v < 0.0 => (i < 0.0 & i > i1)) &
    ((v < v2)  => (i < i2))          &
    ((v > v3)  => (i > i3))
```

The algorithm goes through the following steps:

1. The input, which syntactically takes the form of a single formula (the verification condition, preceeded by `let` clauses introducing the component specifications), is parsed. This yields an abstract representation of the verification condition.
2. Beta reduction is used to eliminate all `let` bindings, that is, to replace locally defined names with their definitions and symbolically evaluate all applications.
3. Any existential quantifiers present in the hypothesis of the formula are moved outwards to the top level (where they become universal quantifiers). In turn, all universals are rewritten as negated existentials.
4. The body of the formula is converted to disjunctive normal form and the existential quantifiers are distributed over the disjunction.
5. Each separate disjunct (which itself corresponds to a conjunction of literals, each one of which is a linear inequality) is converted to a standard linear-programmming matrix form.
6. Each such set of matrices is submitted for analysis to a standard Fortran linear-programming library routine (routine E04MBF in the NAG library). This routine quickly determines whether the set of inequalities has a feasible solution.

If none of the sets of inequalities is found to have a feasible solution, then the original verification condition is true. Conversely, if a feasible solution is found, then this solution provides a counterexample demonstrating the falsity of the verification condition.

In addition to the steps outlined above, the algorithm also incorporates a number of optimisations. For instance, in step (5), a quick test for the presence of opposing literals (that is, pairs of the form $\ldots \wedge (x > a) \wedge \ldots \wedge (x < a) \wedge \ldots$) allows many subformulae to be eliminated without the need for submission to the linear-programming routine.

5   Conclusions

The overall aim of the work described here has been to extend existing predicate-logic based methods used for specifying and verifying systems defined at the "ideal digital" level of abstraction down to the analog level. This allows digital systems that (perhaps in order to gain speed or economise on power consumption) have been implemented in part at the analog level to be specified and reasoned about with the same degree of certainty and transparency as those implemented wholly at the digital level.

The method described is straightforward; it simply involves using predicates to characterise the behaviour of analog components in terms of the voltages and currents at their terminals. A central tenet of the approach is that these characterisations should be *conservative* approximations to the actual behaviours. Based on detailed studies, in particular of implementations of TTL technology, it has been found that these approximations can be remarkably weak and yet still capture enough of the essential behavioural properties of an analog device to allow successful verification at the digital level of abstraction. Perhaps this should come as no surprise; after all, digital designers habitually reason at the analog level simply in terms of devices being 'on' or 'off' and of voltage levels being 'high' or 'low', and so on.

An algorithm for checking the verification conditions that arise with this approach has been programmed and has been used to verify, automatically, simple TTL circuits. The computational time the algorithm takes has, however, been found to increase sharply with circuit complexity. At present it is not clear whether the computational task is an inherently hard one or whether a different algorithm would be less sensitive to circuit size. Irrespective of the algorithm used, one approach to limit the computational time is to partition an implementation into a set of smaller ones by introducing intermediate specifications (in effect, lemmas). This seems to be the approach the human designer uses when confronted with a complex implementation.

Acknowledgments This work was sponsored by ESPRIT (under Grant 8533) and by the UK EPSRC (under grant GR/J/78105).

References

[1] Frederic Benhamon and Alain Colmerauer, editors. *Constraint Logic Programming.* MIT Press, 1993.

[2] Eugene C. Freuder and Alan K. Mackworth, editors. *Constraint-based reasoning.* MIT/Elsevier, 1994.

[3] Keith Hanna. Reasoning about Real Digital Circuits. In *Proc Higher Order Logic Theorem Proving and its Applications.* Springer-Verlag, 1994.

[4] D. A. Hodges and H. G. Jackson. *Analysis and design of digital integrated circuits.* McGraw-Hill, 1983.

A Timed Automaton-Based Method for Accurate Computation of Circuit Delay in the Presence of Cross-Talk

S. Taşıran[1] *, S. P. Khatri[1] *, S. Yovine[2] **, R. K. Brayton[1], and A. Sangiovanni-Vincentelli[1]

[1] Department of Electrical Engineering and Computer Sciences,
University of California at Berkeley
[2] VERIMAG, and
California-PATH, University of California at Berkeley

Abstract. We present a timed automaton-based method for accurate computation of the delays of combinational circuits. In our method, circuits are represented as networks of timed automata, one per circuit element. The state space of the network represents the evolution of the circuit over time and delay is computed by performing a symbolic traversal of this state space.

Based on the topological structure of the circuit, a partitioning of the network and a corresponding conjunctively decomposed OBDD representation of the state space is derived. The delay computation algorithm operates on this decomposed representation and, on a class of circuits, obtains performance orders of magnitude better than a non-specialized traversal algorithm.

We demonstrate the use of timed automata for accurate modeling of gate delay and cross-talk. We introduce a gate delay model which accurately represents transistor level delays. We also construct a timed automaton that models delay variations due to cross-talk for two capacitively coupled wires.

On a benchmark circuit, our algorithm achieves accuracy very close to that of a transistor level circuit simulator. We show that our algorithm is a powerful and accurate timing analyzer, with a cost significantly lower than transistor level circuit simulators, and an accuracy much higher than that of traditional timing analysis methods.

1 Introduction

The computation of delay for combinational circuits is a well-studied problem which, until recently, was considered solved. Efficient exact methods have been devised for computing the delays of acyclic combinational circuits ([DKM93], [LB94], [MSBS93], [YH95] among others). However, with the feature sizes of integrated circuits shrinking to sub-micron levels, some of the underlying assumptions of existing delay analysis methods are no longer valid. First, higher clock speeds require more accurate modeling of circuit delay, therefore, more

---

* Supported by SRC under grants DC-324-026 and DC-324-040.
** Supported by NSF under grant ECS 972514

sophisticated gate delay models are needed. For instance, for complex dynamic logic gates, gate delays depend on the relative timing of the input signals, as well as their values. Second, circuit level effects such as capacitive coupling between wires (also referred to as *cross-talk*) need to be taken into account. As a result, it is now commonly accepted that existing methods for modeling and computing delays are inadequate for deep sub-micron circuits (See, for instance, [CWS97,TKB97] and other related papers in the TAU '97 Workshop). Existing timing analysis schemes do not use sufficiently accurate gate delay models, and do not account for cross-talk.

Timed automata have been used to model the delay characteristics of gates and circuits [MP95,TAKB96,TB97,FK95]. Previously, timed automaton-based techniques have been restricted to the analysis of asynchronous circuits because they are generally smaller than synchronous circuits and require a more detailed temporal analysis than delay computation. Today's synchronous circuits with sub-micron feature sizes place stronger demands on timing analysis tools, and this new setting makes the expressiveness of timed automata desirable. Timed automata allow the use of very general gate delay models - including those in which a unique delay can be specified for every pair of input vectors. As a result, effects of cross-talk on delay can be incorporated. One contribution of this paper is the introduction of a new and accurate gate delay model, which efficiently accounts for input sequence dependent delays.

For large portions of a typical design, a less powerful modeling framework suffices. However, the analysis of certain critical portions of a high performance design must be performed with great accuracy, which is currently provided only by transistor level simulators. Exhaustive simulation of these sections of the circuit, on the other hand, is not computationally feasible because of the exponential number of possible input patterns. In this study, we present a timed automaton-based delay computation method. We represent input waveforms and circuits by networks of timed automata, with each automaton modeling a circuit element or input waveform. The state space of this network describes the possible behaviors of the circuit over time. Delay computation is then posed as a variant of the state-space traversal problem. Because all input patterns are covered, unlike simulation, the circuit delay computed by this algorithm is guaranteed to be correct.

State-space traversal of timed automata is a PSPACE-complete problem, and, like other traversal-based methods, it suffers from the state-space explosion problem for systems with large numbers of components. However, by exploiting certain topological properties of combinational circuits and the state spaces of the corresponding timed automata, we devise heuristics which enable us to handle systems that are much larger than could be handled with non-specialized traversal methods (Section 5.2).

The paper is organized as follows. In Section 2 we introduce timed automata. Section 3 discusses how circuits and waveforms are represented using timed automata. Section 4 describes our algorithm for computing maximum delays. Section 5 presents experimental results and contrasts our method with other approaches for delay computation. Section 6 summarizes our work, and discusses avenues for future research.

2 Timed automata

2.1 Notation

Let X be a finite set of real-valued variables. An X-*valuation* ν assigns a non-negative real value $\nu(x)$ to each variable $x \in X$. An X-predicate φ is a positive Boolean combination of constraints of the form $x \diamond k$, where k is a nonnegative integer constant, $x \in X$ is a variable, and $\diamond$ is one of the following: $\leq, \geq, =$. Let P be a finite set of variables, each ranging over a finite type. A P-valuation ξ is an assignment of values to variables in P. A P-event is a pair $\langle \xi, \xi' \rangle$ consisting of P-valuations ξ and ξ' denoting the old and the new values of the variables in P. A P-predicate is a Boolean predicate on ξ and ξ'.

2.2 Timed Automata

A *timed automaton* A is a tuple $\langle S, S_0, O, I, X, \alpha, \mu, E \rangle$, where

- S is the finite set of locations, and $S_0 \subseteq S$ is the set of initial locations.
- O is the set of output variables, each ranging over a finite domain. An *output* of A is an O-valuation.
- I is the set of input variables, each ranging over a finite type. An *input* of A is an I-valuation, an *input-event* is an I-event, and an *input-predicate* is an I-predicate. An *observation* of A is a $(I \cup O)$-valuation, and an *observation-event* of A is a $(I \cup O)$-event.
- X is the finite set of real-valued variables, called *timers*.
- α is the invariant function that assigns an X-predicate $\alpha(s)$ to each location $s \in S$.
- μ is the output function that assigns the output $\mu(s)$ to each location $s \in S$.
- E is the finite set of edges. Each edge e is a tuple $\langle s, t, \varphi, \chi, R \rangle$ consisting of the source location s, the target location t, the X-predicate φ, the input-predicate χ, and a subset of clocks $R \subseteq X$ that specifies the timers to be reset to 0 after the edge is taken.

A *state* σ of A is a pair $\langle s, \nu \rangle$ containing the location $s \in S$ and the X-valuation $\nu \in \alpha(s)$. The set of all states of A is denoted by Σ_A. The state $\langle s, \nu \rangle$ is initial if $s \in S_0$ and $\nu(x) = 0$ for all $x \in X$. Consider a state $\sigma = \langle s, \nu \rangle$ of the timed automaton A and a time increment δ. The automaton A can *wait* for δ in state σ, written $wait(\sigma, \delta)$, iff for all $0 \leq \delta' \leq \delta$, $(\nu + \delta') \models \alpha(s)$. A *timed event* γ of the timed automaton A is a tuple $\langle \delta, \xi, \xi' \rangle$ consisting of a non-negative real-valued increment δ and the observation-event $\langle \xi, \xi' \rangle$. Such an event means that the automaton can wait for the time period δ if $\delta > 0$ and then update its output variables from $\xi(O)$ to $\xi'(O)$ while the environment is updating the input variables from $\xi(I)$ to $\xi'(I)$. The set of all timed events of A is denoted Γ_A.

The timed automaton A corresponds to a labeled transition system over the state-space Σ_A with labels from Γ_A. For states $\sigma = \langle s, \nu \rangle$ and $\tau = \langle t, \mu \rangle$ in Σ_A, and a timed event $\gamma = \langle \delta, \xi, \xi' \rangle$ in Γ_A, define $\sigma \xrightarrow{\gamma} \tau$ iff $\xi(O) = \mu(s)$, $\xi'(O) = \mu(t)$, $wait(\sigma, \delta)$, and there exists an edge $\langle s, t, \varphi, \chi, R \rangle$ such that $(\nu + \delta) \models \varphi$, $\langle \xi, \xi' \rangle \models$

χ, and μ is obtained from ν by applying R as explained above. A *timed event sequence* $\overline{\gamma} = \gamma_0, \gamma_1, ..., \gamma_{k-1}$ is a finite sequence of timed events $\gamma_i = \langle \delta_i, \xi_i, \xi_i' \rangle$ such that $\xi_{i+1} = \xi_i'$ for $0 \le i < k - 1$. A *run* of A on a timed event sequence $\overline{\gamma}$ is a sequence of states $\sigma_0, \sigma_1, \sigma_2, ..., \sigma_k$ such that $\sigma_0 \xrightarrow{\gamma_0} \sigma_1 \xrightarrow{\gamma_1} \sigma_2 \xrightarrow{\gamma_2} ... \xrightarrow{\gamma_{k-1}} \sigma_k$ in A. The timed event sequence $\overline{\gamma}$ is called a *trace* of A if there exists a run in A on $\overline{\gamma}$ starting from an initial state.

Composition of Timed Automata A timed automaton representing a circuit is obtained by *composing* the timed automata representing each component. The composition operation is analogous to that for FSMs: the product of the discrete state spaces is taken, and the set of timer variables for the composition consists of the union of the timers of the components. The composition of timed automata A and B is denoted by $A\|B$. For a formal definition of composition for timed automata, please see [TAKB96].

3 Modeling Circuits and Waveforms

3.1 Modeling Sets of Waveforms

The maximum delays of combinational circuits are typically computed for the following two cases:

- *Floating-mode delay*: Inputs are allowed to change their values arbitrarily until time 0. After time 0 all inputs remain stable.
- *Two vector delay*: Inputs (and all intermediate nodes of the circuit) are stable until time 0. At time 0 the inputs may switch to new values and must remain stable thereafter.

The sets of input waveforms for both cases can be represented concisely by timed automata. Figure 1 shows the timed automaton for the two-vector delay model. i_{old} and i_{new} represent the vectors of primary inputs before and after time 0 ($t = 0$) respectively. i_{old} and i_{new} are selected non-deterministically, which enables the automaton in the figure to represent all input vector pairs.

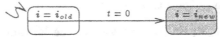

Fig. 1. Input Vector Automaton (Two-vector Delay Model)

It is also straightforward to model different arrival times at different primary inputs, asynchronous inputs, etc., with timed automata using extra timers.

3.2 Modeling Combinational Gates

Timed automata offer a great deal of expressive power for representing gate delay models. In the past, they have been used to model gates [MP95], [TAKB96], [TB97], [FK95] in the verification of asynchronous circuits and library elements. For example, in [TB97], a timed automaton representations for the inertial delay model was described. This model accounts for the possibility that short input pulses may not be reflected at a gate's output.

The gate delay models employed by conventional timing analysis methods have been somewhat simplistic and ad-hoc. The most primitive of these is the

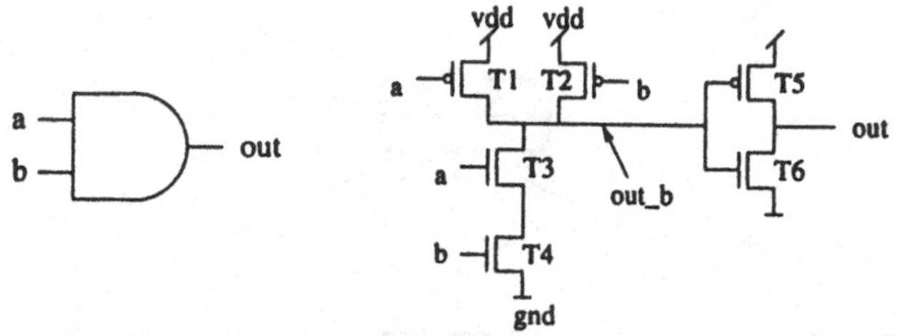

Fig. 2. AND gate: Transistor Level Description

unit delay model, where each gate is assumed to have a delay of one time unit. This was supplanted by the *fixed* delay model, where gates can have different but constant delays. Both of these models are too coarse as they do not allow for variations in gate delay. The *min-max* delay model addresses this problem to some degree by allowing variations in the delay of a gate: transitions at a gate's input are reflected at its output with a delay in the range $[d_{min}, d_{max}]$. In the *pin-delay* model, different input-to-output delays can be specified for different input pins of a gate. Neither the min-max model nor the pin-delay model are powerful enough for an accurate analysis because they can not express the dependency of gate delay on specific input values.

In practice gate delays have a strong dependence on the input vectors applied. This is illustrated by means of the static CMOS AND gate shown in Figure 2. For this gate, under the input sequence $a = b = 1 \rightarrow a = 1, b = 0$, let the gate delay be $t^f_{11 \rightarrow 10}$. The superscript f (r) indicates that the output is falling (rising), and the subscript represents the applied input vector sequence. Similarly, under the input sequence $a = b = 1 \rightarrow a = 0, b = 0$, let the gate delay be $t^f_{11 \rightarrow 00}$. In the second case, the delay of the gate is smaller. This is because, in the latter case, after the inputs transition, both transistors T1 and T2 are on, effectively doubling the current to charge the capacitance of node out_b. This results in a faster falling transition. In the first case, only transistor T2 is on after input b has switched, hence the falling transition is slower. The pin delay model does not distinguish between these two input sequences, and hence is not accurate enough for high performance designs. Some delay models used in the past allow the specification of separate rising and falling delays for the gate output. This is a useful property to model, but in the absence of input sequence dependence of the output delay, this model alone is not very useful.

Timed automata provide a powerful and uniform framework in which gate delay models of varying sophistication can be specified. One has the flexibility to assign a distinct delay for every possible sequence of input vectors (for an n input gate, there are $2^n \cdot (2^n - 1)$ of these). However, in practice, there are only a few sequences of input vectors that result in distinct delays. Many sequences of input vector transitions result in the same gate delay, and we make use of this property to simplify the timed automaton of a gate. For example, in Figure 2, the delays $t^f_{11 \rightarrow 01}$ and $t^f_{11 \rightarrow 10}$ have the same value, and so do the delays $t^r_{01 \rightarrow 11}$, $t^r_{10 \rightarrow 11}$, and $t^r_{00 \rightarrow 11}$. Therefore we group together these delays in our timed automaton model for the AND gate, resulting in a simpler delay model. In this way, our model incorporates input sequence dependent delays without becoming prohibitively large. A delay model in which different vector pairs are assigned distinct delays

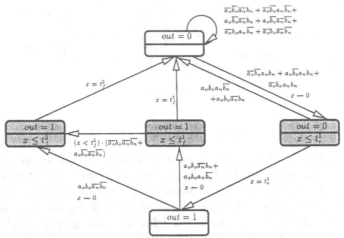

Fig. 3. Timed Automaton for AND gate in Figure 2

was previously introduced by [FK94]. However, this model requires a significantly larger number of transitions to model delays.[1]

The timed automaton for an AND gate is shown in Figure 3. The inputs to the AND gate are a and b. a_o and a_n represent the old and new values of the input a respectively. Similarly for input b. If the input $a = b = 1$ is followed by $a = 1, b = 0$, then this condition is represented by $a_o b_o a_n \overline{b_n}$. Each oblong box in Figure 3 represents a *location*. "Transient" locations are shaded. For each location, the value of the gate output is listed on top. The invariant condition associated with the location is listed below its output value. If $a_o = a_n$ and $b_o = b_n$, we stay in the same location by means of a self-loop arc (not shown). Also, if an input changes and then returns to its original value while the automaton is still in its temporary location (referred to as a *glitch*), we return to the starting location. These glitching arcs are not drawn for ease of readability. For example, if we are in the location corresponding to $out = 0$, we remain in this location if condition $\overline{a_o} b_o a_n \overline{b_n}$ occurs. This is because an AND gate has a 0 output under both the old and new input vectors. If, from the location corresponding to $out = 0$, condition $\overline{a_o} b_o a_n b_n$ occurs, then timer x is set to 0, and a transition is made to a temporary location where $out = 0$. After a delay of t_r^1, a final transition is made to a location where $out = 1$. This corresponds to the gate output changing from 0 to 1 after a delay of t_r^1, when inputs change from $a = 0, b = 1$ to $a = 1, b = 1$. However, if $a = 1, b = 1$ is followed by $a = 1, b = 0$ and $x < t_r^1$, the automaton returns to the location with $out = 0$.

When inputs change from $a = 1, b = 1$, depending on whether one or both inputs change to a 0 value, two delays are possible for this gate. If only one of the inputs changes to a 0, then the delay of the gate is t_f^1. If both inputs change to a 0, then the delay of the gate is t_f^2. As discussed earlier in this section, $t_f^1 > t_f^2$, because t_f^2 corresponds to the case where the current charging node out_b of Figure 2 is doubled. In either case, the timed automaton makes a transition to a temporary location, after setting timer x to 0. From this temporary location,

---

[1] For a 3-input NAND gate, the authors of [FK94] state that 109 transitions would be required in their gate model. In our model, only four distinct transitions are required.

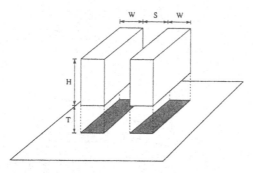

Fig. 4. Two Wires on a VLSI Integrated Circuit

it makes the final transition to the location corresponding to *out* = 0, after the appropriate delay. In case both inputs do not *simultaneously* change to a 0 value, the timed automaton makes a transition to the temporary location corresponding to one input change, from which it makes a transition to the temporary location corresponding to two input changes, if the second input arrives before t_f^2.

Timed automata for other gates are constructed similarly. In the following, we refer to the electrical node in the circuit which is at ground potential as *gnd*, and the node which is at supply potential as *vdd*. For the transistor level representation of any gate, suppose there are n paths from the evaluation (i.e., output) node to *vdd*, and m paths from the evaluation node to *gnd*. In Figure 2, the evaluation node is out_b. Assuming that all paths to *vdd* and *gnd* have the same effective size of transistors, the timed automaton for the gate will have n distinct rising delays, and m distinct falling delays. In general, for a k input gate, each of the $2^k \cdot (2^k - 1)$ input sequences give rise to distinct delays at the output, as was suggested in [CL95]. However, in practice, the delays are tightly clustered around the $n + m$ distinct values we use.

For a general gate, we first determine the values of m and n. After this, we compute each distinct rise and fall delay by means of SPICE [N75], a transistor-level simulator, which gives us the exact delay taking into account the transistor level net-list of the gate. All SPICE simulations in this paper use 0.1 μm transistor models. Interconnections are assumed to be made of copper, and are assumed to correspond to a 0.1 μm fabrication process.

3.3 Modeling Cross-Talk between Wires

As the minimum feature size of VLSI fabrication processes decrease, certain electrical phenomena become significant. Figure 4 shows a graphical view of two wires on a integrated circuit. In most VLSI processes, $W = S$. As minimum feature sizes decrease, W (and S) decrease linearly, but T, the distance between wires on different metal layers decreases sub-linearly. As a result, the ratio of the capacitance between a wire and its neighboring wire to the capacitance between a wire and wires on other metal layers increases with diminishing feature sizes [NTRS97].

One of the effects of this increased capacitance to neighboring wires is a large variation in the delay of the wire. If the neighboring wires switch in the same direction as the wire of interest, the delay of the wire is decreased, and if they switch in the opposite direction, its delay increases. The effect of the transition

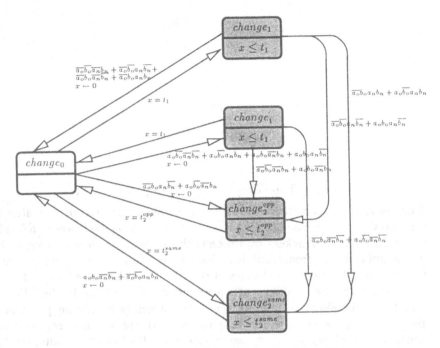

Fig. 5. Timed Automaton Model for Delay of a Pair of Wires

activity of a wire on its neighboring wires is referred to as *cross-talk*. A SPICE simulation of three neighboring wires of length 150 μm in a 0.1 μm process shows a 2:1 variation in the delay of the center wire due to cross-talk. In the set-up for this experiment, the lines were driven by inverters, whose P devices were 4.5 times larger than a minimum P device, and whose N devices were 3 times larger than a minimum N device. The parasitics for this configuration of wires were determined using a 3-dimensional parasitic extractor called SPACE [SPC]. VLSI processes with a 0.1 μm feature size are still a few years away from production, but the above simulation indicates that delay variation effects due to cross-talk will be a major problem in the future. For this reason, it is becoming increasingly important for circuit timing analyzers to incorporate the effect of cross-talk.

We model the effect of cross-talk between two wires by the timed automaton shown in Figure 5. In this figure, the location $change_0$ represents the condition when both wires have a stable value. From this location, if one of the wires switches, a transition is made to temporary location $change_1$ or $change_1'$. While in these temporary locations, if the other wire switches, an transition is made to a location which models both wires switching. There are two such locations, $change_2^{same}$, which models the two wires switching in the same direction, and $change_2^{opp}$ which models the two wires switching in opposite directions. The delay associated with $change_1$ and $change_1'$ is t_1. The delays associated with $change_2^{same}$ and $change_2^{opp}$ are t_2^{same} and t_2^{opp} respectively. These delays are determined by running SPICE on the physical configuration of wires encountered. In general, $t_2^{same} \leq t_1 \leq t_2^{opp}$.

4 Delay Computation with Timed Automata

We pose the delay computation problem as follows: Given a combinational circuit, described as an interconnection of circuit components, and a set of primary input waveforms, we want to determine the latest time that primary outputs become stable. The set of input waveforms is represented by a timed automaton I as described in section 3.1. The circuit is described as an interconnection of timed automata, $G_1, ..., G_n$. The delay parameters of these are determined by SPICE simulation of the transistor-level circuit description, as described in the example in Figure 2. The delay computation problem is then formally stated in the following manner: Let

$$F = (I \parallel G_1 \parallel G_2 \parallel ... \parallel G_n)$$

represent the evolution of the circuit over time for the primary input waveforms described by I. Let us denote by $G_{o_1}, ..., G_{o_m}$ the circuit components whose outputs are primary outputs of the circuit. For each j, define E_{o_j} to be the edges $\langle s_{o_j}, t_{o_j}, \varphi, \chi, R \rangle$ of G_{o_j} for which the primary outputs at s_{o_j} and t_{o_j} are different. The delay of the circuit is the latest time that edge in some E_{o_j} can be traversed in F. Denote by E_{switch} the set of edges of F whose projection onto some G_{o_j} lies in E_{o_j}, and let $S_{switch} \subseteq \Sigma_F$ be the set of states of F that have an outgoing E_{switch} transition. The goal is then to compute the largest time elapsed on paths from initial states of F to S_{switch}. The most straightforward way to obtain this information is to traverse the state-space of F. However, the size of this state-space is often very large. Automata networks describing circuits have certain special properties, which enable significant improvements to the traversal techniques. The rest of this section elaborates on these properties and how they are used for making state-space exploration more efficient.

4.1 A Region Automaton with Integer Delays

Traversal techniques for the dense state-spaces of timed automata can be broadly classified into two categories. The first class of methods use a set of inequalities to represent a convex subset of the timer space, and state sets are represented by pairing locations with sets of such convex subsets (See, [LPW97], for instance). These methods are not suitable for our purposes because they represent locations explicitly, and the boolean component (therefore the number of locations) of the automaton F is often considerably large. Our delay computation technique is based on the second category of methods, called "region automaton"-based methods [AD94]. The key feature of these methods is the division of the state-space into a finite number of "regions", each of which is one equivalence class of a relation ("region equivalence") defined on Σ_F. Each region makes up one state of a finite automaton, called the region automaton, and a transition relation is defined on the regions such that the essential information about F is captured. This finite automaton can then be analyzed using OBDD-based methods. In many cases this enables one to handle large state spaces.

Region equivalence for timed automata as defined in [AD94] has to distinguish between clock valuations which have different orderings of the fractional parts of clock values. This contributes a factor of k to the state-space, where k

is the total number of timers used in F. For a subclass of timed automata, this component of the state space can be eliminated by using the following fact proven in [HMP92]: if a timed automaton uses no strict inequalities in clock predicates, for each run of the automaton, there exists a run that makes transitions only at integer time points, and goes through the same sequence of locations. In fact, such a run can be obtained as follows: Let $0 \leq \epsilon \leq 1$ be arbitrary. For a transition γ, let t_γ denote the time it is taken. If the fractional part of t_γ is $\leq \epsilon$, round t_γ to the nearest integer from below, otherwise, round it up the nearest integer from above. The run obtained in this manner is also a run of the automaton. It follows that, for delay computation purposes, timers can be treated as integer valued variables which increase at the same rate.

In our context, timer predicates refer to the time elapsed between two transitions in node voltages, which are analog waveforms over real-valued time. It does not make any physical sense to specify that the time elapsed between two analog transitions can be any value less than (or greater than) but not equal to c time units. The set of possible elapsed times is simply rounded up to the nearest closed interval with integer end-points. Therefore, combinational circuits can be modeled without strict inequalities and the above integer interpretation for timers can be used. Then, a run $\sigma_0 \xrightarrow{\gamma_0} \sigma_1 \xrightarrow{\gamma_1} \sigma_2 \xrightarrow{\gamma_2} ... \xrightarrow{\gamma_{k-1}} \sigma_k$ can be viewed as an interleaving of *time passage transitions* and *control transitions* as follows: $\sigma_0 \xrightarrow{\delta_0} (\sigma_0 + \delta_0) \xrightarrow{\langle \xi_0, \xi'_0 \rangle} \sigma_1 \xrightarrow{\delta_1} (\sigma_1 + \delta_1) \xrightarrow{\langle \xi_1, \xi'_1 \rangle} ... \xrightarrow{\langle \xi_{k-1}, \xi'_{k-1} \rangle} \sigma_k$, where $\sigma_i + \delta_i$ is shorthand for all timers in σ_i being incremented by $\delta_i \in \mathbb{N}$. Note that time passage transitions with $\delta_i \geq 1$ can be a realized by a sequence of $\delta = 1$ transitions. We have found that taking time steps of one at a time results in more efficient ordered binary decision diagram (OBDD)-based analysis algorithms and makes the formulation of the delay problem easier, as will be seen in Section 4. Clearly, the reachability and delay properties of the automaton remain the same.

With these observations, the timed automaton takes the form of a finite-state machine with states of the form $\sigma = \langle s, \nu \rangle$, where ν is an integer-valued clock evaluation. The transition relation T_A of a timed automaton A is given as $T_A^\delta \cup T_A^c$ where

$$T_A^\delta = \{ \langle \sigma, \xi, \sigma + \delta, \xi \rangle | \ \sigma \in \Sigma_A, \delta = 1, \text{ and } wait(\sigma, \delta) \text{ and } \xi \text{ is an observation} \}$$

represents *the time passage transitions*, where all variables except timers remain constant, and

$$T_A^c = \{ \langle \sigma, \xi, \sigma', \xi' \rangle | \ \sigma, \sigma' \in \Sigma_A, \ \sigma' \text{ is obtained from } \sigma \text{ by the traversal of some edge } e \in E \text{ on observation event } \langle \xi, \xi' \rangle. \}$$

represents the *control transitions*, where the location of the automaton changes. In the rest of the paper, the timed automata will be identified with this discrete transition structure. Our delay computation algorithm is essentially a breadth-first search of this discrete state-space using OBDDs to represent state sets and transition relations.

4.2 The Region Automaton is Acyclic

The transition structure of a timed automaton representing a combinational circuit is acyclic for the following reason: The primary input waveforms that we consider in delay analysis remain constant after a certain point in time (Section 3.1). Combinational circuits are designed to stabilize to a certain final state after the inputs become constant. If F had a cycle, it could be traversed an unbounded number of times, which would point to an instability in the circuit.[2] A cycle in the transition structure of F corresponds to a cycle in each component of F. Observe that all such cycles involve a change in some signal in the circuit. Therefore, such a cycle in F points to oscillations in some circuit nodes. For a correctly designed combinational circuit, this sort of behavior should not occur[3].

While exploring a large state transition graph, the bottleneck is often the size of the representation for the set of explored states. However, if an acyclic state transition graph is traversed in breadth-first manner starting with the set of initial states, one does not need to store all of the traversed states: Denoting by $S^{(k)}$ the set of states that can be reached from the initial states by traversing exactly k edges, it suffices to store $S^{(k)}$ at the kth step of the traversal. This allows a (heuristic) memory-time trade-off: Typically, memory is saved by storing a smaller set, but a state may be visited more than once, which results in re-computation. For an arbitrary system, the amount of re-computation could be prohibitively large. However, the delay of a combinational circuit is bounded by the delay of the longest topological path, which places a polynomial bound on k.

In practice, we found that this approach results in significant savings in memory (See Section 5). Our technique, like many OBDD based methods, is memory limited, and by representing $S^{(k)}$ only, we were able to handle circuits that we could not have handled otherwise.

4.3 The Algorithm

For uniformity of notation, let us rename the automata comprising F to $C_1, ..., C_m$, such that $F = (C_1 \parallel ... \parallel C_m)$. Let $T_1^\delta \cup T_1^c, T_2^\delta \cup T_2^c, \ldots, T_m^\delta \cup T_m^c$ be the transition relations of $C_1, \ldots, C_m$. Recall that the outputs o_j of each timed automaton C_j are a deterministic function of its location given by $o_j = \mu_j(s_j)$. Therefore, for the purposes of traversing the state-space of F, it is possible to express all transition relations in terms of the state variables of the C_j's. Then the transition relation of component j can be expressed solely in terms of state variables in the form $T_j(\sigma_j, \sigma_{\to j}, \sigma_j', \sigma_{\to j}')$. Here $\sigma_{\to j}$ denotes the set of state variables of fan-ins of C_j.

With these, the structure of our delay computation algorithm is described in Figure 6

---

[2] Note that F incorporates automata that generate the primary input waveforms, and is thus a closed system. Therefore, it is possible for every path in F to be traversed.

[3] If the circuit does have unstable behavior, the algorithm that will be described in Section 4.3 will not converge. However, note that the delay of the circuit is bounded from above by the delay of the longest topological path. Using this fact, we can limit the number of iterations, and also determine whether the circuit stabilizes.

$$k \leftarrow 0, \ t \leftarrow 0, \ S^{(0)} = S_1^{(0)} \wedge S_1^{(0)} \wedge ... \wedge S_m^{(0)}$$

repeat

    repeat

        $S^{(k+1)} = \text{ONESTEP}(S^{(k)}, \langle T_1^C, ..., T_m^C \rangle)$

        $k \leftarrow k + 1$

    while $S^{(k)}(\sigma) \neq S^{(k-1)}(\sigma)$

    $S^{(k+1)} = \text{ONESTEP}(S^{(k)}, \langle T_1^\delta, ..., T_m^\delta \rangle)$

    $k \leftarrow k + 1$

    $t \leftarrow t + 1$

while $S^{(k)}(\sigma) \neq S^{(k-1)}(\sigma)$

Fig. 6. Algorithm COMPUTEDELAY

where ONESTEP (described in Figure 7) computes the states reached from $S^{(k)}$ by traversing one edge in the transition relation given by $T = T_1 \wedge T_2 \wedge ... \wedge T_m$

$$\text{ONESTEP}(S^{(k)}, \langle T_1, ..., T_m \rangle)$$

$$S^{(k,k+1)}(\sigma, \sigma') = S^{(k)} \wedge \bigwedge_{i=1}^{m} T_i$$
$$S^{(k+1)}(\sigma') = (\exists \sigma) S^{(k,k+1)}(\sigma, \sigma')$$

Fig. 7. Algorithm ONESTEP

The relation $S^{(k,k+1)}$ represents the outgoing transitions from $S^{(k)}(\sigma)$ as a function of σ where the σ and σ' are variables corresponding to the present and next states, respectively. The algorithm repeats the following loop until the circuit stabilizes: First, control transitions are explored until all states reachable through them are computed, then a time increment of 1 is taken. The delay of the circuit is then the maximum t such that $S^{(k,k+1)}$ includes an edge for which there is a change in a primary output[4]. Note that minimum delay computation, as well as any timed safety property check can easily be incorporated into this scheme.

The practical limitation in applying the algorithm above is the size of the OBDD for $S^{(k)}$. In the next section we present a method for efficiently computing and representing $S^{(k)}$'s.

4.4 A Conjunctively Decomposed Representation

$S^{(k)}(\sigma)$ typically has a large number of variables in its support. In order to get compact (monolithic) OBDD representations, it is necessary to choose a good order of the variables. Heuristically, variables that are strongly correlated must be close to each other in the order, and variables encoding correlated integers must be interleaved. For the problem at hand, it is in general not possible to find a total order that satisfies these constraints, especially because all variables corresponding to timers are correlated with each other [BMPY97]. To deal with this difficulty, we employ a conjunctively decomposed representation for the state sets $S^{(k)}$. The decomposition corresponds to a "slicing" of the circuit in the following manner. The circuit is partitioned into slices $SL_1, ..., SL_p$, where each slice SL_i consists of a set of circuit elements (C_i's). The slices cover the circuit and no C_i belongs to more than one slice. The SL's are ordered topologically, i.e., if $a < b$ no C_i's in SL_b can be in the transitive fan-in of a C_j in SL_a. (See Figure 9 for an example.).

---

[4] It is a simple OBDD operation to check whether $S^{(k,k+1)}$ contains such an edge.

$S^{(k)}$ is then represented by a collection of relations, each corresponding to a slice. We construct $S_1^{(k)}(\sigma_0)$, $S_1^{(k)}(\sigma_0, \sigma_1)$, $S_2^{(k)}(\sigma_1, \sigma_2)$, $\ldots$, $S_p^{(k)}(\sigma_{p-1}, \sigma_p)$ such that each $S_j^{(k)}$ specifies the set of states of SL_j at step k of the algorithm. Note that each $S_j^{(k)}$ has σ_{j-1} in its support in addition to σ_j. Intuitively, $S_j^{(k)}$ specifies what assignments to σ_j are part of $S^{(k)}$ for each assignment to σ_{j-1}, i.e., the correlation between the state variables of adjacent slices is captured. Since each $S_j^{(k)}$ has much fewer variables in its support, the total size of the $S_j^{(k)}$'s is in general much smaller than a monolithic representation for $S^{(k)}$.

For $j = 1$ to p

$$S_j^{(k,k+1)}(\sigma_{j-1}, \sigma_j, \sigma'_{j-1}, \sigma'_j) = \left((\exists \sigma_{j-2}, \sigma'_{j-2}) \, S_{j-1}^{(k,k+1)} \right) \wedge S_j^{(k)} \wedge \bigwedge_{C_i \in SL_j} T_i$$

$$S_j^{(k+1)}(\sigma'_{j-1}, \sigma'_j) = (\exists \sigma_{j-1}, \sigma_j) S_j^{(k,k+1)}$$

Fig. 8. Algorithm ONESTEP'

The modified algorithm described in Figure 8 performs one step of reachability computation operating on one slice at a time. For slice j, slice $j - 1$ provides the inputs, and since the correlation between the states is stored by $S_j^{(k)}$, for each state, the corresponding inputs are supplied. Ideally, for such a decomposed representation, one would like the following equalities to hold: (i) $S^{(k)} = \bigwedge_{j=1}^p S_j^{(k)}$, and (ii) $S_j^{(k)} = (\exists \tau_j) S^{(k)}$, where τ_j is the set of variables not in the support of $S^{(k)}$. This would be the case if, at the kth iteration of the algorithm, the state variables of slice k were only correlated with those of slice $k - 1$. However, for an arbitrary slicing and arbitrary circuit components, state variables in non-adjacent slices may be correlated. For algorithm ONESTEP', it can be proven by induction that $\bigwedge_{j=1}^p S_j^{(k)}$ is a superset of $S^{(k)}$, and thus the slicing based delay computation method is conservative.

In practice, we have not found any cases where the results of the slicing method is different from the results of the monolithic but exact method. We offer the following intuitive explanation for this fact: Combinational circuit elements have bounded memory, and thus, their states are only strongly correlated with the states of other elements close-by. Let us call a circuit component "active" if it is not in a stable state. Suppose that the circuit is sliced in such a way that the topological delay through each slice is roughly equal. Then, at any given time, the active elements are contained in a portion of the circuit consisting of two adjacent slices. Our algorithm performs traversal by sweeping the circuit using a window consisting of two slices. If transitions at any point in time are confined to two adjacent slices, then next-state computation as performed by ONESTEP' is exact. We are now working on an exact algorithm based on this intuition. The algorithm dynamically focuses only on the active portion of the circuit.

5 Experimental Results

As our benchmark circuits, we use n-bit adders built by cascading 2-bit Carry Skip Adder (CSA) blocks (Figure 9). These circuits are known to have false paths, therefore, even in the absence of cross-talk, a topological analysis overapproximates delay.

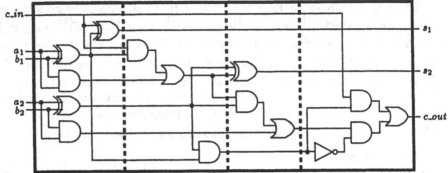

Fig. 9. Two-bit carry-skip adder block. The dashed lines indicate the boundaries between the slices.

We conducted two sets of experiments. The first set, described in Section 5.1 compares the performance and accuracy of conventional delay analysis methods, a circuit simulator and the algorithm described in this paper. The second set, described in Section 5.2 demonstrates that our algorithm for traversing the state-space scales much better than a non-specialized OBDD based method.

5.1 Comparison with conventional methods

For this experiment, we constructed a 4-bit adder using two CSAs (Figure 10). Two configurations of this circuit were used:

K_1: The c_{out} output of CSA1 and primary input A3 were neighboring wires in the circuit layout, resulting in a possible delay variation for both due to cross-talk.

K_2: No cross-talk exists between any wires.

These configurations were meant to be "proof-of-concept" circuits, meant to highlight the difference between various algorithms. For an industrial strength tool, cross-talk information needs to be extracted for the whole circuit from the layout. Here, we performed this only for the pair of wires c_{out} and A3.

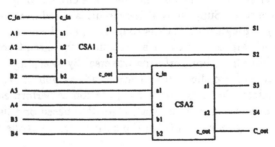

Fig. 10. A 4-bit CSA

Four different algorithms were tested on this circuit:

A_1: *Circuit Simulation:* The circuit was modeled in SPICE, and exact delays were computed by simulating the input vector pairs causing the largest delay. The run-time for exhaustive simulation of all input vector pairs was estimated by multiplying the run-time for one vector pair by the number of possible pairs. Observe that, for this circuit, it was possible to determine the input vector pairs causing worst-case behavior. This is in general not possible, and exhaustive simulation needs to be performed.

A_2: *Our Approach:* We first created timed automata for each gate using the ideas described in Section 3.2. This was done for each of the 4 gate types in the circuit of Figure 9. Similarly, a timed automaton model for cross-talk between two wires was constructed as discussed in section 3.3. The value of different delays was determined by a transistor level simulation, using SPICE. For the gate delay models, the numbers obtained were rounded to the nearest multiple of 5 ps. The maximum delay of the circuit was computed using the algorithm described section 4. Our algorithms were implemented on the verification platform MOCHA [AH+98].

A_3: *"Exact timing analysis":* This is the algorithm described in [MSBS93], implemented on the SIS platform. This algorithm does not account for cross-talk, but reports the delay of the longest true path. Although this method is not truly exact, it is referred to as "exact timing analysis" in the delay computation literature. The implementation we used did not allow input-dependent delays, so worst-case delay parameters obtained from SPICE simulation were used for gates and wires.

A_4: *Topological delay analysis:* This was done using SIS [SSL+92]. This method does not model cross-talk, and does not account for false paths. Again, worst-case delay parameters obtained from SPICE simulation were used for gates and wires.

The input vector sequence which results in the maximum delay for the 4-bit CSA also results in the c_{out} output of CSA1 switching in the opposite direction as the primary input A3. However, these signals are significantly separated in time, so there is no increase in circuit delay due to cross-talk between the wires. An algorithm which is not cross-talk-aware will assign a worst-case delay for each of the wires, and hence estimate a larger delay for configuration K_1 than for configuration K_2, whereas, in reality, the two configurations result in the same delay.

Method	With cross-talk (K_1)		Without cross-talk (K_2)	
	run-time(s)	max delay (ps)	run-time(s)	max delay (ps)
A_1	2.936×10^6	611	3.09×10^6	616
A_2	602	660	617	660
A_3	1.745	770	1.619	740
A_4	0.1	920	0.1	890

Table 1. Experimental Results

The results of these runs are described in table 5.1. SPICE models circuit behavior most closely, and we take the delay that SPICE computes as our reference. Note that the maximum delay that SPICE computes for the two configurations is almost the same, as expected.

The results obtained from A_2 are promising:

- Our method correctly determines that the delay of the circuit is identical whether configuration K_1 or K_2 is used.

- The computed delay is within 10% of the true circuit delay as computed by SPICE. [5] The other two schemes have higher estimates.
- The run-time of the timed-automata scheme is three orders of magnitude less than the estimated run-times for SPICE.

Since algorithm A_3 cannot detect that cross-talk does not actually take place in K_1, it reports a larger delay than for K_2. This is due to the fact that cross-talk in this case was modeled using buffers whose delays are the worst-case delay under cross-talk (t_2^{opp} in Figure 5). Also, the value of delay computed by this scheme for configuration K_2 is larger than that computed by timed automata. This is because the delay model used by exact timing analysis is a pin-delay model, and suffers from the drawback described in section 3.2.

The topological timing analysis scheme (algorithm A_4) has the lowest run-times, but gives the most inaccurate results. False paths in the circuit are not detected, and further, cross-talk is not modeled.

From these results, it is clear that the method we developed is powerful, significantly faster than transistor-level simulation, and much more accurate than other timing analysis methods.

5.2 Comparison with non-specialized traversal

To quantify the improvement brought about by our traversal heuristics, we compared their performance with a non-specialized traversal algorithm. The non-specialized algorithm made use of a partitioned representation of the transition relation while keeping the representation of $S^{(k)}$'s monolithic. The algorithms were run on a family of n-bit carry-skip adders (n-CSA) as described earlier. For this set of experiments, we used the same delay models with a coarser time discretization (1 time unit = approx. 14 ps) in order to demonstrate how the algorithm's performance scales with circuit size. The results presented in Table 2 show that our algorithm scales remarkably well. In contrast, the non-specialized algorithm ran out of space (1GB) for the 4-bit CSA and was not able to complete even with the help of dynamic variable reordering heuristics. Observe that, with the help of our heuristics, we were able to handle models with thousands of OBDD variables.

6 Conclusions and Future Work

We have addressed delay problems due to decreasing minimum feature sizes of VLSI circuits. The need for more accurate timing analysis methods, and also for cross-talk-aware timing analysis was fulfilled by: (i) new gate delay models, and (ii) an accurate, timed automaton-based analysis scheme. The advantages of our approach are:

---

[5] This discrepancy is partly due to the fact that in the timed-automaton models for gates, delays are rounded up to the nearest multiple of 5 ps. The more important reason, however, has to do with the fact that we compute gate delays assuming a nominal loading. In the circuit, each instance of any gate drives a different load. This fact is naturally taken into account by SPICE. If we had incorporated this factor into our delay models, we would have had a larger number of models, but our results would have been closer to SPICE.

no. of CSAs	no. of timers	circuit delay (ps.)	no. of BDD vars	BDD mem.(MB)	CPU time
2	30	634	526	29	2 min
3	45	767	820	58	8 min.
4	60	900	1114	78	13 min.
5	75	1034	1408	85	21 min.
6	90	1167	1702	102*	30 min.
8	120	1440	1180	63*	15 min.
16	240	2480	2362	78*,•	2.8 hr.

Table 2. Experimental Results for n-bit CSA Adders. * denotes dynamic OBDD variable reordering. • denotes a coarser time discretization (1 time unit = approx. 35 ps.). The second column gives the number of timer variables used in modeling each circuit. The fourth and fifth column indicate the number of OBDD variables in the circuit representation and the total memory used by OBDDs.

- Circuit delay is computed much more accurately than algorithms A_3 and A_4 of Section 5.1, since the delay model we use is more realistic. This remains true even in the absence of cross-talk.
- The method models cross-talk between wires in an integrated circuit. This feature is absent from algorithms A_3 and A_4.
- Our preliminary implementation shows reasonable run-times and the run-time of our method is orders of magnitude less than exhaustive circuit simulation. We expect to achieve significant speed-ups with a better implementation

Our method cannot currently handle large circuits. For large circuits, we propose that our scheme be applied to determine the delay of a critical portion by modeling only the path and other nodes in its "electrical neighborhood". By pruning away irrelevant portions of the circuit, our algorithms may be able to analyze the critical portions. We intend to test this conjecture on industrial circuits.

Future work in terms of the computational approach will proceed in two directions: (i) An exact method that dynamically determines the set of active elements and performs computation only on that part of the circuit, and, (ii) the use of an output load dependent delay model for this scheme.

References

[AD94] R. Alur and D.L. Dill. A theory of timed automata. *Theoretical Comp. Sci.*, 126:183–235, 1994.
[AH+98] R. Alur, T.A. Henzinger, F.Y.C. Mang, S. Qadeer, S.K. Rajamani, S. Taşıran MOCHA: Modularity in Model Checking To appear in *Intl. Conf. on Computer-Aided Verification, CAV '98*
[BMPY97] M. Bozga, O. Maler, A. Pnueli, and S. Yovine. Some Progress in the Symbolic Verification of Timed Automata. In *Proceedings of the 9th Intl. Conf. on Computer-Aided Verification, CAV '97*, LNCS 1254, pages 191–201, Springer-Verlag, 1997.
[CL95] B. S. Carlson and S.-J. Lee. Delay optimization of digital CMOS VLSI circuits by transistor reordering. In *IEEE Transactions on Computer-Aided Design of Integrated Circuits and Systems*, vol.14, (no.10), pages 1183–92, Oct. 1995
[CWS97] V. Chandramouli, J. Whittemore, and K. Sakallah. AFTA: A Delay Model for Functional Timing Analysis *Proceedings of the 1997 ACM/IEEE International*

Workshop on Timing Issues in the Specification and Synthesis of Digital Systems,
pages 5–14, 1997, Austin, Texas.

[DKM93] S. Devadas, K. Keutzer and S. Malik. Computation of Floating Mode Delay
in Combinational Circuits: Theory and Algorithms In *IEEE Transactions on
Computer-Aided Design*, 12(12): 1913–1923, December 1993.

[HMP92] Thomas A. Henzinger, Zohar Manna, and Amir Pnueli. What good are
digital clocks? In *Proceedings of the 19th International Colloquium on Automata,
Languages, and Programming (ICALP 1992)*, Lecture Notes in Computer Science
623, Springer-Verlag, 1992, pp. 545-558.

[FK95] J. Frößl, and T. Kropf. Verifying Real-Time Properties of MOS-Transistor
Circuits In *European Design and Test Conference (EDTC)*, pp. 314-319, Paris,
March 1995.

[FK94] J. Frößl, and T. Kropf. A New Model to Uniformly Represent the Function and
Timing of MOS Circuits and its Application to VHDL Simulation In *European
Design and Test Conference (EDTC)*, pp. 343-348, March 1994.

[LB94] W.K.C. Lam, and Robert K. Brayton. *Timed Boolean Functions- A Unified
Formalism for Exact Timing Analysis* ISBN 0-7923-945402, Kluwer Academic
Publishers, 1994.

[LPW97] Kim G. Larsen, Paul Pettersson and Wang Yi. UPPAAL: Status & Developments In *Proceedings of the 9th International Conference on Computer-Aided
Verification, CAV '97*. Haifa, Israel, 22-25 June 1997.

[MP95] O. Maler, A. Pnueli. Timing Analysis of Asynchronous Circuits Using Timed
Automata In *ACM Intl. Workshop on Timing Issues in the Specification and
Synthesis of Digital Systems*, pages 249-257, November 1995.

[M96] K. L. McMillan. A conjunctively decomposed Boolean representation for symbolic model checking. In *Proceedings of the 8th Intl. Conf. on Computer-Aided
Verification, CAV '96*, LNCS 1102, pages 13–25, Springer-Verlag, 1996.

[MSBS93] P. McGeer, A. Saldanha, R. K. Brayton and A. L. Sangiovanni-Vincentelli.
Delay Models and Exact Timing Analysis In *Logic Synthesis and Optimization*,
pages 167–189, T. Sasao, ed., Kluwer Academic Publishers, 1993.

[NTRS97] The National Tecnology Roadmap for Semiconductors, 1997
http://notes.sematech.org/97melec.htm

[N75] L. Nagel SPICE: A Computer Program to Simulate Computer Circuits University of California, Berkeley UCB/ERL Memo M520 May 1995

[SSL+92] E. M. Sentovich, K. J. Singh, L. Lavagno, C. Moon, R. Murgai, A. Saldanha,
H. Savoj, P. R. Stephan, R. K. Brayton, A. L. Sangiovanni-Vincentelli SIS: A
System for Sequential Circuit Synthesis Electronics Research Laboratory, Univ.
of California, Berkeley, CA 94720 UCB/ERL M92/41, May 1992

[TAKB96] S. Taşıran, R. Alur, R. P. Kurshan, and R. K. Brayton. Verifying Abstractions of Timed Systems. In *Proceedings of the 7th Intl. Conf. on Concurrency
Theory, CONCUR '96*, LNCS 1119, pages 546–562, Springer-Verlag, 1996.

[SPC] Physical Design Modeling and Verification Project (SPACE Project)..
http://cas.et.tudelft.nl/research/space.html

[TB97] S. Taşıran and R. K. Brayton. STARI: A Case Study in Compositional and Hierarchical Timing Verification. In *Proceedings of the 9th Intl. Conf. on Computer-
Aided Verification, CAV '97*, LNCS 1254, pages 191-201, Springer-Verlag, 1997.

[TKB97] S. Taşıran, Y. Kukimoto and R. K. Brayton. Computing Delay with Coupling
Using Timed Automata. In *Proceedings of the 1997 ACM/IEEE International
Workshop on Timing Issues in the Specification and Synthesis of Digital Systems*,
pages 232-244, 1997, Austin, Texas.

[YH95] H. Yalcin and John P. Hayes. Hierarchical Timing Analysis Using Conditional
Delays. In *Proc. IEEE/ACM Int'l Conf. on Computer-Aided Design, ICCAD
'95*, pages 371-377, 1995

Maximum Time Separation of Events in Cyclic Systems with Linear and Latest Timing Constraints

Fen Jin[1,*], Henrik Hulgaard[2,**], and Eduard Cerny[1,*]

[1] LASSO, Départment d'Informatique et de Recherche Opérationelle,
Université de Montréal
[2] Department of Information Technology,
Technical University of Denmark

Abstract. The determination of the maximum time separations of events is important in the design, synthesis, and verification of digital systems, especially in interface timing verification. Many researchers have explored solutions to the problem with various restrictions: a) on the type of constraints, and b) on whether the events in the specification are allowed to occur repeatedly. When the events can occur only once, the problem is well solved. There are fewer concrete results for systems where the events can occur repeatedly. We extend the work by Hulgaard et al. for computing the maximum separation of events in cyclic constraint graphs with latest constraints to constraint graphs with both linear and latest constraints.

1 Introduction

The determination of the maximum time separations of events is important in the design, synthesis, and verification of digital systems, especially in interface timing verification. Many researchers have explored solutions to the problem using various restrictions [7, 11, 13, 15–17]: a) on the type of constraints, and b) on whether the events in the specification are allowed to occur repeatedly. When the events can occur only once, the problem is well solved in any combination of timing constraints. The complexity of the algorithms ranges from polynomial for the linear only systems to NP-complete for systems with linear, earliest, and latest constraints (see [7] for a survey). There are fewer concrete results for systems where the events can occur repeatedly (an infinite number of times). To the best of our knowledge, the only work in this field is by Hulgaard et al. [9] for computing the maximum separation of events in cyclic constraint graphs with latest constraints. This algorithm was later extended to systems containing both concurrency and conditional behavior [8]. Here we instead consider more general constraint graphs containing both linear and latest constraints.

* Partially supported by NSERC Canada Grant No. OGP0003879. Experiments were executed on a workstation on loan from the Canadian Microelectronics Corp.
** Financially supported by the Danish Technical Research Council.

The calculation of the maximum time separations between events in cyclic systems can be viewed as finding an infinite number of time separations in acyclic constraint graphs obtained by unfolding the cyclic specification. In [9], the authors solve this problem of determining the maximum over the infinite number of time separations as a matrix closure calculation in an algebra which was specifically developed for this problem. In this paper, we first present an algorithm for computing time separations between events in finite acyclic graphs with linear and latest constraints. Linear combinators bring some difficulties for the computation in cyclic systems: there is not necessarily an order in occurrence times of events related by linear constraints; the occurrence times of events earlier in the unfolding may be determined by those in later unfoldings. In this paper we remove this problem by restricting the constraint graphs in such a way that events can be partitioned into a set of topologically ordered blocks that respect certain properties of causality (see [5] for finite acyclic behaviors). With this restriction we can determine an upper bound on the number of unfoldings one has to consider in order to compute the maximum separation in a cyclic timing constraint graph and thus obtain a practical and exact algorithm.

2 Modeling Temporal Behavior

The temporal behavior of a system is specified using a *timing constraint graph*. Such graphs are similar to event-rule systems [2, 9, 14], but allow both latest and linear constraints. This section defines timing constraint graphs and the maximum separation problem.

2.1 Timing Constraint Graphs

A timing constraint graph is a finite directed graph $G = (V, E)$ whose vertices represent events (e.g., signal transitions on a wire) and whose edges represent dependencies in time between occurrences of the events. Each edge is labeled with an interval $[d, D]$ (with $d \leq D$) and an occurrence index offset, ε. Each edge must either be a *precedence edge* where $0 \leq d$ or a *concurrency edge* where $d < 0$ and $D \geq 0$. Without loss of generality we will assume that ε is either 0 or 1 for all edges in E. We use $u \overset{[d,D],\varepsilon}{\longrightarrow} v$ to denote an edge from u to v labeled with interval $[d, D]$ and occurrence index offset ε. Also, d_{uv}, D_{uv}, and ε_{uv} denotes the lower delay bound, the upper delay bound and the occurrence index offset, respectively, annotated on the edge from vertex u to v.

The timing constraint graph represents a (possibly infinite) set of constraints on the occurrence times of the events. Let v_k to denote the kth occurrence of event v and let $\tau(v_k)$ denote the occurrence time of v_k. An edge $u \overset{[d,D],\varepsilon}{\longrightarrow} v \in E$ represents the set of constraints

$$d \leq \tau(v_k) - \tau(u_{k-\varepsilon}) \leq D \qquad \text{for } k \geq \varepsilon.$$

Each vertex $v \in V$ is associated with either a linear or a latest combinator which determines how the occurrence time of v_k is determined when v has two or more incoming edges. Let preds(v) denote the set of vertices with an edge to v, i.e., preds(v) = $\{u : (u,v) \in E\}$. The two types of combinators are defined as:

Linear (also called conjunctive): If v is a linear combinator then

$$\max_{u \in \text{preds}(v)} \{\tau(u_{k-\varepsilon_{uv}}) + d_{uv}\} \leq \tau(v_k) \leq \min_{u \in \text{preds}(v)} \{\tau(u_{k-\varepsilon_{uv}}) + D_{uv}\}. \quad (1)$$

Latest (also called max): If v is a latest combinator then

$$\max_{u \in \text{preds}(v)} \{\tau(u_{k-\varepsilon_{uv}}) + d_{uv}\} \leq \tau(v_k) \leq \max_{u \in \text{preds}(v)} \{\tau(u_{k-\varepsilon_{uv}}) + D_{uv}\}. \quad (2)$$

All edges incoming to a latest vertex must be precedence edges.

A *timing assignment* τ is a function mapping a vertex $v \in V$ and an occurrence index $k \in \mathbb{N}$ to the occurrence time of v_k (a non-negative real number). A timing assignment τ is *consistent* if for all $v \in V$ and $k \in \mathbb{N}$, $\tau(v_k)$ satisfies the constraints (1) and (2).

A timing constraint graph G is called *cyclic* if there exists a cycle c in G with $\varepsilon(c) > 0$. Conversely, the timing constraint graph is called *acyclic* when all cycles c in G have $\varepsilon(c) = 0$. In an acyclic timing constraint graph, each event occurs only once.

Example 1. The timing diagram specification of a READ cycle of a microprocessor (inspired by MC68360) is shown in Fig. 1. The repeated occurrences of READ cycles on the bus is modeled with the cyclic timing constraint graph in Fig. 2.

2.2 Well-formed Specifications

We restrict our analysis to *well-formed* timing constraint graphs. To a timing constraint graph $G = (V, E)$, we associate an auxiliary directed graph $G_d = (V, E_d)$[1]. For each edge $u \xrightarrow{[d,D],\varepsilon} v \in E$, where v is a latest combinator, there is an edge $v \xrightarrow{-d,-\varepsilon} u \in E_d$. For each edge $u \xrightarrow{[d,D],\varepsilon} v \in E$, where v is a linear combinator, there are two edges in E_d: $u \xrightarrow{D,\varepsilon} v$ (if $D \neq \infty$) and $v \xrightarrow{-d,-\varepsilon} u$ (if $d \neq -\infty$). Let $w(c)$ and $\varepsilon(c)$ denote the sum of delay values (either $-d$ or D) and ε-values, respectively, associated with the edges of a cycle c in G_d. A cycle c is called a *latest cycle* if all vertices on c are latest combinators. A timing constraint graph $G = (V, E)$ is well-formed if the following conditions are all met:

1. For all latest cycles c in G, $\varepsilon(c) > 0$.
2. For all cycles c in G_d, if c is not a latest cycles, then $w(c) \geq 0$ if $\varepsilon(c) = 0$ and $w(c) \cdot \varepsilon(c) > 0$ if $\varepsilon(c) \neq 0$.
3. Let u and v be two vertices on cycles in G_d, then u and v belong to the same strongly connected component (i.e., G_d only contains a single non-trivial strongly connected component).

---

[1] G_d is the dual of the compulsory graph used in [17] for computing time separations in acyclic graphs.

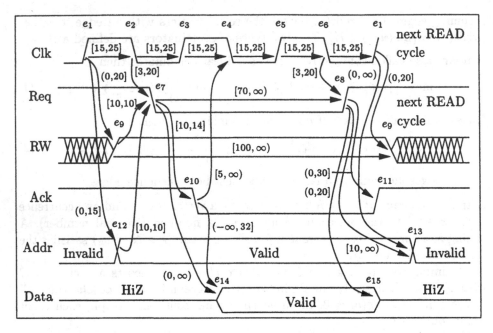

Fig. 1. Timing diagram specification of a READ cycle of a microprocessor (inspired by MC68360). All events are linear combinators except e_7 which is a latest combinator.

2.3 Maximum Separation Problem

The *maximum separation problem* is to determine how far apart in time two given events can occur. More formally, given a well-formed timing constraint graph $G = (V, E)$ and two events $s, t \in V$ (s for "source" and t for "target") and a separation in their occurrence index β, determine the maximum separation Δ,

$$\Delta = \max\{\Delta_k : k \geq \beta_0\}, \tag{3}$$

where $\beta_0 = \max(0, \beta)$. Δ_k is the maximum separation between the $(k - \beta)^{\text{th}}$ occurrence of s and the kth occurrence of t:

$$\Delta_k = \max\{\tau(t_k) - \tau(s_{k-\beta}) : \tau \text{ is a consistent timing assignment}\}. \tag{4}$$

From these definitions it follows that for all consistent timing assignments τ, for all $k \geq \beta_0$, $\tau(t_k) - \tau(s_{k-\beta}) \leq \Delta$. Furthermore, the bound is tight, i.e., there exists a consistent timing assignment τ and some $k \geq \beta_0$ such that $\tau(t_k) - \tau(s_{k-\beta}) = \Delta$. We use $\Delta(s, t)$ to denote the maximum separation between events s and t in an acyclic timing constraint graph, and $\Delta(s, t, \beta)$ to denote the maximum separation between events s and t separated by β in occurrence index in a cyclic graph.

The *minimum separation* δ between events s and t separated in occurrence index by β can be obtained from a maximum separation analysis since $\delta(s, t, \beta) = -\Delta(t, s, -\beta)$.

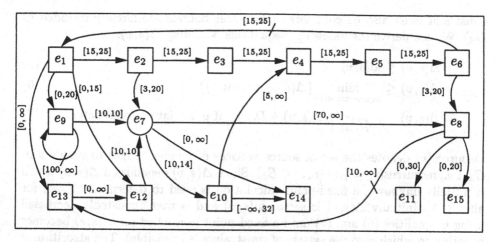

Fig. 2. Cyclic timing constraint graph corresponding to the timing diagram in Fig. 1. Vertices that are latest combinators are draw as circles while linear combinator vertices are drawn as boxes. Edges with $\varepsilon = 1$ are draw with a mark on the edge.

Example 2. In example 1, there are several analyses one may want to perform. For example, in order to ensure that there is no contention on the Ack signal, we need to ensure that the Ack has been deactivated (event e_{11}) by one device before it is activated (event e_{10}) by a possibly different device in the following READ cycle. This corresponds to ensuring that

$$\forall k \geq 0 : \tau(e_{11}, k) \leq \tau(e_{10}, k+1).$$

This property can be checked by determining the maximum separation Δ between e_{10} and e_{11} with $\beta = -1$. If $\Delta > 0$, there is a timing violation since then there exists a consistent timing assignment τ such that for some k

$$\tau(e_{11}, k) - \tau(e_{10}, k+1) = \Delta > 0 \quad \Rightarrow \quad \tau(e_{11}, k) > \tau(e_{10}, k+1).$$

To determine the maximum time separation between s and t in a cyclic timing constraint graph, one has to analyze an infinite number of acyclic constraint graphs, each determining Δ_k for some k. We start by presenting an algorithm for determining Δ_k for a particular value of k.

3 An Algorithm for Acyclic Graphs

Considering a finite acyclic timing constrain graph G, let $sd(u, v)$ denote the shortest path from vertex u to v in G_d. If there is no path from u to v in G_d, we define $sd(u, v)$ to be ∞. It is well known that if G *only* contains linear constraints (i.e., all vertices are linear combinators), the maximum separation between events s and t, $\Delta(s, t)$, is $sd(s, t)$. When the timing constraint graph also contains vertices that are latest combinators, the upper bounds of the latest constraint must also be considered (the lower bounds are already included in G_d). We can derive the following inequalities bounding $\Delta(s, v)$:

$$\Delta(s, v) \leq \text{sd}(s, v) \tag{5}$$

$$\Delta(s, v) \leq \min_{u \in \text{msources}(v)} \{\Delta(s, u) + \text{sd}(u, v)\} \tag{6}$$

$$\Delta(s, v) \leq \max_{u \in \text{preds}(v)} \{\Delta(s, u) + D_{uv}\} \quad \text{if } v \text{ is a latest combinator.} \tag{7}$$

msources(v) denotes the set of source vertices of incident edges to vertex v in G_d, i.e., msources(v) = $\{u : (u, v) \in E_d\}$. Since $\Delta(s, v)$ depends on $\Delta(s, u)$ which is initially unknown, a fixed-point calculation is used to determine $\Delta(s, v)$ for all $v \in V$. Initially $\Delta(s, v)$ is set to sd(s, v) and is then repeatedly decreased using inequalities (6) and (7) until a fixed-point is reached (or $\Delta(s, s)$ becomes negative, in which case the system of constraints is infeasible). The algorithm is shown in Fig. 3.

MaxSep($G = (V, E), s, t$) =
1: Construct G_d.
2: For all $u, v \in V$, compute sd(u, v) in G_d.
3: For all $v \in V$, let $\Delta(s, v) := $ sd(s, v).
4: Repeat
5:      For all $v \in V$
6:          update $\Delta(s, v)$ using inequalities (6) and (7).
7: until $\Delta(s, s) < 0$ or none of the $\Delta(s, v)$ values change.

Fig. 3. McMillan and Dill's algorithm [13] for determining the maximum separation between events s and t in an acyclic timing constraint graph G. If $\Delta(s, s) < 0$ in line 7, the constraint system represented by G is infeasible.

Example 3. Figure 4 shows a simple acyclic timing constrain graph G and the corresponding graph G_d. The computation for determining $\Delta(b, g)$ is:

Event x:	a	b	c	d	e	f	g	h	i
$\Delta(b, x)$, initially	0	0	3	∞	∞	∞	∞	∞	∞
$\Delta(b, x)$, iteration 1	0	0	3	10	20	29	37	25	26
$\Delta(b, x)$, iteration 2	0	0	3	10	20	25	26	25	26

In the third iteration, the fixed-point is reached with $\Delta(b, g) = 26$.

4 Restricted Acyclic Graphs

The algorithm from previous section can be extended to cyclic timing constraint graphs [10]. However, the resulting algorithm is impractical. Here we show how

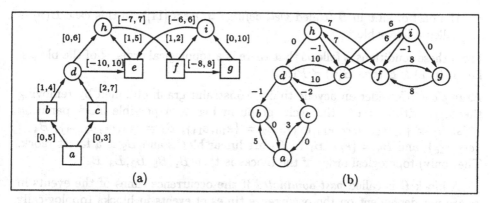

Fig. 4. (a) A simple acyclic timing constraint graph G. The circled vertices are latest combinators while the vertices in boxes are linear combinators. (b) The corresponding auxiliary graph G_d. All edges have $\varepsilon = 0$.

a more practical algorithm can be obtained by restricting the cyclic timing constraint graph. The restriction allows for a practical and exact algorithm for analyzing cyclic timing constraint graphs with both linear and latest combinators.

4.1 Block Partitions

We restrict the acyclic timing constraint graphs to those where events can be partitioned into topologically ordered blocks (see [6,11] for the case of finite constraint systems). Given an acyclic timing constraint graph $G = (V, E)$, a *block* B is simply a nonempty subset of the events, i.e., $B \subseteq V$. A *block partition* P of the graph is a partition of V, i.e., a collection of blocks such that

1. for all $B_1, B_2 \in P$, either $B_1 = B_2$ or $B_1 \cap B_2 = \emptyset$ and
2. $V = \bigcup_{B \in P} B$.

For every block B of a partition P, there are *triggers* and *local constraints*. A trigger of a block is a source event in another block whose sink is in the block, i.e., $\text{trigs}(B) = \{u : (u, v) \in E, u \notin B \text{ and } v \in B\}$. The local constraints of a block are the constraints that either relate events inside the block or relate triggers of the block to events in the block, i.e., the local constraints are $\{u \xrightarrow{[d,D]} v \in E : u, v \in \text{trigs}(B) \cup B\}$. Let $B(v)$ denote the block containing the event v.

We restrict the analysis of acyclic timing constraint graphs to those with a block partition such that:

1. For every event $v \in V$ where v is a latest combinator, $B(v)$ is a singleton, i.e., $B(v) = \{v\}$. In this case $B(v)$ is called a *latest block*.
2. For every event $v \in V$ where v is a linear combinator, all other events in $B(v)$ are also linear combinators. Furthermore, for all vertices $v \in B$ and for all triggers $u \in \text{trigs}(B(v))$, $\text{sd}(v, u) \le 0$, i.e., the triggers must be in

the past of all v in B (called *well-defined triggers* [11]). In this case $B(v)$ is called a linear block.

From these conditions it follows that there is a topological order $\prec$ of the blocks, i.e., if $(u, v) \in E$ then $B(u) \preceq B(v)$.

Example 4. Consider an acyclic timing constraint graph obtained by removing the edges with $\varepsilon = 1$ in the cyclic graph in Fig. 2. A possible block partition P is: $B_1 = \{e_1, e_2, e_3, e_9, e_{12}, e_{13}\}$, $B_2 = \{e_{10}, e_{14}\}$, $B_3 = \{e_4, e_5, e_6, e_8\}$, $B_4 = \{e_{11}, e_{15}\}$, and $B_5 = \{e_7\}$. B_1 to B_4 are linear blocks and B_5 is a latest block. The (only) topological order of the blocks is then B_1, B_5, B_2, B_3, B_4.

A block B is called *past-dominated* if the occurrence times of the events in B are not dependent on the occurrence times of events in blocks topologically after B. More formally, consider an acyclic timing constraint graph G with a block partition $B_1 \prec \cdots \prec B_n$. Let G' be the acyclic timing constraint graph induced by the vertices $V' = \bigcup_{i=1}^{k} B_i$ for some k where $1 \leq k \leq n$. The block B_k is past-dominated, if for any consistent timing assignment τ' for G', there exists a consistent timing assignment τ for G such that for all $v \in V' : \tau(v) = \tau'(v)$.

A block partition P is past-dominated if all the blocks of P are past-dominated. A timing constraint graph that has a past-dominated block partition satisfying the two conditions listed above is called a *causal* system [11].

4.2 The MaxSep Algorithm for Causal Systems

We now simplify the MAXSEP algorithm for causal acyclic timing constraint graphs. Consider an event v in a latest block and let u be a source event of some edge incoming to v. From the construction of G_d, it follows that $u \notin \mathrm{msources}(v)$. Thus, any $u \in \mathrm{msources}(v)$ are in topologically later blocks. As later blocks can not influence the occurrence time of v, (6) can not determine the value of $\Delta(s, v)$ and we can write $\Delta(s, v)$ as

$$\Delta(s, v) = \min\left(\mathrm{sd}(s, v), \max_{u \in \mathrm{preds}(v)}\{\Delta(s, u) + D_{uv}\}\right). \tag{8}$$

Consider now an event v in a linear block. Since $B(v)$ is past-dominated, the occurrence time of v is determined by the occurrence times of the trigger events and the local constraints of $B(v)$ and thus

$$\min_{u \in \mathrm{msources}(v)}\{\Delta(s, u) + \mathrm{sd}(u, v)\} = \min_{u \in \mathrm{trigs}(B(v))}\{\Delta(s, u) + \mathrm{sd}(u, v)\}.$$

For a causal system, it can be shown [4] that

$$\min_{u \in \mathrm{trigs}(B(v))}\{\Delta(s, u) + \mathrm{sd}(u, v)\} = \min_{u \in \mathrm{trigs}(B(v))}\{\Delta(s, u) + \mathrm{sdl}(u, v)\},$$

where $\mathrm{sdl}(u, v)$ is the shortest distance from u to v satisfying the local constraints of $B(v)$. Thus, when $B(v)$ is past-dominated, we can simplify (6) and write $\Delta(s, v)$ as

$$\Delta(s, v) = \min\left(\mathrm{sd}(s, v), \min_{u \in \mathrm{trigs}(B(v))}\{\Delta(s, u) + \mathrm{sdl}(u, v)\}\right). \tag{9}$$

These simplifications allow us to compute $\Delta(s, v)$ in a single iteration using the algorithm in Fig. 3 by determining $\Delta(s, v)$ in topological order of the blocks, see Fig. 5. The runtime of the simplified algorithm is $O(V^3)$ since no fixed-point iterations are needed.

$\text{MaxSep}'(G = (V, E), s, t) =$
1: Construct G_d.
2: For all $u, v \in V$, compute $\text{sd}(u, v)$.
3: Determine a past-dominated block partition P.
4: For all linear blocks $B \in P$, for all $u, v \in B \cup \text{trigs}(B)$, compute $\text{sdl}(u, v)$.
5: For $B \in P$ in topological order
6:    If B is a latest block then let $\{v\} = B$ and set
7:       $\Delta(s, v) := \min\left(\text{sd}(s, v), \max_{u \in \text{preds}(v)}\{\Delta(s, u) + D_{uv}\}\right)$
8:    else (B is a linear block)
9:       For each $v \in B$ set
10:         $\Delta(s, v) := \min\left(\text{sd}(s, v), \min_{u \in \text{trigs}(B(v))}\{\Delta(s, u) + \text{sdl}(u, v)\}\right)$

Fig. 5. Algorithm for determining the maximum separation between events s and t in a causal acyclic timing constraint graph G.

5   Algorithm for Restricted Cyclic Graphs

We now consider the problem of determining the maximum separation in a *cyclic* timing constraint graph. We restrict the analysis to causal constraint graph, i.e., to graphs where each unfolding has a past-dominated block partition (we will only consider identical block partitions in each unfolding). When the constraint graph is causal, we can determine the maximum separation using the simplified MaxSep algorithm in Fig. 5. However, the causality property (the existence of a past-dominated block partition with well-defined triggers) is checked by determining time separations between triggers and between triggers and events in the triggered blocks. Thus, there is an interleaving between checking that the system is past-dominated and the maximum separation analysis. In Sec. 5.3, we give sufficient conditions to check the past-dominated property using the maximum time separation between triggers computed under the assumption that the constraint graph satisfies these conditions.

As mentioned in Sec. 2.3, the maximum separation in a cyclic constraint graph, Δ, is determined as the maximum over Δ_k for all $k \geq \beta_0$. In causal systems, each Δ_k value can be determined from a maximum separation analysis in an *acyclic* timing constraint graph G^k, obtained by unfolding the cyclic graph specification $k + \max(0, -\beta)$ times. That is, Δ_k is the maximum separation between events t_k and $s_{k-\beta}$ in the acyclic timing constraint graph $G^k = (V^k, E^k)$ with vertices

$$V^k = \{v_j : v \in V, 0 \leq j \leq k + \max(0, -\beta)\}$$

and edges

$$E^k = \left\{ u_{j-\varepsilon} \xrightarrow{[d,D]} v_j : u_{j-\varepsilon}, v_j \in V^k \text{ and } u \xrightarrow{[d,D],\varepsilon} v \in E \right\}.$$

We can avoid explicitly analyzing Δ_k for all values of k by exploiting that the behavior of $\Delta(s_{k-\beta}, v_j)$ becomes cyclic for sufficiently large values of k. In the following, we use s (instead of $s_{k-\beta}$) to refer to the source event in G^k.

5.1 Repetitive Behavior

In G^k, the values of $sd(s, v_j)$ for $v_j \in V^k$ are eventually determined by the *maximum ratio cycles* of G_d. A maximum ratio cycle c is a simple cycle in G_d with ratio $w(c)/\varepsilon(c)$ equal to ρ defined by

$$\rho = \max \left\{ \frac{w(c)}{\varepsilon(c)} : c \text{ is a simple cycle in } G_d \text{ with } \varepsilon(c) < 0 \right\}.$$

It is well-known [1, 9, 12] that $sd(s, v_j)$ eventually becomes periodic, i.e., there exists integers k^* and ε^* such that for all $k \geq k^* + \beta_0$

$$sd(s, v_{j+\varepsilon^*}) - sd(s, v_j) = \rho\varepsilon^* \qquad \text{for } 0 \leq j \leq k - \beta - k^* - \varepsilon^*. \tag{10}$$

Example 5. Analyzing G_d for the cyclic timing constraint graph in Fig. 2, we get $\varepsilon^* = 1$, $k^* = 3$, and $\rho = 100$.

In the following we prove that in causal constraint graph, $\Delta(s, v_j)$ exhibit a similar cyclic behavior. This result is key in obtaining an algorithm for computing $\Delta(s, t, \beta)$ in cyclic timing constraint graphs.

Theorem 1. *Let $G = (V, E)$ be a cyclic causal timing constraint graph. Consider determining the maximum time separation from vertex $s_{k-\beta}$ in the graph G^k, where $k \geq k^* + \beta_0$. There exists a finite constant F such that*

$$\Delta(s, v_{j+\varepsilon^*}) - \Delta(s, v_j) = \rho\varepsilon^* \qquad \text{for } F \leq j \leq k - \beta - k^* - \varepsilon^*.$$

To prove this theorem, we need several properties of $\Delta(s, v_j)$ in G^k. In the following, we consider a cyclic causal timing constraint graph $G = (V, E)$. We will assume that E contains a vertex *root* with no incoming constraints. We will analyze the values $\Delta(s, v_j)$ in the graph G^k, where $k \geq k^* + \beta_0$.

A key notion is that of a *shortest path* from s to *root* in G_d^k. A vertex v_j is on a shortest path from s to *root* if and only if $sd(s, root) = sd(s, v_j) + sd(v_j, root)$. First, we show that if an event v_j is on a shortest path from s to *root*, then $\Delta(s, v_j) = sd(s, v_j)$:

Lemma 1. *If a vertex $v_j \in V^k$ is on a shortest path from s to root, then*

$$\Delta(s, v_j) = sd(s, v_j).$$

Proof. Let $\langle v^0, v^1, \ldots, v^n \rangle$ be a path in G^k with $v^0 = root$ and $v^n = s$ such that $\langle v^n, v^{n-1}, \ldots, v^0 \rangle$ is a shortest path from s to $root$ in G_d^k. We prove that $\Delta(s, v^i) = \mathrm{sd}(s, v^i)$ by induction on i. In the base case $i = 0$ and $\Delta(s, root) = \mathrm{sd}(s, root)$ follows from (8). In the inductive step, consider v^i for $i > 0$. From (6) it follows that $\Delta(s, v^{i-1}) \leq \Delta(s, v^i) + \mathrm{sd}(v^i, v^{i-1})$. From the inductive hypothesis we have that $\Delta(s, v^{i-1}) = \mathrm{sd}(s, v^{i-1})$ and thus

$$\mathrm{sd}(s, v^{i-1}) \leq \Delta(s, v^i) + \mathrm{sd}(v^i, v^{i-1}). \tag{11}$$

Since both v^{i-1} and v^i are on a shortest path from s to $root$, v^i must be on a shortest path from s to v^{i-1}, i.e., $\mathrm{sd}(s, v^{i-1}) = \mathrm{sd}(s, v^i) + \mathrm{sd}(v^i, v^{i-1})$ (subpaths of shortest paths are shortest paths). From (11) we get that $\mathrm{sd}(s, v^i) \leq \Delta(s, v^i)$ which combined with (5) yields that $\mathrm{sd}(s, v^i) = \Delta(s, v^i)$. □

Thus, for all vertices v_j on a shortest path from s to $root$ we have that

$$\Delta(s, v_{j+\varepsilon^*}) - \Delta(s, v_j) = \mathrm{sd}(s, v_{j+\varepsilon^*}) - \mathrm{sd}(s, v_j).$$

For such vertices, Theorem 1 immediately follows from (10). If $\Delta(s, v_j)$ is not equal to $\mathrm{sd}(s, v_j)$, then $\Delta(s, v_j)$ must be determined by the upper bound of some latest constraints. For a causal cyclic timing constraint graph $G = (V, E)$, define a cyclic graph $G_D = (V, E_D)$ as follows (G_D is similar to G_d): For each edge $u \xrightarrow{[d,D],\varepsilon} v \in E$, where v is a latest combinator, there is an edge $v \xrightarrow{D,\varepsilon} u \in E_D$. For each edge $u \xrightarrow{[d,D],\varepsilon} v \in E$, where v is a linear combinator, there are two edges in E_d: $u \xrightarrow{\mathrm{sdl}(u,v),\varepsilon} v$ (if $\mathrm{sdl}(u,v) \neq \infty$) and $v \xrightarrow{\mathrm{sdl}(v,u),-\varepsilon} u$ (if $\mathrm{sdl}(v,u) \neq -\infty$). Let $D(c)$ denote the sum of delay values (either D or $\mathrm{sdl}(\cdot, \cdot)$) associated with the edges of a cycle c in G_D.

A cycle c in G_D is said to be *constraining* v_j if $\Delta(s, v_j) < \mathrm{sd}(s, v_j)$ and $\Delta(s, v_j) = \Delta(s, v_{j-\varepsilon(c)}) + D(c)$. In order to bound F in Theorem 1, we need to argue about $\varepsilon(c)$ of constraining cycles. We first show that the weight of a constraining cycle c must be strictly less than $\rho\varepsilon(c)$.

Lemma 2. *Let c be a constraining cycle for $v_j \in V^k$ with $j \leq k - \beta - k^*$. Then*

$$D(c) < \rho\varepsilon(c). \tag{12}$$

Proof. (We will assume that $\varepsilon^* = 1$.) Let u_i be a vertex on a shortest path from s to $root$ such that $B(v_{j-\varepsilon(c)}) \prec B(u_i) \prec B(v_j)$ (such a vertex always exists). As $j \leq k - \beta - k^*$, $u_{i-\varepsilon(c)}$ is also on a shortest path from s to $root$.

$$\Delta(s, v_j) < \mathrm{sd}(s, v_j) \qquad\qquad c \text{ is constraining } v_j.$$
$$\leq \mathrm{sd}(s, u_i) + \mathrm{sd}(u_i, v_j) \qquad\qquad \text{Property of shortest paths.}$$
$$= \mathrm{sd}(s, u_{i-\varepsilon(c)}) + \rho\varepsilon(c) + \mathrm{sd}(u_i, v_j) \qquad \text{From (10).} \tag{13}$$

In a causal constraint graph, $\Delta(s, v_{j-\varepsilon(c)}) \geq \Delta(s, u_{i-\varepsilon(c)}) + \mathrm{sd}(u_{i-\varepsilon(c)}, v_{j-\varepsilon(c)})$, and $\Delta(s, u_{i-\varepsilon(c)}) = \mathrm{sd}(s, u_{i-\varepsilon(c)})$ since $u_{i-\varepsilon(c)}$ is on a shortest path from s to $root$. Combining these two facts, we obtain

$$-\Delta(s, v_{j-\varepsilon(c)}) \leq -\mathrm{sd}(s, u_{i-\varepsilon(c)}) - \mathrm{sd}(u_{i-\varepsilon(c)}, v_{j-\varepsilon(c)}). \tag{14}$$

Adding (13) and (14), it follows that

$$D(c) = \Delta(s, v_j) - \Delta(s, v_{j-\varepsilon(c)}) < \rho\varepsilon(c).$$

since $\mathrm{sd}(u_{i-\varepsilon(c)}, v_{j-\varepsilon(c)}) = \mathrm{sd}(u_i, v_j)$. $\qquad\square$

Lemma 3. *Consider two vertices, u_i and v_j of V^k, such that $j \leq k - \beta - k^*$, u_i has a path to v_j in G^k, and u_i is on a shortest path from s to root. Then there exists a constant $C(u, v, j - i)$ such that*

$$\Delta(s, v_j) - \Delta(s, u_i) \geq C(u, v, j - i). \tag{15}$$

Proof. Let $\langle v^0, v^1, \ldots, v^n \rangle$ be a path from u_i to v_j in G^k, i.e., $v^0 = u_i$ and $v^n = v_j$. We prove that $\Delta(s, v^l) - \Delta(s, v^{l-1}) \geq C^l$ for $1 \leq l \leq n$. The lemma follows by setting $C(u, v, j - i) = \sum_{l=1}^n C^l$. If v^l is a latest vertex, we have from (8) that

$$\Delta(s, v^l) \geq \min\big(\mathrm{sd}(s, v^l), \Delta(s, v^{l-1}) + D_{v^{l-1}v^l}\big).$$

Subtracting $\Delta(s, v^{l-1})$ on both sides, one obtains

$$\Delta(s, v^l) - \Delta(s, v^{l-1}) \geq \min\big(\mathrm{sd}(s, v^l) - \Delta(s, v^{l-1}), D_{v^{l-1}v^l}\big).$$

As $\Delta(s, v^{l-1}) \leq \mathrm{sd}(s, v^{l-1})$ and $\mathrm{sd}(s, v^{l-1}) \leq \mathrm{sd}(s, v^l) + \mathrm{sd}(v^l, v^{l-1})$ it follows that

$$\Delta(s, v^l) - \Delta(s, v^{l-1}) \geq \min\big(-\mathrm{sd}(v^l, v^{l-1}), D_{v^{l-1}v^l}\big) = C^l.$$

If v^l is a linear vertex, we have from (6) that

$$\Delta(s, v^{l-1}) \leq \Delta(s, v^l) + \mathrm{sd}(v^l, v^{l-1})$$

since $v^l \in \mathrm{msources}(v^{l-1})$ and thus

$$\Delta(s, v^l) - \Delta(s, v^{l-1}) \geq -\mathrm{sd}(v^l, v^{l-1}) = C^l.$$

Notice that $\mathrm{sd}(v^l, v^{l-1})$ is a constant, independent on i and j since $i, j \leq k - k^* - \beta$. $\qquad\square$

This lemma allows us to bound the number of unfoldings a constraining cycle can determine the maximum separation.

Lemma 4. *Consider two vertices, u_i and v_j of V^k, such that $j \leq k - \beta - k^*$, u_i has a path to v_j in G^k, and u_i is on a shortest path from s to root. If there is a constraining cycle c in G_D between $v_{j-f\varepsilon(c)}$ and v_j, then*

$$f \leq \frac{\Delta(s, v_{j-f\varepsilon(c)}) - \Delta(s, u_{i-f\varepsilon(c)}) - C(u, v, j - i)}{\varepsilon(c)\rho - D(c)}. \tag{16}$$

Proof. (We will assume $\varepsilon^* = 1$.) From Lemma 3 we know that

$$\Delta(s, v_j) - \Delta(s, u_i) \geq C(u, v, j - i).$$

Since u_i is on a shortest path from s to *root*, from Lemma 1 and (10), it follows that

$$\Delta(s, v_j) \geq \Delta(s, u_{i-f\varepsilon(c)}) + f\varepsilon(c)\rho + C(u, v, j - i). \tag{17}$$

From the definition of a constraining cycle, we have

$$\Delta(s, v_j) = \Delta(s, v_{j-f\varepsilon(c)}) + fD(c),$$

which combined with (17) results in the bound on f as given in (16). □

Thus, eventually the maximum separation $\Delta(s, v_j)$ is determined by those events that are on a shortest path from s to *root*, i.e., $\Delta(s, v_j)$ is determined locally by $\Delta(s, u_i)$ and the constraints from u_i to v_j. Let u_i be an event on a shortest path from s to *root* such that there is a path from u_i to v_j in G^k and define $\Delta'(s, v_j)$ to be the maximum separation determined in the subgraph of G^k induced by the vertices $\{v_l \in V^k : \text{there is a path from } u_i \text{ to } v_l \text{ in } G^k\}$. We have that:

$$\Delta'(s, v_j) = \Delta(s, u_i) + C(u, v, j - i) \quad \text{for } F \leq i, j \leq k - \beta - k^*, \tag{18}$$

where $C(u, v, j - i)$ is a constant. Using u_0 as a reference point and combining this identity with the bound (16) on f, we obtain a bound on F in Theorem 1:

$$F \leq l + \max\left\{ \frac{\Delta(s, v_l) - \Delta'(s, v_l)}{\varepsilon(c)\rho - D(c)} \,\middle|\, v \in c \text{ a simple cycle in } G_D \text{ with } D(c) < \rho \right\},$$

where l is such that there is a path from u_0 to v_l in G^k for all $v \in V$. $\Delta(s, v_l)$ and $\Delta'(s, v_l)$ are determined in the graph G^{k^*+l}. Thus, after F unfoldings, $\Delta(s, v_j)$ is determined entirely by the constraints from u_0, i.e.,

$$\Delta(s, v_j) = \Delta'(s, v_j) \quad \text{when } j \geq F. \tag{19}$$

We can now complete the proof of Theorem 1:

$$\begin{aligned}
\Delta(s, v_{j+\varepsilon^*}) - \Delta(s, v_j) &= \Delta'(s, v_{j+\varepsilon^*}) - \Delta'(s, v_j) && \text{From (19).} \\
&= \Delta(s, u_{i+\varepsilon^*}) - \Delta(s, u_i) && \text{From (18).} \\
&= \text{sd}(s, u_{i+\varepsilon^*}) - \text{sd}(s, u_i) && \text{From Lemma 1.} \\
&= \rho\varepsilon^* && \text{From (10).}
\end{aligned}$$

5.2 An Algorithm for Cyclic Graphs

We now present the algorithm for determining $\Delta(s, t, \beta)$ in a causal cyclic timing constraint graph. In the following we need the notion of a *cutset* of G^k. For a set of vertices $X \subseteq V^k$, let $R(X)$ denote the set of vertices of G^k reachable from a vertex in X. The set of vertices not reachable from any vertex in X is denoted

$\overline{R}(X)$. A cutset C is a finite subset of V^k such that every path from *root* to a vertex in $R(C)$ passes through some vertex in C. Furthermore, we require that C is also a cutset for any larger unfolding, i.e., given a positive integer n, the set $\{v_{j+n} : v_j \in C\}$ must be a cutset for G^{k+n}.

Recall that $\Delta(s, t, \beta)$ is defined as $\max\{\Delta_k : k \geq \beta_0\}$. Using Theorem 1 we show that only a finite number of unfoldings are necessary to determine Δ:

Theorem 2. *Let $G = (V, E)$ be a cyclic causal timing constraint graph. Then*

$$\Delta(s, t, \beta) = \max\{\Delta_k : \beta_0 \leq k \leq k^* + \beta_0 + \varepsilon^* + F\}. \tag{20}$$

Proof. We will relate the values of $\Delta(s, v_j)$ in two different unfoldings, G^k and $G^{k+\varepsilon^*}$ where $k = k^* + \varepsilon^* + F$. Let C be the cutset $\{v_F : v \in V\}$, let $\Delta^k(v_j)$ denote the maximum separation between $s_{k-\beta}$ and v_j in G^k, and let $\Delta^{k+\varepsilon^*}(v_j)$ denote the maximum separation between $s_{k-\beta+\varepsilon^*}$ and v_j in $G^{k+\varepsilon^*}$. From Theorem 1 we have that

$$\Delta^{k+\varepsilon^*}(v_{j+\varepsilon^*}) = \Delta^{k+\varepsilon^*}(v_j) + \rho\varepsilon^* \qquad \text{for all } v_j \in C.$$

Since $\mathrm{sd}(s, v_j)$ in G^k is equal to $\mathrm{sd}(s, v_j) + \rho\varepsilon^*$ in $G^{k+\varepsilon^*}$ for $j \leq F$,

$$\Delta^k(v_j) = \Delta^{k+\varepsilon^*}(v_j) + \rho\varepsilon^* \qquad \text{for all } v_j \in \overline{R}(C) \cup C.$$

Combining these two facts, we get that

$$\Delta^{k+\varepsilon^*}(v_{j+\varepsilon^*}) = \Delta^k(v_j) \qquad \text{for all } v_j \in C.$$

Since $\mathrm{sd}(s, v_j)$ in G^k is equal to $\mathrm{sd}(s, v_{j+\varepsilon^*})$ in $G^{k+\varepsilon^*}$, it follows that

$$\Delta^{k+\varepsilon^*}(v_{j+\varepsilon^*}) = \Delta^k(v_j) \qquad \text{for all } v_j \in R(C)$$

and thus $\Delta_{k+\varepsilon^*} = \Delta_k$. $\qquad\qquad\qquad\square$

This theorem leads to a straightforward algorithm: compute Δ_k for increasing values of k using the MAXSEP' algorithm in Fig. 5 and maximize the results. The efficiency of this algorithm depends on the number of times MAXSEP' is called, i.e., on the values of k^*, ε^*, and F. These numbers all depend on the delay ranges and are not polynomial in the size of the constraint graph. In most realistic constraint graphs $\varepsilon^* = 1$ and F is a small constant. k^* is more of a concern because it can be large if there exists a cycle c in G_d such that $w(c)/\varepsilon(c)$ is almost equal to ρ.

In practice, we can improve the performance of the algorithm by deriving sufficient conditions for when further unfoldings can not result in a larger value of Δ. We observe from (5) that $\mathrm{sd}(s, v_j)$ is an upper bound on $\Delta(s, v_j)$. If $\Delta(s, v_j)$ happens to be equal to $\mathrm{sd}(s, v_j)$ for all vertices in a cutset, there is no reason to consider any further unfoldings since they can only make $\Delta(s, v_j)$ *smaller* and thus can not possibly make Δ larger.

Lemma 5. *Let C be a cutset for G^k where $k \geq \beta_0$. If $\mathrm{sd}(s, v_j) = \Delta(s, v_j)$ for all $v_j \in C$, then*

$$\Delta = \max\{\Delta_j : \beta_0 \leq j \leq k\}.$$

Proof. Consider G^{k+n} for some $n \geq 0$. $\Delta(s, v_{j+n})$ for $v_j \in C$ in G^{k+n} is smaller than or equal to $\mathrm{sd}(s, v_{j+n})$ which is equal to $\mathrm{sd}(s, v_j)$ in G^k, i.e., for all $v_j \in C$, $\Delta(s, v_{j+n})$ in G^{k+n} is smaller than or equal to $\Delta(s, v_j)$ in G^k. Since C is also a cutset for G^{k+n}, it follows from (8) and (9) and the monotonicity of addition, maximization, and minimization that $\Delta_{k+n} \leq \Delta_k$. □

The same observation can be used to determine an upper bound on Δ. Let $\Delta_k^\top$ denote a number such that $\Delta_k^\top \geq \max\{\Delta_j : j \geq k\}$, i.e., $\Delta_k^\top$ is an upper bound for any *further* unfoldings after G^k. From this definition it follows that if $\Delta_k^\top$ is smaller than or equal to $\Delta' = \max\{\Delta_j : \beta_0 \leq j \leq k\}$ then $\Delta = \Delta'$. To compute $\Delta_k^\top$, we use the MAXSEP' algorithm except lines 7 and 10 are not performed for $v_j \in C$ (this makes $\Delta(s, v_j)$ equal to $\mathrm{sd}(s, v_j)$ for all $v_j \in C$). With these optimizations, the algorithm for determining the maximum separation in a cyclic timing constraint graph is as shown in Fig. 6.

CYCLICMAXSEP$(G, s, t, \beta) =$
1: Calculate k^*, ε^*, and F.
2: $k \leftarrow \beta_0 - 1$
3: $\Delta \leftarrow -\infty$
4: **do**
5:     $k \leftarrow k + 1$
6:     $\Delta_k \leftarrow$ MAXSEP'$(G^k, s_{k-\beta}, t_k)$
7:     $\Delta \leftarrow \max\{\Delta, \Delta_k\}$
8:     **if** exists a cutset C in G^k s.t. $\forall v_j \in C : \mathrm{sd}(s_{k-\beta}, v_j) = \Delta(s_{k-\beta}, v_j)$ **then**
9:         **return** Δ
10:    $\Delta_k^\top \leftarrow$ UPPERBOUND$(G^k, s_{k-\beta}, t_k)$
11: **until** $\Delta \geq \Delta_k^\top$ or $k = k^* + \beta_0 + \varepsilon^* + F$
12: **return** Δ.

Fig. 6. Algorithm for determining the maximum separation between events s and t separated by β in occurrence index in a causal cyclic timing constraint graph G.

Example 6. We continue with example 2, i.e., determining the maximum separation in the cyclic timing constraint graph in Fig. 2 between $s = e_{11}$ and $t = e_{10}$ separated in occurrence index by $\beta = -1$. Running the algorithm in Fig. 6 yields that the maximum separation is 0, i.e., for all consistent timing assignments τ,

$$\tau(e_{11}, k) - \tau(e_{10}, k+1) \leq 0 \quad \Rightarrow \quad \tau(e_{11}, k) \leq \tau(e_{10}, k+1).$$

That is, the timing is right on the margin.

5.3 Sufficient Causal Conditions of Cyclic Systems

The algorithm in Fig. 6 assumes that the constraint graph is causal, i.e., that all blocks in the unfolded graph are past-dominated and all blocks have well-defined triggers. The past-dominated property can be checked using the following theorem.

Theorem 3. *A block B in an acyclic timing constraint graph is past-dominated if*

1. *the triggers of B are well-defined (i.e., all events in B occur later than all the triggers of B) and*
2. *$\forall u_1, u_2 \in \text{trigs}(B) : \Delta(u_1, u_2) \leq \Delta_{\text{local}}(u_1, u_2)$, where $\Delta_{\text{local}}(u_1, u_2)$ is the maximum separation when only considering the local constraints of block B.*

Proof. See [3]. □

Notice that if a block satisfy these two conditions, it both has well-defined triggers and is past-dominated. Every latest block is inherently past-dominated since all constraints between triggers and the event in the block are combined using the max operator. We can use the MAXSEP algorithm in Fig. 5 to determine check whether the two conditions in Theorem 3 holds for a linear block in an acyclic constraint graph.

Theorem 4. *Let $G = (V, E)$ be an acyclic timing constraint graph with a block partition P. A linear block $B \in P$ is past-dominated if*

1. *$\forall u \in \text{trigs}(B), v \in B : \Delta_{\text{com}}(v, u) \leq 0$ and*
2. *$\forall u_1, u_2 \in \text{trigs}(B) : \Delta_{\text{com}}(u_1, u_2) \leq \Delta_{\text{local}}(u_1, u_2)$,*

where $\Delta_{\text{com}}(u, v)$ is the maximum time separation computed under the assumption that the system is causal (i.e., the value determined by the algorithm in Fig. 5) and Δ_{local} is the maximum separation when only considering the local constraints of block B.

Proof. We show that the two conditions of the theorem imply those in Theorem 3. Let u and v be two vertices of G. The value computed by the MAXSEP algorithm in Fig. 5, Δ_{com}, is an upper bound on the exact maximum separation, $\Delta(u, v)$. This is because Δ_{com} is determined from (8) and (9) while the real maximum time separation if determined from (5) to (7). Since (5) to (7) contain more terms in the minimization than (8) and (9), and $\text{sdl}(u, v) \geq \text{sd}(u, v)$, we have that

$$\Delta_{\text{com}}(u, v) \geq \Delta(u, v). \tag{21}$$

Combining (21) with condition 1 of the theorem, it follows that

$$\forall u \in \text{trigs}(B), v \in B : \Delta(v, u) \leq \Delta_{\text{com}}(v, u) \leq 0$$

and therefore for all consistent timing assignments τ, $\tau(u) \leq \tau(v)$, i.e., condition 1 on Theorem 3 is satisfied.

Similarly, combining (21) with condition 2 of the theorem, it follows that

$$\forall u_1, u_2 \in \text{trigs}(B) : \Delta(u_1, u_2) \le \Delta_{\text{com}}(u_1, u_2) \le \Delta_{\text{local}}(u_1, u_2),$$

and thus condition 2 of Theorem 3 is satisfied. □

For cyclic timing constraint graphs, we use the CYCLICMAXSEP algorithm in Fig. 6 in a similar manner to check that every block in the unfolded constraint graph G^k is past-dominated.

Theorem 5. *Let $G = (V, E)$ be a cyclic timing constraint graph with a block partition P. A linear block $B \in P$ in G^k for $k \ge \beta_0$ is past-dominated if*

1. $\forall u \in \text{trigs}(B), v \in B : \Delta_{\text{com}}(v, u) \le 0$ and
2. $\forall u_1, u_2 \in \text{trigs}(B) : \Delta_{\text{com}}(u_1, u_2) \le \Delta_{\text{local}}(u_1, u_2),$

where $\Delta_{\text{com}}(u, v)$ is the maximum time separation computed under the assumption that the system is causal (i.e., the value determined by the algorithm in Fig. 6) and Δ_{local} is the maximum separation when only considering the local constraints of block B.

Proof. From the definition of Δ in a cyclic timing constraint graph (3), it follows that

$$\forall k \ge \beta_0 : \Delta_k \le \Delta.$$

Thus, the theorem follows by applying Theorem 4 to each unfolding G^k. □

The number of separation analyses needed to show that a constraint graph is causal can be reduced by the following observation.

Corollary 1. *If all local constraints of a linear block B are precedence constraints, then B has well-defined triggers.*

Example 7. To ensure that the result of the timing analysis in example 6 is valid, we use Theorem 5 to check that the block partition is past-dominated and has well-defined triggers. Only the block $B_2 = \{e_{10}, e_{14}\}$ contains concurrency edges. The only triggers of this block is e_7. Condition 1 of Theorem 5 is satisfied since the results of the following two separation analyses are non-positive.

$$\Delta_{\text{com}}(e_{10}, e_7) = 0$$
$$\Delta_{\text{com}}(e_{14}, e_7) = -10.$$

Condition 2 of the Theorem is trivially satisfied since block B_2 only has a single trigger event.

6 Conclusion

This paper has addressed the problem of determining the maximum separation in time between two events in a cyclic timing constraint graph which contains both linear and latest constraints. We have developed an exact algorithm for solving this problem for the class of causal constraint graphs, i.e., those that have a past-dominated block partition with well-defined triggers. We have stated sufficient conditions for checking causality of the cyclic specification.

References

1. F. Baccelli, G. Cohen, G. J. Olsder, and J.-P. Quadrat. *Synchronization and Linearity*. Wiley Series in Probability and Mathematical Statistics. John Wiley & Sons, New York, 1992.

2. S. M. Burns. *Performance Analysis and Optimization of Asynchronous Circuits*. Ph.D. thesis, California Institute of Technology, 1991. CS-TR-91-1.

3. E. Cerny and K. Khordoc. Interface specifications with conjunctive timing constraints: Realisability and compatibility. In 2nd *AMAST Workshop on Real-Time Systems*, Bordeaux, June 1995.

4. E. Cerny, Y. Wang, and M. Aboulhamid. Scheduling and interface controller synthesis under real-time constraints. In 4th *AMAST Workshop on Real-Time Systems*, Bordeaux, June 1997.

5. P. Girodias. *Interface Timing Verification Using Constrain Logic Programming*. PhD thesis, University of Montreal, 1997.

6. P. Girodias and E. Cerny. Interface timing verification with delay correlation using constrain logic programming. In *Proceedings of the European Design & Test Conference*, Paris, March 1997.

7. P. Girodias, E. Cerny, and W. J. Older. Solving linear, min, and max constraint systems using CLP based on relational interval arithmetic. *Journal of Theoretical Computer Science*, 173:253–281, 1997.

8. H. Hulgaard and S. M. Burns. Bounded delay timing analysis of a class of CSP program. *Formal Methods in System Design*, 11(3):265–294, 1997.

9. H. Hulgaard, S. M. Burns, T. Amon, and G. Borriello. An algorithm for exact bounds on the time separation of events in concurrent systems. *IEEE Transactions on Computers*, 44(11):1306–1317, November 1995.

10. F. Jin and E. Cerny. Time separation of events in cyclic systems with linear and latest timing constraints. Technical report, University of Montreal, 1998.

11. K. Khordoc and E. Cerny. Semantics and verification of action diagrams with linear timing constraints. *ACM Transactions on Design Automation of Electronic Systems*, 3(1), 1998.

12. E. L. Lawler. *Combinatorial Optimization: Networks and Matroids*. Holt, Rinehart and Winston, New York, 1976.

13. K. McMillan and D. L. Dill. Algorithms for interface timing verification. In *IEEE International Conference on Computer Design*, pages 48–51, 1992.

14. C. J. Myers and T. H.-Y. Meng. Synthesis of timed asynchronous circuits. *IEEE Transactions on VLSI Systems*, 1(2):106–119, 1993.

15. P. Vanbekbergen, G. Goossens, and H. D. Man. Specification and analysis of timing constraints in signal transition graphs. In *European Design Automation Conference*, pages 302–306, March 1992.

16. E. A. Walkup and G. Borriello. Interface timing verification with applications to synthesis. In *Proceedings of the Designs Automation Conference*, 1994.

17. T. Y. Yen, A. Ishii, A. Casavant, and W. Wolf. Efficient algorithms for interface timing verification. In *Proceedings of the European Design Automation Conference*, pages 34–39, September 1994.

Using MTBDDs for Composition and Model Checking of Real-Time Systems[1]

Jürgen Ruf and Thomas Kropf

Institut für Rechnerentwurf und Fehlertoleranz (Prof. Dr.-Ing. D. Schmid),
Universität Karlsruhe, Kaiserstr. 12, Geb. 20.20, D-76128 Karlsruhe, Germany
{Juergen.Ruf, Thomas.Kropf}@informatik.uni-karlsruhe.de
http://goethe.ira.uka.de/hvg/cats/raven/

Abstract. In this paper we show that multi-terminal BDDs (MTBDDs) are well suited to represent and manipulate interval based timed transition systems. For many timed verification tasks efficient MTBDD-based algorithms are presented. This comprises the composition of timed structures based on symbolic techniques, heuristics for state variable minimization, and a symbolic model checking algorithm. Experimental results show that in many cases our approach outperforms standard unit-delay approaches and corresponding timed automata models.

1 Introduction

For verifying real-time properties, e.g in the domain of embedded systems, standard model checking is not directly applicable. There, quantized temporal properties have to be verified ("after 231ns something will happen") and the implementation model usually contains timing information like typical system delays etc. Moreover, for real systems typical delay times may vary e.g. due to fabrication tolerances and thus have to be represented by time intervals indicating minimal and maximal bounds.

Various efforts have been undertaken to extend the temporal logic and the proof algorithms to timed systems (i.e. systems containing quantized timing information). Two main approaches have to be distinguished here: those based on timed automata [4] and extensions of symbolic CTL model checking [5]. For both finding efficient algorithms and implementations is an active area of research as real time model checking bears additional challenges compared to standard model checking:

- the model checking algorithms have to cope with time, i.e. with natural or real numbers which makes the application of propositional logic techniques based on ROBDD hard
- adding timing to state transition systems worsens the state space explosion problem, especially if a composition of timed transition systems is necessary and if time intervals are allowed, i.e. non-deterministically varying transition times.

This paper shows how MTBDDs can be used to get symbolic algorithms for all real time verification tasks: model representation, model composition, heuristic minimization of composed structures and model checking. The contributions of this paper are a consistent methodology for using MTBDDs to get symbolic real time verification algorithms, and new composition and model checking algorithms based on symbolic state space representation and traversal techniques. Moreover, our experimental results do not only show that our method outperforms other approaches in many cases but also

---

1. This work has been funded by a german research grant (DFG, SFB 358-C2)

give interesting empirical data how composition of timed systems affects the timing characteristics of the resulting systems.

2 State-of-the-Art

2.1 Timed Automata

A formalism to model real time systems are *timed automata*, developed by Alur and Dill [4]. In timed automata time is represented by clocks carrying real numbers and time passes in the states. A state transition is chosen based on clock predicates on the edges and input events. Specifications are given in TCTL, an extension of CTL. As an arbitrary number of clocks is possible, this is a very powerful approach. Different tools based on this theory have been presented like Kronos[1] or UPPALL[2]. Composition of timed automata is easily possible leading to a new automaton which carries the sum of all the clocks of the original automata.

However, although deriving the composed structure is simple, the state explosion problem is only delayed to the point of model checking. Then in each state the product of the values of all clocks has to be considered. To solve this state explosion problem on-the-fly model checking techniques can be used. However, if specifications are to be proven correct which require the traversal of the complete reachable state space (e.g. mutual reachability of states) the efficiency gain is low. Moreover, the underlying proof algorithm is different from standard CTL fixed point computations and hence it is more difficult to find efficient representations. Thus only recently first efforts have been presented on how to use BDD like techniques for a symbolic state set representation. Therefore, in practice the resulting runtimes may be high especially for model checking composed structures even when using these symbolic techniques [6, 1].

2.2 Extensions to CTL Model Checking

A different approach extends CTL model checking to real-time. It is based on the observation that in many applications the expressive power of timed automata is not necessary. Usually these approaches attribute edges of the transition system with delay times (mostly natural numbers) and allow quantized timing parameters in the temporal operators, leading to CTL extensions like RTCTL (real-time CTL, [7]) or QCTL (quantized CTL, [8]). To retain the efficient BDD representation, delay times are represented by a binary encoding, added to the transition relation or by representing all transitions with a certain delay by a separate transition relation. This can be seen as a special case of timed automata where only one clock carrying natural numbers is allowed which is reset after each state transition. A tool based on this approach is Verus[3]. Recently new timed model checking algorithms based on multi-terminal BDDs (MTBDDs) have been proposed [9, 10]. The advantage of these approaches is that efficient implementation techniques of standard CTL model checking can be used.

Besides the reduced expressiveness compared to timed automata, the main deficiency of these approaches is structure composition. As there is only one global clock the approach of timed automata, where just the clocks of all subsystems are combined,

---

1. [1], http://www.imag.fr/VERIMAG/TEMPORISE/kronos/index-english.html
2. [2], http://www.brics.dk/Projects/UPPALL/
3. [3], http://www.cs.cmu.edu/~modelcheck/verus.html

is not possible, and considerably higher composition effort is necessary. However, once this structure has been computed model checking may be performed again efficiently using standard techniques. Up to now, the problem of finding efficient algorithms for composing timed structures has not been treated in a satisfactory way.

3 Basics

Characteristic functions (CF) represent sets of elements by mapping them to one if they belong to the set and to zero if they are not members of the set: $(\chi_A(s) = 1) \leftrightarrow (s \in A)$ [11]. The advantage of using characteristic functions is that set operations can be efficiently performed by Boolean operations on CFs.

We use *extended characteristic functions* (ECFs) which additionally assign attributes (sets) to set elements: $(\Lambda_A(s) = \alpha) \leftrightarrow (s \in A \wedge attribute(s) = \alpha)$. This leads to a two level representation of sets: the main set and the attribute set. An element is not a member of a main set if its attribute value is the empty set. Set operations may be applied to ECFs, but we have to extend these operations to the attribute sets. E.g. the intersection is defined by: $(\Lambda_A \cap \Lambda_B)(s) := (\Lambda_A(s) \cap \Lambda_B(s))$. In the remainder of this paper we use ECFs to represent sets of natural numbers $\wp(I\!N)$ representing clock values as attributes of states $sc : S \to \wp(I\!N)$.

Another important operation on ECFs is the projection:

Definition 3.1. *Given an ECF mapping tuples of elements to attribute values:* $\Lambda : A_1 \times \ldots \times A_n \to \wp(I\!N)$ *and a binary and associative operation $\Diamond$. The projection*

$$\nabla_\Diamond : [A_i \times (A_1 \times \ldots \times A_{i-1} \times A_i \times A_{i+1} \times \ldots) \to \wp(I\!N)] \to$$
$$(A_1 \times \ldots \times A_{i-1} \times A_{i+1} \times \ldots) \to \wp(I\!N) \tag{1}$$

is defined as $\nabla_\Diamond \; s_i . \Lambda(s_1, \ldots, s_i, \ldots, s_n) := \bigdiamond_{x \in A_i} \Lambda(s_1, \ldots, x, \ldots, s_n)$ \hfill (2)

A projection removes one ECF dimension by $\Diamond$ merging all elements of A_i into a single constant result. For example, if we are interested in the set of all attribute values of Λ, we may use the projection: $V = \nabla_\cup s . \Lambda(s)$. If all A_i are Boolean and $\Diamond$ is the disjunction, we obtain standard existential quantification.

Using ECFs, we can apply operations to the attributes of set elements without altering the set itself. One example is the decrement function which will be used later in the context of real time verification to simultaneously decrement the clock values of a set of states. Applied to an ECF it decrements all attribute values and removes zero values:

$$dec(\Lambda)(s) := \{v - 1 \in I\!N | v \in \Lambda(s)\} \tag{3}$$

As explained in section 7, ECFs can be represented by MTBDDs where the MTBDD leaves carry the attributes. Operations like union, projection and attribute manipulation can then be performed efficiently by usual (MT)BDD "apply" functions.

4 Modeling and Specifying Real-Time Systems

4.1 Modeling Implementations

The basic model for our proof algorithm is the interval structure (IS), i.e. a state transition system where states are labeled with atomic propositions and transitions are labeled with intervals of natural numbers representing the transition times. We assume that an IS has exactly one clock for time measuring. The clock is reset to zero if a state is entered. A state may be left if the current clock value lies within the interval bounds of an outgoing transition. The state must be left if the maximal interval value of all outgoing transitions is reached (Fig. 4.1).

Definition 4.1. *An IS $\mathfrak{S}$ is a tuple $\mathfrak{S} = (P, S, T, L, I)$ with a set of atomic propositions P, a set of states S (i-states), a transition relation between the states $T \subseteq S \times S$ such that every state in S has an successor state, a state labeling function $L{:}S \rightarrow \wp(P)$ and a transition labeling function $I{:}T \rightarrow I\!N \times I\!N$.*

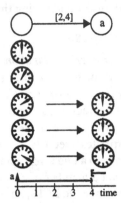

The only difference to standard structures are the transitions which are labeled by the function I with interval bounds.

Definition 4.2. *The maximal state time of an i-state s MaxTime:$S \rightarrow I\!N$ is the upper time interval bound of all outgoing transitions of s, i.e.:*

MaxTime$(s) = max\{t | \exists s'.(s, s') \in T \wedge I(s, s')=[u, t]\}$.

Fig. 4.1. Example IS

Besides the state we also have to consider the currently elapsed time to determine the transition behavior of the system. This is different from a standard CTL structure. Hence the actual state of a system, called *generalized state (g-state)*, is given by a state s and the actual clock value v in this state.

Definition 4.3. *A generalized state (g-state) $g = (s, v)$ is an i-state s associated with a valid clock value v. The set of all g-states in $\mathfrak{S} = (P, S, T, L, I)$ is given by:*

$$G = \{(s, v) | s \in S \wedge 0 \leq v < \text{MaxTime}(s)\} \qquad (4)$$

Definition 4.4. *Given $\mathfrak{S} = (P, S, T, L, I)$ a run r^g is a sequence of g-states $r^g = (g_0, g_1, ...)$. For the sequence $g_i = (s_i, v_i) \in G$ with $g_0 = g$ and for all i it holds that $g_{i+1} = (s_i, v_i + 1)$ with $v_i + 1 < \text{MaxTime}(s_i)$ or $g_{i+1} = (s_{i+1}, 0)$ with $(s_i, s_{i+1}) \in T \wedge v_i + 1 \in I(s_i, s_{i+1})$. We write $r^g(i)$ for the i-th component of r^g.*

4.2 Interval Structures defined in other Semantic Models

For readers familiar with other semantic models and for further clarifying the semantics and expressiveness of ISs we informally present them in terms of unit delay structures and as timed automata.

Unit Delay Models. A unit delay structure $U = (P, S, T, L)$ is defined analogously to ISs with the exception that they have no transition labels. Every transition consumes one time step. A path $p^s = (s_0, s_1, ...)$ is a sequence of states in U with $s_0 = s$ and for all i: $(s_i, s_{i+1}) \in T$. To model ISs with unit delay structures we have to introduce *stutter states* to model the timed transitions.

Definition 4.5. *Given $\mathfrak{S} = (P, S, T, L, I)$, the expanded unit delay structure $U = \text{expand}(\mathfrak{S})$ with $U = (P, S', T', L')$ is defined by (5)*

$$S' = \{(s, v) | s \in S \wedge 0 \leq v < \text{MaxTime}(s)\}$$
$$T' = \{((s, v), (s, v + 1)) | s \in S \wedge 0 < v + 1 < \text{MaxTime}(s)\} \cup$$
$$\{((s, v), (s', 0)) | s, s' \in S \wedge (s, s') \in T \wedge v + 1 \in I(s, s')\} \qquad (5)$$
$$L'((s, v)) = L(s)$$

The first operand of the union in the definition of T' is the connection of the stutter states (time chain) and the second operand corresponds to the connection of the i-states. Fig. 4.2 shows an example of expanding timed transitions with stutter states.

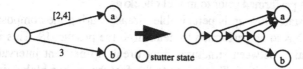

Fig. 4.2. Expansion Semantics of a Timed Structure using Stutter States

The following corollary states the behavioral equivalence of an expanded structure with regard to the original IS.

Corollary 4.6. *Given* $\mathfrak{S} = (P, S, T, L, I)$, *the expanded unit delay structure* $U = (P, S', T', L')$ *and a g-state* $g = (s, v)$. *For every run* r^g *in* $\mathfrak{S}$ *there exists a path* $p^{(s,v)}$ *in* U *with* $\forall i.r^g(i) = p^{(s,v)}(i)$ *and for every path* $p^{(s,v)}$ *in* U *there exists a run* r^g *in* $\mathfrak{S}$ *with* $\forall i.r^g(i) = p^{(s,v)}(i)$.

Timed Automata. ISs may be also modeled by timed automata with one clock as given in Fig. 4.3. The clock has to be reset explicitly at each transition. The maximal state time (see Def. 4.2) has been formalized using a state invariant.

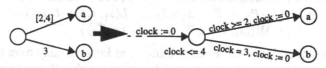

Fig. 4.3. Timed Automaton modeling an IS

4.3 Specifications

The logic we use is an extension of CTL with time scopes and intervals (see [10] for more details). It contains propositional operators ($\neg$, $\wedge$) and temporal operators $EX_{[n]}, EG_{[n]}, EU_{[m,n]}$. Other operators like $EF_{[n]}$ or $AG_{[m,n]}$ are derivable from the three basic operators.

Definition 4.7. *Given the IS* $\mathfrak{S} = (P, S, T, L, I)$, *a g-state* $g=(s, v) \in G$, *the CCTL formulas* φ, ψ *and the natural numbers* $m \in \mathbf{N}$ *and* $n \in \mathbf{N} \cup \{\infty\}$. *The model relation for the bounded temporal operators is defined by:*

$\mathfrak{S}, g \vDash EX_{[m]}\varphi \qquad \Leftrightarrow$ there ex. a run r^g with $\mathfrak{S}, g_m \vDash \varphi$

$\mathfrak{S}, g \vDash EG_{[n]}\varphi \qquad \Leftrightarrow$ there ex. a run r^g s.t. for all $0 \leq i \leq n$ holds $\mathfrak{S}, g_i \vDash \varphi$

$\mathfrak{S}, g \vDash E(\varphi U_{[m,n]}\psi) \Leftrightarrow$ there ex. a run r^g and a i with $m \leq i \leq n$ such that $\mathfrak{S}, g_i \vDash \psi$ and for all $0 \leq j < i$ holds that $\mathfrak{S}, g_i \vDash \varphi$

5 Composition

As systems are usually described in a modular way, different temporal structures representing the modules have to be combined to apply standard model checking techniques. Communication between modules is modeled by signal sharing.

In case of unit delay structures the product structure is obtained by intersecting the state transition relations which only requires the conjunction of the BDDs representing the different structures. The resulting state space is the cross product of the original state spaces and may lead to a state explosion. However, in many cases a large set of unreachable states results due to the composition. Thus a computation of the reachable state set is often performed prior to model checking.

In ISs, this state blow up is perceivable already during the composition as a state space traversal has to be performed to determine the product IS. This is necessary as the communication between structures may break up existent intervals into smaller ones during the composition. This is due to the fact that only a global time (one clock) is available and all interference effects have to be computed explicitly. The advantage however is that once this computation has been completed the following model checking process is very efficient. Moreover building the product structure has to be done only once even if many different properties have to be checked.

5.1 Definitions

Before we present the new algorithms, the composition of structures is formally defined. As we will use it later and as it eases further understanding the composition of unit-delay structures is introduced first.

Definition 5.1. Given $U_1 = (P_1, S_1, T_1, L_1)$ and $U_2 = (P_2, S_2, T_2, L_2)$ (two unit delay structures). The product structure $U = (P, S, T, L)$, written as $U = U_1 \| U_2$ is defined by: $P = P_1 \cup P_2$, $S = S_1 \times S_2$, $L(s_1, s_2) = L_1(s_1) \cup L_2(s_2)$ and $T = \{((s_1, s_2), (s_1', s_2')) | ((s_1, s_1') \in T_1 \wedge (s_2, s_2') \in T_2)\}$.

Defining the composition of ISs is more difficult as intervals may have to be divided into smaller ones due to the composition of transitions with different transition times. A simple example is shown in Fig. 5.1.

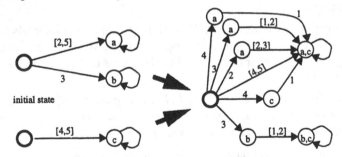

Fig. 5.1. Composition of Interval Transitions

The splitting of the intervals is clearly visible and requires the introduction of additional intermediate states. Moreover it is obvious that for a formal definition of the composition of ISs many different cases have to be taken into account, e.g. all possible interleaving relations of time intervals. To avoid this definition, we rather take another approach and define interval composition using a detour via unit delay structures. Translating ISs into behavioral equivalent unit delay structures has been already treated in Def. 4.5, the composition of unit delay structures was given in Def. 5.1. The single missing function, not defined formally in this paper is reduce which translates unit delay structures into ISs by converting chains of stutter states into single interval transitions.

Definition 5.2. *Given* $\mathfrak{I}_1 = (P_1, S_1, T_1, L_1, I_1)$ *and* $\mathfrak{I}_2 = (P_2, S_2, T_2, L_2, I_2)$. *The composed structure* $\mathfrak{I}$ *is defined as* $\mathfrak{I} = \texttt{reduce}(\texttt{expand}(\mathfrak{I}_1) \, \| \, \texttt{expand}(\mathfrak{I}_2))$. *We write* $\mathfrak{I} = \mathfrak{I}_1 \, \| \, \mathfrak{I}_2$.

Def. 5.2 is only given to formalize the composition of ISs. Though being the basis of our first composition algorithms, it is neither the only nor the best way to efficiently compute $\mathfrak{I}_1 \, \| \, \mathfrak{I}_2$. In the following, we present two composition algorithms where special care is taken to allow symbolic state manipulation techniques and to achieve a compact and efficiently model checkable product structure. In section 5.2 and section 5.3 we describe the basic algorithms. Implementation details showing how to efficiently use MTBDDs are given in section 7.

5.2 Expand-Reduce

The first composition algorithm is a straightforward implementation of Def. 5.2 given in three variants. All variants first expand all ISs to unit delay structures by introducing stutter states as already shown in Fig. 4.2 (see section 7.2 for implementation details of expand and reduce). The resulting structures are then composed according to Def. 5.1 requiring only the conjunction of the respective transition relations. The resulting transition relation is then reduced again by merging state chains into timed transitions.

Three variants for reduce exist. Our first approach applies a technique used in parallel algorithms called pointer jumping. This algorithm (*transitive* in Fig. 5.2[1]) merges two adjacent transitions which have the same original propositions to one transition which carries the sum of the transition times. This procedure is performed symbolically and in parallel on all states of the composed structure. This connection has to be repeated until no adjacent states with equal original propositions exist. In one last step the transitions are extended to the non-stutter successor states.

A second algorithm (*iterative* in Fig. 5.2) directly starts in the initial states and symbolically steps forward in the composed structure in a breadth first search manner. During this operation, it incrementally creates the timed transitions by lengthening the existing transitions to the actual reached states.

A variant of the second algorithm avoids the creation of temporary transitions. This procedure assigns unique indices to the starting states (*indexing* in Fig. 5.2). Using these indices and a counter, the structure is traversed until a non-stutter state is reached. With the help of indices the starting states are identified and as the counter value carries the actual delay time a new transition with this delay time can be added.

5.3 Separate State Space Traversal

Composing based on expand/reduce is intuitive but not very elegant. Therefore we designed another algorithm which steps synchronously through the expanded but not composed structures and successively builds the resulting timed edges of the final IS.

This algorithm starts in the initial states of the structures to be composed and incrementally adds temporary timed edges in the separate structures. If one structure reaches a non-stutter state, all edges in all other structures with the same transition

---

1. The pictures only illustrate the ideas of the various algorithms. The implementation is based exclusively on symbolic ECF operations implemented with MTBDDs as shown in section 7.2. No explicit state enumeration is necessary.

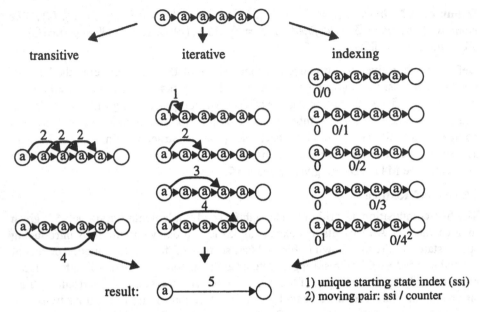

Fig. 5.2. Three reduction algorithms based on Expand/Reduce

time are combined (by conjuncting them) to one new edge in the resulting structure. After this combination, a new temporary transition has to start in all structures. Temporary transitions may be removed, if their counterpart (transitions with same time) is missing in any of the other structure. Due to space limitations we demonstrate the idea of this algorithm by an example only (Fig. 5.3).

It has to be noted that this algorithm also performs a symbolic computation without requiring the explicit enumeration of all states.

6 Real-Time Model Checking

In this section we describe the symbolic model checking algorithm for CCTL and ISs. More details can be found in [10]. We represent sets of g-states symbolically using ECFs by mapping states to valid clock values. The transition relation is represented symbolically by mapping pairs of states to the possible transition times minus one (we start counting at zero, see Fig. 4.1). As in standard model checking, our algorithm builds the syntax diagram of the CCTL formula and computes sets of states in a bottom-up manner for each operator holding the corresponding subformula.

The set of g-states holding the formula $EX_{[1]}\varphi$ corresponds to the set of direct predecessor states of g-states holding φ. The generalized predecessor state g of a g-state $g' = (s', v')$ with a non-zero clock value v may be computed by decrementing the clock value: $g = (s', v' - 1)$. This decrementation may be expanded to sets of states, i.e. to all values of an ECF Λ:

$$\texttt{localpre}(\Lambda) := dec(\Lambda) \qquad (6)$$

We call this operation the local predecessor computation since it computes the predecessors locally on the i-states by decrementing the clock values.

If the clock value of $g' = (s', 0)$ is zero then for all generalized predecessors, $g = (s, v)$ holds (see Def. 4.4): $(s, s') \in T, v + 1 \in I(s, s')$. To compute the predeces-

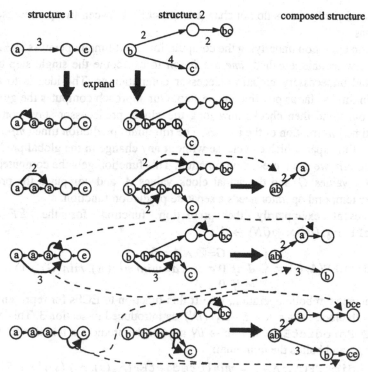

Fig. 5.3. Alternative Structure Composition

sor in this case we first have to select the i-states carrying a zero clock value. In terms of ECFs we define a zero clock extraction $<>$:

$$<\Lambda>(s) := \begin{cases} \mathit{IN} & \text{if } 0 \in \Lambda(s) \\ \varnothing & \text{else} \end{cases} \tag{7}$$

This leads to the global predecessor computation for the transition relation T and the actual g-state set Λ:

$$\texttt{globalpre}(\Lambda) := \nabla_\cup s'.T(s,s') \cap (<\Lambda>(s))\big|_{s \leftarrow s'} \tag{8}$$

Without clocks and using Boolean functions instead of set operations, we obtain the relational product used in symbolic CTL model checking. As the predecessors may be local and global, we compute the EX-operator by:

$$\Lambda_{EX_{[1]}\varphi} = \texttt{globalpre}(\Lambda_\varphi) \cup \texttt{localpre}(\Lambda_\varphi) \tag{9}$$

with the actual set of g-states holding φ given as ECF Λ_φ.

All other timed operators may be computed by using this $EX_{[1]}\varphi$ operator, e.g. $EF_{[n]}\varphi = \varphi \cup EX_{[1]}EF_{[n-1]}\varphi$ [10]. Untimed operators like AG are computed by applying the usual fixed point operations using the $EX_{[1]}\varphi$ operator instead of $EX\varphi$.

6.1 Making Real Time Model Checking more Efficient

If we analyze the predecessor computation, we observe the following:

- The global predecessor computation is more expensive than the local predecessor computation since it takes the complete transition relation into account.

- The global predecessors do not change necessarily between two predecessor computations.
- The basic operation underlying the computation of all timed operators is $EX_{[1]}\varphi$.

We use a new technique called *time prediction* to overcome the single step traversal and to avoid unnecessary global predecessor computations. The idea is to advance "locally" in time as far as possible in one step. Our approach computes the global predecessors once and then checks how long the global predecessor stays constant by using the time information of the i-states. The minimum prediction time (mp) of all i-states is the time span which can elapse without any change in the global predecessor. To compute mp we use a prediction function. This function gets the computed predecessor clock values G and the actual clock values A and computes the prediction time. Every temporal operator needs a separate prediction function.

We present exemplarily the prediction function for the EF-operator predictEF : $\wp(IN) \times \wp(IN) \to IN$:

$$\text{predictEF}(A, G) = \begin{cases} \infty & \text{if } G=\emptyset \wedge A=\emptyset \\ d & \text{if } 0 \notin A \wedge d=min(min(A), min(G)+1) \\ \infty & \text{if } 0 \in A \end{cases} \quad (10)$$

For an efficient symbolic operation, we lift this function to ECFs for representing sets of states with clock values $sc : S \to \wp(IN)$ as introduced in section 3. This extended function ECFpredictEF : $sc \times sc \to IN$ applies the prediction function to the ECFs Λ_A and Λ_{G_p} and builds the minimum:

$$\text{ECFpredictEF}(\Lambda_A, \Lambda_{G_p}) = min\{\text{predictEF}(\Lambda_A(s), \Lambda_{G_p}(s)) \mid s \in S\} \quad (11)$$

After this prediction, the fixpoint iteration of the temporal operators is performed locally mp times in the i-states. E.g. for the EF operator, we execute the algorithm given on the right side, which computes the fixpoint iteration mp times: iterateEF: $\wp(IN) \times \wp(IN) \times IN \to \wp(IN)$.

```
function iterateEF(A, G, mp)
  res = A
  while mp > 0 do
    res = res ∪ G ∪ dec(res)
    mp = mp - 1
  return res
```

We also lift this operation to ECFs: ECFiterateEF : $sc \times sc \times IN \to sc$ with

$$\text{ECFiterateEF}(\Lambda_A, \Lambda_G, mp) = \text{iterateEF}(\Lambda_A(s), \Lambda_G(s), mp) \quad (12)$$

i.e. ECFiterateEF results in an ECF of type sc mapping states to sets of clock values. The complete algorithm for computing checkEF is given on the right side.

The local fixpoint iteration needs $O(mp)$ set operations for execution. We define a technique called *time jumping* which avoids this iterative

```
function checkEF(ECF Λ_A, int step)
  Λ_0 = Λ_S /* state space ECF */
  while (step > 0 && Λ_0 ≠ Λ_A)
    Λ_0 = Λ_A
    Λ_G = globalpre(Λ_A)
    mp = ECFpredictEF(Λ_A, Λ_G)
    if mp > step then mp = step
    Λ_A = ECFiterateEF(Λ_A, Λ_G, mp)
    step = step - mp
  return Λ_A
```

execution. Since this time jump is dependent on the implementation representation of sets of clock values, we only give a formal definition of the EF-time jump here:

$$res = \{v \mid \exists v' \in A.v + mp \geq v'\} \cup \{v \mid \exists v' \in G.v + mp - 1 \geq v'\} \quad (13)$$

We add those new values to the values in A which are smaller than any value in A and have a maximal distance mp to A. Also, new values are added to the G values which are smaller than any value in G and have a maximal distance $mp - 1$ to G. Hence we are able to perform the local fixpoint iteration with one special set operation.

7 Implementation

In this section we describe how ISs are represented using MTBDDs and bitvectors, how algorithms may be realized using MTBDDS and how the algorithms can be further optimized. The various algorithms are described in detail as their efficiency is directly related to finding elegant and simple MTBDD operations.

7.1 Representing Interval Structures

In symbolic CTL model checking, state sets and transition relations are represented by characteristic functions which in turn are represented by ROBDDs [12]. For timed model checking, state sets and relations are – due to the additional timing information – represented by ECFs which can in turn be represented by MTBDDs. MTBDDs are a straightforward extension of ROBDDs where the leaves of the graph carry arbitrary values [13, 14]. All set and state traversal operations are then executed by MTBDD operations. As for ROBDDs there exist functions for applying a terminal defined function to one or two (apply1/2) MTBDDs. The intersection of two ECFs $\Lambda_A(s) \cap \Lambda_B(s)$, presented in section 3, can be computed with apply2(Λ_A, Λ_B, terminal-$\cap$). The function terminal-$\cap$ intersects the sets in the MTBDD leaves. There exist MTBDD operations for projection and for analyzing (analyze1/2) MTBDD leaves (e.g. computing the minimum).

In our implementation we use bitvectors of arbitrary length to represent sets of natural numbers (clock values) in the MTBDD leaves. Thus ISs as given in Def. 4.1 are represented by MTBDDs analogously to representing unit delay structures with ROBDDs. The transition times are captured as attributes of ECFs. Therefore, an MTBDD representing an IS contains two sets of atomic propositions for the current and the next state. Each path in the MTBDD determined by a valuation of these two variable sets corresponds to a transition; the leaf value of this MTBDD path carries the interval value of this transition [10].

7.2 Composition Algorithms

Expand. As structures are represented symbolically, i.e. states are uniquely represented by their labels, we have to introduce new atomic propositions to distinguish the newly added stutter states from the original states. We call these propositions *chain propositions* (encoding propositions c) in contrast to the already existing *original* propositions. First we determine the new chain propositions for the stutter states. Then we descend the MTBDD of the transition relation and replace leaves representing the transition times by ROBDDs representing the stutter states time chains. An illustrative example is given in Fig. 7.1. It has to be noted that the substitution of interval transitions by chains of stutter states only requires a constant number of operations on MTBDD leaves. No state space traversal is required.

Reduce. The main operation of the *transitive* and *iterative* algorithms of section 5.2 (and also of the separate state space traversal of section 5.3) is merging two transitions. This is done by the following operation: $\nabla_{\cup} s''. T_1(s, s')|_{s' \leftarrow s''} \oplus T_2(s, s')|_{s \leftarrow s''}$ where

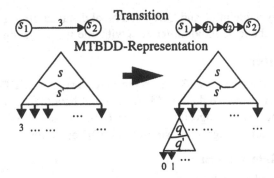

Fig. 7.1. Introducing Stutter States by simple MTBDD Leaf Operations

T_1 contains the starting transitions and T_2 contains the following transitions (it may happen that $T_1 = T_2$, e.g. for building the transitive closure). The starting states of transitions are defined by the atomic propositions in s and the final states of a transition are represented by atomic propositions in s'. s'' represents new temporary atomic propositions identifying the intermediate state. The plus operator builds the addition of all combinations of the set elements and joins them into one set:

$$A \oplus B = \{a + b \mid a \in A, b \in B, A \neq \emptyset, B \neq \emptyset\} \qquad (14)$$

The corresponding MTBDD operations are informally illustrated in Fig. 7.2.

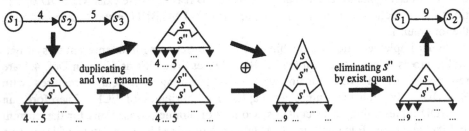

Fig. 7.2. Merging Transitions using simple MTBDD operations

7.3 MTBDD Minimization

The necessity to add additional variables to encode necessary stutter states has turned out to be the performance bottleneck of the composition algorithm as this may considerably increase the number of MTBDD nodes and directly affects the model checking time. The main effect is that the composed structure usually needs considerably less chain propositions compared to the sum of chain propositions in the original but expanded structures. This is due to the fact that after the structure reduction, usually only few additionally states have to be encoded. Those states occur if the original intervals have to be broken up into parts, each break point requiring an additional state. In the extreme – if all original timed transitions can be retained – no chain propositions are required anymore in the reduced composed structure.

In the following section we present two MTBDD based minimization algorithms which heuristically reduce the number of MTBDD nodes. They are not guaranteed to find the optimal solution but only require a negligible amount of computation effort compared to the composition and model checking effort as they directly work on the MTBDD data structures. Many implementation details are given in [15].

Reencoding. The first approach reencodes the chain propositions of the resulting structure. To reencode these additional propositions, we first have to determine how many states carry the same original labels (but different chain labels). For this operation, we take a representation of the state space without timing information. It maps all existing states represented by their original and chain propositions to 1. The chain propositions are removed by projection: $count(o) = \nabla_+ c.S(oc)$ where $c = \{c_0, ..., c_{m-1}\} \subseteq P$ are the chain propositions, $o = \{o_0...o_{k-1}\} \subseteq P$ are the original propositions and S represents the state space ROBDD (Fig. 7.3).

After this computation, we determine the maximum $m = \nabla_{max} o.count(o)$. We encode the state space with $l = \lceil \log_2(m) \rceil$ new atomic propositions by retaining the original atomic propositions. It has to be noted that this state counting does not require a state traversal. Only a few MTBDD operations are needed.

The aim is to substitute the set of old encoding variables c by new propositions $n = \{n_0, ..., n_{l-1}\}$ such that $l \le m$. Therefore, we build an encoding relation C which maps $(o_0...o_{k-1}c_0...c_{m-1}n_0...n_{l-1})$ to 1 where the valuation in n is the new encoding for the valuation in c in the state given by the valuation in o. This encoding relation is created by a special ROBDD operation which is described in more detail in [15].

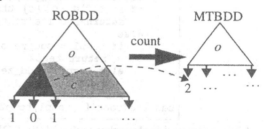

Fig. 7.3. Counting the Number of States with the same original Propositions with one MTBDD Operation

We can compute the new encoded transition relation by an operation similar to the relational product of ROBDDs:

$$\nabla_\cup c, c'. \Big((T(o, c, o', c') \bullet C(o, c, n)) \bullet C(o', c', n') \Big) \qquad (15)$$

The variable sets o', c' and n' represent copies for representing successor states. The operation $\bullet$ selects the valuations of the first argument, if the second argument for this valuation is not the empty set.

If the encoding ROBDDs are given, this procedure reencodes the state space in a constant number of ROBDD/MTBDD operations. This algorithm guarantees a smaller or equal number of chain propositions and therefore a smaller or equal number of MTBDD variables. But it cannot guarantee a smaller number of MTBDD nodes. Especially if the number of old and new chain propositions are similar in size, it may happen that a larger MTBDD results (see also Table 3 in section 8).

Minimal MTBDD Paths. The second minimization technique does not reencode the chain propositions by replacing them with new ones. It rather removes those parts of the MTBDD which are not necessary to distinguish different states. Thus this operation builds equivalence classes of encodings.

Paths in ROBDDs or MTBDDs to non zero leaves correspond to states or sets of states in the state space. On paths with non-zero leaves, all nodes with a zero-successor can be removed (n_i in Fig. 7.4). This operation corresponds to finding the minimal path length in the MTBDD/ROBDD to distinguish different states.

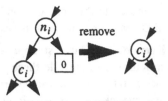

Fig. 7.4. Remove MTBDD nodes

For building the encoding relation we introduce new atomic propositions n in the state space ROBDD, such that every chain proposition is coupled with a new proposition. This ROBDD establishes an identical encoding relation (id) for the variables n and c. The minimal path algorithm works as follows:

```
function minimal_path(bdd f)
  if is_leaf(f) then return f
  else
      temp1 = minimal_path(bdd_then(f))
      temp2 = minimal_path(bdd_else(f))
      if root(f)∈ (o ∪ c) then
         return bdd_ite(root(f),temp1,temp2)
      else
         if temp1 = bdd_zero then
            return temp2
         else temp2 = bdd_zero then
            return temp1
         else return
bdd_ite(root(f),temp1,temp2)
```

After applying this algorithm to the id-ROBDD, we obtain a representation with minimal ROBDD paths in the variables of n. This encoding relation may be combined with the relational product, described above, to compute the reduced transition relation. With the exception of building the id function, this procedure works with a constant number of ROBDD/MTBDD operations.

7.4 Real Time Model Checking

The set of natural numbers necessary to encode the clock values of states and the transition times are represented using bitvectors. With this representation we may extend the *and, or* and *not* operator to the corresponding bitvector operators. To obtain decremented clock values (*dec* function as introduced in section 3) we shift all bitvectors that represent clock values one position to the right.

7.5 Making Real Time Model Checking more efficient

The prediction function as described in section 6.1 is defined over sets of clock values, i.e. it can be seen as a terminal function of MTBDD leaves. Therefore, we are able to apply this function to all corresponding leaves of two MTBDDs:

$$mp = \text{analyze2}(\varphi, \Lambda_{new}, \text{predictEF}, \text{Min}) \qquad (16)$$

Min indicates the analyze function to take the minimum of the subtrees in an MTBDD node during the recursive MTBDD descent. After computing mp we perform the local fixpoint iteration steps with one MTBDD operation: $\text{apply2}(\varphi, \Lambda_{new}, \text{shiftEF})$. More details about the implementation of the time jump technique with bitvector transducers are given in [15].

8 Experimental Results

We demonstrate our algorithms on various examples. First we compare approaches for structure composition and minimization. Afterwards we present results which compare our model checking approach to SMV and KRONOS.

In our first example (Table 1) we compose n ISs, each with two states toggling with a fixed delay (8 for the 1. IS, 12 for the 2. IS, 16 for the 3. IS etc.). The columns *trans*, *iter*, *index* and *sep* indicate the runtimes in seconds of the four composition algorithms of section 5[1]. The next columns denote the number of nodes of the MTBDD representation for the reachable state space and the transition relation, respectively. #stat/trans is the result directly after composition (without minimization); reencode and minpath are the results for the algorithms of section 7.3.

Table 1. First Example

#	trans	iter	index	sep	#stat/trans	reencode	minpath
2	0,1	0,2	0,05	0,05	22/66	3/19	5/(42)[a]19
3	0,6	0,4	0,1	0,1	64/198	3/37	5/(126)37
4	4,4	2,6	1,2	0,6	163/731	16/114	12/(457)119
5	10,4	4,2	1,9	1,4	303/1054	24/145	12/(628)147
6	76,5	52,6	25,5	14,4	676/3484	44/427	20/(2016)327
7	300,5	175,9	78,5	35,2	1858/9464	44/601	20/(5770)501

a. The result in brackets indicate the size of the transition relation in case that only one set (the starting states) of state variables is reencoded. It has to be noted that this still leads to a correctly represented structure.

The second example is similar to the first one, but the structures carry a fixed delay of 7, 12, 18 and 25, i.e. we have a less homogeneous timing. This directly affects the results as shown in Table 2.

Table 2. Second Example

#	trans	iter	index	sep	#stat/trans	reencode	minpath
2	0.3	0.3	0.1	0.1	56/182	32/116	24/(140)99
3	3.2	1.6	0.8	0.7	253/872	54/234	45/(513)339
4	427.6	87.0	48.9	30.7	930/4687	706/4644	267/(3519)2398

In Table 3, equivalent toggling structures with 2 states, each with an interval range [9, 12], are composed.

Table 3. Third Example

#	iterative	separate	#stat/trans	reencode	minpath
2	0.8	1.4	83/346	44/313	34/(315)257
3	58.6	14.7	299/1782	302/4724	227/(1733)1822
4	13737.6	56.9	519/5323	2461/58350	420/(5173)5511
5	> 50000.0	383.3	739/12654	23899/765322	613/(12569)13158

---

1. All experiments have been performed on an ULTRASparc Station with 166MHz and 196MB.

The fourth example shows a system of one writer and n readers on a shared memory. The access of the writer and the readers to the shared memory is synchronized by signals. Writing/reading to the shared memory is mutually exclusive. Writing may consume between 30 and 60 time units, reading between 20 and 40 (Table 4).

Table 4. Shared Memory Model

#	trans	iter	index	#stat/trans	reencode	minpath
2	20.3	1.8	1.1	412/1531	99/355	72/(834)162
3	57.4	7.1	4.7	1050/12188	102/944	95/(2871)783
4	175.4	24.5	19.3	2320/71582	105/972	118/(14278)1146
5	492.6	77.4	66.0	4854/280219	108/1000	141/(41920)1547
6	1332.1	245.5	229.8	9916/954956	111/1028	164/(100759)1988

Comparing the results of minimization, it is visible that no single algorithm outperforms the others. Moreover, reencoding may even worsen the result, i.e. does not minimize at all! However, if minimal_path is used with a reencoding of the starting state set of the transition relation only, we always get a reduction.

Fig. 8.1 shows the effect of the composition on the delay times and the interval widths of the shared memory example with 6 readers and one writer. Together the 7 modules have 22 transitions, the composed structure carries 105,084 transitions. On the left (right) side of the y-axis the distribution before (after) composition is given. The delay times (ranging from 1 to 60) in the left picture are an enumeration of all transitions (e.g. an interval [3, 5] leads to three transitions of value 3, 4 and 5; the interval widths in the right part range from 1 to 31. It is obvious that the composition considerably reduces the interval widths resulting in many single delay values. This effect usually occurs for all real-time verification approaches, however, in our approach this effect is explicitly visible.

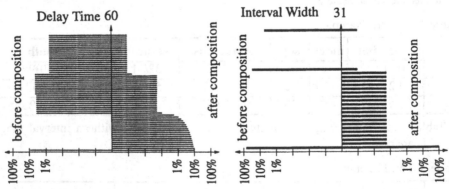

Fig. 8.1. Relative Delay and Interval Width Distribution before and after Composition

Fig. 8.2 shows the results of the model checking approach presented in section 6. The example is the priority inheritance protocol [10]. Modeling the protocol was performed using the techniques presented in section 4.2, the timed specification could be encoded in a straightforward way in the respective temporal logic.

We have compared our algorithm to SMV and KRONOS. Our examples could not be treated directly with UPPAAL as the specification language is not sufficiently expressive to state mutual time-bounded reachability properties of states. This is necessary for the specification of the properties to be verified. We have omitted the experimental results of the VERUS system, as the underlying model checking are the standard algorithms as used in SMV. Thus, in principle, the results do not differ from our SMV results. In Fig. 8.2 CCTL and CCTL+ denote the runtime without and with time prediction.

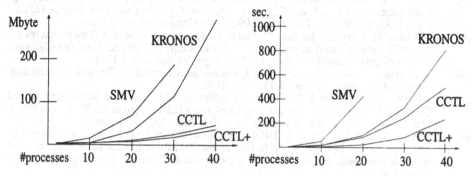

Fig. 8.2. Comparing Model Checking Results (Memory and Runtime)

9 Conclusion

In this paper we have presented algorithms based on symbolic state set representation and traversal. All aspects of real time verification are covered: model representation, model composition, heuristic minimization of composed structures and model checking. Although building the composition of timed structures is expensive (especially if time intervals are supported and used as in our case) our new composition and minimization algorithms perform well due to the symbolic state set representation and manipulation using MTBDDs. Many steps of the algorithms can be performed by simple and inexpensive MTBDD operations. Minimization often leads to a considerable reduction of the product structure which in turn reduces the model checking effort. It was also shown that MTBDD-based model checking outperforms the corresponding SMV and KRONOS verification.

The aim of the presented work was to set up a framework to study empirical results of composing temporal structures. Thus we deliberately have not (yet) used abstraction techniques [see e.g. 16] which are complementary to our work and will be studied in the future. We currently implement a language front-end for our real-time verification environment RAVEN which will be based on a real-time extension of ESTEREL. RAVEN is currently being integrated as part of our verification system C@S[1] [17].

10 References

[1]   M. Bozga, O. Maler, A. Pnueli, and S. Yovine. Some progress in the symbolic verification of timed automata. In O. Grumberg, editor, *Conference on Computer Aided Verification (CAV)*, volume 1254 of *Lecture Notes in Computer Science*, pages 179–190. Springer Verlag, June 1997.

---

1. http://goethe.ira.uka.de/hvg/cats/

[2]   K. Larsen, P. Pettersson, and W. Yi. UPPAAL: Status & developments. In O. Grumberg, editor, *Conference on Computer Aided Verification (CAV)*, volume 1254 of *Lecture Notes in Computer Science*, pages 456–459. Springer Verlag, June 1997.

[3]   S. Campos, E. Clarke, and M. Minea. The verus tool: A quantitative approach to the formal verification of real-time systems. In O. Grumberg, editor, *Conference on Computer Aided Verification (CAV)*, volume 1254 of *Lecture Notes in Computer Science*, pages 452–455. Springer Verlag, June 1997.

[4]   R. Alur, C. Courcoubetics, and D. Dill. Model Checking for Real-Time Systems. In *IEEE Symposium on Logic in Computer Science (LICS)*, pages 414–425, Washington, D.C., June 1990. IEEE Computer Society Press.

[5]   J. Burch, E. Clarke, K. McMillan, D. Dill, and L. Hwang. Symbolic Model Checking: 1020 States and Beyond. In *IEEE Symposium on Logic in Computer Science (LICS)*, pages 1–33, Washington, D.C., June 1990. IEEE Computer Society Press.

[6]   E. Asarin, M. Bozga, A. Kerbrat, O. Maler, M. Pnueli, and A. Rasse. Data structures for the verification of timed automata. In O. Maler, editor, *Hybrid and Real-Time Systems*, pages 346–360, Grenoble, France, 1997. Springer Verlag, LNCS 1201.

[7]   E. Emerson, A. Mok, A. Sistla, and J. Srinivasan. Quantitative Temporal Reasoning. *Journal of Real-Time Systems*, 4:331–352, 1992.

[8]   J. Frößl, J. Gerlach, and T. Kropf. An Efficient Algorithm for Real-Time Model Checking. In *European Design and Test Conference (EDTC)*, pages 15–21, Paris, France, March 1996. IEEE Computer Society Press (Los Alamitos, California).

[9]   T. Kropf and J. Ruf. Using MTBDDs for discrete timed symbolic model checking. Technical Report SFB358-C2-5/96, Universität Karlsruhe, Institut für Rechnerentwurf und Fehlertoleranz, August 1996. ftp://goethe.ira.uka.de/pub/hvg/techreports/SFB358-C2-5-96.ps.gz.

[10]   J. Ruf and T. Kropf. Symbolic model checking for a discrete clocked temporal logic with intervals. In E. Cerny and D. Probst, editors, *Conference on Correct Hardware Design and Verification Methods (CHARME)*, pages 146–166, Montreal, Canada, October 1997. IFIP WG 10.5, Chapman and Hall.

[11]   J. Lipson, editor. *Elements of Algebra and Algebraic Computing*. The Benjamin/Cummings Publishing Company, Inc., 1981.

[12]   R. Bryant. Graph-Based Algorithms for Boolean Function Manipulation. *IEEE Transactions on Computers*, C-35(8):677–691, August 1986.

[13]   E. Clarke, K. McMillian, X. Zhao, M. Fujita, and J.-Y. Yang. Spectral Transforms for large Boolean Functions with Application to Technologie Mapping. In *ACM/IEEE Design Automation Conference (DAC)*, pages 54–60, Dallas, TX, June 1993.

[14]   R. Bahar, E. Frohm, C. Gaona, G. Hachtel, E. Macii, A. Pardo, and F. Somenzi. Algebraic Decision Diagrams and Their Applications. In *IEEE/ACM International Conference on Computer Aided Design (ICCAD)*, pages 188–191, Santa Clara, California, November 1993. ACM/IEEE, IEEE Computer Society Press.

[15]   J. Ruf and T. Kropf. Using MTBDDs for composition and model checking of real-time systems. Technical Report SFB358-C2-1/98, Universität Karlsruhe, Institut für Rechnerentwurf und Fehlertoleranz, January 1998. ftp://goethe.ira.uka.de/pub/hvg/techreports/SFB358-C2-1-98.ps.gz.

[16]   S. Graf and H. Saidi. Construction of abstract state graphs with PVS. In O. Grumberg, editor, *Conference on Computer Aided Verification (CAV)*, volume 1254 of *Lecture Notes in Computer Scienece*, pages 72–83. Springer Verlag, June 1997.

[17]   K. Schneider and T. Kropf. A unified approach for combining different formalisms for hardware verification. In M. Srivas and A. Camilleri, editors, *International Conference on Formal Methods in Computer Aided Design (FMCAD)*, volume 1166 of *Lecture Notes in Computer Science*, pages 202–217, Palo Alto, USA, November 1996. Springer Verlag.

Formal Methods in CAD from an Industrial Perspective

Carl-Johan H. Seger
Strategic CAD Labs
Intel Corporation

Abstract. Symbolic trajectory evaluation (STE) is a model checking approach for efficient circuit verification. In the talk we follow the development of STE; from an ad-hoc use of symbolic simulation to a rich theory of model checking algorithms, inference rules, and abstraction verification techniques. In parallel, we discuss the development of a practical verification system based on STE with an emphasis on the impact on the system by actual use in an industrial setting.

A Methodology for Automated Verification of Synthesized RTL Designs and Its Integration with a High-Level Synthesis Tool *

Nazanin Mansouri and Ranga Vemuri

Digital Design Environments Laboratory
ECECS Department
University of Cincinnati
Cincinnati, Ohio 45221-0030
nmansour@ececs.uc.edu,ranga.vemuri@uc.edu
http://www.ececs.uc.edu/~ddel

Abstract. High-level synthesis tools generate RTL designs from algorithmic behavioral specifications and consist of well defined tasks. Widely
used algorithms for these tasks retain the overall control flow structure of
the behavioral specification allowing limited code motion. Further, HLS
algorithms are oblivious to the mathematical properties of arithmetic
and logic operators, selecting and sharing RTL library modules solely
based on matching uninterpreted function symbols and constants. This
paper reports a verification methodology that effectively exploits these
features to achieve efficient and fully automated verification of synthesized designs and its incorporation in a relatively mature HLS tool.
In the proposed methodology, a correctness condition generator is tightly
integrated with the HLS tool to automatically generate (1) formal specifications of the behavior and the RTL design, (2) the correctness lemmas
establishing equivalence between them, and (3) their proof scripts that
can be submitted to a higher-order logic proof checker without further
human interaction.

1 Introduction

High-level synthesis systems generate register-transfer level (RTL) designs from
algorithmic behavioral specifications (Figure 1). The RTL design consists of a
data path and a controller. The data path consists of component instances selected from an RTL component library. The controller is a finite-state machine
(FSM) description subject to down-stream FSM synthesis. High-level synthesis
process begins by compiling the behavior specification into a control-data flow
graph (CDFG) representation. The CDFG typically consists of operator nodes

---

* This work is sponsored in part by DARPA and monitored by US Army Ft. Huachuca
under contract number DABT63-96-C-0051.

representing the arithmetic and logical operators in the specification and control nodes representing the control flow operations in the specification. The goal of the high-level synthesis system is to bind the operator nodes to arithmetic-logic units (ALUs), specification variables and data dependencies to registers and interconnect units, and control nodes to states in the controller FSM such that the user constraints on speed and cost (area) are met.

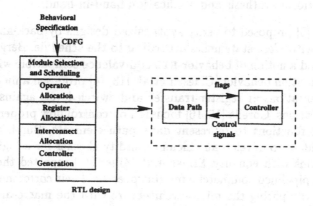

Fig. 1. High-Level Synthesis Process

In this paper we present a methodology to automatically generate the correctness conditions and their proof scripts as a byproduct of synthesis. Our verification strategy exploits the features of common high-level synthesis algorithms. We argue that the high-level synthesis tools can easily produce information about behavior-to-RTL state and variable binding which can be effectively exploited during verification. Further, we show that in the verification of automatically synthesized designs, functions and constant symbols can be left uninterpreted leading to efficient verification based mostly on rewriting strategies and that the necessary rewriting steps can be automatically generated by making use of the variable and state binding information. We have implemented this method in the context of a relatively mature high-level synthesis system and verified several synthesized designs using the PVS proof checker.

In Section 2, we review related research. In Section 3, we motivate the basis for the proposed technique through an example. In Section 4, we discuss the proposed method in a more formal setting. In Section 5, we present the correctness condition generator integrated with the high-level synthesis system. In Section 6, we present further enhancements to the proposed method, eliminating some of the assumptions made earlier. Section 7 discusses the implementation issues and concluding remarks on ongoing work are presented in Section 8.

2   Related Research

Several authors have proposed techniques for verification of synthesized designs. Devadas et al. [3] proposed methods to verify equivalence between sequential machines at RTL and logic levels. Transformational approach to hardware synthesis has been pioneered by Johnson [7]. Vemuri [8] and Feldbusch and Kumar [9] proposed converting RTL implementation to a normal form . However, transformation into normal form seems to be possible only for restricted classes of designs. Rajan [10] addressed the same question using theorem proving. Eisnbiegler and Kumar [12] proposed tight integration of high-level synthesis with theorem proving to perform synthesis and verification hand-in-hand.

Corella et al. [5] proposed to verify synthesized designs by back-annotating the specification with *clock* statements according to the schedule. Bergamaschi and Raje [6] defined a notion of behavior-RTL equivalence compatible with the specification's simulation behavior. Claesen et al. [13, 14] proposed a method to compare implementations at register-transfer and switch levels against signal flow graph specifications. Corella [15, 16] focussed on control flow properties by using uninterpreted functions to represent data path elements. Burch and Dill [17] used similar ideas to develop an efficient validity checker for a logic of uninterpreted functions with equality. Srivas and Miller [18] reported the verification of AAMP5, a pipelined commercial microprocessor. Their correctness conditions are based on comparing the micro-architecture with the macro-architecture at *visible* states.

In general, all the proposed methods for formal verification of correctness of synthesized designs take advantage of the features and limitations of the synthesis and optimization algorithms which generate design structures in a particular style. These methods resort to syntactic or symbolic comparisons avoiding the combinatorial explosion problem involved in introducing more general notion of correctness based on, for example, boolean equivalence. Our method also takes advantage of the nature of the HLS algorithms. In addition, our method is fully integrated with an HLS tool such that both the correctness theorems and their proofs are automatically produced as auxiliary outcomes of the synthesis process. These proofs are then readily processed by a higher-order logic proof checker.

3 Motivation

Consider the simple behavioral specification shown in Figure 2.a and its representation as a *behavioral automaton* in Figure 2.b. The behavior automaton [19] is a finite state automaton which consists of a set of states and a set of transitions among these states. There is a unique *start* state, BS_0. A state may be an *assignment* state or a *conditional* state. An assignment state is annotated with one or more assignment statements and has exactly one outgoing transition to a *next* state. A conditional state is not annotated with any assignment statements and has exactly two outgoing transitions, one of which is labeled with the condition v and the other is labeled with the condition $\neg v$, where v is a specification variable.

Figure 3 shows a register level design generated by an HLS system. The data path consists of a number of registers, one ALU, two buses and a number of

variables max, sum, val, grt;

max := 0;
sum := 0;

repeat
 val := read_input();
 sum := sum + val;
 grt := val > max;
 if grt **then** max := val;
forever;

(a) **Algorithmic Specification**   (b) **Behavioral Automaton**

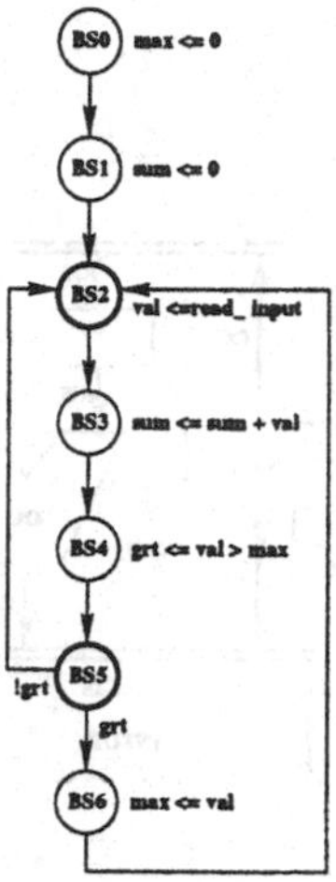

Fig. 2. A Behavior Specification

wires. Each data path component may have one or more control signals. For registers, we assume that the 'load' control is active 'high.' The controller is a finite state machine which interacts with the data path through *control signals* and *flags*. Controller has a unique start state DS_0. There are two types of states: *assignment* and *conditional*. An assignment state is annotated with the control signals that must be asserted 'high' in that state and has exactly one outgoing transition. We assume that the remaining control signals must be asserted 'low' in that state. A conditional state has exactly two outgoing transitions; one of these transitions is annotated with the condition f and the other is annotated with the condition $\neg f$, where f is a status flag from the data path. [1] Conditional states have no control signal annotations implying that all control signals must be held 'low'. In particular, this means that no data path register will be loaded in conditional states.

Our technique of determining the correctness of the RTL design depends upon (symbolically) comparing the values of certain critical specification variables in certain critical behavioral states with those of certain critical RTL registers in certain critical controller states. In our illustrative example, we compare the values of *sum*, *max*, *val* and *grt* with those of the registers with the same names in the following pairs of states: $\langle BS_0, DS_0 \rangle$, $\langle BS_2, DS_1 \rangle$, $\langle BS_5, DS_6 \rangle$. Determining the right pairs of variables and registers to compare and the pairs of states at which to compare requires help from the synthesis tool. However, further insight into the synthesis process reveals that such help is easily obtainable.

---

[1] In this paper we assume conditional states with exactly two outgoing transitions in both the behavior automaton and the RTL controller. This is easily generalizable to include conditional transitions based on multiple flags.

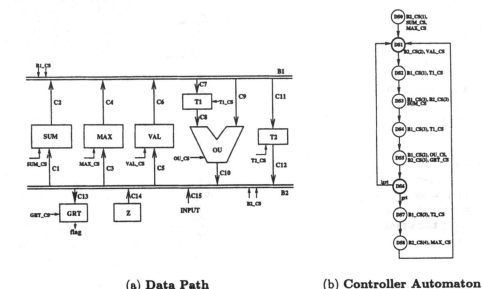

(a) **Data Path**   (b) **Controller Automaton**

Fig. 3. Example of a register level design generated by a high-level synthesis system

The HLS system generates a control-data flow graph (CDFG) from the behavioral specification and goes through the steps of scheduling, function unit allocation, register allocation, interconnect allocation and controller generation. High-level synthesis tools are oblivious to the mathematical properties of the operators used in the behavioral specification. Further, while scheduling the CDFG, HLS tools do not perform code motion across conditional and iterative statement boundaries. Such behavioral transformations are performed prior to the commencement of scheduling in a preprocessing step called *behavior transformation* [1, 2]. Proving the correctness of such general behavioral transformations involves techniques similar to those used in program verification. McFarland studied various behavioral transformations and proposed the algebra of behavioral expressions as means to specify and verify their correctness [4]. In this paper, we assume that the behavioral specification has already been transformed into the desired algorithmic form before synthesis commences. Once submitted to the synthesis system, it is not subjected to further behavioral transformation. High-level synthesis process is then predominantly concerned with constraint-satisfying sharing among resources (ALUs, registers and interconnections). To facilitate resource-sharing, HLS tool performs scheduling which permits time sharing of resources whose life times do not overlap across the scheduled time-scale.

Scheduling process may be viewed as code motion across a time scale. Operations may be scheduled at any time-step as long as the data and control dependencies are not violated. That is, if an operator is data-dependent on a source operator, it can only be scheduled after the source operator is scheduled. Similarly, if an operator is control dependent on a control operator then it is not scheduled until after the control operator is scheduled. All scheduling algorithms in high-level synthesis assume that control operators introduce sequential control flow points into the CDFG being scheduled [1, 2]. This ensures that the control flow branches in the behavior specification are preserved and no new control flow branches are introduced. Scheduling, thus, is the process of implicit code motion possibly involving introduction of additional temporary variables in order to explore the design-space to determine a constraint-satisfying time-area tradeoff point.

Operator, register and interconnect allocation algorithms, which follow scheduling, are typically based on clique partitioning or graph coloring following life-cycle analysis of the scheduled flow graph. Operator, register and interconnect allocation algorithms perform no code motion and do not alter the control flow. Their focus is essentially on resource sharing to meet cost (area) constraints.

The final step in the HLS process is controller generation [1]. A finite-state controller is generated by creating a state corresponding to each control-step in the scheduled flow graph and by creating a transition between each control-step boundary. Operation nodes in the CDFG are turned into assignment states in the controller where control signals to effect the corresponding register transfer in the data path are asserted. Conditional nodes in the CDFG are turned into conditional states predicated on the appropriate flags.

Outline of the Method: This discussion suggests that the following states in the behavior specification should be marked as *critical states*: (1) conditional states – states with more than one outgoing transition, (2) join states – states with more than one incoming transition; (3) output states – states that write to output ports; (4) input states – states that read from input ports, and, (5) the start state. These states introduce control flow dependencies into the CDFG and the HLS system never moves code across these states. Further, since, specification code between a pair of these states is subject to scheduling and hence to possible code motion during high-level synthesis, no other states can be marked as being critical. The following variables in the behavior specification should be marked as *critical variables* in the specification: (1) input variables, (2) output variables, (3) any variable that has a live value across a critical state. [2] The last category of critical variables are easily identified during life-time analysis that precedes register allocation in high-level synthesis. The critical variables are preserved by the HLS tool and manifest in the RTL design in the form of critical registers and the critical states in the behavior automaton are preserved by the HLS tool and manifest in the form of critical states in the controller automaton. Therefore, conditional, join, input, output and start states in the controller together form the critical states of the controller. We assume that all critical states are reachable from the start states respectively in the behavior automaton and the RTL controller, that is, any dead-code has been eliminated prior to high-level synthesis .

---

[2] Note that these include any variable used as the deciding variable in transitioning from a conditional state.

How do we identify the correspondence between critical variables and critical registers and between the behavioral critical states and the RTL critical states? We can count on the HLS tool to produce such correspondences in the form of auxiliary information as a byproduct of the synthesis process by maintaining links with the elements of the behavior specification throughout the synthesis process. Many authors, for example Thomas et al. [20], have described detailed methods to maintain the binding information during the synthesis process. *Binding* refers to the final assignment of behavioral operators, variables and data/control dependencies to RTL ALUs, registers and interconnect units.

In particular, we assume that the high-level synthesis tool can generate the mapping between critical variables in the specification and registers in the RTL data path and the mapping between the critical states in the behavior automaton and critical states in the RTL controller. $B_r : CR_b \times CR_d = \{\langle max, MAX\_reg \rangle,$ $\langle sum, SUM\_reg \rangle, \langle val, VAL\_reg \rangle, \langle grt, GRT\_reg \rangle\}$ defines the critical register binding and $B_s : CS_b \times CS_d = \{\langle BS_0, DS_0 \rangle, \langle BS_2, DS_1 \rangle, \langle BS_5, DS_6 \rangle\}$ defines the critical state binding for the example shown in Figures 2 and 3.

We define a *critical path* in a behavioral automaton as a directed path from a critical state s_1 to another critical state s_2 without traversing through any other critical state. s_1 is called the first or originating state of this critical path and s_2 is called its last or terminating state. Since we assumed that states in the behavioral automaton have exactly one or two outgoing transitions, there can be at most two critical paths between any pair of critical states. Critical paths in the RTL controller are similarly defined. $B_p = CP_b \times CP_d = \{\langle bcp_1, dcp_1 \rangle, \langle bcp_2, dcp_2 \rangle, \langle bcp_3, dcp_3 \rangle, \langle bcp_4, dcp_4 \rangle\}$ is the critical path binding for our example, where $bcp_1, bcp_1, bcp_1, bcp_1, dcp_1, dcp_1, dcp_1,$ and dcp_1 the critical paths of the behavior automaton and controller are equal to $[BS_0, BS_1, BS_2],$ $[BS_2, BS_3, BS_4, BS_5],$ $[BS_5, BS_2],$ $[BS_5, BS_6, BS_2],$ $[DS_0, DS_1],$ $[DS_1, DS_2,$ $DS_3, DS_4, DS_5, DS_6],$ $[DS_6, DS_1]$ and $[DS_6, DS_7, DS_8, DS_1]$ respectively.

In order to show that the RTL correctly implements the value transfers in the behavioral specification, we need to symbolically show that the values in critical variables match values in corresponding pairs of critical registers at the critical states, provided that they match at the start states.

We accomplish this by showing that for each critical path between a pair of critical states, if the critical variables match the critical registers at the originating states, then they will match them at the terminating states. This involves computation and comparison of the symbolic values of the various critical registers and variables along corresponding pairs of critical paths in the behavior automaton and the RTL controller. In the following section, we will present the key elements of this discussion in a more formal setting.

4 Formalization of the Verification Technique

Let R_b be the set of variables in the behavioral specification. Let $CR_b \subseteq R_b$ be the set of *critical variables*. Let S_b be the set of states in the behavioral automaton and the X_b be the set of transitions among these states. Let $S0_b$ be the unique start state of the behavioral automaton. Let $CS_b \subseteq S_b$ be the set of critical states. We assume that every critical state is reachable from the start state. Let CP_b be the set of critical paths among these critical states. For any critical path $p \in CP_b$, $F_b(p)$ and $L_b(p)$ denote the originating and terminating states of p.

Let R_d be the set of registers in the register level data path. Let $CR_d \subseteq R_d$ be the set of critical registers in the data path. Let S_d be the set of states in the controller and X_d be the set of state transitions. Let $S0_d$ be the unique start state of the controller. Let $CS_d \subseteq S_d$ be the set of critical states in the controller and let CP_d be the set of critical paths among these states. We assume that every critical state is reachable from the start state. For any critical path $p \in CP_d$, $F_d(p)$ and $L_d(p)$ denote the originating and terminating states of p.

Following the previous discussion, we postulate that the high-level synthesis tool can produce, as a byproduct of the synthesis process, the following two

mappings (called bindings in the synthesis terminology): $B_r : CR_b \rightarrow CR_d$, is the *critical register binding*, and, $B_s : CS_b \rightarrow CS_d$ is the *critical state binding*. The start state of the behavior is always mapped to the start state of the controller, that is, $B_s(S0_b) = S0_d$.

From B_s we can easily derive another mapping $B_p : CP_b \rightarrow CP_d$ which is the *critical path binding*. A critical path $p_b \in CP_b$ is mapped to critical path $p_d \in CP_d$ if and only if their originating and terminating states are mapped by B_s and the transition conditions, if any, on the outgoing transitions of their originating states match. More formally, $B_p(p_b) = p_d$ if and only if $B_s(F_b(p_b)) = F_d(p_d)$, $B_s(L_b(p_b)) = L_d(p_d)$, and if v ($\neg v$) is the condition variable annotation on the originating transition of p_b then $B_r(v)$ ($\neg B_r(v)$) is the condition register annotation on the originating transition of p_d. This ensures that if p_b is traversed in the behavior, p_d will be traversed in the RTL controller.

An *execution path* in the behavior is a finite sequence of critical states $[s_1, s_2, \cdots, s_i, s_{i+1}, \cdots]$ such that the first state in the sequence is the start state $S0_b$ and any two successive states in the sequence form a critical path, that is, $\forall i > 1, \exists p \in CP_b : s_i = F_b(p)$ and $s_{i+1} = L_b(p)$. Execution path in the RTL controller is similarly defined. Let EP_b denote the set of all possible behavioral execution paths and EP_d denote the set of all possible RTL execution paths. We can construct an execution path binding, $B_e : EP_b \rightarrow EP_d$ using the critical state binding as follows: If $e = \{s_1, s_2, \cdots, s_i, s_{i+1}, \cdots\}$ is an execution path in the behavior automaton, then $B_e(e) = \{B_s(s_1), B_s(s_2), \cdots, B_s(s_i), B_s(s_{i+1}), \cdots\}$. The last state in an execution path e is called the *termination state* of e and denoted by $T_b(e)$ for the behavior automaton and by $T_d(e)$ for the RTL controller.

For the purposes of defining co-execution equivalence, we postulate an uninterpreted domain of *values*, V. These values can be 'stored' in behavioral variables as well as RTL registers. We postulate two functions for assigning values to critical variables and critical registers: $V_b : EP_b \times CR_b \rightarrow V$ determines the value of a critical variable r_b when the behavioral automaton traversed the execution path e and reached the state $T_b(e)$. $V_d : EP_d \times CR_d \rightarrow V$ is similarly defined. In the next section we show axiomatic definitions of V_b and V_d suitable for symbolic manipulation, automatically generated from behavioral specifications and RTL descriptions respectively.

We are now ready to define various equivalence relationships between the behavior and the RTL design. We say that the initial state $S0_b$ in behavior automaton is equivalent to the initial state $S0_d$ in the controller provided $\forall r \in CR_b, V_b([S0_b], r) = V_d([S0_d], B_r(r))$. We denote initial state equivalence by $S0_b \equiv S0_d$.

We say that a critical state s in the behavior and the RTL critical state $B_s(s)$ are *equivalent* to each other provided $S0_b \equiv S0_d \Rightarrow \forall e \in EP_b : s = T_b(e), \forall r \in CR_b : V_b(e, r) = V_d(B_e(e), B_r(r))$. We denote state equivalence by $s \equiv B_s(s)$.

We say that a behavioral execution path e is *equivalent* to the RTL execution path $B_e(e)$ provided $S0_b \equiv S0_d \Rightarrow \forall s \in e, s \equiv B_s(s)$. We denote execution path equivalence by $e \equiv B_e(e)$.

We say that the RTL design is equivalent to the behavior specification provided $\forall e \in EP_b, e \equiv B_e(e)$.

We say a behavioral critical path p is equivalent to the RTL critical path $B_p(p)$ provided $F_b(p) \equiv F_d(B_p(p)) \Rightarrow L_b(p) \equiv L_b(B_p(p))$. We denote critical path equivalence by $p \equiv B_p(p)$.

We claim that critical path equivalence implies execution path equivalence per the following theorem, offered here without proof:

Theorem: If every critical path in the behavior is equivalent to the RTL critical path to which it is bound during the synthesis process, then the RTL design is equivalent to the behavior specification. Formally, $\forall p \in CP_b, p \equiv B_p(p) \Rightarrow \forall e \in EP_b, e \equiv B_e(e)$.

The proof of this statement is straight forward and follows from the fact that, in both behavioral automaton and the RTL controller, the critical states are reachable from the respective start states. The proofs can be done by induction on the length of the execution path.

5 Correctness Condition Generator

We assume that the operating environment of the designs ensures that $S0_b \equiv S0_d$. Typically, the environment ensures that all the data and control registers

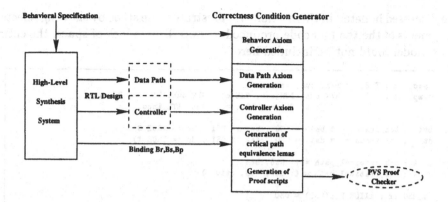

Fig. 4. Stages of Correctness Condition Generation

are reset at the start states. The goal of our proof effort is to show that each critical path in the behavior is equivalent to its corresponding critical path in the structure. We determine this by using symbolic term rewriting in a higher-order logic theorem prover.

```
equivalent_states : [beh_state,rtl_state -> bool]

beh_trans : TYPE+ = [# bs1 : beh_state, bs2 : beh_state , bc : bool #]
beh_critical_path : TYPE+ = list[beh_trans]
Val_b : [spec_var, beh_state -> value]

flag : [rtl_state -> bool]

beh_transition : PRED[beh_trans]
source_beh_trans(t : beh_trans) : beh_state = bs1(t)
target_beh_trans(t : beh_trans) : beh_state = bs2(t)

B_s : [beh_state -> rtl_state]
B_r : [spec_var -> comp_out]
B_p : [beh_critical_path -> rtl_critical_path]

First_b(cp : beh_critical_path) : beh_state = source_beh_trans(car(cp))

Last_b(cp  : beh_critical_path) : beh_state = target_beh_trans(car(reverse(cp)))

beh_transition_condition(cp : beh_critical_path) : bool = beh_transition(car(cp))

% similar declarations for RTL design go here.
```

Fig. 5. Some Basic Definitions

Our proof effort is carried out in the PVS theorem prover environment [11]. We modified the high-level synthesis system DSS [23] to generate the three bindings B_r, B_s and B_p. These bindings along with the behavior specification and the RTL design are the inputs to the Correctness Condition Generator (CCG) shown in Figure 4. CCG generates the following theories in five steps: (1) behavior axioms; (2) data path axioms; (3) controller axioms; (4) critical path correctness lemmas; and, (5) proof scripts for each correctness lemma. These five steps will be discussed in detail in this section. We illustrate these steps by showing selected fragments of the the PVS code produced by CCG. Due to lack of space, the entire PVS model could not be included here.

```
spec_var : TYPE+ = max, sum, val, grt
comp_out : TYPE+ = MAX_out, SUM_out, VAL_out, GRT_out, B1_out, B2_out,
                   OU_out, T1_out, T2_out, Z_out, RTL_input

bt1  : beh_trans = (# bs1 := BSO , bs2 := BS1 , bc := TRUE #)
dt1  : rtl_trans = (# ds1 := DSO , ds2 := DS1 , dc := TRUE #)

bcp4 : beh_critical_path = (: bt6, bt8 :)
dcp4 : rtl_critical_path = (: dt7, dt9, dt10 :)

Bs_BSO_ax : AXIOM B_s(BSO) = DSO
Br_max_ax  : AXIOM B_r(max) = MAX_out
Bp_bcp4_ax : AXIOM B_p(bcp4) = dcp4

state_equivalence : AXIOM (FORALL (bs : beh_state, ds : rtl_state):

             equivalent_states(bs,ds)
               IFF
             (ds = B_s(bs) AND
             Val_b(max,bs) = Val_d(MAX_out,ds) AND
             Val_b(sum,bs) = Val_d(SUM_out,ds) AND
             Val_b(val,bs) = Val_d(VAL_out,ds) AND
             Val_b(grt,bs) = Val_d(GRT_out,ds)))
```

Fig. 6. Some Basic Design-Specific Declarations

At first, the CCG generates a set of general definitions and a set of declarations specific to the design under investigation. Figure 5 shows some of the basic definitions. These definitions are used in all the equivalence checking exercises. Figure 6 shows some of the design-specific elements of the PVS model. This set of declarations and axioms define information about the specification and RTL design. This information includes specification variables and the RTL component declarations, critical path specifications for both behavior and RTL automata, and bindings B_r, B_s and B_p.

```
trans_bs5_bs6_ax : AXIOM beh_transition(bt6) = (Val_b(grt,BS5) = ONE)

bs5_bs6_ax : AXIOM  beh_transition(bt6)
        IMPLIES
            (Val_b(max,BS6) = Val_b(max,BS5) AND
               Val_b(sum,BS6) = Val_b(sum,BS5) AND
                 Val_b(val,BS6) = Val_b(val,BS5) AND
                   Val_b(grt,BS6) = Val_b(grt,BS5))

bs6_bs2_ax : AXIOM beh_transition(bt8)
        IMPLIES
            (Val_b(max,BS2) = Val_b(val,BS6) AND
               Val_b(sum,BS2) = Val_b(sum,BS6) AND
                 Val_b(val,BS2) = Val_b(val,BS6) AND
                   Val_b(grt,BS2) = Val_b(grt,BS6))
```

Fig. 7. Axioms for Some Behavior Transitions

5.1 Behavior Axiom Generation

This step examines the behavior specification, written in a simple subset of VHDL in our case, and converts it into a series of axioms that collectively specify the value transfers in the behavior. For each state transition in the behavior design one axiom is generated. This axiom specifies the value of each specification variable, at the destination state of the transition, in terms of the value of all the specification variables prior to transition. Figure 7 shows the axioms for the critical path $[BS_5, BS_6, BS_2]$. This critical path consists of two transitions $t6 = \langle BS_5, BS_6 \rangle$ and $t8 = \langle BS_6, BS_2 \rangle$. For each transition, one axiom is generated.

5.2 Data Path Axiom Generation

A pre-existing library of axioms defines the behavior of each type of RTL component. The input of a component can be the output of some other component or a primary input. Each library axiom specifies the input-output relation of a component, at each state. The value at the output of a component at a particular state is defined in terms of the data and control inputs of the component at that state, or its output at a previous state in the case of sequential components. Figure 8 shows the axiomatic behavior description for the register and bus components. These axioms are not design specific, and are included with the specification of all designs.

```
register        : [comp_out, comp_out, signal[bool] -> bool]
bus             : [arr[comp_out], comp_out, arr[signal[bool]] -> bool]

reg_ax : AXIOM (FORALL (c_in        : comp_out,
                        c_out        : comp_out,
                        load         : signal[bool],
                        s1           : rtl_state,
                        s2           : rtl_state):
    register(c_in,c_out,load) AND (s2 = next_state(s1)
        IMPLIES
    IF (load(s2)) THEN
        Val_d(c_out,s2) = Val_d(c_in,s2)
    ELSE
        Val_d(c_out,s2) = Val_d(c_out,s1)
    ENDIF))

bus_ax  : AXIOM (FORALL (c_ins       : arr[comp_out],
                         c_out        : comp_out,
                         wrs          : arr[signal[bool]],
                         ds           : rtl_state,
                         index        : nat) :
    bus(c_ins,c_out, wrs)
        IMPLIES
    (wrs(index)(ds) => (Val_d(c_out,ds) = Val_d(c_ins(index),ds))))
```

Fig. 8. Axioms for Some RTL Library Components

At the second stage of correctness condition generation, the data-path of the synthesized RTL design is modeled as a PVS theory [22]. For each component of the data-path, an axiom is generated, which specifies its type, and its interface with the rest of the components. The data-path axioms together with the the component library axioms define the behavior and interface of each individual component. Also, the interconnection of the control inputs of the RTL components with the controller, and the interconnection of the flags from the data-path to the controller are specified in this theory. Axiomatic specifications of some of the data-path components of our example design are shown in Figure 9.

```
GRT_cs, SUM_cs, MAX_cs, VAL_cs, T1_cs, T2_cs, OU_cs :  signal[bool]

B1_wrs      : arr[signal[bool]]
B1_ins      : arr[comp_out]

grt_cs_ax : AXIOM (FORALL (s : rtl_state) : GRT_cs(s) = Control_Signal(s,0))

b1_wrs_ax1 : AXIOM (FORALL (s : rtl_state) : B1_wrs(1)(s) = Control_Signal(s,7))
b1_ins_ax1 : AXIOM (B1_ins(1) = SUM_out)

flag_ax : AXIOM (FORALL (s : rtl_state) : flag(s) IFF (Val_d(GRT_out,s) = ONE))

grt_ax : AXIOM register(B2_out,GRT_out,GRT_cs)
z_ax : AXIOM constant_register(Z_out,ZERO)
ou_ax : AXIOM alu(T1_out,B1_out,OU_out,OU_cs)
b1_ax : AXIOM bus(B1_ins,B1_out,B1_wrs)
```

Fig. 9. Axioms for Some Data Path Components

5.3 Controller Axiom Generation

At this step of correctness condition generation, a constructive model of the RTL controller is generated. Two functions *next_state*, and *Control_Signal* are extracted from the controller and converted to PVS functions. The function *next_state*, defines the next state in terms of the current state and the values of the flags from the data-path to the controller at current state. The function *Control_Signal* defines the values of control signals from the controller to data-path at each state of the controller. Since the PVS specification of these functions are straight forward, they are not presented here.

5.4 Generation of Critical Path Equivalence Lemmas

For each pair of the behavior-RTL critical paths which are bound by the function B_p, a set of lemmas are generated. A general lemma states that if the initial states of the pair of the critical paths are equivalent, their final states should be equivalent. The instantiation of this statement for the behavior critical path $bcp_4 = [BS_5, BS_6, BS_2]$ and its RTL counterpart $dcp_4 = B_p(bcp_4) = [DS_6, DS_7, DS_8, DS_1]$ is the lemma eq_cp4 in Figure 10. A set of sub-lemmas (one sub-lemma for each specification variable v) state that if the initial states of the pair of critical paths are equivalent, then the values stored in the specification variable v and its RTL counterpart $B_r(v)$ at the final states of the critical paths will be equivalent. The proof of these sub-lemmas together complete the proof of the main lemma for the critical paths. An instance of such a lemma for the specification variable *max* is the lemma *eq_cp4_l1* in Figure 10. The equivalence of the behavior specification and the RTL design is established by proof of these lemmas for all the critical paths in the design.

```
eq_cp4_l1 : LEMMA
   (equivalent_states(First_b(bcp4),First_d(B_p(bcp4))) AND
   beh_transition_condition(bcp4) AND
   rtl_transition_condition(B_p(bcp4)))
      IMPLIES
   Val_b(max,Last_b(bcp4)) = Val_d(MAX_out,Last_d(B_p(bcp4)))

eq_cp4 : LEMMA
   (equivalent_states(First_b(bcp4),First_d(B_p(bcp4))) AND
   beh_transition_condition(bcp4) AND
   rtl_transition_condition(B_p(bcp4)))
      IMPLIES
   equivalent_states(Last_b(bcp4),Last_d(B_p(bcp4)))
```

Fig. 10. Some Critical Path Equivalence Lemmas

5.5 Generation of Proof Scripts

The generation of the proof scripts is the most elaborate stage in CCG. In this stage all the information about the design is processed and the rules for proving each lemma are produced. Proof scripts are generated making use of the axioms and definitions generated in the previous stages. These proofs are then subjected to verification by the PVS proof checker. These proofs make extensive use of symbolic rewriting, involving instantiation of definitions, axioms and other proven lemmas. A portion of the proof script for the lemma *eq_cp4_l1* is shown in Figure 11.

6 Further Enhancements

In the preceding sections, for clarity of discussion, we made several assumptions in presenting the method of behavior-RTL equivalence checking. Some of these assumptions are easily alleviated. In particular, the correctness condition generator we implemented incorporates the following enhancements to the basic technique discussed in the previous sections:

1. *Freedom of the Scheduler:* We assumed earlier that the CDFG scheduler does not move behavioral assignments and operators across critical nodes. However, some schedulers, including the DSS scheduler, do move operators and assignments across input/output statements as long as no data dependency is violated. This is handled by masking the criticality of certain variables (registers) at certain critical states. Essentially, a critical variable whose instance is moved by the scheduler across a critical state is masked at that state and the register to which that variable is bound is also masked at the

```
(AUTO-REWRITE-THEORY "des" :ALWAYS? T)
(FLATTEN)
(LEMMA "bs6_bs2_ax")
(LEMMA "bs5_bs6_ax")
(LEMMA " btc_bcp4_ax")
(PROP)
(LEMMA "Bp_bcp4_ax")
(REPLACE -1)
(EXPAND "beh_transition_condition")
(EXPAND "rtl_transition_condition")
(EXPAND "First_b")
(EXPAND "First_d")
(EXPAND "Last_b")
(EXPAND "Last_d")
(ASSERT)
(REPEAT (EXPAND "reverse"))
(REPEAT (EXPAND "append"))
(LEMMA "trans_ds6_ds7_ax")
(LEMMA "flag_ax" ("s" "DS6"))
(PROP)
(LEMMA "max_ax")
(LEMMA "reg_ax" ("c_in" "B2_out" "c_out" "MAX_out" "load" "MAX_cs" "s1" "DS8" "s2" "DS1"))
(ASSERT)
(EXPAND "Control_Signal")
(ASSERT)
....
(ASSERT)
```

Fig. 11. Proof Scripts for a Correctness Theorem

corresponding RTL state. For these masked variables (registers), comparison is turned-off at that state. This masking is taken into account by the CCG when correctness lemmas are generated. (Although not necessary for the DSS scheduler, this technique may be further extended in the context of schedulers that may move operations across conditional/join node boundaries.)

2. *Value Based Register Allocation:* DSS register allocation algorithms perform a mixture of carrier based and value based register allocations. For variables that may contain live values across condition or loop boundaries, carrier-based allocation is assumed; that is, all values stored in the variable are allocated to the same register. For variables whose life-times never cross a condition or loop boundary, value based allocation is performed, where different values in the variable may be allocated to different registers. Earlier in the paper we assumed carrier based allocation. Value based allocation is handled by generating new names for each instance (value) of those variables. Further, some variables subjected to carrier based allocation may have instances (values) that are not live across critical states. The scheduler is free to eliminate these instances by chaining of ALUs or to bind these instances to registers other than the one to which the critical instances of the variable are bound. This implies that some variables are critical at certain states and are not critical at other states. Such situations are also handled by the criticality masking technique discussed above.

3. *Commuting ALU Inputs During Interconnect Allocation:* Contrary to the assumption made earlier, the DSS interconnect allocation algorithms do allow exploitation of commutativity of certain operators (addition, multiplication, etc.) in order to reduce interconnect cost by eliminating multiplexers. This is handled by keeping track of the places where commutativity is exploited, generating commutativity axioms for those (uninterpreted) functions, and instantiating these axioms during the proof.

4. *Folding Operators into Registers:* As in the case of many synthesis tools, DSS permits folding certain types of operators such as addition/subtraction-by-1 and divide/multiply-by-2 into register operations like increment/decrement, and shift right/left. This is handled in a straight-forward way by generating appropriate register axioms.

7 Implementation and Results

The method discussed in the previous sections has been implemented in a correctness condition generator module integrated with the DSS [23] high level synthesis system. DSS has been in development for about ten years and is relatively mature. DSS accepts behavioral specifications in VHDL and generates RTL designs also in VHDL. Using parallel synthesis algorithms, DSS searches through vast regions of design space [24]. DSS uses enhancements of force-directed list scheduling [25, 26] and a hierarchical clique partitioning algorithm for register allocation [27]. DSS has been used to generate numerous designs both in the university and industry and has been throughly tested using systematic benchmark development, test generation and simulation [28]. In addition, as a byproduct of the synthesis process, DSS automatically generates control flow properties in CTL logic [29, 30] for verification by the SMV model checker [31].

Figure 4 shows the integration of the correctness condition generator (CCG) with the DSS system as explained in the previous section. The CCG component of DSS is highly experimental to help us determine how much of the verification effort can be automated and further develop the techniques discussed in this paper. A major limitation of the verification condition generator currently is that it can handle a much smaller subset of VHDL than that can be synthesized by DSS. The modified DSS system with this generator produces a PVS file containing declarative specifications of the behavior and data path and constructive specifications of the controller. In addition, it produces all of the critical path equivalence lemmas and proof scripts to prove these lemmas. The PVS theories generated are not necessarily very elegant, but are amenable to completely automated verification. PVS system is used to execute these scripts automatically. No manual interaction is necessary to conduct the proof and inspection is necessary only in the event of a failure. In experiments with more than ten RTL designs generated by DSS we registered CCG execution times under one minute for all designs.

8 Discussion and Ongoing Work

This paper presented an automated generator of specifications, correctness lemmas and proofs for verification of synthesized RTL designs using theorem proving. The generator is tightly integrated with a synthesis system.

We believe that it is possible to relax the notion of critical states such that critical states partition the state transition graphs into acyclic subgraphs. In such an approach, symbolic rewriting would encompass conditional branches (that are not associated with loops) as well. This is similar to the approach taken by Claesen et al. [13] in the SFG-tracing methodology. However, in contrast to the signal flow graphs used to represent data-dominated DSP-style computations, we wish to be able to handle behavior automaton rich in conditional control flow constructs.

Acknowledgements - The authors like to thank Adrian Nunez-Aldana, Narendra Narasimhan, Elena Teica and Rajesh Radhakrishnan for their valuable help in reviewing this paper.

References

1. G. De Micheli, "Synthesis and Optimization of Digital Circuits", McGraw-Hill, 1994.
2. D.E. Thomas et al., "Algorithmic and Register Transfer Level Synthesis: The System Architect's Workbench", Kluwer Academic Publishers, 1990.
3. Srinivas Devadas, Hi-Keung Tony Ma, Richard Newton, "On Verification of Sequential Machines at Differing Levels of Abstraction", IEEE Transactions on Computer-Aided Design, June 1988.
4. Michael McFarland, "An Abstract Model of Behavior for Hardware Descriptions", IEEE Transactions on Computers, July 1983.
5. F. Corella, R. Camposano, R. Bergamaschi, M. Payer, "Verification of Synchronous Sequential Circuits Obtained from Algorithmic Specifications," Proc. Intl. Workshop on Formal Methods in VLSI Design, Miami, 1991.
6. Reinaldo A. Bergamaschi, Salil Raje, "Observable Time Windows: Verifying The Results of High-Level Synthesis", IEEE Design & Test of Computers", May 1997.
7. Steven Johnson, "Synthesis of Digital Designs from Recursion Equations", MIT Press, Cambridge, 1984.
8. , Ranga Vemuri, "On the Notion of Normal Form Register-Level Structures and Its Applications in Design-Space Exploration ", European Design Automation Conference, March 1990.
9. F. Feldbusch, R. Kumar, "Verification of Synthesized Circuits at Register Transfer Level with Flow Graphs," Proc. IEEE EDAC Conf., pp. 22-26, 1991.
10. Sreeranga Rajan, "Correctness Transformations in High Level Synthesis: Formal Verification", Proceedings of the International Conference on Computer Hardware Description Languages, Japan, August 1995.
11. N. Shankar, S. Owre and J. M. Rushby, "The PVS Proof Checker: A Reference Manual (Beta Release)", March 1993.
12. Dirk Eisnbiegler, Ramayya Kumar, "Formally Embedding Existing High Level Synthesis Algorithms", "Correct Hardware Design and Verification Methods", Germany, October 1995.

13. Luc Claesen, Mark Genoe, Eric Verlind, Frank Proesmans, Hugo De Man, "SFG-Tracing: A Methodology of Design for Verifiability", Proceedings of Advanced Workshop on Correct Hardware Design Methodologies, North-Holland, 1991.
14. Luc Claesen, Frank Proesmans, Eric Verlind, Hugo De Man, "SFG-Tracing: A Methodology for the Automatic Verification of MOS Transistor Level Implementations from High-Level Behavioral Specifications", Proc. Intl. Workshop on Formal Methods in VLSI Design, Miami, 1991.
15. Francisco Corella, "Automated High-Level Verification Against Clocked Algorithmic Specifications," Proc. Computer Hardware Description Languages and Their Applications, April 1993.
16. Francisco Corella, "Automated Verification of Behavioral Equivalence for Microprocessors", Research Report, IBM Research division, T.J. Watson Research Center, 1992.
17. Jerry R. Burch and David L. Dill, "Automatic Verification of Pipelined Microprocessor Control", Proceedings of Computer-Aided Verification, July 1994.
18. M. K. Srivas and S. P. Miller, "Formal Verification of the AAMP5 Microprocessor," Chapter 7 in Industrial Applications of Formal Verification.
19. A. Takach and W. Wolf, "Scheduling Constraint Generation for Communicating Processes", Princeton University, November 1993.
20. D. E. Thomas, R. L. Blackburn, and J. V. Rajan, "Linking the Behavioral and Structural Domains of Representation for Digital System Design", IEEE Trans. CAD, vol. CAD-6, pp. 103-110, January 1987.
21. F. J. Kurdahi, A. C. Parker, "REAL: A Program for REgister ALlocation", Proceedings of the 24th ACM/IEEE Design Automation Conference, pp. 210-215, 1987.
22. S. Owre, N. Shankar, J. M. Rushby, "The PVS Specification Language (Beta Release)", June 1993.
23. J. Roy, N. Kumar, R. Dutta, R. Vemuri, "DSS: A Distributed High-Level Synthesis System", IEEE Design and Test of Computers, June 1992.
24. R. Dutta, J. Roy, R. Vemuri, "Distributed Design Space Exploration for High-Level Synthesis Systems", 29th Design Automation Conference, pp. 644-650, 1992.
25. Sriram Govindarajan, Ranga Vemuri, "Dynamic Bounding of Successor Force Computations in the Force Directed List Scheduling Algorithm", International Conference on Computer Design (ICCD), October 1997.
26. Sriram Govindarajan, Ranga Vemuri, Cone-Based Clustering Heuristic for List Scheduling Algorithms, Proceedings of the European Design and Test Conference, pp. 456-462., (ED&TC), March 1997.
27. Srinivas Katkoori, Jay Roy, Ranga Vemuri, "A Hierarchical Register Optimization Algorithm for Behavioral Synthesis", Proceedings of International Conference on VLSI Design, pp. 126-134, Banglore, India 1996.
28. R. Vemuri et al, "Experiences in Functional Validation of a High Level Synthesis System", 30th ACM/IEEE Design Automation Conference, 1993.
29. E.M. Clarke, E.A. Emerson, A.P. Sistla, "Automatic Verification of Finite-State Concurrent Systems using Temporal Logic Specifications", ACM Trans. Prog. Lang. Syst., pp. 244-263, 1986.
30. N. Narasimhan, R. Vemuri, "Specification of Control Flow Properties for Verification of Synthesized VHDL Designs", "Proceedings of Formal Methods in CAD", pp. 327-345, Springer-Verlag, November 1996.
31. Kenneth L. McMillan, "Symbolic Model Checking: An Approach to the State Explosion Problem" Carnegie Mellon University, 1992.

Combined Formal Post– and Presynthesis Verification in High Level Synthesis*

Thomas Lock[1], Michael Mendler[1], and Matthias Mutz[2]

[1] University of Passau, D–94030 Passau, Germany
Chair for Computer Architectures, Prof. Dr. Ing. W. Grass
Tel.: +49-851-509-3044
Fax: +49-851-509-3042
EMail: {lock,mendler}@fmi.uni-passau.de
[2] SICAN Braunschweig GmbH
Richard–Wagner–Straße 1, D–38106 Braunschweig, Germany
Tel.: +49-531-3807-226
Fax: +49-531-3807-299
EMail: MMutz@sican-bs.de

Abstract. We propose a formal framework based on higher order logic theorem proving as a support for high level synthesis. We suppose that the design process starts from a behavioural or functional specification in terms of a VHDL description. It produces a structural description at the register transfer level. We propose a method for proving the correctness of synthesis results combining the advantages of presynthesis and postsynthesis verification. To perform the postsynthesis task automatically and efficiently correctness–implying properties of an intermediate synthesis result are checked.

1 Introduction

Behavioural synthesis produces a register transfer structure from a behavioural specification. It is generally decomposed into six major transformations: generation of an intermediate format, scheduling, allocation, binding, architecture generation, and transformation into a textual (VHDL) representation. A lot of behavioural synthesis tools have been developed and described in the literature. Discussions of many tools and concepts can be found in [GDW+92], [MLD92] and [JDKR92]. Existing tools may differ from several points of view like application domains, the underlying architectural models, the amount of user controlhoro and interaction with the synthesis algorithms, as well as the supported VHDL subset and its interpretation.

The use of synthesis tools in itself does not guarantee the correctness of the results (see [GMT94]). Although they may be carefully engineered and thoroughly tested, state–of–the–art systems are too complicated for the user to be sure that the algorithms work correctly in all cases. In particular, this is true when sophisticated optimizations are involved or when designers are allowed to

---

* This work is supported by the *Deutsche Forschungsgemeinschaft* under project no. GR 1006/2–1 (FORVERTIS)

interact with the tool and to influence some of the design steps themselves. Several papers have therefore considered the problem of formally verifying results of synthesis tools.

There are basically two ways to do this verification task: presynthesis and postsynthesis verification ([KBES96]). Pure presynthesis approaches are described in [FFFH89, BBL89, CGJ96]. More recent works are [BF96, WM95, WM96, BTC96] Further references can be found in [KBES96]. To pick one example, in [BEK96] a presynthesis verification is proposed where high level synthesis is done by consecutive applications of formally correct design steps. The advantage of presynthesis verification is that the verification task is solved for a specific tool and independently from concrete problem instances to be synthesized. The major disadvantage is that this kind of support requires complete information about the internal workings of the tool, which is unrealistic for commercial systems. Thus, formal verification of high level synthesis, in general, will always require some amount of design dependent postsynthesis verification.

Postsynthesis verification compares formal interpretations of corresponding input and output descriptions of a synthesis tool. Works concerning this methodology are [Pie89, BS89, Cha89, BEF+89, SP89]. Newer works are [TC96, Pie95, BGMW95, NFG+95, CBSB96]. Again further references can be found in [KBES96]. In the postsynthesis approach the algorithms used in the tool for the different transformation steps do not need to be considered. But in general this comparison is difficult and has to be made at least partially with user interaction. The problem is that most of the design decisions taken by the synthesis tool, which are important for verification, are lost in the synthesis result and must be reengineered with a lot of effort. This means, that postsynthesis verification cannot reasonably be done without some minimal amount of presynthesis information.

In view of the problems with pure presynthesis and pure postsynthesis verification we follow a meet–in–the–middle approach that does neither completely ignore the internals of a synthesis tool like the postsynthesis method nor completely depend on them as the presynthesis method. Instead we factorize the verification task into a presynthesis part that relies on the common features of most behavioural synthesis tools and a postsynthesis part that deals with the remaining design choices resolved by executing a particular tool on a particular input description. Indeed, we found that the tool independent presynthesis verification work that can be done for a large class of systems adhering to some common principles makes the remaining postsynthesis verification of high level synthesis solvable in an automatic and efficient way.

What is the common principle underlying all behavioural synthesis tools? High level synthesis is the task of clustering objects into groups so that a given objective function is optimized with respect to design constraints (see [GDW+92]). As a result of high level synthesis, variables, operations, and logical connections are clustered so that each group is mapped in a storage element, a functional unit or an interconnection unit of the resulting design. Also operations are clustered into groups so that all operations of the same group are executed in the same control step. For our approach the only restriction on the synthesis tool is that it makes available these clustering functions as an intermediate result.

Although its format depends on the considered tool, the clustering functions are always contained, implicitly or explicity, in most systems.

In our context we are not interested in optimality but in correctness. We introduce properties that must be fulfilled for a mapping to be correct. We have proven a correctness lemma as a tool independent presynthesis task showing that the clustering result is correct with regard to the input if it meets these properties. The remaining postsynthesis verification of the textual representation with regard to the clustering result is not hard to automate.

Our approach combines the advantages of postsynthesis verification with the advantages of presynthesis verification. In the first place our method can be added like any other postsynthesis verification to a behavioural synthesis tool without knowing about its internals or even changing it. In the second place the postsynthesis part of our method can be efficiently automated because proving the fulfillment of the properties is much simpler than the usual postsynthesis verification. So no manual work has to be done after synthesis.

2 High Level Synthesis

High level synthesis (HLS) produces as an intermediate result mappings from objects of the behavioural description to objects of the transfer description. As a final result a transfer description in — in our case — VHDL is generated. In this paper we only consider basic blocks. These are pure data flow descriptions containing no control flow elements like loops or conditions.

In a data flow description values represented by data flow variables are combined by data flow operations. Data flow oriented HLS translates data flow descriptions into transfer descriptions. In transfer descriptions values are transferred (via buses or by multiplexors, depending on the target architecture) from registers to modules performing the required operation. The new computed values are then transferred back to registers. The allocation part of HLS determines the required modules, buses or multiplexors, and registers. Then data flow objects (that is operations and variables) are bound to those objects of the transfer description. The scheduling figures out when each necessary module operation is to be executed.

We use higher order logic theorem proving performed in the HOL system (see [GM93]) to create verification support. Higher order logic has turned out to be a good formalism for specifying and verifying hardware. It has also been applied to verify incremental synthesis steps (see [FFFH89] and [Lar95]).

To formally reason about the data flow description, the intermediate result, and the final result, the textual representations have to be formally captured. The following sections deal with this problem.

2.1 Formal Capture of the Data Flow Description

A data flow description without control elements is given as a linear VHDL program. VHDL variables correspond to data flow variables and VHDL signals correspond to data flow input and output variables. Input variables never take the result of an operation, output variables are never used as a source of an operation. Figure 1 shows a VHDL description of an example and its interpretation as a generalized data flow graph (see [Jon93]). In the data flow graph the operations are shown in circles and oval boxes while the variables are represented by squared boxes. For simplicity it is assumed that operations have at most two input ports, referred to as *IN1*, *IN2*, and exactly one output port, referred to as *OUT*.

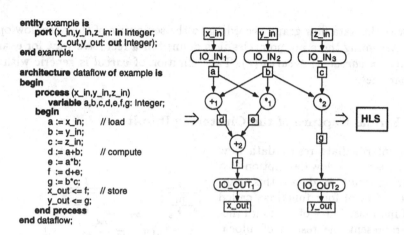

```
entity example is
    port (x_in,y_in,z_in: in Integer;
          x_out,y_out: out Integer);
end example;

architecture dataflow of example is
begin
    process (x_in,y_in,z_in)
        variable a,b,c,d,e,f,g: Integer;
    begin
        a := x_in;      // load
        b := y_in;
        c := z_in;
        d := a+b;       // compute
        e := a*b;
        f := d+e;
        g := b*c;
        x_out <= f;     // store
        y_out <= g;
    end process
end dataflow;
```

Fig. 1. Data flow VHDL specification and data flow graph interpretation

The data flow is represented by a function *flow*. The HOL definition of the function is generated in the following way: parsing the VHDL text builds an abstract syntax tree. Semantic functions generate the definition of function *flow* according to the abstract syntax tree.

The parsing functions are generated with the tool CLaReT (see [Bou95] and [Bou96]). CLaReT needs the grammar of the text for which the functions should be generated. In future CLaReT will also generate the 'semantic function' but at the moment we define them manually. The deep embedding approach discussed in [Tas95] manually defines abstract syntax of a VHDL subset in HOL. There, syntactic functions are also defined manually. Abstract and concrete syntax are related by inspection. CLaReT directly supports concrete syntax definitions given as grammars in BNF style. A similar approach is used in [RK95], which also supports an axiomatic style of semantic definitions. We decided to use CLaReT because its denotational style of semantic definitions is well suited for our application.

The function *flow* for the above data flow example is given by the definition in Table 1. For example the value at the second input port *IN2* of the data flow operation $+_1$ is contained in the data flow variable b: $flow(IN2, +_1) = b$. The definition of the function *flow* is consistent with the VHDL simulation model (see [DB95]).

The semantics of the data flow description is given by the function *varval*. For any assignment of values to the data flow input variables (here x_in, y_in, and z_in), the function returns their initial values. For all other variables it

Table 1. Definition of function *flow* for data flow example

flow	IO_IN_1	IO_IN_2	IO_IN_3	$+_1$	$+_2$	$*_1$	$*_2$	IO_OUT_1	IO_OUT_2
IN1	x_in	y_in	z_in	a	d	a	b	f	g
IN2	$\perp$	$\perp$	$\perp$	b	e	b	c	$\perp$	$\perp$
OUT	a	b	c	d	f	e	g	x_out	y_out

evaluates the data flow graph according to the semantics of the data flow operations. Assuming there are no cycles it is defined in a recursive way, for example $varval(f) = varval(d) + varval(e)$. The definition of *varval* is generic with *flow* as a parameter.

2.2 Formal Capture of the Clustering Result

As an intermediate result data flow HLS maps a data flow description to a clustering result. In detail, this mapping consists of four functions called HLS functions. Three of the four functions represent the results of allocation: *mall* for modules, *ball* for connections and *rall* for registers. Another function *osch* formalizes the result of operation scheduling. Figure 2 gives an example for the four functions. The appropriate definitions of the four clustering functions are extracted out of the textual intermediate result given by the synthesis tool.

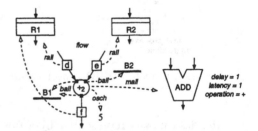

Fig. 2. Example application of the four HLS functions

This is done in the same way as above for the definition of function *flow*: parsing the intermediate result builds an abstract syntax tree. Semantic functions generate function definitions for the four functions from the abstract syntax tree. For our example a suitable set of functions *osch*, *mall*, *ball*, and *rall* are given in Tables 2 and 3.

The semantics of the intermediate result is given by a transfer desription consisting of tuples which determine sources, sinks, execution times etc. of transfers. It is generated by instantiating generic predicates whose definitions depend on the supported tool. Their generic parameters are the HLS functions. An example for generic transfer predicates will be given in Sect. 2.4.

2.3 Formal Capture of the Final Result

Once more CLaReT–generated functions are used to parse the output of the synthesis tool and to build an abstract syntax tree. Semantic functions process the syntax tree and generate a transfer description according to the definitions of transfer predicates (see Sect. 2.4).

Table 2. Definitions of functions *ball*, *osch*, and *mall*

ball	IO_IN_1	IO_IN_2	IO_IN_3	$+_1$	$+_2$	$*_1$	$*_2$	IO_OUT_1	IO_OUT_2
IN1	$\perp$	$\perp$	$\perp$	B1	B1	B1	B1	B1	B1
IN2	$\perp$	$\perp$	$\perp$	B2	B2	B2	B2	$\perp$	$\perp$
OUT	B1	B2	B1	B1	B1	B2	B1	$\perp$	$\perp$
osch	1	2	3	3	5	3	4	7	6
mall	M_IN	M_IN	M_IN	ADD	ADD	MULT	MULT	M_OUT	M_OUT

Table 3. Definition of function *rall*

	x_in	y_in	z_in	a	b	c	d	e	f	g	x_out	y_out
rall	$\perp$	$\perp$	$\perp$	R1	R2	R1	R1	R2	R1	R1	$\perp$	$\perp$

2.4 Semantics of the Clustering and Final Result

The semantics of the clustering result and the tool–generated output are both defined by a transfer decription. The form of the transfer description must be specified in accordance with the format of the tool's output description. A transfer description is determined by transfer predicates and consists of the tuples fulfilling them. The output format reflects the target architecture used during synthesis. The following is based on one possible such architecture and gives an example for the appropriate transfer predicate definitions.

The operation principle for the target architecture divides up every clock cycle into six different phases. These are two successive read phases, a compute phase, two successive write phases and one store phase. In the first read phase (rA) values of output ports of registers are transferred to buses. In the next read phase (rB) values from buses are transferred to input ports of modules. In the compute phase (cM) new values are computed by the modules. In the first write phase (wA) values are transferred from output ports of modules to buses and in the second write phase (wB) values are transferred from buses to input ports of registers. In the last phase of a clock cycle — the store phase (sR) — values at input ports of registers are stored into the registers. The operation principle is shown in Fig. 3.

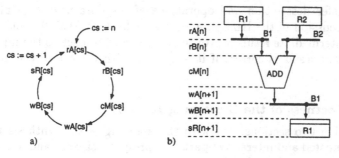

Fig. 3. Control step cycle and corresponding transfers

Corresponding to each clock cycle phase where transfers are required (rA, rB, wA, wB) there is a generic predicate definition. For a transfer in phase rA there is a predicate $R\_to\_B(r, b, cs)$. Here r gives the register that is the source of the transfer, b gives the bus which is the sink of the transfer and cs gives the control step in which the transfer has to be carried out. Similarly the predicate $B\_to\_M(b, m, in, cs)$ reflects a transfer in phase rB of control step cs from the bus b to the inport in of module m. In the same way there are two further predicates corresponding to phases wA and wB. A generic predicate definition is, for example

$$R\_to\_M(r, b, m, in, cs) = \exists o. (osch(o) = cs) \wedge (r = rall(flow(o, in))) \wedge$$
$$(b = ball(o, in)) \wedge (m = mall(o)) \text{ , where}$$

$$R\_to\_B(r, b, cs) = \exists m, in. R\_to\_M(r, b, m, in, cs) \quad \text{and}$$
$$B\_to\_M(b, m, in, cs) = \exists r. R\_to\_M(r, b, m, in, cs) \text{ .}$$

3 Correctness

Using the predicates defined in Sect. 2.4 a semantic function $regval$ can be assigned to the intermediate result. $regval(r, cs)$ gives the value of register r in control step cs. Let $invar : VARS \to BOOL$ and $outvar : VARS \to BOOL$ be predicate functions that indicate whether a variable $v \in VARS$ is an input resp. an output variable. Let $varinit$ give the value of an input variable and let $reginit$ be a function that gives the initial value of a register. Then the condition for correctness of the intermediate result according to the data flow description can be defined as:

$$CORRECTNESS = (\forall v : invar \ v \Rightarrow (varinit \ v = reginit(rall \ v))) \Rightarrow$$
$$(\forall v : outvar \ v \Rightarrow (varval \ v = regval(rall \ v, cs\_max)))$$

where cs_max is the last control step. This condition cannot be checked automatically and efficiently after synthesis. Without additional information this would be a very complex task because terms mixed of arithmetic and logic components must be compared. An example of such an approach is [CPVD90].

We exploit the fact that there is a direct relation between a data flow description and the intermediate result to split the verification task into two parts. This close relation results from the HLS paradigm followed by HLS tools. An example for this close relation is the strong correspondence between modules of the transfer description and operations of the data flow description. For a detailed discussion of the HLS paradigm we refer to [GDW+92] and [Ach93].

If the intermediate result is proven correct with respect to the input description, the final result is correct if it is correct with respect to the intermediate result.

3.1 Correctness of the Clustering Result

The following two sections deal with the splitting of the synthesis task into a machine–assisted and interactive part as a presynthesis task and an automatical one as a postsynthesis task.

Presynthesis Task We have devised correctness properties such that if an intermediate result matches these properties it is correct with respect to the data flow description. The first four of these properties are that neither in the read phase nor in the write phase of a control step more than one source (output port or bus) is connected to a sink (bus or input port). The next two properties are that the modules perform the data flow operation bound to them and their operands are available in time. The following property is that the liveness intervals of variables assigned to the same register do not overlap. This property allows us to prove a relationship between the values of the data flow variables at specific steps of data flow computation and the register contents at specific control steps. The last property states that output values are not overwritten after being assigned to their respective registers. We did the presynthesis work that consists of proving the 'Correctness Lemma' (see [LM97]). It says that the properties guarantee correctness of the intermediate result. Formally, this lemma is:

$$NO\_READ\_MODULE\_CONFLICTS \wedge NO\_READ\_BUS\_CONFLICTS\wedge$$
$$NO\_WRITE\_REGISTER\_CONFLICTS \wedge NO\_WRITE\_BUS\_CONFLICTS\wedge$$
$$OPS\_ON\_MODULES \wedge OPERANDS\_IN\_TIME\wedge$$
$$NO\_REGISTER\_OVERLAPS\wedge$$
$$OUTPUTS\_STABLE \Rightarrow$$
$$CORRECTNESS.$$

Precise definitions of the single properties can be found Sect. 4.

Postsynthesis Task The postsynthesis work only consists of proving the properties mentioned in Sect. 3.1 in the intermediate result, so that the 'Correctness Lemma' can be instantiated. The proof procedure is implemented as a tactic in HOL ([Mut]). The basic procedure is as follows. For a concrete intermediate synthesis result, all quantifications range over finite domains. They are replaced by disjunctions and conjunctions by rewriting with theorems provided by enumeration type definition. The property definitions are rewritten with the *flow*, *osch*, *mall*, *ball*, and *rall* definitions. This rewriting reduces all comparisons of non boolean subterms to boolean constants. Rewriting with boolean simplification rules proves the property theorem. If the rewriting steps do not succeed, the tactic fails. That means that either the HLS tool did not work correctly or it did not follow the supposed HLS paradigm. If the properties can be proven, the correctness lemma is instantiated in order to obtain the correctness theorem. The main benefit of such a verification support lies not so much in the debugging of the HLS tool by the tool developer, for which it may be used as well, but in increasing the confidence in the correctness of the synthesis results obtained by the user of the tool.

Figure 4 illustrates the design flow above the dashed line and the attached postsynthesis verification flow below the dashed line. All actions depicted by this figure are performed automatically. The left half of the verification flow shows the verification of the intermediate result.

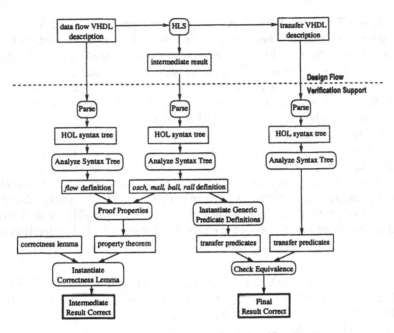

Fig. 4. Design flow and postsynthesis verification support

3.2 Correctness of the Final Result

The right half of the verification flow in Fig. 4 illustrates the verification of the final result. To get the form of the transfer description introduced in Sect. 2.4 out of the final result (transfer description in VHDL) the VHDL text is parsed using CLaReT generated functions. Semantic functions extract a set of tuples matching a particular format (see Sect. 2.4) from the resulting abstract syntax tree. For correctness all tuples must fulfill the (with the definitions of *flow*, ... and *rall*) instantiated generic predicates (see Sect. 2.4) and no fulfilling tuple may be left out. This is the Equivalence Check seen in Fig. 4.

In a complete postsynthesis verification flow, which cannot exploit the scheduling and allocation result (the intermediate result), the data flow description and the tool–generated transfer description would have to be related directly. So our approach drastically simplifies the verification task.

4 Experimental Results

We are using the HOL 90.7 version based on SML 0.93 running on a SUN Sparc 20 workstation with 128 MB of main memory under SunOS 5.6. The proof script for the presynthesis verification part ('Correctness Lemma') consists of about 850 lines of SML code and is executed in about 4 seconds.

With this platform we have performed the postsynthesis verification procedure for several examples. The proof script contains proofs of lemmas at HOL level as well as as meta level decision procedures. It consists of about 2000 lines of SML code.

To estimate the computation time needed to perform the check for a given correctness property we have defined a separate complexity function for the algorithm checking each property. The parameters to the functions are the number of data flow level objects (variables and operations) and transfer level objects (modules, registers, ports and buses). The values are roughly based on the number and depth of quantifications the property definitions include, and is divided by 10000. For example the formula $\forall op \in OPS : P(op) \wedge \exists v, v' \in VARS : Q(v, v')$ would have the complexity value $(|OPS| + |VARS|^2)/10000$.

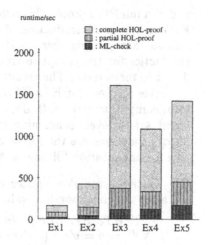

Fig. 5. Results of the five different safety levels

In Table 4 we present five examples where the sum of the complexity values for all properties and the parameters to the complexity functions are given. The first one is a working example synthesized by the AMICAL system [Rah95]. The second one is part of the IKS pseudo code (Inverse Kinematic Solution, [LS89]) whereby the scheduling and allocation result was manually generated from the microcode and an implementation on logic level. This one is the example used in Sects. 2.1 – 2.3. The third example consists of a basic block of a differential equation solver which was processed by the HLS system SUSAN ([Ach93]). The fourth example is a differential equation solver too, taken from the HLSynth92 benchmarks (known as the HAL example, available at [NCS]). It was scheduled by hand. The fifth example is the same, but with a better scheduling, also done manually.

Table 4. Complexity values for selected examples

	complexity value	OPS	VARS	MODULES	BUSES	REGISTERS
EXAMPLE1	8.3084	6	8	2	4	4
EXAMPLE2	18.3999	9	12	4	3	3
EXAMPLE3	88.1776	12	16	3	4	5
EXAMPLE4	174.7764	10	15	4	4	7
EXAMPLE5	346.1365	10	15	4	6	7

Our complexity value (cpx-value) has turned out to give a good estimate of the computation time needed to perform a property and to perform the whole task. In Fig. 5 the five examples are ordered by their total complexity value, which is simply the sum of the cpx-values for each property, for simplicity all taken with the same weight.

For performing the verification task, we consider three safety levels in more detail: a) a check only based on SML decision procedures, b) a partial HOL proof, where some conversions used by the theorem prover are introduced as axioms, and c) a full HOL proof. This reflects a tradeoff between performance and safety. For example the introduction of conversions as axioms is certainly less safe but quicker than proving them as theorems. To achieve a quite safe check still first the tactics for the complete HOL proof are developed. Then these tactics are 'lifted to meta level'. The results for the five examples are shown in Fig. 5.

Table 5 shows for EXAMPLE3 how the single properties contribute to the total computation time. It also clearly demonstrates the difference between different safety levels concerning the performance. The properties and the complexity functions are the following. The table refers to the given abbreviations. The formal notation '∃!' means 'there is one and not more than one':

OS 'Outputs stable': a register associated with an output variable is not written after the corresponding value is stored

$$OUTPUTS\_STABLE = \forall v : outvar(v) \Rightarrow$$
$$\forall v' : \neg(v = v') \wedge (rall(v) = rall(v')) \Rightarrow (cs\_def(v') < cs\_def(v))$$

cpx: $|VARS|^2$

OIT 'Operands in time': each value is stored into a register before any operation using it is scheduled

$$OPERANDS\_IN\_TIME = \forall op, p : ((p = IN1) \vee (p = IN2)) \Rightarrow$$
$$(cs\_def((flow(op))(p)) < osch(op))$$

cpx: $|OPS| * |PORTS|$

NRO 'No register overlaps': no value in a register is overwritten before its last use by an operation

$$NO\_REGISTER\_OVERLAPS = \forall op, op', p :$$
$$let\ v = (flow(op))(p)\ and\ v' = (flow(op'))(OUT)\ in$$
$$((p = IN1) \vee (p = IN2)) \wedge \neg(v = v') \wedge (rall(v) = rall(v')) \Rightarrow$$
$$let\ n = cs\_def(v')\ in$$
$$(cs\_def(v)) \vee (n \geq osch(op))$$

cpx: $|OPS|^2 * |PORTS|$

OOM 'Operations on module': modules correctly implement the operations they have to execute

$$OPS\_ON\_MODULES = \forall op : opfn\ op = modfn\ (mall(op))$$

cpx: $|OPS|$

NRBC 'No read bus conflicts': not more than one register writes to a bus in read phase

$$NO\_READ\_BUS\_CONFLICTS = \forall b, cs :$$
$$(\exists! r : R\_to\_B(r, b, cs)) \vee \neg(\exists r : R\_to\_B(r, b, cs))$$

cpx: $|REGISTERS|^2 * |MODULES|^2 * |BUSES| * |OPS|^2$

NRMC 'No read module conflicts': no module reads from more than one bus in read phase

$$NO\_READ\_MODULE\_CONFLICTS = \forall m, p, cs :$$
$$(\exists! b : B\_to\_M(b, m, p, cs)) \vee \neg(\exists b : B\_to\_M(b, m, p, cs))$$

cpx: $|REGISTERS|^2 * |MODULES| * |PORTS| * |BUSES|^2 * |OPS|^2$

NWBC 'No write bus conflicts': not more than one module writes to a bus in write phase

$NO\_WRITE\_BUS\_CONFLICTS = \forall b, cs :$
$\quad (\exists! m : M\_to\_B(m, b, cs)) \lor \neg (\exists m : M\_to\_B(m, b, cs))$

cpx: $|REGISTERS|^2 * |MODULES|^2 * |BUSES| * |OPS|^2$

NWRC 'No write register conflict': no register reads from more than one bus in write phase

$NO\_WRITE\_REGISTER\_CONFLICTS = \forall r, cs :$
$\quad (\exists! b : B\_to\_R(b, r, cs)) \lor \neg (\exists b : B\_to\_R(b, r, cs))$

cpx: $|MODULES|^2 * |REGISTERS| * |BUSES|^2 * |OPS|^2$

Table 5. Complexity values and computation times (in sec CPU) at different safety levels for the correctness properties

EXAMPLE3	$\sum$	OS	OIT	NRO	OOM	NRBC	NRMC	NWBC	NWRC
complexity value	88.2	0.026	0.004	0.043	0.001	12.96	51.84	12.96	10.37
ML check	113.69	0.26	0.45	10.56	0.04	26.66	45.9	15.52	12.95
partial HOL proof	404.82	19.25	11.11	83.05	2.5	53.73	108.88	36.48	89.82
total HOL proof	1619.59	36.1	30.85	585.67	2.21	153.64	371.06	88.99	349.66

Table 5 clearly demonstrates the advantage of using ML decision procedures.

5 Summary and Outlook

We presented a methodology to support existing VHDL based high level synthesis tools with automatic formal postsynthesis verification support. By using an intermediate result containing the clustering generated by an HLS tool during scheduling and allocation no direct relation between input and output description has to be proven. Instead, in two much simpler steps first the correctness of the intermediate result is guaranteed by detecting certain correctness properties automatically. In a presynthesis proof (interactively constructed) these properties are shown to be sufficient for its correctness. In the postsynthesis part only these correctness conditions need to be verified for any given intermediate result. In a second step the intermediate result is associated with a transfer description whose format and architectural information have to be determined according to the supported tool. For correctness this transfer description is compared with the one extracted of the tool–generated output description, which can be done automatically, too.

So far we only considered pure data flow descriptions as HLS input. We are currently extending our approach to control flow dominated descriptions (see [LM98]). They are interpreted as a control graphs whose edges are annotated with data flow graphs. Each of these can be verified with the developed methodology. This can be done provided the structure of the basic blocks in the input description is preserved. One more property has to be found to guarantee the correct connection of the annotated data flow graphs.

Control driven HLS tools (path based synthesis) do not preserve the structure of basic blocks. Our idea is to interpret such tools as first changing the structure of basic blocks in a preprocessing step and then allocating and scheduling each basic block. The preprocessing step will generate an intermediate result containing path information as simple HLS does for scheduling and allocation. So a similar methodology for verifying the preprocessing step will be developed. In this way HLS tools that optimize the input description before processing it (optimization at source code level) will be verified.

References

[Ach93]   H. Achatz. SUSAN: System for universal scheduling and allocation. In *SASIMI Synthesis and Simulation Meeting and International Interchange*, pages 138–44, 1993.

[BBL89]   D. A. Basin, G. M. Brown, and M. E. Leeser. Formally verified synthesis of combinational CMOS circuits. In L. Claesen, editor, *Applied Formal Methods for Correct VLSI Design*, pages 251–260. Organizing Commitee of the IMEC–IFIP–Workshop, November 1989.

[BEF+89]   A. Bartsch, H. Eveking, H.-J. Faerber, M. Kelelatchew, J. Pinder, and U. Schellin. LOVERT — a logic verifier of register–transfer level description. In L. Claesen, editor, *Applied Formal Methods for Correct VLSI Design*, pages 522–531. Organizing Commitee of the IMEC–IFIP–Workshop, November 1989.

[BEK96]   C. Blumenröhr, D. Eisenbiegler, and R. Kumar. Implementation issues about the embedding of existing high level synthesis algorithms in HOL. In Joakim Wright, editor, *Theorem proving in higher order logics*, pages 165–172. Springer, August 1996. LNCS 1125.

[BF96]   D. Basin and S. Friedrich. Modelling a hardware synthesis methodology in Isabelle. In G. Goos, J. Hartmanis, and J. van Leeuwen, editors, *Theorem Proving in Higher Order Logics*, pages 33–50. Springer, August 1996. LNCS 1125.

[BGMW95]   H. Barringer, G. Gough, B. Monahan, and A. Williams. Formal support for the ELLA hardware description language. In Paulo E. Camurati and Hans Eveking, editors, *Formal Methods in Computer Aided Design*, pages 225–245. Springer, July 1995. LNCS 987.

[Bou95]   R. J. Boulton. CLaReT user's manual. Technical report, University of Cambridge, Computer Laboratory, 1995.

[Bou96]   R. J. Boulton. A single language for specifying abstract syntax trees, lexical analysis, parsing and pretty printing. Technical report, University of Cambridge, Computer Laboratory, 1996.

[BS89]   D. Borrione and A. Salem. Proving an on-line multiplier with OBJ3 and TACHE: A practical experience. In L. Claesen, editor, *Applied Formal Methods for Correct VLSI Design*, pages 271–280. Organizing Commitee of the IMEC–IFIP–Workshop, November 1989.

[BTC96]   Bhaskar Bose, M. Esen Tunar, and Venkatesh Choppella. A tutorial on digital design derivation using DRS. In Mandayam Srivas and Albert Camilleri, editors, *Formal Methods in Computer Aided Design*, pages 270–274. Springer, September 1996. LNCS 1166.

[CBSB96] F. J. Cantu, A. Bundy, A. Smaill, and D. Basin. Experiments in automating hardware verification using inductive proof planning. In Mandayam Srivas and Albert Camilleri, editors, *Formal Methods in Computer Aided Design*, pages 94–108. Springer, September 1996. LNCS 1166.

[CGJ96] Solange Coupet-Grimal and Line Jakubier. Coq and hardware verification: A case study. In G. Goos, J. Hartmanis, and J. van Leeuwen, editors, *Theorem Proving in Higher Order Logics*, pages 125–140. Springer, August 1996. LNCS 1125.

[Cha89] Sergio R. Ramirez Chavez. Formal proof of the cascading property of a parallel sorting circuit. In L. Claesen, editor, *Applied Formal Methods for Correct VLSI Design*, pages 338–346. Organizing Commitee of the IMEC–IFIP–Workshop, November 1989.

[CPVD90] L. Claesen, F. Proesmans, E. Verlind, and H. DeMan. SFG–tracing: A formal verification methodology. In *ACM–SIGDA Workshop*, February 1990. Miami.

[DB95] C. Delgado Kloos and P. T. Breuer, editors. *Formal Semantics for VHDL*. Kluwer Academic Publishers, 1995.

[FFFH89] S. Finn, M. P. Fourman, M. Francis, and R. Harris. Formal system design — interactive synthesis based on computer–assisted formal reasoning. In L. Claesen, editor, *Workshop on Applied Formal Methods for Correct VLSI Design*, pages 97–110, 1989. North Holland.

[GDW+92] D. Gajski, N. Dutt, A. Wu, S. Lin, et al. *High Level Synthesis: Introduction to Chip and System Design*. Kluwer Academic Publishers, 1992.

[GM93] M. J. C. Gordon and T. F. Melham. *Introduction to HOL: A Theorem Proving Environment for Higher Order Logic*. Cambridge University Press, 1993.

[GMT94] W. Grass, M. Mutz, and W.-D. Tiedemann. High level synthesis based on formal methods. In *EUROMICRO '94*, pages 83–91, 1994. Liverpool (Great Britain).

[JDKR92] A. A. Jerraya, H. Ding, P. Kission, and M. Rahmouni. *Behavioral Synthesis and Component Reuse with VHDL*. Kluwer Academic Publishers, 1992.

[Jon93] G. G. de Jong. *Generalized Data Flow Graphs — Theory and Applications*. PhD thesis, Eindhoven, 1993. Cip–Gegevens Koninklijke Bibliotheek, Den Haag.

[KBES96] R. Kumar, C. Blumenröhr, D. Eisenbiegler, and D. Schmid. Formal synthesis in circuit design — a classification and survey. In Mandayam Srivas and Albert Camilleri, editors, *Formal Methods in Computer Aided Design*, pages 294–309. Springer, September 1996. LNCS 1166.

[Lar95] M. Larsson. An engineering approach to formal digital system design. *The Computer Journal*, 38(2):101–10, 1995.

[LM97] Th. Lock and M. Mutz. FORVERTIS: Projektbericht. Technical report, Universität Passau, Lehrstuhl für Rechnerstrukturen, Oktober 1997.

[LM98] Th. Lock and M. Mendler. Formale Modellierung von kontrollflußdominierten High–Level–Synthese–Eingabebeschreibungen zur Verifikation von Ergebnissen kontrollflußgesteuerter Einplanungsverfahren. In F. J. Rammig and W. Müller, editors, *Methoden und Beschreibungssprachen zur Modellierung und Verifikation von Schaltungen und Systemen*, pages 75–84. GI/ITG/GMM, HNI–Verlagsschriftenreihe, März 1998. Paderborn.

[LS89] S. S. Leung and M. A. Shanblatt. *ASIC System Design with VHDL: A Paradigm*. Kluwer Academic Publishers, 1989.

[MLD92]   P. Michel, U. Lauther, and P. Duzy. *The Synthesis Approach to Digital System Design.* Kluwer Academic Publishers, 1992.

[Mut]   M. Mutz. Presynthesis proof and postsynthesis tactic for pure dataflow descriptions in high level synthesis. http://acrux.fmi.uni-passau.de/~lock/pre-post-proof.html.

[NCS]   The benchmark archives at CBL. http://www.cbl.ncsu.edu/www/-benchmarks/.

[NFG+95] R. De Nicola, A. Fantechi, S. Gnesi, S. Larosa, and G. Ristori. Verifying hardware components with JACK. In Paulo E. Camurati and Hans Eveking, editors, *Formal Methods in Computer Aided Design*, pages 246–260. Springer, July 1995. LNCS 987.

[Pie89]   Laurance Pierre. The formal proof of sequential circuits described in CASCADE using the boyer–moore theorem proover. In L. Claesen, editor, *Applied Formal Methods for Correct VLSI Design*, pages 365–384. Organizing Commitee of the IMEC–IFIP–Workshop, November 1989.

[Pie95]   Laurence Pierre. Describing and verifying synchronous circuits with the boyer–moore theorem proover. In Paulo E. Camurati and Hans Eveking, editors, *Formal Methods in Computer Aided Design*, pages 35–55. Springer, July 1995. LNCS 987.

[Rah95]   M. Rahmouni. AMICAL: Interactive behavioral synthesis based on VHDL for control–flow dominated systems. *Journal of the Brazilian Computer Society*, November 1995.

[RK95]   R. Reetz and T. Kropf. *A Flow Graph Semantics of VHDL: A Basis for Hardware Verification with VHDL*, pages 205–38. Kluwer Academic Publishers, 1995.

[SP89]   H. Simonis and T. Le Provost. Verification in CHIP: Benchmark results. In L. Claesen, editor, *Applied Formal Methods for Correct VLSI Design*, pages 570–574. Organizing Commitee of the IMEC–IFIP–Workshop, November 1989.

[Tas95]   J. P. Van Tassel. *An Operational Semantics for a Subset of VHDL*, pages 71–106. Kluwer Academic Publishers, 1995.

[TC96]   S. Tabar and P. Curzon. A comparison of MDG and HOL for hardware-verification. In G. Goos, J. Hartmanis, and J. van Leeuwen, editors, *Theorem Proving in Higher Order Logics*, pages 415–430. Springer, August 1996. LNCS 1125.

[WM95]   Li-Guo Wang and Michael Mendler. Formal design of a class of computers. In Paulo E. Camurati and Hans Eveking, editors, *Formal Methods in Computer Aided Design*, pages 84–102. Springer, July 1995. LNCS 987.

[WM96]   L. G. Wang and M. Mendler. Abstraction of hardware construction. In G. Dowek, J. Heering, B. Möller, and K. Meinke, editors, *Higher-Order Algebra, Logic, and Term Rewriting, HOA'95*, pages 264–287. Springer, 1996. LNCS 1074.

Formalization and Proof of a Solution to the PCI 2.1 Bus Transaction Ordering Problem*

Abdel Mokkedem, Ravi Hosabettu, and Ganesh Gopalakrishnan

Department of Computer Science, University of Utah,
Salt Lake City, UT 84112-9205
mokkedem,hosabett,ganesh@cs.utah.edu

Abstract. The transaction ordering problem of the original PCI 2.1 standard bus specification violates the desired correctness property of maintaining the so called 'Producer/Consumer' relationship between writers and readers. In [3], a correction to this ordering problem was proposed and informally proved (called the "HP solution" here). In this paper, we present a formalization of the PCI 2.1 protocol in PVS. We formalize the fact that with Local Master ID added to the protocol no completion stealing is possible and the Producer/Consumer property is provided even in the presence of multiple readers. The state of our proofs leading to this result, as well as some of the much needed enhancements to theorem-proving frameworks that will greatly facilitate similar proofs, are also elaborated.

1 Introduction

Modern system I/O busses such as the PCI are defined with the primary goal of establishing industry-wide open standards. Given the considerable complexity of these standards, only when developed with sufficient rigor can such standards allow peripheral VLSI chip manufacturers to *independently* develop *compatible* I/O devices and bus bridges. Current industrial practices are, however, such that rigor is not reflected either in the development or the documentation of these standards. As a result, open standards often contain significant inconsistencies, resulting in costly work-arounds and interfacing problems. Formal methods based on theorem proving can play a crucial role in the development of these standards by providing the precise and expressive notation of higher order logic to characterize entire families of devices, allowing the precise statement and proof of crucial correctness properties, and forming a solid framework for reliable future extensions to standards.

In this paper, we demonstrate how a modern theorem prover, namely PVS [11], can be used to carry out such a formalization for a significant portion of the PCI 2.1 standard. Our main contribution is a PVS-based formalization of a solution to the PCI 2.1 transaction ordering problem proposed, as well as informally

* Supported in part by DARPA under contract #DABT6396C0094 (Utah Verifier), and NSF MIP MIP-9321836

proved, in [3, 12]. Briefly, there are two problems with the PCI 2.1 standard: *deadlock*, arising from inadvertently disallowing certain transaction re-orderings; and *incorrect behavior*, arising from the so called "completion stealing". More specifically, [4, 3] show that the PCI 2.1 standard allows implementations that deadlock, by not allowing certain transactions to bypass others in bus channels (as will be elaborated later). They also point out that the PCI 2.1 standard allows a crucial correctness property—namely Producer/Consumer—to be violated. Briefly, Producer/Consumer requires that if a writer writes data in one location and then writes a flag announcing the availability of the data, a reader which reads the flag true can subsequently read the data location and be guaranteed that it receives the data that was written. [4, 3] show that this criterion is violated because PCI 2.1 does not maintain master IDs, and consequently third-party bus masters can inadvertently steal a "completion token" intended for another requester (as will be elaborated later). The solution proposed in [4, 3] involves using reserved pins in the PCI 2.1 interface to implement *Local Master IDs*. This solution has been widely discussed in the PCI mailing list `pci-sig-request@znyx.com` as "the Hewlett-Packard solution". In [3, 12], an intuitive manual proof is given for the fact that this solution guarantees Producer/Consumer by avoiding completion stealing.

Our main contributions are as follows. First, we use a more detailed model of the PCI bus which is closer to the actual standard. Next, we build a reusable theory hierarchy which will allow transaction oriented interconnect structures and busses to be easily specified in future projects of this kind (we have invested considerable effort in building this theory hierarchy for instance, by extending recently proposed graph libraries [1]). We identify the necessary conditions which characterize well-formed PCI networks and mechanically prove crucial properties pertaining to the PCI bus, including providing a formal specification of *all the graph-configurations* to be considered in proving Producer/Consumer. We formulate and prove many interesting properties about the PCI protocol which are subsequently used in the proof of the absence of completion stealing and the proof of Producer/Consumer. Most of the mechanical proofs leading upto these results have been finished, with the remaining effort estimated to be one person month. We also report on some of the much needed enhancements to theorem-proving systems without which proofs of this nature will be too tedious to finish in reasonable times—as has been our experience.

Related Work: Modeling modern coherent memory busses [6], as well as the verification of coherence and memory models [13] has been done in the context of PVS. Although I/O busses have been studied [9], the examination of entire I/O bus standards in the context of modern theorem provers has, to the best of our knowledge, not been done. While model-checking tools have been used to debug many bus protocols (e.g. [2]), formal examination of entire bus standards cannot adequately be performed in their context. Our work is believed to be one of the few that addresses, in the context of a powerful theorem prover, a modern as well as widely used I/O standard.

2 The PCI Local Bus

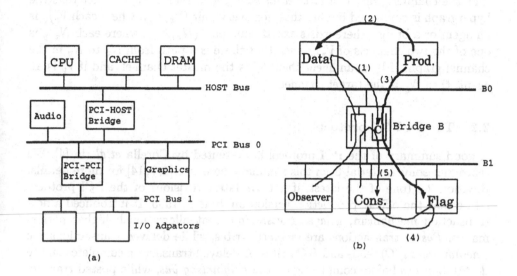

Fig. 1. (a) A PCI acyclic network and (b) Producer-Consumer violation.

The PCI Local Bus has been defined with the primary goal of establishing an industry open standard for system I/O buses that addresses the performance issues associated with peripheral functions with high bandwidth requirements. PCI is a high performance, 32-bit or 64-bit bus with multiplexed address and data lines. It is intended for use as an interconnect mechanism between highly integrated peripheral controller components, peripheral add-in boards, and processor/memory systems. Beginning with the next section, we present an informal description of the PCI bus.

2.1 Topology

Figure 1 (a) shows a typical acyclic network connecting several PCI busses and a Host bus through *bridges*. Primarily, a bridge interfaces one bus to another and limits the loading of each bus. A bridge between two busses B and B' has two *opposite channels* corresponding to the two direction of the traffic. One of the channels has B as its *in-bus* and B' as its *out-bus*, while the other has B' as its *in-bus* and B as its *out-bus*. Similarly, every agent has two opposite channels, called the *master channel* and the *target channel*. The hypergraph defined by the PCI network topology, which has bridges and agents as vertices, gives rise to a directed graph that has the *channels of bridges* and *channels of agents* as nodes. There is an *edge* from a channel N_1 to a channel N_2, written $N_1 \to N_2$, if the out-bus B of N_1 is the same as the in-bus of N_2. We call this directed graph the *channel graph*. Note that $N_1 \to N_2$ implies $N_2' \to N_1'$, where N_1' and N_2' are the opposite channels of N_1 and N_2 respectively.

The acyclicity of the hypergraph defined by the PCI network topology implies that the channel graph is a *directed acyclic graph*. Moreover, the fact that the hypergraph is connected implies that, for every pair (V_x, V_y), where each $V_{x/y}$ is an agent or a bridge, there exists exactly one pair (N_x, N_y), where each $N_{x/y}$ is one of the two channels of $V_{x/y}$, such that there is a path from N_x to N_y in the channel graph. If V_x is an agent then N_x is the master channel, and if V_y is an agent, then N_y is the target channel.

2.2   The PCI 2.1 protocol

A good summary of the PCI protocol is presented by Corella *et. al.* in [4]. We reproduce some material from this summary here; see [4, 7, 14] for more details. Revision 2.1 (one of the latest, if not *the* latest revision) of the PCI protocol [7] became the official production version on June 1, 1995. It introduced a new transaction mechanism, *delayed transaction*, that allows much higher performance. *Posted* transactions are *memory* writes, while delayed transactions are (*memory and I/O*) reads and *I/O* writes. A delayed transaction completes on the *destination bus* before completing on the *originating bus*, while posted transactions complete on the *originating bus* before completing on the *destination bus*. One advantage of a delayed transaction is that the bus is not held in wait states while completing an access to a slow device.

Transaction propagation: A transaction is issued by an agent, the *master* of the transaction, and specifies an address which uniquely determines another agent, the *target* of the transaction. A *posted* transaction propagates from the *originating bus* of the transaction to the *destination bus*. A *delayed* transaction propagates from the originating bus to the destination bus, and then the *completion* of the transaction travels back from the destination to the originating bus. The completion carries the data, in the case of a delayed read, or the termination status (normal or abnormal) in the case of a delayed write. The address of the transaction uniquely determines the target of the transaction and hence the path that the transaction must follow.

As a (*global*) transaction propagates it causes one or more (*local*) subtransactions to be issued on one or more buses that separate the master from the target of the global transaction. Each of those subtransactions has a local master, which may be the master of the global transaction or a bridge acting on its behalf, and a local target, which may be the target of the global transaction or a bridge acting on its behalf.

For a *posted* subtransaction, the local target may either *complete* it (in which case the subtransaction propagates to the next level of bus interconnect) or request the local master to retry. In this case, the local master must reissue the posted subtransaction until it is completed. The local target then latches the transaction information, creating a Posted Memory Write (PWM, or "P entry"). For a *delayed* subtransaction, the local target may either *ignore* it or latch it creating a *DRR entry* (in case of a Delayed Read Request) or a *DWR entry* (in the case of a Delayed Write Request). In the latter case, an "R entry" (referring both to DRR and DWR) that specifies the address of the transaction (and in the case of a write, also the data) is created. The local master of a delayed transaction must reissue the transaction until it is completed. *Meanwhile*, the local target tries to obtain a completion corresponding to every R entry. When the completion is obtained, a "C entry" (a Delayed Read Completion, *DRC*, or a Delayed Write Completion, *DWC*) is created at the local target. The C entry specifies the address and the data (in case of DRC) and the address and the termination status (normal or abnormal) (in case of DWC). Once a C entry is created at the local target, the delayed subtransaction completes when *issued again* by the local master.

Ordering rules: PCI bridge channels must *disallow* certain re-orderings between P, R, and C entries in order to preserve desired semantics such as the Producer/Consumer relationship, while at the same time *permitting* certain re-ordering in order to avoid deadlocks. For instance, not allowing C entries to pass R enties can cause deadlocks [4]. Table 1 specifies these orderings (this incorporates the fix proposed in [4] and is different from the PCI standard specification). In [4] a DWC entry is not allowed to pass a P entry because both DRC and DWC entries are treated as one common class called "C entries". As we maintain the distinction between DRC and DWC (as per the PCI standard), we can actually specify the detail that DWCs can pass P, thus achieving our formalization with respect to a more accurate bus model.

Row pass Col.?	P	R	C
P	No	Yes	Yes
R	No	Y/N	Y/N
DRC	No	Yes	Y/N
DWC	Y/N	Yes	Y/N

Yes: Must be allowed to pass.
No: Can not be allowed pass.
Y/N: The bridge designer
    may choose either way.

Table 1. Ordering rules in a Bridge.

R and C discarding rules: ¿From the point of view of forward progress, the PCI protocol allows R entries to be discarded under certain circumstances. A bridge is allowed to discard an R entry *only* from the time it is enqueued till it has been attempted on the destination bus. This is because, as said earlier, an attempt on the destination bus *may* or *may not* result in the latching of the request. Taking the conservative position that it *does* get latched, a request is *committed* once it is attempted *at least once on the bus*. Thus, committed R entries may not be discarded; they must be repeated until they complete.

A bridge is allowed to discard a C entry in two cases: (i) when the completion corresponds to a read to a pre-fetchable address or to a Memory Read, (ii) when the master has not repeated the corresponding Delayed Request within 2^{15} clock cycles. In our PVS model of the protocol we model a weaker rule which allows a C entry to be discarded if it is not the oldest C entry in the channel.

States, events and transitions: At any time t, each channel contains a collection of *entries* $(E_i)_{0 \leq i \leq n}, n \geq 0$. An entry is a tuple (ϵ, α, τ), where ϵ is the transaction type (P, R, or C), α is the transaction address, and τ is the data (in the case of a read) or the status (in the case of a write) carried by the completion. As said earlier, a channel functions as a FIFO buffer except for the re-orderings permitted by Table 1. The committed R entries present in a bridge channel N at time t comprise the *retry set* of N at time t. At the level of abstraction we consider, state transitions are caused by the following *events* [4]:

- A P *event* that *occurs on bus* B; A P event is triggered by a P entry E in a bridge channel or master channel N with out-bus B. If N' is the target channel specified by the address α of E, there exists a path $N = N_0, N_1, \ldots, N_n = N'$ from N to N' in the channel graph. A P event is an abstraction of a posted bus transaction on bus B. There are *two kinds* of P events:

• A *P_retry* event, which has no effect other than marking the P entry as *committed.*

• A *P_completion* event, which deletes the entry E from N_0, and creates a P entry E' in N_1 with the same parameters. In the case where $N_1 = N'$, the corresponding global posted transaction completes and the entry E' is created in the target channel of the addressed agent (i.e. when N'). In fact, we use the target channel of an agent to store all the transactions completed in that agent for specification purposes (in other words, a "history" of all transactions completed is kept to assist in the specification and proof).

- A D *event* that occurs on bus B; A D event is *triggered* by an R entry E in a channel N with out-bus B. If N is a bridge channel, such an event may happen only if there are no P entries older than E in N. (This is to abide by the orderings in Table 1.) If N is a bridge channel and E is not in the *retry set* of N, then E is added to the retry set of N and the event is said to be an *R_commit event* that commits the entry E. If N' is the target channel specified by the address α, there exists a path $N = N_0, N_1, \ldots, N_n = N'$ from N to N' in the channel graph. A D event is an abstraction of a delayed transaction on bus B. There are *two kinds* of D events:

• A *D_retry* event which can in turn be a *D_noop* event which has no effect, or a *D_latch* event, which creates an R entry E' in N_1 with the same parameters as E. A *D_latch* event can only happen if $N_1 \neq N'$, E' is not already latched in N_1 and the channel N_1' opposite to N_1 does not contain a C entry with the same parameters as E.

- A *D_completion* event, which is enabled only if (i) $N_1 = N'$, or (ii) $N_1 \neq N'$, the channel N_1' opposite to N_1 contains a C entry E'' with the same parameters as E and there is no P entry in N_1' older than E'' (in the case of DRC only). Its effect on the state is to remove E'' from N_1' in case (ii) and to create a C entry E''' with the same parameters in the channel N_0' opposite to N_0 and to remove E from N_0. This is the mechanism by which completions for transactions travel from the target of the transaction back to the master. Also, we record, as a 'history' of all completed transactions, all R entries that reach the final destination and all C entries that reach the original master.

3 Transaction ordering problem and the HP proposal

Overview of the transaction ordering problem: In the PCI protocol, a C entry is matched against an R entry based on the Address and Byte Enable fields. The identity of the master that issued the transaction cannot be used in the comparison, because *the Master ID information is not maintained in the PCI 2.1 standard.* Thus, the local target does not know which master issued the R entry that resulted in a C entry. A completion of a delayed transaction may thus be given to a transaction initiated by a different master, resulting in completion stealing, causing violation of the Producer/Consumer relationship [3, 12]. To recap, the Producer/Consumer relationship says that if a master, called the Producer, writes to a location called Data, and then to a location called Flag, and if a second master, called the Consumer, sees the value written to Flag by the Producer and then reads Data, the Consumer is guaranteed to see the value written to Data by the Producer no matter where Producer, Consumer, Data and Flag are placed in an arbitrary PCI bus network. Figure 1 (b) describes one possible scenario which violates this rule. A third master called the Observer reads Data and the completion C containing the Data value meant for this read is waiting in the bridge B. A Producer, then, writes a new value to Data followed by writing to Flag and a Consumer then reads the new Flag and attempts to read Data. Since the Master ID information is not maintained, the bridge returns the completion C meant for the Observer. Since C contains the old value of Data, Producer/Consumer is violated. Many variants of this scenario are described in `pci-sig-request@znyx.com` [12], but they all relate to completion stealing. The problem is also documented in Section 3.11 of the PCI Local Bus Specification Revision 2.1, pages 116–117; however the severity of the problem is under-estimated there.

HP proposal: The solution proposed by HP [12] has two aspects: (i) using uncommitted hardware resources, *local master-ids are implemented,* and (ii) the requirement that bridges and multi-functions devices have only *one* outstanding transaction at a time for a given address. The reasons for the second requirement are a bit too involved, but basically allowing mutiple outstanding transactions again invites failure due to completion-stealing—essentially, R entries bearing local master-IDs can get "aliased" after one hop through a common bridge.

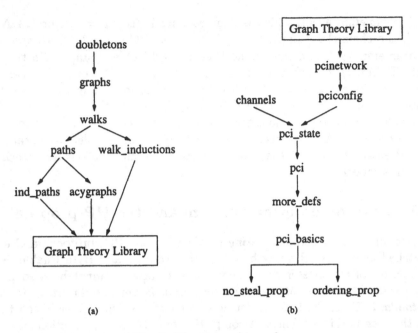

Fig. 2. The theory hierarchy

Our PVS model of PCI protocol includes these two components. Our contribution is towards formally proving that (i) and (ii) together provide a complete solution to the transaction ordering problem for any acyclic network of PCI buses.

4 Formalization of the PCI Protocol in PVS

We now describe, in stages, how our PVS specification was constructed. Details not presented here, including the proofs completed so far, are available on the web [8].

4.1 The Model of PCI Network

The graph theory we used to model an arbitrary acyclic network of PCI busses is based on a modification of "A PVS Graph Theory Library" from NASA [1]. Figure 2(a) shows the graph theory hierarchy. Since the channel graph of a PCI network is a directed acyclic graph, we modified `doubleton` to be an ordered pair (used to define an edge) and then defined `graph` as directed graphs. A `walk` is a sequence of vertices connected by edges and a `path` is walk where no vertex appears twice. `ind_paths` provides the definition of independent paths used later in the proofs and `walk_inductions` defines induction schemes for walks. Finally an acyclic graph is defined as a graph where every walk is a path (in `acygraphs`). While most of the properties proved in the NASA library remained true, some needed modifications and we added some of our own.

The initial fragment of `pcinetwork` theory is shown in $\boxed{1}$. A PCI device (an agent or a bridge) is represented by a natural number and Nodes defines the type of the channels of these devices. The type acynet defines acyclic graphs over Nodes.

```
pcinetwork: THEORY                                                    1
BEGIN
    Agent_count, Bridge_count: posnat % The number of agents and bridges
    MasterId: TYPE = nat
    PCI_Device: TYPE = {n:MasterId | n < Agent_count+Bridge_count}
    Chanls: TYPE = { MASTER, TARGET }
    Nodes: TYPE = [# name: PCI_Device, chan: Chanls #]
    acynet: TYPE = acygraph[Nodes]
    opposite(nd:Nodes): Nodes = (# name:= name(nd),
                           chan:= IF (chan(nd)=MASTER) THEN TARGET
                                  ELSE MASTER ENDIF #)
```

However, not every acyclic graph over Nodes represents a valid network of PCI busses. For example, there cannot be any edge coming into the master channel of an agent. We have identified six conditions on a *well-formed* network of PCI busses:

1) To every edge, there is an opposite edge:

```
all_opposite_exists(net:acynet):bool = FORALL(nd1,nd2:Nodes):      2
    edge?(net)(nd1,nd2) IMPLIES edge?(net)(opposite(nd2),opposite(nd1))
```

2) For all agents, the master channels are only sources and target channels are only destinations:

```
agents_edges_ok(net:acynet):bool = FORALL(nd1:Nodes):              3
    is_agent?(nd1) IMPLIES
    (IF chan(nd1) = MASTER THEN
        (NOT (EXISTS (nd2:Nodes): edge?(net)(nd2,nd1)))
     ELSE NOT (EXISTS (nd2:Nodes): edge?(net)(nd1,nd2)) ENDIF)
```

3) There is a path from the master channel of every agent to the target channel of every other agent:

```
master_target_connected(net:acynet):bool = FORALL(nd1,nd2:Nodes):  4
    (is_agent?(nd1) AND is_agent?(nd2) AND chan(nd1) = MASTER
         AND chan(nd2)=TARGET) IMPLIES
    (EXISTS (w:prewalk[Nodes]): path_from?(net,w,nd1,nd2))
```

4) The two channels of a bridge are not connected:

```
no_self_loop(net:acynet):bool =  FORALL(nd1,nd2:Nodes):            5
(edge?(net)(nd1,nd2) and name(nd1) = name(nd2)) IMPLIES is_agent?(nd1)
```

5) All the channels driving a PCI bus are connected "properly":

```
bus_connections_ok(net:acynet):bool = FORALL(nd1,nd2,nd3:Nodes):   6
(edge?(net)(nd2, nd1) AND edge?(net)(nd3, nd1) AND nd2 /= nd3) IMPLIES
(edge?(net)(nd2, opposite(nd3)) AND edge?(net)(nd3,opposite(nd2)))
```

6) Every channel has a local master id and for any two channels driving the same PCI bus, the local master id's are different:

```
lmid_map : [Nodes -> nat]                                                7

lmid_ok(net:acynet):bool = FORALL(nd1,nd2:Nodes):
    (edge?(net)(nd1,nd2) and name(nd1) /= name(nd2))
        IMPLIES lmid_map(nd1) /= lmid_map(opposite(nd2))
```

Finally, a well-formed PCI network is an acyclic graph over Nodes that satisfies all the above predicates.

4.2   The Properties of PCI Network

We have proved many interesting properties about our PCI network model. The three very crucial properties that will be used in the proofs later are discussed below:

1. For every path p in the network, there exists another path p' in the opposite direction, with the nodes in p' being the opposite nodes of the corresponding nodes in p. The proof of this property uses the well-formedness conditions stated before and the induction scheme for walks.

2. There is a unique path between any two nodes in the network, if any. Its proof uses the acyclicity property of the network, the induction scheme for walks and the well-formedness conditions stated before.

3. The third property captures all the possible configurations of the four nodes corresponding to the Producer *Prod*, the Consumer *Cons*, Data *Data* and Flag *Flag*. The two *trivial* cases are when *Prod* and *Cons* are the same and when *Data* and *Flag* are the same. All non-trivial configurations showing the paths *Prod* → *Data*, *Cons* → *Data*, *Prod* → *Flag* and *Cons* → *Flag*—all of which are guaranteed to exist by the third well-formedness condition—are shown in Figure 3. In this figure, there are two significant points called the *hit* (h) and the *take-off* (t) points. The hit point is where the path from *Cons* to *Data* hits the path from *Prod* to *Data*, while the take-off point is where the path from *Prod* to *Flag* leaves the path from *Prod* to *Data*. It is proved that the hit point exists in all cases, the segments *Prod* → h and *Cons* → h are disjoint and the segment h → *Data* is common to both the paths *Prod* → *Data* and *Cons* → *Data*. (A similar remark applies for the takeoff point too). Figure 3(a) illustrates the case when t is ahead of or same as h. Figure 3(b) illustrates the case when t is before h by at least two edges. (The two dotted line segments show the paths in the opposite direction corresponding to the two paths shown by the solid line segments. See the first property discussed earlier.) In this case, the path from *Flag* to *Cons* is made up of the segment *Flag* → t, the segment t → h, and the segment h → *Cons*. Figure 3(c) illustrates the case when t is before h, and is directly connected (by an edge). In this case, there exist points 1 and 2 connected by an edge and the path from *Cons* to *Flag* is made up of the segment *Cons* → 1, the edge 1 to 2 and the segment 2 → *Flag*. We have mechanically proved that these cases *completely cover* all possible configurations of *Prod*, *Cons*, *Data* and *Flag* (in the pciconfig theory).

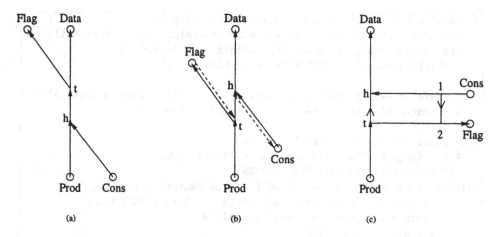

Fig. 3. The possible configurations of *Prod, Cons, Data* and *Flag*.

4.3 The PCI Protocol Formalization

The `pci_state` and the pci theories contain the formal specification of the PCI protocol including all the transitions discussed in Section 2. Recall that each node in our network model corresponds to a channel of a PCI device. The theory `channels` parametrized by the type of all PCI entries (`Entries`, shown in $\boxed{8}$) provides some operations on channels like addition/deletion of entries. An entry in our model includes all the transaction fields discussed in Section 2 augmented by (i) a local master ID 'lmid' introduced in the HP proposal and (ii) auxmid and `tid` introduced by us and (iii) a `committed` field which indicates if the entry has been tried on the out-bus at least once. Each agent has a unique global master ID (auxmid) associated with it and every transaction issued by it has a transaction ID (`tid`) associated. These two fields are never used by the protocol and are introduced since they simplify the specification of Producer/Consumer. `addr_map` maps (associates) an address to an agent. The state of the network (shown in $\boxed{8}$) specifies the contents of every channel, the value of the next transaction ID for every agent and a record of the committed and completed transactions at every agent (discussed later).

```
Entries : TYPE = [# trans : Trans, addr: Addr, data: Data, lmid: nat│ 8
                    auxmid: nat, tid: nat, committed: bool #]

State : TYPE = [# channel: [Nodes -> channel[Entries]],
                   nextTid: [Agent_nodes -> nat],
                   trans_history: [Agent_nodes -> channel[CC_pair]] #]
```

We now give the definition of one of the transactions described in Section 2.2.

```
C_ready?(e:Entries,mid:nat,ch:channel[Entries]):bool =                    | 9
  (trans(e) = DRR AND EXISTS (i:below[length(ch)]): (trans(nth(ch,i))=DRC
   AND addr(nth(ch,i))=addr(e) AND mid(nth(ch,i))= mid AND
   (FORALL (j:below[i]): NOT P_Entry?(nth(ch,j)))))
                      OR
  (trans(e) = DWR AND EXISTS (i:below[length(ch)]): (trans(nth(ch,i))=DWC
   AND addr(nth(ch,i))=addr(e) AND mid(nth(ch,i))=mid))


D_completion(nd:Nodes,s:State) : State =
 IF NOT (is_agent?(nd) AND chan(nd) = TARGET) THEN
   IF DR_exists?(channel(s)(nd)) THEN
(a) LET i= choose_DR(channel(s)(nd)), e=nth(channel(s)(nd),i),
        target = (# name:=addr_map(addr(e)), chan:=TARGET #),
        succ = path_thru(net, nd, target) IN
     IF (succ = target) THEN
     LET new_ce = complete(e) IN
       IF (is_agent?(nd) AND chan(nd) = MASTER) THEN
       LET cc_ent= (# entry:=e , flag:=Completed #),
           cc_pair= (# request:=cc_ent, completion:=new_ce #) IN
        s WITH [(channel):=channel(s) WITH
                [(nd):=del(channel(s)(nd),i),
                 (opposite(nd)):=enq(channel(s)(opposite(nd)),new_ce),
                 (target):=enq(channel(s)(target),
                                     e WITH [(mid):= lmid(nd)])],
               (trans_history):=trans_history(s) WITH
                 [(nd):= enq(trans_history(s)(nd),cc_pair)]]
       ELSE s WITH [(channel):= "Exactly same changes as above" ENDIF
(b)    ELSIF C_ready?(e,lmid(nd),channel(s)(opposite(succ))) THEN
       LET c_index = C_ready?(e,lmid(nd),channel(s)(opposite(succ))),
           c_entry = nth(channel(s)(opposite(succ)),c_index),
           new_ce = c_entry WITH [(mid) := mid(e)] IN
       IF (is_agent?(nd) AND chan(nd) = MASTER) THEN
       LET cc_ent= (# entry:=e , flag:=Completed #),
           cc_pair= (# request:=cc_ent, completion:=new_ce #) IN
        s WITH [(channel):=channel(s) WITH
                [(nd):=del(channel(s)(nd),i),
                (opposite(succ)):=del(channel(s)(opposite(succ)),c_index),
                 (opposite(nd)):=enq(channel(s)(opposite(nd)),new_ce)],
                (trans_history):=trans_history(s) WITH
                 [(nd):=enq(trans_history(s)(nd),cc_pair)]]
       ELSE s WITH [(channel):= "Exactly same changes as above" ENDIF
      ELSE s ENDIF
    ELSE s ENDIF
  ELSE s ENDIF
```

The D_completion transaction (shown in 9) obtains the completion for a R
entry present in a node nd. The R entry is selected such that there is no older
P entries and no other committed R entries for the same address in the channel
(choose_DR predicate on line (a)). Depending on the address specified in the
selected R entry, the final target as well as the successor of nd along the path

to that target is determined. Now there are two cases: (i) the successor of nd is the final target in which case a new completion entry is created or (ii) there is a completion ready for the R entry in the opposite channel of the successor. In case (ii) the local master ID of nd is used in determining if a completion is ready for its R entry (predicate C_ready? on line (b)). Also, this completion is allowed to pass a P entry in case it is a Delayed Write Completion. In both cases, the R entry is removed from nd and the completion obtained is put in opposite(nd). When the original master obtains the completion of a R entry, that global delayed transaction is complete.

As pointed out in Section 2, we record all the transaction entries that have reached an agent in its target channel. Accordingly, the D_completion event records the R entry in the target channel of the final destination and the C entry in the target channel of the original master. In addition, every agent records a history of all transactions committing and completing on its out-bus (trans_history field of the state). In the case of a delayed transaction completing, the completion obtained is recorded too. As explained in subsequent sections, these are used for specification purposes only.

5 Formalization of Producer/Consumer

Our formulation of Producer/Consumer relies on: (i) the absence of completion stealing, and (ii) the existence of a partial order which constrains read and write transactions to complete in the right order. Producer/Consumer requires that the "data" read by the consumer be the same as the "new data" written by the producer. While (ii) only guarantees that the transactions complete in the right order, (i) is necessary to meet the "new data" requirement.

Definition 1. *A PCI based architecture provides the Producer/Consumer memory model if (i) there is no completion stealing and (ii) for every two agents Prod and Cons (which may or may not be identical), for every two addresses Data and Flag (which may or may not be identical): (1), (2), (3), (4) and (5)* **implies** *(6), where:*

(1) *the agent Prod issues two write transactions: W_{Data} (a PMW or DWR transaction to the address Data) followed by W_{Flag} (a PMW or DWR transaction to the address Flag),*

(2) *the agent Cons issues two Delayed Read Request transactions R_{Flag} followed by R_{Data},*

(3) *W_{Flag} is committed on the originating bus after the completion of W_{Data} on the originating bus,*

(4) *R_{Data} is committed on the originating bus after the completion of R_{Flag} on the originating bus,*

(5) *R_{Flag} is completed on the destination bus after W_{Flag} completes on the destination bus, and*

(6) *R_{Data} is completed on the destination bus after W_{Data} completes on the destination bus.*

It is quite simple and intuitive to conclude from (i) and (ii) that the value read by R_{Data} is the same value written by W_{Data} if there is no intervening write to $Data$ between W_{Data} and R_{Data}, even when multiple readers are allowed. Notice that our definition has no other constraints on the environment in which transactions W_{Data}, W_{Flag}, R_{Flag} and R_{Data} are executed. In particular, both W_{Data} and W_{Flag} can be Delayed Writes too. Hence, this is a more general definition of Producer/Consumer.

To prove the above memory model for PCI in PVS, we specified two properties No_steal_prop and Ordering_prop encoding the conditions (i) and (ii) of the above definition respectively. No_steal_prop states that when a delayed transaction is completed by a master, the R entry and the C entry corresponding to the completion obtained should satisfy the trans_match predicate shown in $\boxed{10}$. If a delayed request is serviced either with a completion meant for another master's transaction or with a completion meant for another transaction of the same master, this property is violated:

```
trans_match(e1:R_Entries,e2:C_Entries): bool =                        10
  ((trans(e1)=DRR and trans(e2)=DRC) OR (trans(e1)=DWR and trans(e2)=DWC))
AND addr(e1)=addr(e2) AND auxmid(e1)=auxmid(e2) AND tid(e1)=tid(e2)

No_steal_prop(s:State): bool = FORALL (nd:Agent_master,cc_pair:CC_pair):
  member(cc_pair,trans_history(s)(nd)) AND
  R_Entry?(entry(request(cc_pair))) AND flag(request(cc_pair))=Completed
  IMPLIES trans_match(entry(request(cc_pair)),completion(cc_pair))
```

The formal definition of Ordering_prop, shown in $\boxed{11}$, mirrors part (ii) of the definition 1 given earlier.

```
Ordering_prop(s: State): bool = FORALL(Prod:Agent_master,            11
            Cons:Agent_target,(WD,WF:W_Entries),(RF,RD:DRR_Entries)):
  LET Flag = (# name:= addr_map(addr(RF)), chan:= TARGET #),
      Data = (# name:= addr_map(addr(RD)), chan:= TARGET #)
  IN (addr(WD) = addr(RD)  AND addr(WF) = addr(RF) AND
    sameAgent(WD,WF,Prod) AND tid(WD) < tid(WF)      AND
    sameAgent(RD,RF,Cons) AND tid(RF) < tid(RD) AND
    (orig_commit_cmp?(WF,trans_history(s)(Prod))                       (i)
      IMPLIES (orig_completed?(WD,trans_history(s)(Prod))
              AND In_ch_order(WD,WF,trans_history(s)(Prod)))) AND
    (orig_commit_cmp?(RD,trans_history(s)(Cons))                       (ii)
      IMPLIES (orig_completed?(RF,trans_history(s)(Cons))
              AND In_ch_order(RF,RD,trans_history(s)(Cons)))) AND
    (cmp_in(RF,channel(s)(Flag))                                       (iii)
      IMPLIES (cmp_in(WF,channel(s)(Flag))
              AND In_ch_order(WF,RF,channel(s)(Flag)))))
    IMPLIES (cmp_in(RD,channel(s)(Data))
          IMPLIES (cmp_in(WD,channel(s)(Data))
                  AND In_ch_order(WD,RD,channel(s)(Data))))
```

The following predicates (defined in more_defs theory. See Figure 2(b).) are used in the specification of Ordering_prop. Two entries are issued from the same agent (sameAgent predicate) if their auxmid is the same as the global mid assigned to the agent. The predicate `orig_completed?` tests if a request is completed on the originating bus, `orig_committed?` tests if it is committed on the originating bus and `orig_commit_cmp?` tests if either of these is true. The predicate `cmp_in` tests if an entry is in the specified target channel. Finally, the predicate `In_ch_ord` tells if entry e1 is "before" entry e2 in a given channel.

Note that the definitions in 10 and 11 use the two auxiliary fields `auxmid` and `tid` introduced by us as part of the definition of an entry, the "history" of the completed transactions maintained in the target channels of the agents and the "history" of the committed/completed transactions maintained at the agent nodes.

6 Proof of Producer/Consumer

In this section, we sketch the proofs of `No_steal_prop` and `Ordering_prop` properties and also list some of the auxiliary invariants and lemmas that were needed during the proofs. A property is shown to be an invariant by proving that it is true in the initial state and is preserved by all transitions. The proofs mentioned in this section are in progress.

6.1 Auxiliary Invariants

The auxiliary invariants listed here establish some basic properties about the PCI protocol (`pci_basics` theory in Figure 2(b)) and were needed for the proofs of the main properties. These include:

• All entries in the master channel of an agent have the same `auxmid` as the global mid assigned to that agent.

• The next transaction id of an agent is always greater than the id of all the transactions issued by it earlier.

• For any entry present in a channel, there exits a path in the network leading to the correct destination specified by the address of that entry.

• A transaction cannot be completed without being committed.

• If a C entry is present in a channel, then it is unique (as determined by `addr` and `lmid` fields) and the corresponding R entry is already removed from its opposite channel.

• At any node, only one R entry for a particular address may be committed.

• If a delayed transaction is completed on the originating bus, it is completed on the destination bus too.

Some of the other invariants needed will be discussed in later subsections. In addition to these invariants, we needed several lemmas specifying the correctness of the channel operations like addition/deletion of entries.

6.2 Proof of No_steal_prop

The proof of this property essentially requires us to prove that R and C entries are "correctly" matched by a D_completion event. This fact is captured by the predicate `RC_match` shown in (12):

```
RC_match(s:State): bool = FORALL (nd1:Nodes):                              12
   LET re = nth(channel(s)(nd1), choose_DR(channel(s)(nd1))),
       succ = path_thru(net, nd1,
                           (# name := addr_map(addr(re)),chan := TARGET #)),
       ce = nth(channel(s)(opposite(succ)),
                C_ready(re, lmid(nd1), channel(s)(opposite(succ)))) IN
   DR_exists?(channel(s)(nd1)) AND
   NOT (succ = (# name := addr_map(addr(re)),chan := TARGET #)) AND
   C_ready?(re, lmid(nd1), channel(s)(opposite(succ)))
   IMPLIES (auxmid(re)=auxmid(ce) AND tid(re)=tid(ce) AND committed(re))
```

The D-completion event chooses an R entry re in a node nd1 and matches it against a C entry ce present in the opposite channel of its successor node succ (along the path to the final destination). The matched C entry ce contains the same lmid as the local mid assigned to nd1. We are required to prove that R and C entries thus matched belong to the same transaction *i.e.*, have the same auxmid and tid. This relies on the fact that only one delayed transaction for a particular address is committed in any node and that all nodes driving the same PCI bus have different local mids (well-formedness condition lmid_ok).

6.3   Proof of Ordering_prop

We sketch the proof of this property for the case when WD is a posted transaction(refer to [11]). The proof for the case when WD is a Delayed Write Request is trivial since the premise ensures that WF is not even committed on the originating bus before WD completes on the originating bus, which guarantees that WD has already reached the target before WF is committed. (A delayed transaction completes on the destination bus before completing on the originating bus).

Assume RD has reached its destination Data. It must, then, have completed on the originating bus (at node Cons). From the invariants proved earlier and the condition on line (ii) in [11], RF must have completed at its destination bus and the corresponding completion CF must have reached the node that issued RF (*i.e.*, Cons). From (iii) in [11], it follows that WF has reached its destination Flag. In this scenario, we prove that WD must have reached its destination Data by a case analysis on the possible configurations (See Figure 3):

Case (a) Since WF has reached the Flag, WD entry must be ahead of t. This is because WF can not pass the posted transaction WD and hence must push it past t. Now RD (which is issued after CF has reached Cons which in turn implies that it is issued after WF has reached Flag) will push WD to its destination Data. These "pushing" arguments are formulated and proved as invariants and rely crucially on the the fact that there is a unique path between any two nodes in the channel graph, if any and the ordering rules of the PCI protocol.

Case (b) This is the most complex case where all three entries—WF, CF, RD—"push" WD. Before WF reaches Flag, it would have pushed WD past t, before CF reaches Cons, it would have pushed WD past h and finally before RD reaches Data, it would have pushed WD to its destination.

Case (c) Similar to case (a). When WF reaches Flag, it would have pushed WD past t which means that WD is in the segment h → Data (since t → h is an edge). Now RD will push WD to its final destination.

Note that limit cases (d) Prod and Cons are the same nodes and (e) Data and Flag are the same nodes are special cases of case (a).

7 Conclusions

Using PVS we have formalized that the PCI 2.1 protocol extended by local master ID provides the Producer/Consumer memory model. Due to technical difficulties, we have not established the Producer/Consumer relationship at the time of writing, but believe that we are well on our way. Our proofs are generic in the sense that they cover any acyclic network of a finite number of PCI buses. We have shown that theorem proving technology is mature enough today to be able to tackle such a challenging problem. The proofs make extensive use of induction and several fundamental theorems from the acyclic directed graph theory we have proved in PVS. Although the expertise we accumulated in both PVS and formal verification in the last few years [5, 6, 10] was crucial to realize this work in a reasonable time (5 person months so far) we believe that significant progress has been made in the theorem proving technology contributing to practical use of tools like PVS in the formal verification of commercial protocols. However, this progress should continue from different perspectives in order to make such tool accessible by a larger community. We believe that a user-interface providing suitable functionalities including basic graphical operations such that zoom, shrink, expand, etc. of nodes in the graphical representation of the proof would be very helpful. Also, it is unacceptable that the user has to cut-and-paste a thousand-line proof and apply it step by step to a similar subgoal in an adjacent node in the proof tree. A mouse-drag from the proved node to an unproved node that is similar should be enough. Such situations arose many times in our work. In addition, the proof manager in PVS should be smarter in naming the formulas. The current naming scheme is a serious impediment to automatic proof reuse, especially when proofs are very long. A local update of a proved property[1] should not affect the global proof of the original property: one should be able to easily reuse the proof corresponding to the un-modified part.

References

1. Ricky W. Butler and Jon A. Sjogren. A PVS Graph Theory Library. Technical Report Memorandum, NASA Langly Research Center, December 1997. http://atb-www.larc.nasa.gov/ftp/larc/PVS-library.

---

[1] For example it is very common to add a conjunct to the assumptions or the conclusions of a sequent.

2. E. M. Clarke, O. Grumberg, H. Hiraishi, S. Jha, D. E. Long, K. L. McMillan, and L. A. Ness. Verification of the futurebus+ cache coherence protocol. In L. Claesen, editor, *Eleventh International Symposium on Computer Hardware Description Languages and their Applications*. North-Holland, April 1993.

3. Francisco Corella. Verifying memory ordering model of I/O systems. In *Invited talk at Computer Hardware Description Languages 1997, Toledo*, Spain, April 1997.

4. Francisco Corella, Robert Shaw, and Cui Zhang. A formal proof of absence of deadlock for any acyclic network of PCI buses. In *Computer Hardware Description Languages*, 1997.

5. Rajnish Ghughal, Abdel Mokkedem, Ratan Nalumasu, and Ganesh Gopalakrishnan. Using "test model-checking" to verify the runway-pa8000 memory model. In *Tenth Annual ACM Symposium On Parallel Algorithms And Architectures*, pages 231–239, Puerto Vallarta, Mexico, June 1998. ACM Press.

6. G. Gopalakrishnan, R. Ghughal, R. Hosabettu, A. Mokkedem, and R. Nalumasu. Formal modeling and validation applied to a commercial coherent bus: A case study. In Hon F. Li and David K. Probst, editors, *CHARME*, Montreal, Canada, 1997.

7. PCI Special Interest Group. PCI Local Bus Specification, Revision 2.1, June 1995.

8. A. Mokkedem. Verification of PCI 2.1 Local Bus in PVS. http://www.cs.utah.edu/~mokkedem/pvs/pvs.html.

9. Vijay Nagasamy, Sreeranga Rajan, and Preeti R. Panda. Fiber channel protocol: Formal specification and verification. In *Sixth Annual Silicon Valley Networking Conference*, 1997.

10. Ratan Nalumasu, Rajnish Ghughal, Abdel Mokkdem, and Ganesh Gopalakrishnan. The 'test model-checking' approach to the verification of formal memory models of multiprocessors. In Alan J. Hu and Moshe Y. Vardi, editors, *Computer Aided Verification*, volume 1427 of *LNCS*, pages 464–476, Vancouver, BC, Canada, June 1998. Springer-Verlag.

11. Sam Owre, John Rushby, Natarajan Shankar, and Friedrich von Henke. Formal verification for fault-tolerant architectures: Prolegomena to the design of PVS. *IEEE Transactions on Software Engineering*, 21(2):107–125, February 1995.

12. F. Corella. Hewlett Packard. Proposal to fix ordering problem in PCI 2.1, 1996. http://www.pcisig.com/reflector/thrd8.html#00706.

13. S. Park and D. L. Dill. Protocol verification by aggregation of distributed transactions. In Rajeev Alur and Thomas A. Henzinger, editors, *Computer Aided Verification*, volume 1102 of *Lecture Notes in Computer Science*, pages 300–309, New Brunswick, NJ, USA, July 1996. Springer-Verlag.

14. Edward Solari and George Willse. *PCI Hardwarde and Software Architecture & Design*. Annabooks, 3rd edition edition, December 1996. ISBN 0-929392-32-9.

A Performance Study
of BDD-Based Model Checking

Bwolen Yang[1], Randal E. Bryant[1], David R. O'Hallaron[1],
Armin Biere[1], Olivier Coudert[2], Geert Janssen[3],
Rajeev K. Ranjan[2], and Fabio Somenzi[4]

[1] Carnegie Mellon University, Pittsburgh PA 15213, USA
[2] Synopsys Inc., Mountain View CA 94043, USA
[3] Eindhoven University of Technology, 5600 MB Eindhoven, Netherlands
[4] University of Colorado, Boulder CO 90309, USA

Abstract. We present a study of the computational aspects of model
checking based on binary decision diagrams (BDDs). By using a trace-
based evaluation framework, we are able to generate realistic benchmarks
and perform this evaluation collaboratively across several different BDD
packages. This collaboration has resulted in significant performance improvements and in the discovery of several interesting characteristics of
model checking computations. One of the main conclusions of this work
is that the BDD computations in model checking and in building BDDs
for the outputs of combinational circuits have fundamentally different
performance characteristics. The systematic evaluation has also uncovered several open issues that suggest new research directions. We hope
that the evaluation methodology used in this study will help lay the
foundation for future evaluation of BDD-based algorithms.

1 Introduction

The binary decision diagram (BDD) has been shown to be a powerful tool in
formal verification. Since Bryant's original publication of BDD algorithms [7],
there has been a great deal of research in the area [8, 9]. One of the most
powerful applications of BDDs has been to symbolic model checking, used to
formally verify digital circuits and other finite state systems. Characterizations
and comparisons of new BDD-based algorithms have historically been based on
two sets of benchmark circuits: ISCAS85 [6] and ISCAS89 [5]. There has been
little work on characterizing the computational aspects of BDD-based model
checking.

There are two qualitative differences between building BDD representations
for combinational circuits versus model checking. The first difference is that for
combinational circuits, the *output BDDs* (BDD representations for the circuit
outputs) are built and then are only used for constant-time equivalence checking.
In contrast, a model checker first builds the BDD representations for the system transition relation, and then performs a series of fixed point computations

analyzing the state space of the system. In doing so, it is solving PSPACE-complete problems. Another difference is that BDD construction algorithms for combinational circuit operations have polynomial complexity [7], while the key operations in model checking are NP-hard [16]. These differences indicate that results based on combinational circuit benchmarks may not accurately characterize BDD computations in model checking.

This paper introduces a new methodology for the systematic evaluation of BDD computations, and then applies this methodology to gain a better understanding of the computational aspects of model checking. The evaluation is a collaborative effort among many BDD package designers. As results of this evaluation, we have significantly improved model checking performance, and have identified some open problems and new research directions.

The evaluation methodology is based on a trace-driven framework where execution traces are recorded from verification tools and then replayed on several BDD packages. In this study, the benchmark consists of 16 execution traces from the Symbolic Model Verifier (SMV) [16]. For comparison with combinational circuits, we also studied 4 circuit traces derived from the ISCAS85 benchmark. The other part of our evaluation methodology is a set of platform independent metrics. Throughout this study, we have identified useful metrics to measure work, space, and memory locality.

This systematic and collaborative evaluation methodology has led to better understanding of the effects of cache size and garbage collection frequency, and has also resulted in significant performance improvement for model checking computations. Systematic evaluation also uncovered vast differences in the computational characteristics of model checking and combinational circuits. These differences include the effects of the cache size, the garbage collection frequency, the complement edge representation [1], and the memory locality of the breadth-first BDD packages. For the difficult issue of dynamic variable reordering, we introduce some methodologies for studying the effects of variable reordering algorithms and initial variable orders.

It is important to note that the results in this study are obtained based on a very small sample of all possible BDD-based model checking computations. Thus, in the subsequent sections, most of the results are presented as hypotheses along with their supporting evidence. These results are not conclusive. Instead, they raise a number of interesting issues and suggest new research directions.

The rest of this paper is organized as follows. We first present a brief overview of BDDs and relevant BDD algorithms (Sec. 2) and then describe the experimental setup for the study (Sec. 3). This is followed by three sections of experimental results. First, we report the findings without dynamic variable reordering (Sec. 4). Then, we present the results on dynamic variable reordering algorithms and the effects of initial variable orders (Sec. 5). Third, we present results that may be generally helpful in studying or improving BDD packages (Sec. 6). After these result sections, we discuss some unresolved issues (Sec. 7) and then wrap up with related work (Sec. 8) and concluding remarks (Sec. 9).

2 Overview

This section gives a brief overview of BDDs and pertinent BDD algorithms. Detailed descriptions can be found in [7] and [16].

2.1 BDD Basics

A BDD is a directed acyclic graph (DAG) representation of a Boolean function where equivalent Boolean sub-expressions are uniquely represented. Due to this uniqueness property, a BDD can be exponentially more compact than its corresponding truth table representation. One criterion for guaranteeing the uniqueness of the BDD representation is that all the BDDs constructed must follow the same variable order. The choice of this variable order can have a significant impact on the size of the BDD graph.

BDD construction is a memoization-based dynamic programming algorithm. Due to the large number of distinct subproblems, a cache, known as the *computed cache*, is used instead of a memoization table. Given a Boolean operation, the construction of its BDD representation consists of two main phases. In the top-down *expansion phase*, the Boolean operation is recursively decomposed into subproblems based on the Shannon decomposition. In the bottom-up *reduction phase*, the result of each subproblem is put into the canonical form. The uniqueness of the result's representation is enforced by hash tables known as *unique tables*. The new subproblems are generally recursively solved in a depth-first order as in Bryant's original BDD publication [7]. Recently, there has been some work that tries to exploit memory locality by using a breadth-first order [2, 18, 19, 21, 26].

Before moving on, we first define some terminology. We will refer to the Boolean operations issued by a user of a BDD package as the *top-level operations* to distinguish them from *sub-operations* (subproblems) generated internally by the Shannon expansion process. A BDD node is *reachable* if it is in some BDDs that external users have references to. As external users free references to BDDs, some BDD nodes may no longer be reachable. We will refer to these nodes as *unreachable* BDD nodes. Note that unreachable BDD nodes can still be referenced within a BDD package by either the unique tables or the computed cache. Some of these unreachable BDD nodes may become reachable again if they end up being the results for new subproblems. When a reachable BDD node becomes unreachable, we say a *death* has occurred. Similarly, when an unreachable BDD node becomes reachable again, we say a *rebirth* has occurred. We define the *death rate* as the number of deaths over the number of subproblems (*time*) and define the *rebirth rate* as the fraction of the unreachable nodes that become reachable again, i.e., the number of rebirths over the number of deaths.

2.2 Common Implementation Features

Modern BDD packages typically share the following common implementation features based on [4, 22]. The BDD construction is based on depth-first traversal.

The unique tables are hash tables with the hash collisions resolved by chaining. A separate unique table is associated with each variable to facilitate the dynamic variable reordering process. The computed cache is a hash-based direct mapped (1-way associative) cache. BDD nodes support *complement edges* where, for each edge, an extra bit is used to indicate whether or not the target function should be inverted. Garbage collection of unreachable BDD nodes is based on reference counting and the reclaimed unreachable nodes are maintained in a *free-list* for later reuse. Garbage collection is invoked when the percentage of the unreachable BDD nodes exceeds a preset threshold.

As the variable order can have significant impact on the size of a BDD graph, dynamic variable reordering is an essential part of all modern BDD packages. The dynamic variable reordering algorithms are generally based on *sifting* or *window permutation* algorithms [22]. Typically, when a variable reordering algorithm is invoked, all top-level operations that are currently being processed are aborted. When the variable reordering algorithm terminates, these aborted operations are restarted from the beginning.

2.3 Model Checking and Relational Product

There are two popular BDD-based algorithms for computing state transitions: one is based on applying the *relational product* operator (also known as *AndExists* or *and-smooth*) on the transition relations and the state sets [10]; the other is based on applying the *constrain* operator to Boolean functional vectors [11, 12].

The benchmarks in this study are based on SMV, which uses the relational product operation. This operation computes "$\exists v. f \wedge g$" and is used to compute the set of states by the forward or the backward state transitions. It has been proven to be NP-hard [16]. Figure 1 shows a typical BDD algorithm for computing the relational product operation. This algorithm is structurally very similar to the BDD-based algorithm for the *AND* Boolean operation. The main difference (lines 5–11) is that when the top variable (τ) needs to be quantified, a new BDD operation ($OR(r_0, r_1)$) is generated. Due to this additional recursion, the worst case complexity of this algorithm is exponential in the graph size of the input arguments.

3 Setup

3.1 Benchmark

The benchmark used in this study is a set of execution traces gathered from the Symbolic Model Verifier (SMV) [16] from Carnegie Mellon University. The traces were gathered by recording BDD function calls made during the execution of SMV. To facilitate the porting process for different packages, we only recorded a set of the key Boolean operations and discarded all word-level operations. The coverage of this selected set of BDD operations is greater than 95% of the total SMV execution time for all but one case (*abp11*) which spends 21% of CPU time in the word-level functions constructing the transition relation.

```
RP(v, f, g)
        /* compute relational product: ∃v.f ∧ g */
1    if (terminal case) return result
2    if the result of (RP, v, f, g) is cached, return the result
3    let τ be the top variable of f and g
4    r₀ ← RP(v, f|τ←0, g|τ←0)   /* Shannon expansion on 0-cofactors */
5    if (τ ∈ v)   /* existential quantification on τ ≡ OR(r₀, RP(v, f|τ←1, g|τ←1)) */
6        if (r₀ == true)   /* OR(true, RP(v, f|τ←1, g|τ←1)) ≡ true */
7            r ← true.
8        else
9            r₁ ← RP(v, f|τ←1, g|τ←1)   /* Shannon expansion on 1-cofactors */
10           r ← OR(r₀, r₁)
11   else
12       r₁ ← RP(v, f|τ←1, g|τ←1)   /* Shannon expansion on 1-cofactors */
13       r ← reduced, unique BDD node for (τ, r₀, r₁)
14   cache the result of this operation
15   return r
```

Fig. 1. A typical relational product algorithm.

A side effect of recording only a subset of BDD operations is that the construction process of some BDDs is skipped, and these BDDs might be needed later by some of the selected operations. Thus in the trace file, these BDDs need to be reconstructed before their first reference. This reconstruction is performed bottom-up using the *If-Then-Else* operation. This process is based on the property that each BDD node (v_i, $child_0$, $child_1$) essentially represents the Boolean function "If v_i then $child_1$ else $child_0$".

For this study, we have selected 16 SMV models to generate the traces. The following is a brief description of these models along with their sources.

abp11: alternating bit protocol.
Source: Armin Biere, Universität Karlsruhe.

dartes: communication protocol of an Ada program.
dpd75: dining philosophers protocol.
ftp3: file transfer protocol.
furnace17: remote furnace program.
key10: keyboard/screen interaction protocol in a window manager.
mmgt20: distributed memory manager protocol.
over12: automated highway system overtake protocol.
Source: James Corbett, University of Hawaii.

dme2-16: distributed mutual exclusion protocol.
Source: SMV distribution, Carnegie Mellon University.

futurebus: futurebus cache coherence protocol.
Source: Somesh Jha, Carnegie Mellon University.

motor-stuck: batch-reactor system model.
valves-gates: batch-reactor system model.
Source: Adam Turk, Carnegie Mellon University.

phone-async: asynchronous model of a simple telephone system.
phone-sync-CW: synchronous model of a telephone system with call
waiting.
Source: Malte Plath and Mark Ryan, University of Birmingham,
Great Britain.

tcas: traffic alert and collision system for airplanes.
Source: William Chan, University of Washington.

tomasulo: a buggy model of the Tomasulo algorithm for instruction
scheduling in superscalar processors.
Source: Yunshan Zhu, Carnegie Mellon University.

As we studied and improved on the model checking computations during the
course of the study, we compared their performance with the BDD construction
of combinational circuit outputs. For this comparison, we used the ISCAS85
benchmark circuits as the representative circuits. We chose these benchmarks
because they are perhaps the most popular benchmarks used for BDD performance evaluations. The ISCAS85 circuits were converted into the same format
as the model checking traces. The variable orders used were generated by the
order-dfs in SIS [24]. We excluded cases that were either too small (< 5 CPU
seconds) or too large (> 1 GBytes of memory requirement). Based on this criteria, we were left with two circuits — C2670 and C3540. To obtain more circuits,
we derived 13-bit and 14-bit integer multipliers, based on the C6288, which we
refer to as C6288-13 and C6288-14. For the multipliers, the variable order is
$a_{n-1} \prec a_{n-2} \prec ... \prec a_0 \prec b_{n-1} \prec b_{n-2} \prec ... \prec b_0$, where $A = \sum_{i=0}^{n-1} 2^i a_i$ and
$B = \sum_{i=0}^{n-1} 2^i b_i$ are the two n-bit input operands to the multiplier.

Figure 2 quantifies the sizes of the traces we used in the study. The statistic
"*# of BDD Vars*" is the number of BDD variables used. The statistic "*Min. #
of Ops*" is the minimum number of sub-operations (or subproblems) needed for
the computation. This statistic characterizes the minimum amount of work for
each trace. It was gathered using a BDD package with a complete cache and no
garbage collection. Thus, this statistic represents the minimum number of sub-
operations needed for a typical BDD package. Due to insufficient memory, there
are 4 cases (*futurebus, phone-sync-CW, tcas, tomasulo*) for which we were not
able to collect this statistic. For these cases, the results shown are the minimum
across all the packages used in the study. These results are marked with the
"$<$" symbol. The third statistic, "*Peak # of Live BDDs*", represents the peak
number of reachable BDD nodes during the execution. It provides a lower bound

on the memory required to execute the corresponding trace. Note that neither "*Min. # of Ops*" nor "*Peak # of Live BDDs*" reflects the effects of the dynamic variable reordering process.

Trace	# of BDD Vars	Min. # of Ops ($\times 10^6$)	Peak # of Live BDDs ($\times 10^3$)
abp11	122	116	53
dartes	198	6	468
dme2-16	586	106	905
dpd75	600	41	1719
ftp3	100	132	763
furnace17	184	30	2109
futurebus	348	< 10270	4473
key10	140	91	626
mmgt20	264	35	1113
motors-stuck	172	29	325
over12	174	58	3008
phone-async	86	329	1446
phone-sync-CW	88	< 3803	22829
tcas	292	< 1323	19921
tomasulo	212	< 1497	26944
valves-gates	172	44	433
c2670	233	15	4363
c3540	50	57	7775
c6288-13	26	60	3378
c6288-14	28	178	9662

Fig. 2. Sizes of the benchmark traces. "# of BDD Vars" is the number of BDD variables. "*Min. # of Ops*" is the minimum number of sub-operations which characterizes work. "*Peak # of Live BDDs*" is the maximum number of reachable BDD nodes, which characterizes the minimum memory requirement.

3.2 BDD Packages

The following is a list of the BDD packages used in the study. For each BDD package, we note how it differs from the common implementation described in Sec. 2.2. Although many of these BDD packages contain a wide variety of useful features, only those pertinent to the study are described in this section.

ABCD (Author: Armin Biere)

ABCD [3] is an experimental BDD package based on the classical depth-first traversal. Interesting features include mark-and-sweep based garbage collection, the integration of BDD nodes with the BDD unique table by using open addressing, and index-based (instead of pointer-based) references to

BDD nodes. These techniques reduce the BDD node size by half (2 machine words instead of 4). In addition, to avoid clustering in open addressing, ABCD uses a quadratic probe sequence for the hashing collision resolution.

CAL (Authors: Rajeev Ranjan and Jagesh Sanghavi)

CAL [20] is a publicly available BDD package based on breadth-first traversal to exploit memory locality. The garbage collection algorithm is based on reference-counting with memory compaction. To increase locality of reference, each BDD node contains the indices of its cofactor nodes. To keep the node size to 4 machine words, bit tagging is used to store and retrieve the value of the reference count of a node. For this study, the relational product operation is based on the depth-first traversal with the quantification step (line 7 in Fig. 1) computed using the breadth-first traversal.

CUDD (Author: Fabio Somenzi)

CUDD [25] is a publicly available BDD package based on depth-first traversal. In CUDD, the reference counts of the nodes are kept up-to-date throughout the computation. To counter the impact on performance of these updates when many nodes are freed and reclaimed, CUDD enqueues the requests for updates and performs them only if they are still valid when they are extracted from the queue. The growth of the tables in CUDD is determined by a reward policy. For instance, the cache grows if the hit rate is high. CUDD partially sorts the free list during garbage collection to improve memory locality. Another distinguishing feature is that CUDD contains a suite of heuristics for dynamic variable reordering.

EHV (Author: Geert Janssen)

EHV [14] is a publicly available BDD package based on depth-first traversal. The main differences from the common implementation are additional support for inverted inputs [17] and provisions for user data to be attached to a BDD node. The latter feature allows intermediate results to be stored in the BDD nodes, which in turn, removes the need to use separate computed caches for some special BDD operations. This feature incurs a memory overhead of 2 extra machine words per BDD node.

PBF (Authors: Bwolen Yang and Yirng-An Chen)

PBF [26] is an experimental BDD package based on partial breadth-first traversal. The partial breadth-first traversal along with per-variable memory managers and the memory-compacting mark-and-sweep garbage collector are used to exploit memory locality. The partial breadth-first traversal also bounds the breadth-first expansion to avoid the potential excessive memory overhead of a full breadth-first expansion.

TiGeR (Authors: Olivier Coudert, Jean C. Madre and Herve Touati)

TiGeR [13] is a commercial BDD package based on the depth-first approach. Interesting features include the segmentation of the computed caches and the garbage collection algorithm. In TiGeR, each operation type has its own cache. This allows the caches to be tuned independently. For this study, the caches for the non-polynomial operations such as relational product are set to be about four times as sparse as the caches for the polynomial operations. TiGeR's garbage collection algorithm is different from typical garbage collection algorithms in two ways: the free-list is sorted to maintain memory locality, and the memory compaction is performed when memory resources become critical.

3.3 Evaluation Process

The performance study was carried out in two phases. The first phase studied performance issues in BDD construction without variable reordering. The second phase focused on the dynamic variable reordering computation. The evaluation process was iterative, with the study evolving dynamically as new issues were raised and new insights gained. Based on the results from each iteration, we collaboratively tried to identify the performance issues and possible improvements. Each BDD package designer then incorporated and validated the suggested improvements. During this iterative process, we also tried to hypothesize the characteristics of the computation and design new experiments to test these hypotheses.

4 Phase 1 Results: No Variable Reordering

Figure 3 presents the overall performance improvements for Phase 1 with dynamic variable reordering disabled. There are 6 packages and 16 model checking traces, for a total of 96 cases. Figure 3(a) categorizes the results for these cases based on speedups. Note that the speedups are plotted in a cumulative fashion; i.e., the $> x$ column represents the total number of cases with speedups greater than x. Figure 3(b) presents a comparison between the initial timing results (when we first started the study) and the current timing results (after the authors made changes to their packages based on insights gained from previous iterations). The n/a results represent cases where results could not be obtained.

Initially, 19 cases did not complete because of implementation bugs or memory limits. Currently, 13 of these 19 cases now complete (the *new* cases in the figures). The other 6 cases still do not complete within the the resource limit of 8 hours and 900 MBytes (the *failed* cases in the figures). There is one case (the *bad* case in the charts) that initially completed, but now does not complete within the memory limit.

Figure 3(a) shows that significant speedups have been obtained for many cases. Most notably, 22 cases have speedups greater than an order of magnitude (the > 10 column), and 6 out of these 22 cases actually achieve speedups greater than two orders of magnitude (the > 100 column)!

Figure 3(b) shows that significant speedups have been obtained mostly from the small to medium traces, although some of the larger traces have achieved speedups greater than 3. Another interesting point is that the *new* cases (those that initially failed but are now doable) range across small to large traces.

Overall, for the 76 cases where the comparison could be made, the total CPU time was reduced from 554,949 seconds to 127,786 seconds — a speedup of 4.34. Another interesting overall statistic is that initially none of the 6 BDD packages could complete all 16 traces, but currently 3 BDD packages can complete all of them.

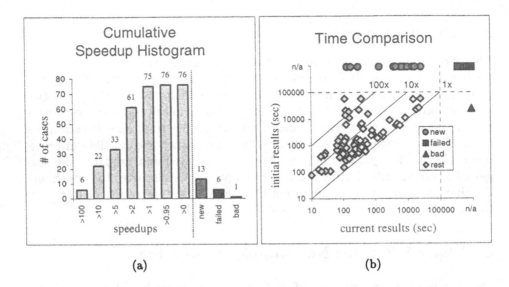

(a)   (b)

Fig. 3. Overall results. The *new* cases represent the number of cases that failed initially and are now doable. The *failed* cases represent those that currently still exceed the limits of 8 CPU hours and 900 MBytes. The *bad* shows the case that finished initially, but cannot complete currently. The *rest* are the remaining cases. (a) Results shown as histograms. For the 76 cases where both the initial and the current results are available, the speedup results are shown in a cumulative fashion; i.e., the $> x$ column represents the total number of cases with speedups greater than x. (b) Time comparison (in seconds) between the initial and the current results. n/a represents results that are not available due to resource limits.

The remainder of this section presents results on a series of experiments that characterize the computational aspects of the BDD traces. We first present results on two aspects with significant performance impact — computed cache size and garbage collection frequency. Then we present results on the effects of the complement edge representation. Finally, we give results on memory locality issues for the breadth-first based traversal.

4.1 Computed Cache Size

We have found that dramatic performance improvements are possible by using a larger computed cache. To study the impact of the computed cache, we performed some experiments and arrived at the following two hypotheses.

Hypothesis 1 *Model checking computations have a large number of repeated subproblems across the top-level operations. On the other hand, combinational circuit computations generally have far fewer such repeated subproblems.*

Experiment: Measure the minimum number of subproblems needed by using a complete cache (denoted CC-NO-GC). Compare this with the same setup but with the cache flushed between top-level operations (denoted CC-GC). For both cases, BDD-node garbage collection is disabled.

Result: Figure 4 shows the results of this experiment. Note that the results for the four largest model checking traces are not available due to insufficient memory.

These results show that for model checking traces, there are indeed many subproblems repeated across the top-level operations. For 8 traces, the ratio of the number of operations in CC-GC over the number of operations in CC-NO-GC is greater than 10. In contrast, this ratio is less than 2 for building output BDDs for the ISCAS85 circuits. For model checking computations, since subproblems can be repeated further apart in time, a larger cache is crucial.

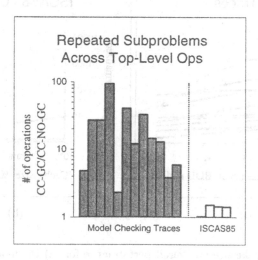

Fig. 4. Performance measurement on the frequency of repeated subproblems across the top-level operations. CC-GC denotes the case in which the cache is flushed between the top-level operations. CC-NO-GC denotes the case in which the cache is never flushed. In both cases, a complete cache is maintained within a top-level operation and BDD-node garbage collection is disabled. For four model checking traces, the results are not available (and are not shown) due to insufficient memory.

Hypothesis 2 *The computed cache is more important for model checking than for combinational circuits.*

Experiment: Vary the cache size as a percentage of the number of BDD nodes and collect the statistics on the number of subproblems generated to measure the effect of the cache size. In this experiment, the cache sizes vary from 10%

to 80% of the number of BDD nodes. The cache replacement policy used is FIFO (first-in-first-out).

Results: Figure 5 plots the results of this experiment. Each curve represents the result for a trace with varying cache sizes. The "# of Ops" statistic is normalized over the minimum number of operations necessary (i.e., the CC-NO-GC results). Note that for the four largest model checking traces, the results are not available due to insufficient memory.

These results clearly show that the cache size can have much more significant effects on the model checking computations than on building BDDs for the ISCAS85 circuit outputs.

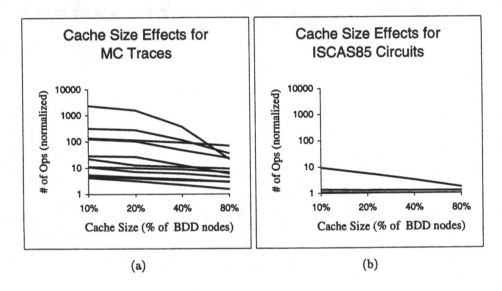

(a)   (b)

Fig. 5. Effects of cache size on overall performance for (a) the model checking traces and (b) the ISCAS85 circuits. The cache size is set to be a percentage of the number of BDD nodes. The number of operations (subproblems) is normalized to the minimum number of subproblems necessary (i.e., the CC-NO-GC results).

4.2 Garbage Collection Frequency

The other source of significant performance improvement is the reduction of the garbage collection frequency. We have found that for the model checking traces, the rate at which reachable BDD nodes become unreachable (death rate) and the rate at which unreachable BDD nodes become reachable (rebirth rate) can be quite high. This leads to the following conclusions:

- Garbage collection should occur less frequently.

- Garbage collection should not be triggered solely based on the percentage of the unreachable nodes.
- For reference-counting based garbage collection algorithms, maintaining accurate reference counts all the time may incur non-negligible overhead.

Hypothesis 3 *Model checking computations can have very high death and rebirth rates, whereas combinational circuit computations have very low death and rebirth rates.*

Experiment: Measure the death and rebirth rates for the model checking traces and the ISCAS85 circuits.

Results: Figure 6(a) plots the ratio of the total number of deaths over the total number of sub-operations. The number of sub-operations is used to represent *time*. This chart shows that the death rates for the model checking traces can vary considerably. In 5 cases, the number of deaths is higher than the number of sub-operations (i.e., death rate is greater than 1). In contrast, the death rates of the ISCAS85 circuits are all less than 0.3.

That the death rates exceed 1 is quite unexpected. To explain the significance of this result, we digress briefly to describe the process of BDD nodes becoming unreachable (death) and then becoming reachable again (rebirth). When a BDD node become unreachable, its children can also become unreachable if this BDD node is its children's only reference. Thus, it is possible that when a BDD node become unreachable, a large number of its descendants also become unreachable. Similarly, if an unreachable BDD node becomes reachable again, a large number of its unreachable descendants can also become reachable. Other than rebirth, the only way the number of reachable nodes can increase is when a sub-operation creates a new BDD node as its result. As each sub-operation can produce at most one new BDD node, a death rate of greater than 1 can only occur when the corresponding rebirth rate is also very high. In general, high death rate coupled with high rebirth rate indicates that many nodes are toggling between being reachable and being unreachable. Thus, for reference-counting based garbage collection algorithms, maintaining accurate reference count all the time may incur significant overhead. This problem can be addressed by using a bounded-size queue to delay the reference-count updates until the queue overflows.

Figure 6(b) plots the ratio of the total number of rebirths over the total number of deaths. Since garbage collection is enabled in these runs and does reclaim unreachable nodes, the rebirth rates shown may be lower than without garbage collection. This figure shows that the rebirth rates for the model checking traces are generally very high — 8 out of 16 cases have rebirth rates greater than 80%. In comparison, the rebirth rate for the ISCAS85 circuits are all less than 30%.

The high rebirth rates indicate that garbage collection for the model checking traces should be delayed as long as possible. There are two reasons for this: first, since a large number of unreachable nodes do become reachable again, garbage collection will not be very effective in reducing the memory usage. Second, the high rebirth rate may result in repeated subproblems involving the currently unreachable nodes. By garbage collecting these unreachable nodes, their corresponding computed cache entries must also be cleared. Thus, garbage collection may greatly increase the number of recomputations of identical subproblems.

The high rebirth rates and the potentially high death rates also suggest that the garbage collection algorithm should not be triggered based solely on the percentage of the dead nodes, as with the classical BDD packages.

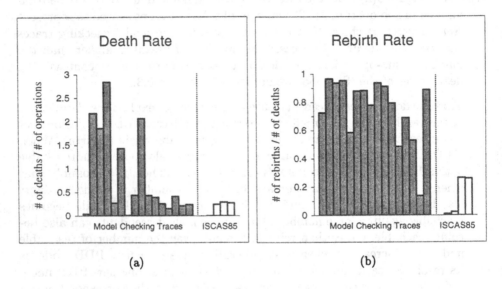

(a) (b)

Fig. 6. (a) Rate at which BDD nodes become unreachable (death). (b) Rate at which unreachable BDD nodes become reachable again (rebirth).

4.3 Effects of the Complement Edge

The complement edge representation [1] has been found to be somewhat useful in reducing both the space and time required to build the output BDDs for the ISCAS85 circuits [17]. In the following experiments, we study the effects of the complement edge on the model checking traces and compare it with the results for the ISCAS85 circuits.

Hypothesis 4 *The complement edge representation can significantly reduce the amount of work for combinational circuit computations, but not for model checking computations. However, in general, it has little impact on memory usage.*

Experiment: Measure and compare the number of subproblems (amount of work) and the resulting graph sizes (memory usage) generated from two BDD packages — one with and the other without the complement-edge feature. For the graph size measurements, sum the resulting BDD graph sizes of all top-level operations. Note that since two packages are used, minor differences in the number of operations can occur due to different garbage collection and caching algorithms.

Results: Figure 7(a) shows that the complement edges have no significant effect for model checking traces. In contrast, for the ISCAS85 circuits, the ratio of the no-complement-edge results over with-complement-edge results ranges from 1.75 to 2.00. Figure 7(b) shows that the complement edges have no significant effect on the BDD graph sizes in any of the benchmark traces.

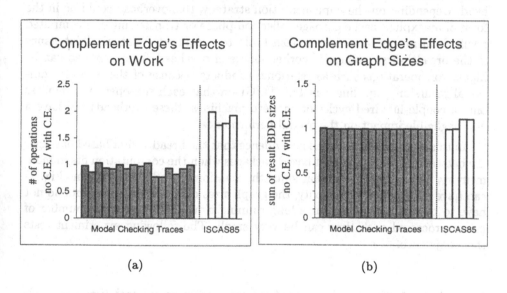

(a)   (b)

Fig. 7. Effects of the complement edge representation on (a) number of the operations and (b) graph sizes.

4.4 Memory Locality for Breadth-First BDD Construction

In recent years, a number of researchers have proposed breadth-first BDD construction to exploit memory locality [2, 18, 19, 21, 26]. The basic idea is that for each expansion phase, all sub-operations of the same variable are processed together. Similarly, for each reduction phase, all BDD nodes of the same variable are produced together. Note that even though this *levelized* access pattern is slightly different from the traditional notion of breadth-first traversal, we will continue to refer to this pattern as breadth-first to be consistent with previous work. Based on this structured access pattern, we can exploit memory locality by using per-variable memory managers and per-variable breadth-first queues to cluster the nodes of the same variable together. This clustering is beneficial only if many nodes are processed for each breadth-first queue during each expansion and reduction phase.

The breadth-first approach does have some performance drawbacks (at least in the two packages we studied). The breadth-first expansion usually has higher memory overhead. In terms of running time, one drawback is in the implementation of the cache. In the breadth-first approach, the sub-operations are explicitly represented as *operator nodes* and the uniqueness of these nodes is ensured by using a hash table with chaining for collision resolution. Accesses to this hash table are inherently slower than accesses to the direct mapped (1-way associative) computed cache used in the depth-first approaches. Furthermore, handling of the computed and yet-to-be-computed operator nodes adds even more overhead. Depending on the implementation strategy, this overhead could be in the form of an explicit cache garbage collection phase or transferring of a computed result from an operator node's hash table to a computed cache. Maintenance of the breadth-first queues is another source of overhead. This overhead can be higher for operations such as relational products because of the possible additional recursion (e.g., line 7 in Fig. 1). Given that each sub-operation requires only a couple hundred cycles on modern machines, these overheads can have a non-negligible impact on the overall performance.

In this study, we have found no evidence that the breadth-first based packages are better than the depth-first based packages when the computation fits in main memory. Our conjecture is that since the relational product algorithm (Fig. 1) can have exponential complexity, the graph sizes of the BDD arguments do not have to be very large to incur a long running time. As a result, the number of nodes processed each time can be very small. The following experiment tests this conjecture.

Hypothesis 5 *For our test cases, few nodes are processed each time a breadth-first queue is visited. For the same amount of total work, combinational circuit computations have much better "breadth-first" locality than model checking computations.*

Experiment: Measure the number of sub-operations processed each time a breadth-first queue is visited. Then compute the maximum, mean, and standard deviation of the results. Note that these calculations do not include the cases where the queues are empty since they have no impact on the memory locality issue.

Result: Figure 8 shows the statistics for this experiment. The top part of the table shows the results for the model checking traces. The bottom part shows the results for the ISCAS85 circuits. We have also included the *"Average / Total # of Ops"* column to show the results for the average number of sub-operations processed per pass, normalized against the total amount of work performed.

The results show that on average, 10 out of 16 model checking traces processed less than 300 sub-operations (less than one 8-KByte memory page) in each pass. Overall, the average number of sub-operations in a breadth-first queue is at most 4685, which is less than 16 memory pages (128 KBytes). This number is quite small given that hundreds of MBytes of total memory are used. This shows that for these traces, the breadth-first approaches are not very effective in clustering accesses.

Trace	# of Ops Processed per Queue Visit			Average / Total # of Ops $(\times 10^{-6})$
	Average	Max.	Std. Dev.	
abp11	228	41108	86.43	1.86
dartes	27	969	12.53	3.56
dme2-16	34	8122	17.22	0.31
dpd75	15	186	4.75	0.32
ftp3	1562	149792	63.11	8.80
furnace17	75	131071	42.40	2.38
futurebus	2176	207797	76.50	0.23
key10	155	31594	48.23	1.70
mmgt20	66	4741	21.67	1.73
motors-stuck	11	41712	50.14	0.39
over12	282	28582	55.60	3.32
phone-async	1497	175532	87.95	3.53
phone-sync-CW	1176	186937	80.83	0.19
tcas	1566	228907	69.86	1.16
tomasulo	2719	182582	71.20	1.95
valves-gates	25	51039	70.41	0.55
c2670	3816	147488	71.18	204.65
c3540	1971	219090	45.49	34.87
c6288-13	4594	229902	24.92	69.52
c6288-14	4685	237494	42.29	23.59

Fig. 8. Statistics for memory locality in the breadth-first approach.

Another interesting result is that the maximum number of nodes in the queues is quite large and is generally more than 100 standard deviations away from the average. This result suggests that some depth-first and breadth-first hybrid (perhaps as an extension to what is done in the CAL package) may obtain further performance improvements.

The result for *"Average / Total # of Ops"* clearly shows that for the same amount of work, the ISCAS85 computations have much better locality for the breadth-first approaches. Thus, for a comparable level of *"breadth-first"* locality, model checking applications might need to be much larger than the combinational circuit applications.

We have also studied the effects of the breadth-first approach's memory locality when the computations do not fit in the main memory. This experiment was performed by varying the size of the physical memory. The results show that the breadth-first based packages are significantly better only for the three largest cases (largest in terms of memory usage). The results are not very conclusive because as an artifact of this BDD study, the participating BDD packages tend to use a lot more memory than they did before the study began, and furthermore, since these BDD packages generally do not adjust memory usage based on the actual physical memory sizes and page fault rates, the results are heavily influenced by excessive memory usage. Thus, they do not accurately reflect the effects of the memory locality of the breadth-first approach.

5 Phase 2 Results: Dynamic Variable Reordering

Dynamic variable reordering is inherently difficult for many reasons. First, there is a tradeoff between time spent in variable reordering and the total elapsed time. Second, small changes in the triggering and termination criteria may have significant impact in both the space and time requirements. Another difficulty is that because the space of possible variable orders is so huge and variable reordering algorithms tend to be very expensive, many machines are required to perform a comprehensive study. Due to these inherent difficulties and lack of resources, we were only able to obtain very preliminary results and have performed only one round of evaluation.

For this phase, only the CAL, CUDD, EHV, and TiGeR BDD packages were used, since the ABCD and PBF packages have no support for dynamic variable reordering. There are 4 packages and 16 traces, for a total of 64 cases. Figure 9 presents the timing results for these 64 cases. In this figure, the cases that did not complete within the resource limits are marked with n/a. The speedup lines ranging from 0.01x to 100x are included to help classify the performance results.

Figure 9(a) compares running time with and without dynamic variable reordering. With dynamic variable reordering enabled, 19 cases do not finish within the resource limits. Six of these 19 cases also cannot finish without variable reordering (the *failed* cases in Fig. 9(a)). Thirteen of these 19 cases are doable without dynamic variable reordering enabled (the *bad* cases in Fig. 9(a)). There is one case that does not finish without dynamic variable reordering, but finishes with dynamic variable reordering enabled (the *new* in Fig. 9(a)). The remaining 45 cases are marked as the *rest* in Fig. 9(a). These results show that given reasonably good initial orders (e.g., those provided by the original authors of these SMV models), dynamic variable reordering generally slows down the computation. This slowdown may be partially caused by the cache flushing in the dynamic variable reordering phase; i.e., given the importance of the computed cache, cache flushing can increase the number of repeated subproblems.

To evaluate the quality of the orders produced, we used the final orders produced by the dynamic variable reordering algorithms as new initial orders and reran the traces without dynamic variable reordering. Then we compared these results with the results obtained using the original initial order and also without dynamic variable reordering. This comparison is one good way of evaluating the quality of the variable reordering algorithms since in practice, good initial variable orders are often obtained by iteratively feeding back the resulting variable orders from the previous variable reordering runs.

Figure 9(b) plots the results for this experiment. The y-axis represents the cases using the original initial variable orders. The x-axis represents the cases where the final variable orders produced by the dynamic variable reordering algorithms are used as the initial variable orders. In this figure, the cases that finished using the original initial orders but failed using the new initial orders are marked as the *bad* and the remaining cases are marked as the *rest*. The results show that improvements can still be made from the original variable orders. A few cases even achieved a speedup of over 10.

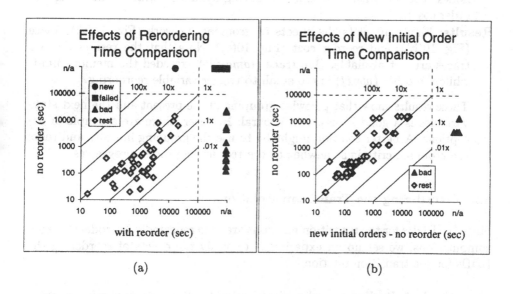

(a)   (b)

Fig. 9. Overall results for variable reordering. The *failed* cases represent those that always exceed the resource limits. The *bad* cases represent those that are originally doable but failed with the new setup. The *rest* represent the remaining cases. (a) Timing comparison between with and without dynamic variable reordering. (b) Timing comparison between original initial variable orders and new initial variable orders. The new initial variable orders are obtained from the final variable orders produced by the dynamic variable reordering algorithms. For results in (b), dynamic variable reordering is disabled.

The remainder of this section presents results of a limited set of experiments for characterizing dynamic variable reordering. We first present the results on two heuristics for dynamic variable reordering. Then we present results on sensitivity of dynamic variable reordering to the initial variable orders. For these experiments, only the CUDD package is used. Note that the results in this section are very limited in scope and are far from being conclusive. Our intent is to suggest new research directions for dynamic variable reordering.

5.1 Present and Next State Variable Grouping

We set up an experiment to study the effects of *variable grouping*, where the grouped variables are always kept adjacent to each other.

Hypothesis 6 *Pairwise grouping of present state variables with their corresponding next state variables is generally beneficial for dynamic variable reordering.*

Experiment: Measure the effects of this grouping on the number of subproblems (work), maximum number of live BDD nodes (space), and number of nodes swapped with their children during dynamic variable reordering (reorder cost).

Results: Figure 10 plots the effects of grouping on work (Fig. 10(a)), space (Fig. 10(b)), and reorder cost (Fig. 10(c)). Note that the results for two traces are not available. One trace (*tomasulo*) exceeded the memory limit, while the other (*abp11*) is too small to trigger variable reordering.

These results show that pairwise grouping of the present and the next state variables is a good heuristic in general. However, there are a couple of exceptions. A better solution might be to use the grouping initially and relax the grouping criteria somewhat as the reordering process progresses.

5.2 Reordering the Transition Relations

Since the BDDs for the transition relations are used repeatedly in model checking computations, we set up an experiment to study the effects of reordering the BDDs for the transition relations.

Hypothesis 7 *Finding a good variable order for the transition relation is an effective heuristic for improving overall performance.*

Experiment: Reorder variables once, immediately after the BDDs for the transition relations are built, and measure the effect on the number of subproblems (work), maximum number of live BDD nodes (space), and number of nodes swapped with their children during dynamic variable reordering (reorder cost).

Effects of Grouping

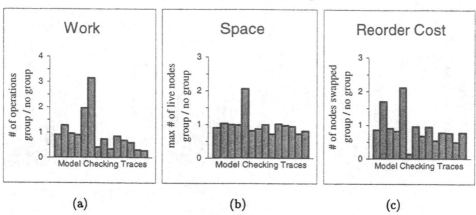

Fig. 10. Effects of pairwise grouping of the current and next state variables on (a) the number of subproblems, (b) the number of maximum live BDD nodes, and (c) the amount of work in performing dynamic variable reordering.

Results: Figure 11 plots the results of this experiment on work (Fig. 11(a)), space (Fig. 11(b)), and reorder cost (Fig. 11(c)). The results are normalized against the results from automatic dynamic variable reordering for comparison purposes. Note that the results for two traces are not available. With automatic dynamic variable reordering, one trace (*tomasulo*) exceeded the memory limit, while the other (*abp11*) is too small to trigger variable reordering.

The results show that reordering once, immediately after the construction of transition relations' BDDs generally works well in reducing the number of subproblems (Fig. 11(a)). This heuristic's effects on the maximum number of live BDD nodes is mixed (Fig. 11(b)). Figure 11(c) shows that this heuristic's reordering cost is generally much lower than automatic dynamic variable reordering. Overall, the the number of variable reordering for automatic dynamic variable reordering is 5.75 times the variable reordering frequency using this heuristic. These results are not strong enough to support our hypothesis as cache flushing may be the main factor for the effects on the number of subproblems. However, it does provide an indication that the automatic dynamic variable reordering algorithm may be invoking the variable reordering process too frequently.

5.3 Effects of Initial Variable Orders

In this section, we study the effects of initial variable orders on BDD construction with and without dynamic variable reordering. We generate a suite of initial variable orders by perturbing a set of good initial orders. In the following, we

Effects of Reordering Transition Relations

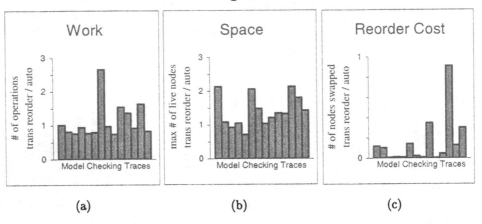

Fig. 11. Effects of variable reordering the transition relations on (a) the number of subproblems, (b) the number of maximum live BDD nodes, and (c) the amount of work in performing variable reordering. For comparison purposes, all results are normalized against the results for automatic dynamic variable reordering.

describe this experimental setup in detail and then present some hypotheses along with supporting evidence.

Experimental Setup

The first step is the selection of good initial variable orders — one for each model checking trace. The quality of an initial variable order is evaluated by the running time using this order without dynamic variable reordering.

Once the best initial variable order is selected, we perturb it based on two perturbation parameters: the probability (p), which is the probability that a variable will be moved, and the distance (d), which controls how far a variable may move. The perturbation algorithm used is shown in Figure 12. Initially, each variable is assigned a weight corresponding to its variable order (line 1). If this variable is chosen (with the probability of p) to be perturbed (by the distance parameter d), then we change its weight by δw, where δw is chosen randomly from the range $[-d, d]$ (lines 3-5). At the end, the perturbed variable order is determined by sorting the variables based on their final weights (line 6). This algorithm has the property that on average, p fraction of the BDD variables are perturbed and each variable's final variable order is at most $2d$ away from its initial order. Another property is that the perturbation pair ($p = 1, d = \infty$) essentially produces a completely random variable order.

Since randomness is involved in the perturbation algorithm, to gain better statistical significance, we generate multiple initial variable orders for each pair of perturbation parameters (p, d). For each trace, if we study n_p different perturbation probabilities, n_d different perturbation distances, and k initial orders for each perturbation pair, we will generate a total of $k n_p n_d$ different initial variable

perturb_order($v[n]$, p, d)
    /* perturb the variable order with probability p and distance d.
      $v[\ ]$ is an array of n variables sorted based on decreasing
      variable order precedence. */
1   for $(0 \le i < n)$ $w[i] \leftarrow i$ /* initialize weight */
2   for $(0 \le i < n)$ /* for each variable, with probability p, perturb its weight. */
3   With probability p do
4            $\delta w \leftarrow$ randomly choose an integer from $[-d, d]$
5            $w[i] \leftarrow w[i] + \delta w$
6   sort variables in array $v[\ ]$ based on increasing weight $w[\ ]$
7   return $v[\ ]$

Fig. 12. Variable-order perturbation algorithm.

orders. For each initial variable order, we compare the results with and without dynamic variable reordering enabled. Thus, for each trace, there will be $2kn_p n_d$ runs. Due to lack of time and machine resources, we were only able to complete this experiment for one very small trace — *abp11*.

The perturbed initial variable orders were generated from the best initial variable ordering we found for *abp11*. Using this order, the *abp11* trace can be executed (with dynamic variable reordering disabled) using 12.69 seconds of CPU time and 127 MBytes of memory on a 248 MHz UltraSparc II. This initial order and its results are used as the base case for this experiment. Using this base case, we set the time limit of each run to 1624.32 seconds (128 times the base case) and 500 MBytes of memory.

For the perturbation parameters, we let p range from 0.1 to 1.0 with an increment of 0.1. Since *abp11* has 122 BDD variables, we let d range from 10 to 100 with an increment of 10 and added the case for $d = \infty$. These choices result in 110 perturbations pairs (with $n_p = 10$ and $n_d = 11$). For each perturbation pair, we generate 10 initial variable orders ($k = 10$). Thus, there are a total of 1100 initial variable orders and 2200 runs.

Results for *abp11*

Hypothesis 8 *Dynamic variable reordering improves the performance of model checking computations.*

Supporting Results: Figure 13 plots the number of cases that did not complete within various time limits for runs with and without dynamic variable reordering. For these runs, the memory limit is fixed at 500 MBytes. The time limits in this plot are normalized to the base case of 12.69 seconds and are plotted in log scale.

The results clearly show that given enough time, the cases with dynamic variable reordering perform better. Overall, with a time limit of 128 times the base case, only 10.1% of cases with dynamic variable reordering exceeded the resource limits. In comparison, 67.6% of cases without dynamic variable reordering failed to complete.

Note that for the time limit of 2 times the base case (the > 2x case in the chart), the results with dynamic variable reordering is worse. This reflects the fact that dynamic variable reordering can be expensive. As the time limit increases, the number of unfinished cases for with dynamic variable reordering drops more quickly until at about 32 times the base case. After this point, the number of unfinished cases for both with and without dynamic variable reordering appear to be decreasing at about the same rate.

Another interesting result is that none of the cases takes less time to complete than the base case of 12.69 seconds (i.e., the > 1x results are both 1100). This result indicates that the initial variable order of our base case is indeed a very good variable order.

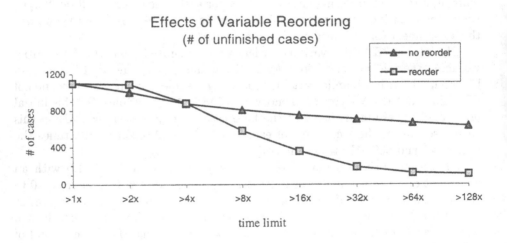

Fig. 13. Effects of variable reordering on *abp11*. This chart plots the number of unfinished cases for various time limits. The time limits are normalized to the base case of 12.69 seconds. The memory limit is set at 500 MBytes.

To better understand the impact of the perturbations on running time, we analyzed the distribution of these results (in Fig. 13) across the perturbation space and formed the following hypothesis.

Hypothesis 9 *The dynamic variable reordering algorithm performs "unnecessary" work when it is already dealing with reasonably good variable orders. Overall, given enough time, dynamic variable reordering is effective in recovering from poor initial variable order.*

Supporting Results: Figure 14(a) shows the results with a time limit of 4 times the base case of 12.69 seconds. These plots show that when there are small perturbations ($p = 0.1$ or $d = 10$), we are better off without dynamic variable reordering. However, for higher levels of perturbations, the cases with dynamic variable reordering usually does a little better.

Figures 14(b) and 14(c) show the results with time limits of 32 and 128 times, respectively, the base case. Note that since 128 times is the maximum time limit we studied, Fig. 14(c) also represents the distribution of the cases that did not complete at all for this study. These results clearly show that given enough time, the cases with dynamic variable reordering perform much better.

Hypothesis 10 *The quality of initial variable order affects the space and time requirements, with or without dynamic variable reordering.*

Supporting Results: Figure 15 classifies the unfinished cases into memory-out (Fig. 15(a)) or timed-out (Fig. 15(b)). For clarity, we repeated the plots for the total number of unfinished cases (memory-out plus timed-out results) in Fig. 15(c). It is important to note that because the BDD packages used in this study still do not adapt *very well* upon exceeding memory limits, memory-out cases should be interpreted as indications of high memory pressure instead of that these cases inherently do not fit within the memory limit.

The results show that levels of perturbation directly influence the time and memory requirement. With a very high level of perturbation, most of the unfinished cases are due to exceeding the memory limit of 500 MBytes (the upper-left triangular regions in Fig. 15(a)). For a moderate level of perturbation, most of the unfinished cases are due to the time limit (the diagonal bands from the lower-left to the upper-right in Fig. 15(b)).

Note that the results in Fig. 15 are not very monotonic; i.e., the results are not necessarily worse with a larger degree of perturbation. This leads to the next hypothesis.

Hypothesis 11 *The effects of the dynamic variable reordering algorithm and the initial variable orders are very chaotic.*

Supporting Results: Fig. 16 plots the standard deviation of running time normalized against average running time. For the cases that cannot complete within the resource limits, they are included as if they use exactly the time limit. Note that as an artifact of this calculation, when all 10 variants of a perturbation pair exceed the resource limits, the standard deviation is 0. In particular, without variable reordering, none of the cases can be completed in the highly perturbed region (upper-left triangular region in Fig 15(c)) and thus these results are all shown as 0 in the chart.

The results show that the standard deviations are generally greater than the average time (i.e., with the normalized result of > 1). This finding partially

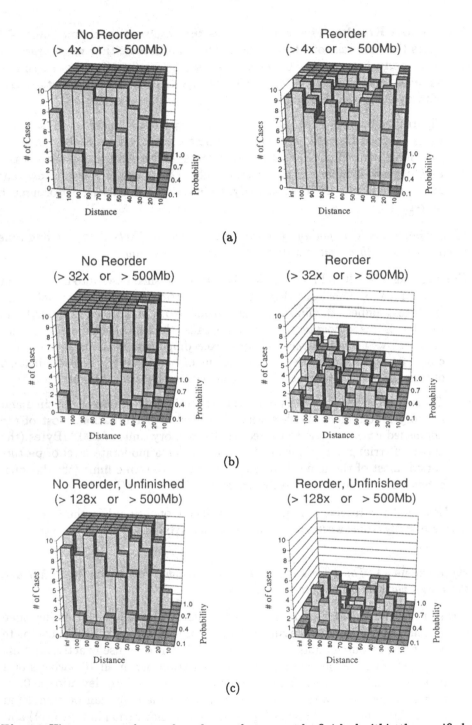

Fig. 14. Histograms on the number of cases that cannot be finished within the specified resource limits. For all cases, the memory limit is set at 500 MBytes. The time limit varies from (a) 4 times, (b) 32 times, to (c) 128 times the base case of 12.69 seconds.

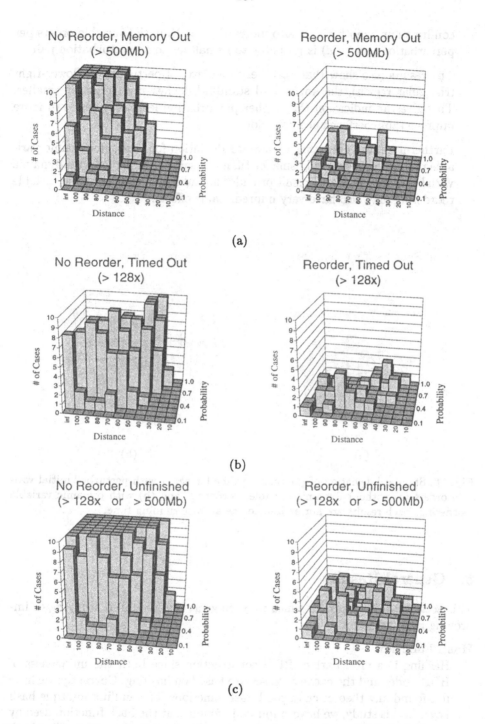

Fig. 15. Breakdown on the cases that cannot be finished. (a) memory-out cases, (b) timed-out cases, (c) total number of unfinished cases.

confirms our hypothesis. It also indicates that 10 initial variable orders per perturbation pair (p, d) is probably too small for some perturbation pairs.

The results also show that with very low level of perturbation (lower-right triangular region), the normalized standard deviation is generally smaller. This gives an indication that higher perturbation level may result in more unpredictable performance behavior.

Furthermore, the normalized standard deviation for without dynamic variable reordering is generally smaller than the same statistic for with dynamic variable reordering. This result provides an indication that dynamic variable reordering may also have very unpredictable effects.

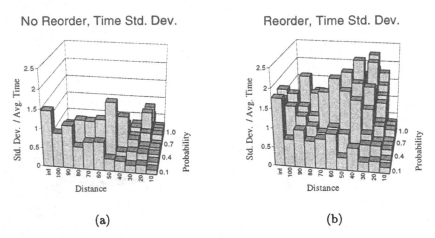

(a)   (b)

Fig. 16. Standard deviation of the running time for *abp11* with perturbed initial variable orders (a) without dynamic variable reordering, and (b) with dynamic variable reordering. Each results are normalized to the average running time.

6 General Results

This section presents results which may be generally helpful in studying or improving BDD packages.

Hash Function

Hashing is a vital part of BDD construction since both the uniqueness of BDD nodes and the cache accesses are based on hashing. Currently, we have not found any theoretically good hash functions for handling multiple hash keys. In this study, we have empirically found that the hash function used by the TiGeR BDD package worked well in distributing the nodes. This hash function is of the form

$$H(k_1, k_2) = ((k_1 p_1 + k_2) p_2)/2^{w-n}$$

where k's are the hash keys, p's are sufficiently large primes, w is the number of bits in an integer, and 2^n is the size of the hash table. Note that division by 2^{w-n} is used to extract the n most significant bits and is implemented by right shifting $(w - n)$ bits.

The basic idea is to distribute and combine the bits in the hash keys to the higher order bits by using integer multiplications, and then to extract the result from the high order bits. The power-of-2 hash table size is used to avoid the more expensive *modulus* operation. Some small speedups have been observed using this hash function. One pitfall is that for backward compatibility reasons, some compilers might generate a function call to compute integer multiplication, which can cause significant performance degradation (up to a factor of 2). In these cases, architecture-specific compiler flags can be used to ensure the integer-multiplier hardware is used instead.

Caching Strategy

Given the importance of cache, a natural question is: *Can we cache more intelligently?* One heuristic, used in CUDD, is that the cache is accessed only if at least one of the arguments has a reference count greater than 1. This technique is based on the fact that if all arguments have reference counts of 1, then this subproblem is not likely to be repeated within the current top-level operation. In fact, if a complete cache is used, this subproblem will not be repeated within the same top-level operation. Using this technique, CUDD is able to reduce the number of cache lookups by up to half, with a total time reduction of up to 40%.

Relational Product Algorithm

The relational product algorithm in Fig. 1 can be further improved. The new optimizations are based on the following derivations. Let r_0 be the result of the 0-cofactors (line 4 in Fig. 1), v be the set of variables to be quantified, and h be any Boolean function, then

$$r_0 \vee (\exists v.r_0 \wedge h) = r_0 \vee (r_0 \wedge \exists v.h) = r_0$$

and

$$r_0 \vee (\exists v.(\neg r_0) \wedge h) = r_0 \vee ((\neg r_0) \wedge \exists v.h) = r_0 \vee \exists v.h$$

The validity comes from the fact that r_0 does not depend on the variables in v. Based on these equations, we can add the following optimizations (between line 7 and line 8 in Fig. 1) to the relational product algorithm:

7.1	else if $(r_0 == f\|_{\tau \leftarrow 1})$ or $(r_0 == g\|_{\tau \leftarrow 1})$
7.2	$r \leftarrow r_0$
7.3	else if $(r_0 == \neg f\|_{\tau \leftarrow 1})$
7.4	$r \leftarrow r_0 \vee (\exists v.g\|_{\tau \leftarrow 1})$
7.5	else if $(r_0 == \neg g\|_{\tau \leftarrow 1})$
7.6	$r \leftarrow r_0 \vee (\exists v.f\|_{\tau \leftarrow 1})$

In general, these optimizations only slightly reduces the number of subproblems, with the exception of the *futurebus* trace, where the number of subproblems is reduced by over 20%.

BDD Package Comparisons

In comparing BDD packages, one fairness question is often raised: *Is it fair to compare the performance of a bare-bones experimental BDD package with a more complete public domain BDD package?* This question arises particularly when one package supports dynamic variable reordering, while the other does not. This is an issue because supporting dynamic variable reordering requires additional data structures and indirection overheads to the computation for BDD construction. To partially answer this question, we studied a package with and without its support for variable reordering in place. Our preliminary results show that the additional overhead to support dynamic variable reordering has no measurable performance impact. This may be due to the fact that BDD computation is so memory intensive, a couple additional non-memory intensive operations can be scheduled either by the hardware or the compiler without any measurable performance penalty.

Cache Hit Rate

The computed cache hit rate is not a reliable measure of overall performance. In fact, it can be shown that when the cache hit rate is less than 49%, a cache miss can actually result in a higher hit rate. This is because a cache miss generates more subproblems and these subproblems' results could have already been computed and are still in cache.

Platform Independent Metrics

Throughout this study, we have found several useful machine-independent metrics for characterizing the BDD computations. These metrics are:

- the *number of subproblems* as a measure for work,
- the *maximum number of live nodes* as a measure for the lower bound on memory requirement,
- the *number of subproblems processed for each breadth-first queue visit* to reflect the possibility of exploiting memory locality using the breadth-first traversal, and
- the *number of nodes swapped with their children during dynamic variable reordering* as a measure of the amount of work performed in dynamic variable reordering.

7 Issues and Open Questions

Cache Size Management

In this study, we have found that the size of the compute cache can have a significant impact on model checking computations. Given that BDD computations are very memory intensive, there is an inherent conflict between using a larger cache for better performance and using a smaller cache to conserve memory usage. For BDD packages that maintain multiple compute caches, there are additional conflicts as these caches will compete with each other for the memory resources. As the problem sizes get larger, finding a good dynamic cache management algorithm will become more and more important for building an efficient BDD package.

Garbage Collection Triggering Algorithm

Another dynamic memory management issue is the frequency of garbage collection. The results in Fig. 6(b) clearly suggest that delaying garbage collection can be very beneficial. Again, this is a space and time tradeoff issue. One possibility is to invoke garbage collection when the percentage of unreachable nodes is high and the rebirth rate is low. Note that for BDD packages that do not maintain reference counts, the rebirth rate statistic is not readily available and thus a different strategy is needed.

Resource Awareness

Given the importance of space and time tradeoff, a commercial strength BDD package not only needs to know when to gobble up the memory to reduce computation time, it should also be able to free up space under resource contention. This contention could come from different parts of the same tool chain or from a completely different job. One way to deal with this issue is for BDD packages to become more aware of the environment, in particular, the available physical memory, various memory limits, and the page fault rate. This information is readily available to the users of modern operating systems. Several of the BDD packages used in this study already have some limited form of resource awareness. However, this problem is still not well understood and probably cannot be easily studied using the trace-driven framework.

Cross Top-Level Sharing

For the model checking traces, why are there so many subproblems repeated across the top-level operations? We have two conjectures. First, there is quite a bit of symmetry in some of these SMV models. These inherent symmetries are *somehow* captured by the BDD representation. If so, it might be more effective to use higher level algorithms to exploit the symmetries in the models. The other conjecture is that the same BDDs for the transition relations are used repeatedly throughout model checking in the fixed-point computations. This repeated use of the same set of BDDs increases the likelihood of the same subproblems being repeated across top-level operations. At this point, we do not know how to validate these conjectures. To better understand this property, one starting point would be to identify how far apart are these cross top-level repeated subproblems; i.e., is it within one state transition, within one fixed-point computation, within one temporal logic operator, or across different temporal logic operators?

Breadth-First's Memory Locality

In this study, we have found no evidence that breadth-first based techniques have any advantage when the computation fits in the main memory. An interesting question would be: *As the BDD graph sizes get much larger, is there going to be a crossover point where the breadth-first packages will be significantly better?* If so, another issue would be finding a good depth-first and breadth-first hybrid to get the best of both worlds.

Inconsistent Cross Platform Results

Inconsistency in timing results across machines is yet another unresolved issue in this study. More specifically, for some BDD packages, the CPU-time results on a UltraSparc II machine are up to twice as long as the corresponding results on a PentiumPro, while for other BDD packages, the differences are not so significant. Similar inconsistencies are also observed in the Sentovich study [23]. A related performance discrepancy is that for the depth-first based packages, the garbage collection cost for UltraSparc II is generally twice as high as that of PentiumPro. However, for the breadth-first based packages, the garbage collection performances between these two machines are much closer. In particular, for one breadth-first based package, the ratio is very close to 1. This discrepancy may be a reflection of the memory locality of these BDD packages. To test this conjecture, we have performed a set of simple tests using synthetic workloads. Unfortunately, the results did not confirm this hypothesis. However, the results of this test do indicate that our PentiumPro machine appears to have a better memory hierarchy than our UltraSparc II machine. A better understanding of this issue can probably shed some light on how to improve memory locality for BDD computations.

Pointer- vs. Index-Based References

Another issue is that within the next ten years, machines with memory sizes greater than 4 GBytes are going to become common. Thus the size of a pointer (i.e., memory address) will increase from 32 to 64 bits. Since most BDD packages today are pointer-based, the memory usage will double on 64-bit machines. One way to reduce this extra memory overhead is to use integer indices instead of pointers to reference BDDs as in the case of the ABCD package. One possible drawback of an index-based technique is that an extra level of indirection is introduced for each reference. However, since ABCD's results are generally among the best in this study, this provides a positive indication that the index-based approach may be a feasible solution to this impending memory overhead problem.

Computed Cache Flushing in Dynamic Variable Reordering

In Sec. 5, we showed that dynamic variable reordering can generally slow down the entire computation when given a reasonably good initial variable order. Since the computed cache is typically flushed when dynamic variable reordering takes place, it would be interesting to study what percentage of the slowdown is caused by an increase in the amount of work (number of subproblems) due to cache flushing. If this percentage is high, then another interesting issue would be in finding a good way to incorporate the cache performance as a parameter for controlling dynamic variable reordering frequency.

8 Related Work

In [23], Sentovich presented a BDD study comparing the performance of several BDD packages. Her study covered building output BDDs for combinational circuits, computing reachability of sequential circuits, and variable reordering.

In [15], Manne *et al.* performed a BDD study examining the memory locality issues for several BDD packages. This work compares the hardware cache miss rates, TLB miss rates, and page fault rates in building the output BDDs for combinational circuits.

In contrast to the Sentovich study, our study focuses in characterizing the BDD computations instead of doing a performance comparison of BDD packages. In contrast to the Manne study, our work uses platform independent metrics for performance evaluation instead of hardware specific metrics. Both types of metrics are equally valid and complementary. Our study also differs from these two prior studies in that our performance evaluation is based on the execution of a model checker instead of benchmark circuits.

9 Summary and Conclusions

By applying a new evaluation methodology, we have not only achieved significant performance improvements, we have also identified many interesting characteristics of model checking computations. For example, we have confirmed that model checking and combinational circuit computations have fundamentally different performance characteristics. These differences include the effects of the cache size, the garbage collection frequency, the complement edge representation, and the memory locality for the breadth-first BDD packages. For dynamic variable reordering, we have introduced some new methodologies for studying the effects of variable reordering algorithms and initial variable orders. From these experiments, we have uncovered a number of open problems and future research directions.

As this study is very limited in scope, especially for the dynamic variable reordering phase, further validations of the hypotheses are necessary. It would be especially interesting to repeat the same experiments on execution traces from other BDD-based tools.

The results obtained in this study clearly demonstrate the usefulness of systematic performance characterization and validate our evaluation methodology. We hope that the trace-drive framework and the machine-independent metrics will help lay the foundation for future benchmark collection and performance-characterization methodology.

Acknowledgement

We thank Yirng-An Chen for providing the ISCAS85 circuits in the trace format and providing the software which forms the core of our evaluation tool.

We thank Claudson F. Bornstein and Henry R. Rowley for numerous discussions on experimental setups and data presentation. We are grateful to Edmund M. Clarke for providing additional computing resources. This research is sponsored in part by the Defense Advanced Research Projects Agency (DARPA) and Rome Laboratory, Air Force Materiel Command, USAF, under agreement number F30602-96-1-0287, in part by the National Science Foundation under Grant CMS-9318163, in part by DARPA under contract number DABT63-96-C-0071, in part by Cadence Design Systems, and in part by SRC contract 96-DJ-560. This work utilized Silicon Graphics Origin 2000 machines from the National Center for Supercomputing Applications at Urbana-Champaign.

References

[1] AKERS, S. B. Functional testing with binary decision diagrams. In *Proceedings of Eighth Annual International Conference on Fault-Tolerant Computing* (June 1978), pp. 75–82.

[2] ASHAR, R., AND CHEONG, M. Efficient breadth-first manipulation of binary decision diagrams. In *Proceedings of the International Conference on Computer-Aided Design* (November 1994), pp. 622–627.

[3] BIERE, A. ABCD: an experimental BDD library, 1998. http://iseran.ira.uka.de/~armin/abcd/.

[4] BRACE, K., RUDELL, R., AND BRYANT, R. E. Efficient implementation of a BDD package. In *Proceedings of the 27th ACM/IEEE Design Automation Conference* (June 1990), pp. 40–45.

[5] BRGLEZ, F., BRYAN, D., AND KOZMISKI, K. Combinational profiles of sequential benchmark circuits. In *1989 International Symposium on Circuits And Systems* (May 1989), pp. 1924–1934.

[6] BRGLEZ, F., AND FUJIWARA, H. A neutral netlist of 10 combinational benchmark circuits and a target translator in Fortran. In *1985 International Symposium on Circuits And Systems* (June 1985). Partially described in F. Brglez, P. Pownall, R. Hum. Accelerated ATPG and Fault Grading via Testability Analysis. In *1985 International Symposium on circuits and Systems*, pages 695-698, June 1985.

[7] BRYANT, R. E. Graph-based algorithms for Boolean function manipulation. *IEEE Transactions on Computers C-35*, 8 (August 1986), 677–691.

[8] BRYANT, R. E. Symbolic Boolean manipulation with ordered binary decision diagrams. *ACM Computing Surveys 24*, 3 (September 1992), 293–318.

[9] BRYANT, R. E. Binary decision diagrams and beyond: Enabling technologies for formal verification. In *Proceedings of the International Conference on Computer-Aided Design* (November 1995), pp. 236–243.

[10] BURCH, J. R., CLARKE, E. M., LONG, D. E., MCMILLAN, K. L., AND DILL, D. L. Symbolic model checking for sequential circuit verification. *IEEE Transactions on Computer-Aided Design of Integrated Circuits and Systems 13*, 4 (April 1994), 401–424.

[11] COUDERT, O., BERTHET, C., AND MADRE, J. C. Verification of sequential machines using Boolean functional vectors. In *Proceedings of the IFIP International Workshop on Applied Formal Methods for Correct VLSI Design* (November 1989), pp. 179–196.

[12] COUDERT, O., AND MADRE, J. C. A unified framework for the formal verification of circuits. In *Proceedings of the International Conference on Computer-Aided Design* (Feb 1990), pp. 126–129.

[13] COUDERT, O., MADRE, J. C., AND TOUATI, H. *TiGeR Version 1.0 User Guide*. Digital Paris Research Lab, December 1993.

[14] JANSSEN, G. *The Eindhoven BDD Package*. University of Eindhoven. Anonymous FTP address: ftp://ftp.ics.ele.tue.nl/pub/users/geert/bdd.tar.gz.

[15] MANNE, S., GRUNWALD, D., AND SOMENZI, F. Remembrance of things past: Locality and memory in BDDs. In *Proceedings of the 34th ACM/IEEE Design Automation Conference* (June 1997), pp. 196–201.

[16] MCMILLAN, K. L. *Symbolic Model Checking*. Kluwer Academic Publishers, 1993.

[17] MINATO, S., ISHIURA, N., AND JAJIMA, S. Shared binary decision diagram with attributed edges for efficient Boolean function manipulation. In *Proceedings of the 27th ACM/IEEE Design Automation Conference* (June 1990), pp. 52–57.

[18] OCHI, H., ISHIURA, N., AND YAJIMA, S. Breadth-first manipulation of SBDD of Boolean functions for vector processing. In *Proceedings of the 28th ACM/IEEE Design Automation Conference* (June 1991), pp. 413–416.

[19] OCHI, H., YASUOKA, K., AND YAJIMA, S. Breadth-first manipulation of very large binary-decision diagrams. In *Proceedings of the International Conference on Computer-Aided Design* (November 1993), pp. 48–55.

[20] RANJAN, R. K., AND SANGHAVI, J. CAL-2.0: Breadth-first manipulation based BDD library. Public software. University of California, Berkeley, CA, June 1997. http://www-cad.eecs.berkeley.edu/Research/cal_bdd/.

[21] RANJAN, R. K., SANGHAVI, J. V., BRAYTON, R. K., AND SANGIOVANNI-VINCENTELLI, A. High performance BDD package based on exploiting memory hierarchy. In *Proceedings of the 33rd ACM/IEEE Design Automation Conference* (June 1996), pp. 635–640.

[22] RUDELL, R. Dynamic variable ordering for ordered binary decision diagrams. In *Proceedings of the International Conference on Computer-Aided Design* (November 1993), pp. 139–144.

[23] SENTOVICH, E. M. A brief study of BDD package performance. In *Proceedings of the Formal Methods on Computer-Aided Design* (November 1996), pp. 389–403.

[24] SENTOVICH, E. M., SINGH, K. J., LAVAGNO, L., MOON, C., MURGAI, R., SALDANHA, A., SAVOJ, H., STEPHAN, P. R., BRAYTON, R. K., AND SANGIOVANNI-VINCENTELLI., A. L. SIS: A system for sequential circuit synthesis. Tech. Rep. UCB/ERL M92/41, Electronics Research Lab, University of California, May 1992.

[25] SOMENZI, F. CUDD: CU decision diagram package. Public software. University of Colorado, Boulder, CO, April 1997. http://vlsi.colorado.edu/~fabio/.

[26] YANG, B., CHEN, Y.-A., BRYANT, R. E., AND O'HALLARON, D. R. Space- and time-efficient BDD construction via working set control. In *1998 Proceedings of Asia and South Pacific Design Automation Conference* (Feb 1998), pp. 423–432.

Symbolic Model Checking Visualization

Gila Kamhi, Limor Fix, Ziv Binyamini
Design Technology
Intel, Haifa, Israel
gkamhi@iil.intel.com

Abstract

The industrial deployment of formal verification technology that has not yet reached its maturity is extremely difficult. The more automated and therefore most widely used formal verification technology, symbolic model checking, has a severe problem of *limited capacity*. The capacity limitation reflects itself in long, and most importantly, unpredictable run time duration which results in low productivity.

In this paper, we demonstrate how we integrated techniques from the fields of *algorithm animation*, *performance monitoring*, and *knowledge engineering* in order to boost two major problem areas of model checking: the capacity limitation and low productivity. We have developed a prototype, called *Palette*, in order to demonstrate these concepts.

Palette uses visualization techniques to give insight to the execution of the symbolic model checking algorithms. Furthermore, it tracks the progress of the verification run and enables mid-course analysis of the status. Palette makes a step forward in estimation of the run time duration by predicting the amount of work done in some selected execution tasks, and informing the ones that are to be executed in the future. An additional important goal of Palette is to assist in building and automating the model checking usage methodology, and consequently to reduce the need for user expertise and intervention. In this aspect, Palette is a light-weight expert system for model checking. It can determine which algorithms are not efficient and reason on their failure. It can also advice on how to make a verification task complete successfully.

1 Introduction

Post-silicon functional bugs in hardware design are very expensive both in time and money; therefore, there is a strong need to raise the quality of the hardware design before manufacture. This can be achieved by exhaustive simulation which is very costly in terms of CPU time and can very rarely be achieved in practice. The inability of simulation to guarantee the absence of design errors has prompted the use of formal verification technology in industrial practice.

Hardware design validation via formal verification technology gives very high confidence in the correctness of the specifications being verified. However, the current formal verification tools are still at their infancy. The theory behind these tools has been investigated over the last couple of decades, but only in the last five years practical tools have emerged. The main barriers to the intensive utilization of formal verification tools at their current maturity level are: capacity limitation and low productivity [2].

Two approaches can be distinguished in the formal verification technology; model checking and theorem proving. The more automated formal verification technology, i.e., model checking, has a severe problem of *limited capacity*. The size of the model and the complexity of the properties that can be verified by model checking is still too limited. Researchers in academia and industry are developing new algorithms to improve the efficiency of model checking. However, most of the innovations and breakthroughs in hardware verification are unfortunately not applicable to all design domains; for example, a verification technique which provides good results for data path intensive circuits may perform very poorly for control intensive designs and vice versa. Therefore, a variety of algorithms that address different design domains have to be integrated into the model checking engine.

The capacity limitation of model checking technology results in an additional obstacle - *low productivity*. To effectively use symbolic model checking based tools requires a lot of user interaction and guidance. Moreover, the integration of many different algorithms that do not apply to all design domains introduces several distinct flows of operation to the tools and, furthermore, complicates their usage. In summary, currently even the more automated formal verification tools are far from being push-button. A well-defined usage methodology is needed to lead the user on how and when to use the distinct flows and options of the formal verification system. Furthermore, we witness that a detailed user guide that describes the usage methodology is not sufficient to convey this information to the user.

In this paper, we introduce a prototype, Palette (Profile, AnaLyzE and Track Tool) that has been developed to address two major problem areas of symbolic model checking: the capacity and the productivity limitation. Palette makes a successful attempt to ease the development and usage of model checking. It exposes and analyzes the behavior of the model checking engine by visualizing the progress of its internal algorithms. An additional important goal of Palette is to assist in building and automating the formal verification usage methodology and reduce the user expertise needed to successfully use the model checking tools. To achieve this goal, Palette acts as a lightweight expert system for model checking. It can determine which algorithms are not efficient and reason on their failure. It can also advice on how to make a verification task complete successfully.

Palette is being used in Intel by multiple user groups (i.e., formal verification development groups, application engineers, formal verification end-users) to verify the next-generation micro-processors. Model checking expert system role of Palette enables efficient knowledge and expertise transfer between different user groups. For example, understanding the formal verification technology enables hardware engineers to actively influence the formal verification development. Furthermore, Palette is helps in introducing the novice formal verification end-users to the new technology.

We believe the techniques used in Palette are not only specific to symbolic model checking but may be used in different areas of design technology (e.g., logic synthesis) or applications that require a lot of user intervention and expertise.

This paper is organized as follows: Section 2 contains a brief overview of Palette. Section 3 summarizes the features of Palette. This section describes the role of Palette to profile and analyze new algorithms, and demonstrates how Palette can be used to track the progress of the run of symbolic model checking. Additionally, it outlines the use of Palette to automate the model checking expertise. Finally, in Section 4 we present future directions.

2 Palette (Profile, AnaLyze, and Track Tool)

New techniques and algorithms are constantly being developed to increase the capacity of symbolic model checking. Their integration into a single tool introduces several distinct flows of operation and thus complicates the usage. Furthermore, the capacity limitation of symbolic model checking reflects itself in unpredictable run times. A tool with never-ending run times and run times in units of hours is in high risk of low productivity. Therefore, we face the challenge of building a methodology around the different flows to ensure their best usage by the verification end-user.

Palette attempts to enhance the productivity of symbolic model checking engines[1] by providing a framework to: 1) Profile the model checking performance; 2) AnaLyze the internal algorithms; 3) Track the progress of the verification run; 4) Build and automate the model checking usage methodology.

Palette addresses multiple audiences: the model checker developer, the application engineer of the formal verification tool, and the formal verification end-user. All of the features of Palette may interest all the user groups. However, some features interest most specific audiences.

The first two goals of Palette (i.e., profiling and analysis), mainly meet the needs of the symbolic model checker developer. Profiling and analysis enable the ranking and rating of new algorithmic enhancements; therefore, they are prerequisites of any significant work that needs to be done to improve the core technology. The profiling and analysis are done by gathering and computing selected metrics during (or after) the run. The internal algorithms of the model checking module are decomposed into a hierarchical structure of tasks and sub-tasks, and performance quantifiers are computed separately for each of these tasks. For example, some potential symbolic model checking tasks of interest may be *reachable states build*, *counter-example generation,* and *fair states build.* Similarly, an example to a performance quantifier is the *number of BDD* [4] *nodes allocated* during the execution of a task.

The third goal of Palette (i.e., tracking), serves the needs of both the end-user, and the developer. The ability to track the progress of execution will help the user to figure out how to control the run. Similarly, understanding the progress of the run will help the developer improve the internal algorithms. Palette tracks the progress by displaying which tasks have already successfully terminated and which tasks are yet to be executed. Moreover, for selected tasks[2] which are currently being

---

1. Palette is integrated to Intel's symbolic model checking engine [1].

2. Currently, Palette is not able to predict the percentage of work done on all major tasks of model checking. The estimation is done only on the *transition relation build* stage.

executed, Palette provides an educated estimate on their progress, namely, the percentage of the work already done in these tasks. For example, 60% of the *transition relation build* task has already been completed.

The more ambitious goals of Palette, i.e., building and automating usage methodology, reduces the need for end-user expertise on formal verification. Palette provides these abilities through a wizard service. During (or after) the run, the user may activate the wizard that will provide advice for improving the performance. This advice is either of the form of suggesting specific values to the control switches of the symbolic model checking engine or of the form "kill the run and start again with the following variable order".

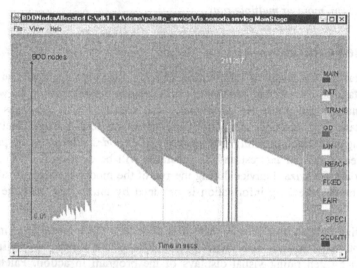

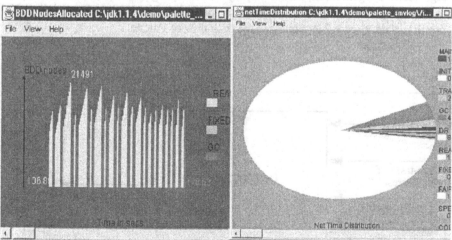

Figure 1: Three Palette views profiling and analyzing the same model checking run. The different colors correspond to the different model checking tasks of interest. The top view displays the number of BDD nodes allocated by sub-tasks of model checking run. The left-most view at the bottom displays the BDD nodes allocated when the task of interest is reachable states computation instead of whole model checking run. The right-most view displays the run-time distribution among sub-tasks of the model checking run.

In order to be an efficient profiling, analysis, and tracking aid, and furthermore, a potential expert system, Palette is designed according to the following guidelines:

- The run-time information is kept in a *modifiable, extendable* and *object-oriented data model.* New stages of interest in symbolic model checking algorithms and new profiling metrics can be easily defined and added.

- Palette has a user-friendly *graphical interface.* The profiling information that it generates can be displayed graphically by means of display widgets (e.g., graphs, bar-charts, pie-graphs).

- Palette can provide *interactively* real-time profiling information and can additionally be used to *post-process* the run.

- Palette can *monitor multiple runs.*

3 Palette Features

Palette enables two levels of information filtering. The user specifies the model checking task of interest (e.g., *reachable states computation*) that he or she wants to profile, analyze and track. He additionally can specify which sub-tasks of the selected task (e.g., *garbage collection*) are of no interest. Lastly, Palette user specifies against which profiling metrics (e.g., *BDD nodes allocated*) the progress of the selected task of interest and its sub-tasks will be displayed. The user can optionally ask for wizard advice during the run of the model checking or after the run. The model checking information is obtained by interacting with the model checker.

The profiling data on long runs is huge. It is difficult if not impossible to interpret huge amounts of textual data. Therefore, Palette offers visual insight into the algorithms by generating visual displays of the program in action. Palette uses mainly graphical widgets as bar-charts, pie-graphs, polygon charts to display profiling data. Figure 2 summarizes the data flow of Palette.

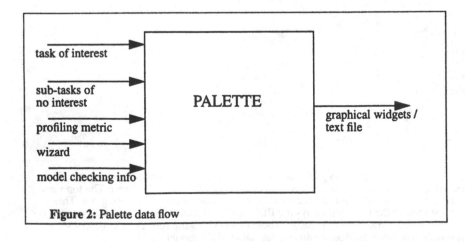

Figure 2: Palette data flow

Section 3.1 describes the role of Palette to profile and analyze new algorithms. Section 3.2 demonstrates how Palette can be used to track the progress of the run of symbolic model checking. Finally, Section 3.3 outlines the use of Palette to automate the model checking expertise.

3.1 Profiling and Analysis of Algorithms

It is essential to be able to accurately measure the performance of internal symbolic model-checking algorithms, in order to improve their quality, and predictability.

Palette predefines which tasks of a symbolic model checking system (e.g., *reachable states computation*, *transition relation build*, and *counter-example generation*) are worthy to be profiled, analyzed, and tracked. In addition, it enables the quantification of the tasks of interest using selected metrics (e.g., *number of BDD nodes allocated, number of memory bytes allocated, the size of transition relation in units of BDD nodes*). The architecture of Palette enables the definition of new tasks of interest and changing the granularity of the predefined ones. Similarly, new profiling metrics can easily be added [1]to the system.

Palette monitors the performance by gathering and displaying data on selected metrics during the run. The model checker developer can use the feedback from the profiling metrics to improve the internal algorithms. Feed-forward of data enables mid-course analysis and making decisions to try different algorithms or adjusting the parameters of the current ones.

Palette facilitates the profiling, analysis and tracking of the model checking process by giving the end-user three dimensions of freedom: 1) task of interest 2) sub-tasks of the selected task 3) profiling metric. It visualizes the progress of the task of interest and its selected sub-tasks against the selected profiling metric (See Figure 2).

Palette offers visual insight into the algorithms by generating visual displays of the program in action. It uses multiple views to illustrate more than one aspect of major tasks. Each view is easy to comprehend in isolation, and the composition of several views is more informative than the sum of their individual contributions [2].

The run time data can be observed in batch or interactive mode; i.e., Palette enables both real-time and post-processing of the data. In addition, it enables profiling multiple runs in one session. Monitoring of multiple runs is useful for ranking and rating of new algorithms and techniques.

Figure 3. illustrates an abstract view of a display of Palette of hierarchical structure of tasks and subtasks of execution. In this example, *task1* is the task of interest. *task2*, *task3*, *task4*, *task6*, and *task7* are sub-tasks of *task1*, and *task5* is a sub-task

---

1. Currently, only the Palette developer is responsible for the addition of new metrics and tasks.

2. These visualization techniques have been used and recommended in other algorithm animation systems [12].

of *task4*. The tasks are measured against the selected profiling metric along the time axis. To concretize the example and map it to symbolic model checking tasks, suppose the profiling metric is the *total number of BDD nodes allocated*, and *task1* is the *main stage of execution*, *task2*, *task5*, and *task6* are instances of *garbage collection* task and *task3*, *task4*, and *task7* correspond respectively to *transition relation build*, *reachable states build*, and *fair states build* tasks.

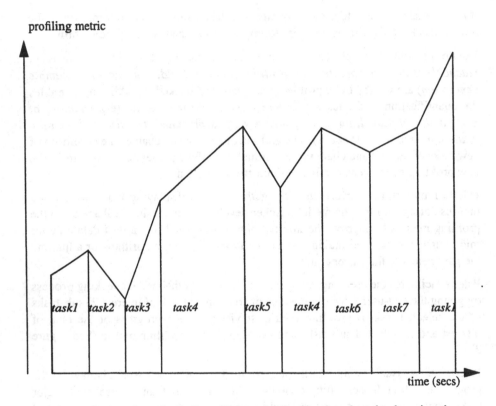

Figure 3: An abstract view of a display of Palette that profiles the selected task against the selected profiling metric

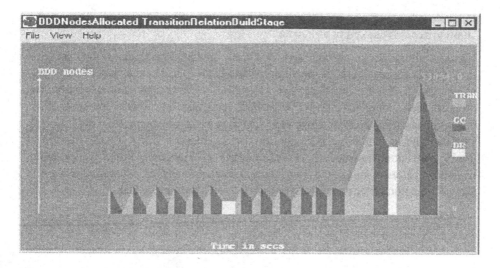

Figure 4: A display of Palette that profiles the *Transition Relation Build* (TRANS) task against the profiling metric BDD nodes allocated along the time axis. *Garbage Collection* (GC) and *Dynamic Reordering* (DR) are sub-tasks of *Transition Relation Build*.

3.2 Tracking the Progress

Symbolic model checking is a technique that, given a state-transition graph and a temporal logic formula, determines which states satisfy the formula. In symbolic model checking systems, BDDs [4] and extensions to BDDs, are the building blocks used to represent the transition relations and sets of states. The verification process is performed iteratively on these BDDs. The major parameter that effects the size of BDDs is the number of state variables (i.e. inputs and sequential devices) in the state-transition graph. Although the number of variables gives an indication on the size and complexity of the BDDs and indirectly the complexity of the symbolic model verification task, the verification time and the number of variables in the pruned [5] model are not directly proportional. The run time of the model checking module may take a few seconds to verify a pruned model that has three hundred variables and, on the other hand, several hours for a pruned model that has only eighty variables. Palette attempts to bring to light the mystery of symbolic model checking by visualizing its progress. The main initiative is to be able to predict the run-time duration. It displays to the user which tasks of execution are successfully completed, which have not yet started, and which are in progress. Moreover, for tasks which are currently being executed, Palette provides an educated estimation on their progress, namely, the percentage of the work done already in these tasks.

A symbolic model checking expert can decide based on the progress information whether to terminate the run and try rerunning with different parameters or not.

Palette automates these expert decisions by providing a wizard mode. Section 3.3 summarizes this capability of Palette.

Figure 5. illustrates an abstract view of Palette tracking the status of the execution of *Task1*. In this example, the profiling metric is the *status* of the task. *Task1* is still active. *Task2*, *Task3*, and *Task4* are sub-tasks of *Task1* and their execution has already been completed. *Task5* is still in progress and ninety percent of its execution is estimated to be completed. *Task6* is yet to be executed. In order to map this example to symbolic model checking tasks, suppose *Task1*, *Task2*, *Task5*, and *Task6* correspond respectively to *main stage of execution*, *initial states build*, *transition relation build*, and *specification evaluation* tasks; whereas, *Task3* and *Task4* correspond to instances of *dynamic variable reordering* task.

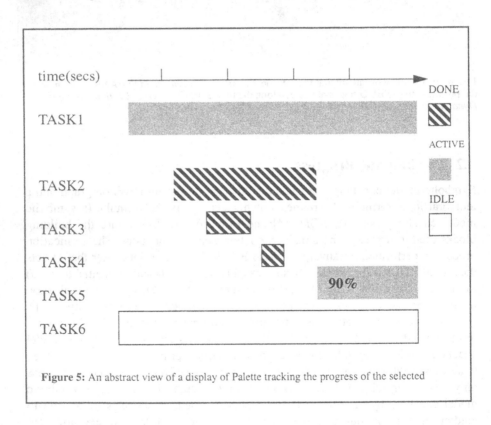

Figure 5: An abstract view of a display of Palette tracking the progress of the selected

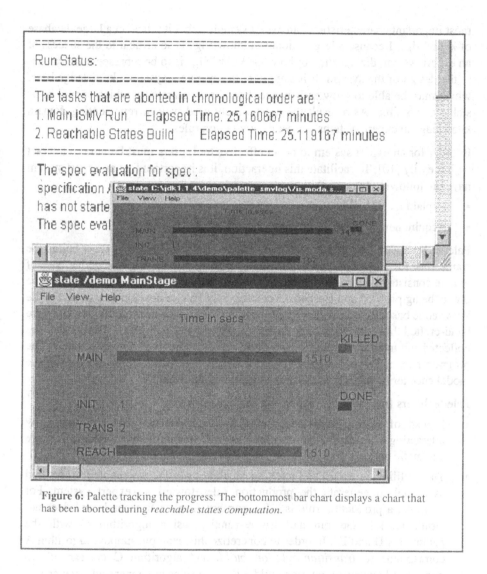

Figure 6: Palette tracking the progress. The bottommost bar chart displays a chart that has been aborted during *reachable states computation*.

3.3 Automation of the Expertise

Formal verification requires a great deal of expert knowledge. It can be performed by experts who have accumulated the required knowledge. Tools that can automate the formal verification expertise will be very useful since there is usually a shortage of qualified human experts.

Tools that can automate human expertise are called expert systems and the construction of them is referred to as knowledge engineering [6,7,8,9,10,11]. The most important characteristic of an expert system is that it relies on a large database of knowledge. Because a large amount of knowledge is so critical to the success of an expert system, the question of how that knowledge is to be represented is critical to the design of the system. It is important to separate a system's knowledge base, which must be able to grow and change, from its program part, which should be as stable as possible. As a result the most widely used way of representing domain knowledge in expert systems is a set of production rules.

In order for an expert system to be an effective tool, users must be able to interact with it easily [10]. To facilitate this interaction, it is important that an expert system have the following two capabilities:

- Explain its reasoning.
- Acquire new knowledge and modification of old knowledge.

Palette, by attempting to automate symbolic model checking expertise, builds the framework for a symbolic model checking expert system. The knowledge base of Palette consists of a set of production rules related with the model checking engine that is being profiled and tracked. Currently, only Palette developers can access the knowledge base. Based on the model checking experience new production rules are hand-crafted by end-users and formal verification developers. These rules are collected and integrated to the knowledge base of Palette by Palette developers. The program part of Palette's expert system activates these rules during the run of the model checker by request of the user to get wizard advice.

Palette differs from standard expert systems in two ways:

- Instead of getting knowledge from the user, Palette gets information by interacting with the model checking engine. In this manner, Palette is more automatic and requires less user intervention than standard expert systems.
- The profiling, analysis, and tracking capabilities of Palette enable it to be used as a tool to generate the production rules toward an expert system. For example, a production rule is "IF algorithm A takes B seconds and does not terminate, kill the run and try re-running using algorithm C with the parameters D and E". In order to concretize this example suppose algorithm A corresponds to *transition relation build*, and algorithm C corresponds to *partitioned transition relation* build. Once the rules are generated, they can be stored in the knowledge base of Palette.

Palette is an expert system in the sense that it can advice on how to make the verification task complete successfully with the current formal verification system with no algorithmic changes. Furthermore, it can determine which algorithms are not efficient (i.e., take too long) and reason on the failure.

Palette builds the framework for an efficient symbolic model checking expert system. Its knowledge database is object oriented, easily extendable and modifiable. Palette gets all of its information on the status of the model checking by interacting with the model checking engine and decides which production rules to apply based on this information; therefore, it is automatic and does not require user intervention. Additionally, it can reason on its advice by providing visual information on the run.

Model checking expert system role of Palette enables efficient knowledge and expertise transfer between different user groups. For example, understanding the formal verification technology enables hardware engineers to actively influence the formal verification development. Furthermore, Palette helps in introducing the novice formal verification end-users to the new technology.

4 Conclusions and Future Work

In this paper, we have introduced a prototype, Palette, that has been developed to ease the development and usage of symbolic model checking based verification in industrial practice.

Palette uses visualization techniques that have already been proven to be useful in other algorithm animation systems [12,13,14,15,16] to give insight to the symbolic model checking algorithms. A major contribution of Palette is in showing the usefulness of visualization techniques for improving the quality, productivity, and predictability of symbolic model checking systems. Palette, additionally, identifies which tasks of symbolic model checking algorithms are worthy to be profiled, analyzed, and tracked and quantifies them using selected metrics.

Run time duration prediction is a major bottleneck that endangers the productivity of model checking in practice. Palette makes a step forward in estimation of the run time duration by predicting the amount of work done in some selected execution tasks, and informing the ones that are to be executed in the future. Further work could solidify the concept of predicting the run time duration. We believe that integrating machine learning techniques [17] to Palette will improve the accuracy of the tool to estimate the run time duration.

The novelty of Palette in comparison to other algorithm animation and performance monitoring systems is in its ability to help create and automate symbolic model checking methodology and expertise. In this respect, Palette unifies three apparently separate fields of computer science: 1) algorithm animation 2) performance monitoring 3) knowledge engineering (i.e., expert systems).

Palette facilitates the utilization of the maximum capacity of the formal verification tool by providing expert advice on which action to take during a verification run. By storing all the formal verification expert rules in its knowledge base and providing them on demand, Palette transfers the expert knowledge among different user groups, and speeds up the ramp-up of novice end-users to the formal verification technology. Palette's knowledge base is continuously being enhanced as new rules on when and where to apply different options of symbolic model checking are hand-crafted.

Palette is being used in Intel by multiple user groups (i.e., formal verification development groups, application engineers, formal verification end-users) to verify the next-generation micro-processors. Increasing use of Palette by different user groups is encouraging and is an indication on Palette's accomplishment in achieving its ultimate goal, namely, to address two major problem areas of model checking: the capacity limitation and the low productivity.

5 Acknowledgments

Barukh Ziv, Roni Rosner, Ze'ev Shtadler, Aviad Cohen, Andreas Tiemeyer, Sela Mador-Haim, Micha Moffie and Osnat Weissberg have helped refine the ideas presented in this paper. We thank Avi Puder and Eyal Ackerman for their current effort in the support and future development of Palette.

6 References

[1] G.Kamhi, O.Weissberg, L.Fix, Z.Binyamini, Z.Shtadler. Automatic Datapath Extraction for Efficient Usage of HDDs. In Proceedings of the 9th International Conference on Computer-Aided Verification, CAV'97, Haifa, Israel, June 1997, LNCS 1254.

[2] K.L.McMillan. Fitting Formal Methods into the Design Cycle, In Proceedings of the 31st Design Automation Conference, June 1994.

[3] K.L.McMillan. Symbolic Model Checking, Kluwer Academic Publishers.

[4] R.E.Bryant. Graph-based Algorithms for Boolean Function Manipulation. IEEE Transactions on Computers, c-35(8), pp. 677-691, Aug. 1986

[5] I.Beer, S.Ben-David, C.Eisner, and Avner Landver. Rulebase: an industry-oriented formal verification tool. In Proceedings of the 33rd Design Automation Conference. IEEE Computer Society Press, June 1996.

[6] E.A. Feigenbaum. The Art of Artificial Intelligence: Themes and Case Studies in Knowledge Engineering, In Proceedings of IJCAI 5, 1977.

[7] Stefik, Aikins, Balzer, Benoit, Birnbaum, Hayes Roth, Sacerdoti. "The Organization of Expert Systems", Artificial Intelligence, Vol. 18, pp. 135-173, Mar. 1982

[8] Sridharan, N.S., "Artificial Intelligence 11, Special Issue on Applications to the Sciences and Medicine", 1978

[9] Shortliffe, Buchanan, Feigenbaum, "Knowledge Engineering for Medical Decision Making: A Review of Computer-based Clinical Decision Aids", Proceedings of IEEE, Vol. 67, pp. 1207-1224, 1979

[10] C.A. Kulikowski, "Artificial Intelligence Methods and Systems for Medical Consultation", IEEE Transactions on Pattern Analysis and Machine Intelligence, Sept. 1980

[11] E.Rich, Artificial Intelligence. MC Graw Hill Series.

[12] M.H.Brown, J.Hershberger, "Color and Sound in Algorithm Animation", IEEE Computer, December 1992.

[13] M.H.Brown, R.Sedgewik, "Techniques for Algorithm Animation", IEEE Software, Vol. 2, No. 1, Jan 1985, pp. 28-39

[14] M.H.Brown, J. Hershberger, "Zeus: A System for Algorithm Animation", In Proceedings of IEEE 1991 Workshop on Visual Languages

[15] J.T. "Tango: A Framework and System for Algorithm Animation", IEEE Computer, Sept. 1990

[16] M.H.Brown, "Exploring Algorithms Using Balsa-II, IEEE Computer, May 1988

[17] R.S.Michalski, R.L. Chilausky, "Learning by Being Told and Learning from Examples: an experimental comparison of the two methods of knowledge acquisition in the context of developing an expert system for soybean disease diagnosis", In "Policy Analysis and Information Systems", Special Issue on Knowledge Acquisition and Induction, No.2, 1980

Input Elimination and Abstraction
in Model Checking

Sela Mador-Haim and Limor Fix

Future CAD technologies
Intel
Haifa, Israel
{selam,lfix}@iil.intel.com

Abstract. Symbolic model checking, while gaining success in the industry as a valuable tool for finding hardware design bugs, is still severely limited with respect to the size of the verifiable designs. This limitation is due to the non-linear memory consumption of the data structure (namely, BDD and its variants) used to represent the model and the explored states. *Input elimination* is a known method that reduces the size of the model by existential quantification of the inputs. In this paper, we improve this technique in several dimensions: we present a novel re-encoding of the model that results in a much large set of quantifiable inputs, we introduce a new greedy algorithm for early quantification of the inputs during the transition relation build, and we suggest a new algorithm to reconstruct the input values in an error trace. *Model abstraction* is a semi-automatic method that requires the user to provide an *abstraction mapping*, and can dramatically reduce the size of models with large data-path. We show that data abstraction can be *reduced to* input elimination using few simple manipulations of the hardware netlist description. Model abstraction is a well-known technique and our contribution is a novel technique that generates the minimal transition relation with respect to a given abstraction mapping.

1. Introduction

The limited capacity of model checking is still a major problem. Our experience in hardware formal verification have taught us that useful and intuitive specifications usually describe modules which exceed the capacity of the model checking tools. Currently, we deal with this problem via the "divide and concur" approach using hierarchical and modular verification [Lon93, GL94, McM97]. However, these techniques make the verification task more complex and time consuming and therefore we must continue to develop the capacity of the core model checking technology.

The compilation of a hardware module into a finite-state-machine produces two types of state signals: signals that represent the sequential elements of the hardware module and signals that represent the inputs of the hardware module. In many cases, the number of inputs is relatively large, 30-50%, out of the total number of state signals. Moreover, we use static pruning techniques in order to reduce the size of the model and these pruning techniques often transform sequential elements of the model into inputs, resulting in an even larger number of input signals. A known technique to increase the capacity of model checking is to existentially quantify out these inputs [CCLQ97, HKB96]. However, earlier applications of this technique rely on a number of implicit simplifying assumptions:

1. A simplified model is assumed, where next-state functions do not depend on next state values of primary input signals. This prevents us from using edge-triggered (flip-flop) or level sensitive latches.
2. Similarly, the properties we want to check cannot depend on these inputs.
3. In an assume-guarantee framework, we cannot use assumptions which depend on next state values of inputs.
4. No fairness constraints depend on the inputs.
5. Values of inputs do not appear in error traces.

We explicitly address all the above issues, by identifying three major tasks for input elimination: (a) the identification and maximization the number of the quantifiable inputs, (b) adapting the early quantification algorithm for the specific needs of input elimination and (c) reconstructing the values of eliminated inputs in counter example generation.

Intuitively, a signal sig can be existentially quantified out if either its current value, sig, or its next state value, sig', is constrained by the transition relation, but not both of them. Following this guideline only few inputs are quantifiable, since in hardware design many sequential elements are defined in terms of both the current and the next value of the inputs. We propose techniques to re-encode the latches, the control signals and the auxiliary specification signals, hence increasing up to 3 times the number of inputs that can be quantified out, in several test cases. To our knowledge, no such re-encoding was presented before.

Early quantification is used in two contexts within model checking: in the image computation [BCL91, GB94, RABPP95] and while building the transition relation [HKB96]. While most of the work on early quantification focuses on its use in image computation, we are using early quantification in the building of the transition relation. The advantage of quantifying the inputs early while building the transition relation and not in image computation is that the quantification of each input is done only once. This is especially important since quantification is an expensive operation over BDDs. In addition, the quantification done during the image computation is inherently limited to "linear" or "mostly linear" order. That is, the partitions of the transition relation are incrementally conjuncted with the current frontier, which refer to most variables. However, in the building of the transition relation we can afford to conjunct any set of partitions (in any order) before the quantification. As a consequence, the quantification of each input is often performed on smaller BDDs.

[HKB96] also addresses the use of early quantification for building the transition relation. Their quantification order is based on a weight function that estimates the support of the resulting BDD. We use a similar weight function; however, while they choose at each iteration the best pair of partitions to be conjuncted and then quantify the corresponding inputs we primarily choose at each iteration the best input signals to be quantified and then we conjunct together the corresponding partitions, possibly more than two. Our experimental results show that, on average, our quantification algorithm keeps the intermediate BDDs relatively small, that is, in the same order of magnitude as the resulting BDD.

When a counter-example is generated by the model checker the user usually wants to examine a particular subset of variables that would explain the violating sequence of events. In many cases, it is important to show the particular values of inputs that make each transition step possible. Therefore, it is necessary to reconstruct the values of inputs that were quantified out. We enhance the counter example generation algorithm in order to address this problem. To our knowledge no previous work addresses this issue.

Another contribution of this work is in identifying input elimination as a particular application of the more general technique of *model abstraction* [CGL92]. Unlike the general case, input elimination has two advantages. First, it is completely automatic, as it does not rely on the user providing an abstraction mapping. Second, being a bisimulation, it preserves all CTL* formulas, while general abstraction preserves only ACTL* (universal CTL*) formulas. In particular, this means that no loss in accuracy occurs during input elimination, and thus preventing either false negative or false positive results.

Furthermore, we show how the general case of model abstraction can be reduced to input elimination, by transforming variables to be eliminated into inputs. This allows us to reuse some of the techniques developed for input elimination, such as early quantification. This is particularly important for building the abstract transition relation without building the concrete one. This problem is also addressed in [CGL92], where only an approximation of the transition relation is built by pushing the quantifiers inwards and applying the quantification over the elementary partitions.

Our approach is more accurate in the sense that it builds the minimal transition relation with respect to the given abstraction mapping, and thus reducing the risk of getting false negatives. This is important since the user is given full responsibility over the amount of information which is lost, as this can occur only due to the abstraction mapping, and not due to the approximation while building the transition relation.

The paper is organized as follows. In Section 2, we present the theory of abstraction and input elimination. Section 3 contains heuristics for maximizing the number of quantifiable inputs. In Section 4 we introduce a new algorithm for early quantification. Section 5 discusses reconstruction of input values in error traces. In Section 6 we describe model abstraction and its reduction to input elimination. Experimental results are summarized in Section 7, and Section 8 concludes.

2. Preliminaries

A *labeled transition system*, LTS, is a tuple $M = <S,S_0,R, L>$ where S is a set of states, $S_0 \subseteq S$ is a set of initial states, $R \subseteq S{\times}S$ is a transition relation and $L: S \to 2^{AP}$ is a labeling function mapping every state in S to a subset of AP, where AP is a finite set of propositions. In our framework, a state s in S is an evaluation of the set of all inputs and memory elements of the hardware circuit that is formally modeled by the LTS.

A *trace* of a labeled transition system M is an infinite sequence of states $\pi = s_0$, $s_1, \dots$ where $s_0 \in S_0$ and for all i > 0, $(s_{i-1},s_i) \in R$. Let *TRACE(M)* denote the set of all traces of M.

Definition 1. Given two labeled transition systems $M = <S,S_0,R,L>$ and $M' = <S',S_0',R',L'>$ and corresponding proposition sets AP and AP′ such that $AP \supseteq AP'$, a relation $H \subseteq S{\times}S'$ is a *simulation relation* iff for every s and s′, such that H(s,s′), the following conditions hold:
1. $L(s) \cap AP' = L'(s')$
2. $\forall s_1. R(s,s_1)$ implies $\exists s_1'. (R'(s',s_1') \wedge H(s_1,s_1'))$

We say that M′ *simulates* M if there exists a simulation relation H such that for every initial state s_0 in M there is an initial state s_0' in M′ for which $H(s_0,s_0')$. It can be proved [CGL96] that if M′ simulates M, then for every ACTL* formula f, $M' \models f$ implies $M \models f$.

Lemma 1. *Simulation* is a preorder (reflexive and transitive) on the set of LTS.

Definition 2. Given two labeled transition systems, M and M′, with the same set of atomic propositions AP, a relation $B \subseteq S{\times}S'$ is a *bisimulation relation* iff for all s and s′, such that B(s,s′) the following conditions hold:
1. $L(s) = L'(s')$
2. $\forall s_1 R(s,s_1)$ implies $\exists s_1'. (R'(s',s_1') \wedge B(s_1,s_1'))$
3. $\forall s_1' R'(s',s_1')$ implies $\exists s_1. (R(s,s_1) \wedge B(s_1,s_1'))$

The structures M and M′ are *bisimular* if there exists a bisimulation relation B such that for every initial state $s_0 \in S_0$ in M there is an initial state $s_0' \in S_0'$ in M′ for which $B(s_0,s_0')$ and conversely, for every initial state $s_0' \in S_0'$ there is an initial state $s_0 \in S_0$ for which $B(s_0,s_0')$. It can be proved [CGL92] that if M and M′ are bisimular, then for every CTL* formula f, $M \models f$ iff $M' \models f$.

A particular class of simulation relations are obtained from an *abstraction mapping* h:S → S' by taking R = {(s,h(s)) | s ∈ S}.

Definition 3. A labeled transition system M' simulates M under an abstraction mapping h: S → S' when:
1. $\exists$s. (h(s) = s' $\wedge$ s $\in$ S$_0$) $\Rightarrow$ s' $\in$ S$_0'$
2. $\exists$s$_1\exists$s$_2$. (h(s$_1$) = s$_1'$ $\wedge$ h(s$_2$) = s$_2'$ $\wedge$ R(s$_1$,s$_2$)) $\Rightarrow$ R'(s$_1'$,s$_2'$)
3. L(s) $\cap$ AP' = L'(h(s))

The system M' defined above is also known as an *abstract approximation* of M [CGL92].

2.1 Elimination of free and constrained inputs

The modeling of a hardware requires to represent not only the internal state of the system, but also its primary inputs. Thus we extend the notion of labeled transition system such that every transition is labeled with an *input-state i* that assigns values to the set of all primary inputs of the model. The transition s→ s' means that the system in state s with input-state i will move in one step to the state s'.

Usually, a hardware model is part of a larger model, i.e., its environment, and therefore the possible input sequences are constrained by the environment. A property on the model may hold only when legal inputs are provided, and be violated on illegal inputs. The *assume-guarantee* paradigm addresses this problem by proving that a property is *guaranteed* only under a certain *assumption* on the environment.

Such an assumption is modeled by constraining the possible inputs provided by the environment. Thus, we define a *constrained-input transition system* as follows.

Definition 4. An *constrained-input transition system* can be described as M = <I,S,S$_0$,R,L,C>,
1. I is the set of input-states.
2. S is the set of states.
3. S$_0$ $\subseteq$ S is the set of initial states.
4. R $\subseteq$ S×I×S is the transition relation.
5. L : S → 2^{AP} is the labeling function.
6. C $\subseteq$ S×I is a relation where C(s,i) is true iff input i is legal in state s.
   The transition s→s' *occurs* with given input-state i only if (s,i,s') ∈R.

In order to apply the results of the previous section we can map an input driven transition system into a labeled transition system by hiding the inputs in the machine state.

Theorem 1. Any CTL* formula over AP holds for M=< I,S,S$_0$,R,L,C> iff it holds for M*=< S*,S$_0$*,R*, L*>, where:

1. $S^* \equiv I \times S$.
2. $S_0^* \subseteq S^*$, $\quad \forall i\ (i,s) \in S_0^* \Leftrightarrow s \in S_0$
3. $R^* \subseteq S^* \times S^*$, $\quad \forall i_1, i: R^*((i,s),(i_1,s_1)) \Leftrightarrow R(s,i,s_1) \wedge C(s,i)$
4. $L^*\ (<i,s>) = L(s)$.

Note that we did not conjoin C with the initial state, nor conjoin R with C over the next states, as might be expected if we wish to avoid reaching states that does not satisfy C. Nevertheless, we prevent further transitions from these states. Considering that we restrict our system only to infinite paths, all such states are removed from the system, and hence the additional conjunctions are redundant in the context of infinite paths.

The inclusion of inputs as part of the machine state may significantly increase the size of the transition relation. In order to address this problem we introduce a second labeled transition system, that is derived from the initial system M by existentially quantifying the inputs in the transition relation.

Theorem 2. Any CTL* formula with no occurrences of the inputs holds for M* iff it holds for $M' = <S',S_0',R', L'>$, where:
1. $S' = S$.
2. $s \in S_0' \Leftrightarrow s \in S_0 \wedge \exists i.\ C(i,s)$.
3. $R'(s,s_1) \Leftrightarrow \exists i\ R(i,s,s_1) \wedge C(i,s)$
4. $L' = L$

Proof: It is sufficient to prove that the models M* and M' are bisimular.
The following lemma identifies a bisimulation relation between the two models:

Lemma 2. The relation $B = \{((i,s),s) \mid s \in S\}$ is a bisimulation between M* and M'.

Proof: step (1) in definition 2 is trivial from the definition of $L^*(i,s)$ and L'.
To prove step (2), we note that $R^*((i,s),(i_1,s_1))$ implies that $R(s,i,s_1)$ holds, and using the definition of R' we conclude that $R'(s,s_1)$ holds. $B((i_1,s_1),s_1)$ is true, and thus this part is proved.
Proving step (3), similarly, from $R'(s,s_1)$, its definition implies that there is some i for which $R(s,i,s_1)$, and then R* definition implies $R^*((i,s),(i_1,s))$ holds for all i_1. Also $B((i_1,s_1),s_1)$ is true for an arbitrary i_1. Q.E.D.
Now to complete the proof that M* and M' are bisimulation equivalent, we need to show that for any initial state in M* there is a bisimular initial state in M' and vice-versa. This is trivially implied from the definitions of S_0^* and S_0'. Q.E.D

As a consequence of Theorems 1 and 2, we conclude that any deterministic system with free inputs $M = <I,S,S_0,R,L>$ we can be transformed into a non-deterministic one without inputs, such that the transition preserves the correctness of all CTL* formulas.

3. Netlist remodeling for maximizing quantifiable inputs

The previous section has introduced the technique of input elimination in the context of labeled transition systems. As real hardware systems are rather modeled as netlists, we need to map them into the above framework.

Definition 5. A netlist is a representation of a design at the *structural* level [DeM94]. Using the "next state" functions notation, a netlist can be formally represented as the tuple (V,Init,f,C), where:
1. $V = \{v_i\}_{i=1..n}$ is the set of all variables in the model. A variable is either an input or a memory element.
2. Init $\equiv \{Init_v\}_{v \in V}$ is a set of boolean relations defining the legal initial values of each variable v.
3. $f \equiv \{f_v\}_{v \in V}$ is a set of next-state functions, defined below.
4. C: is a boolean relation, representing the invariant of the system, which is the set of combinational constraints on the inputs.

Let variable v_i denote a memory element in the netlist. For each v_i, f_{vi} is a function that defines the value of v_i in the next state, based on the current values of all variables and the next-state values of the variables (except v_i itself):

$f_{vi}: 2^V \times 2^{V-\{v_i\}} \to$ Boolean

The transition relation F_i for a given variable v_i is:

$F_i \equiv (next(v_i) = f_{vi}(V,next(V)))$

where $next(v_i)$ is the next value of v_i. If v_i is a primary input, F_i then defined as: $F_i \equiv$ True.

A *labeled transition system* can be constructed from the above description in the following way:
1. $S = 2^V$
2. $S_0 \equiv \wedge_{i=1..n} Init_{vi}$. Thus, $S_0 \subseteq S$
3. $R \equiv (\wedge_{i=1..n} F_i) \wedge C$. Thus, $R \subseteq S \times S$
4. $L \equiv Id(s)$, where Id is the identity function $Id(s) = s$, $s \in S$.

3.1 Remodeling of latches

The above definition constructs an LTS from next state functions. However, to apply the results of Section 0 (i.e., existential quantification of inputs), we need to construct a *constrained input transition system* from next-state functions, by distinguishing between input and non-input variables. We need to redefine the next-state function f_{vi}, by introducing the *input-state* $I = i_1 \times i_2 \times ... i_m$:

(*) $f_{vi} : S \times I \times S \to$ Boolean

Also $R \equiv (\wedge_{i=1..n} F_i)$ is now of the type $S \times I \times S$. The most important property of (*) is that all inputs must appear only in the current state in all next-state functions.

In practice, however, the above statement is not generally true for all primary inputs of the model. The form of the next state function for each sequential element depends on the type of that element. Two distinct types of sequential elements are used in common hardware design - one is *edge-triggered* latches, and the other is *level-sensitive* latches.

The form of the next-state function for an *edge triggered* latch is:

$F_i \equiv ($ next$(v_i) = $ if $(!clock \wedge$ next$(clock))$ then in else v_i),

for a *rising edge*. A *falling edge* latch is similarly defined.

From the definition above we learn that the only variable whose next-state value affects f_i is *clock*. Therefore, we choose *clock* to be part of S and not part of I (*clock* will not be quantified out).

The case for *level sensitive* latches, however, is a bit more difficult.

An active-high (active-low) level-sensitive latch stores the current input value if the clock is high (low) at the current state and retains its previous value in the other case. This is expressed using a next-state function of the following form:

$F_i \equiv ($ next$(v_i) = $ if (next$(clock)$) then next(in) else v_i)

In this case, the function uses the next value of *in* that can be an input, or a combinational function of several inputs. But according to (*), next values of inputs cannot be used in next-state functions, and therefore we need to find another way to encode *level sensitive* latches, so that the maximal number of inputs could be included in I and thus be quantified out.

The solution we suggest is to translate each *level sensitive* latch into a combination of a next-state function and a constraint. The next state function has the form:

$Fi \equiv ($next$(v_i) = $ if (next$(clock)$) then *random* else v_i)

where *random* represent a nondeterministic choice,
And the constraint is:

$C_i \equiv$ if $(clock)$ then $v_i = in$

3.2 Re-encoding control signals

As discussed in the previous section, primary input signals that drive clocks and enable latches cannot be existentially quantified. The same applies to additional control signals, such as asynchronous sets and resets. These signals are modeled as combinational functions of one or more variables. In some designs a large number of input signals may be driving a single control signal of a latch, and in such case none of these inputs will be eliminated. It may be worthwhile to replace the control signal with a new variable, whose definition is identical to that of the control signal. At the price of adding one new variable, several inputs can now be eliminated, as they no longer directly drive the control of the latch.

Things get more complicated, as each primary input may be driving several control signals, while each control signal may be affected by several input signals. We need to estimate for each control signal if it is worthwhile to convert it into a variable.

We propose an algorithm for choosing which control signals to convert to variables, by calculating the ratio between the number of inputs that drive certain control signals, and the number of control signals those inputs drive, which works as follows:

(a) calculate the weight of each control signal as $w = \sum_i 1/t_i$ where i ranges over all primary input signals that drive that control signal, and t_i is the number of control signals driven by the primary input i.

(b) convert all control signals with weight greater than 1.0 to variables, and then update the weights.

(c) repeat steps (a) and (b), until all weights are less than or equal to 1.0.

3.3 Re-encoding of Specifications

A limitation of the input elimination technique described in Section 0, is that it preserves only CTL* formulas defined over non-input variables of the model. However, experience shows that we may sometimes need to verify formulas which do refer to free inputs. The obvious solution in such case would be to avoid elimination of primary inputs referred by CTL* formulas. However, there is an alternative solution that may give a better tradeoff.

Each CTL* formula is built from one or several atomic (combinational) sub-formulas. Each sub-formula referring to primary input signals can be converted into a variable. In case the number of the added variables is smaller than the number of the primary inputs involved in the formula, we do this translation.

For example, suppose we want to prove that $Always_{1 \leq j \leq 10} \wedge: i_j = v_j$.

We can introduce a new variable v_k defined as $v_k := (i_1 = v_1) \wedge (i_2 = v_2) \wedge \dots (i_{10} = v_{10})$ and check that $AG\ v_k$ while quantifying out the free inputs $i_1, \dots, i_{10}$.

An interesting special case is that of LTL formulas. In case of LTL formulas which refer to free inputs, we can eliminate all inputs at the price of adding only one variable, based on the tableau construction of [CGH94]. In that paper, an LTL formula is verified with a CTL model checker by translating the formula into an automaton and a corresponding CTL formula. One signal of this automaton is used to indicate if the LTL formula is correct and the CTL formula only refers to this signal. We convert this signal into a variable and thus the free inputs referred to in the original LTL formula can be quantified out.

3.4 Fairness constraints

So far we have not considered models with fairness constraints. Intuitively, a path (trace) is said to be *fair* with respect to a set of fairness constraints if each constraint holds infinitely often along the path. The path quantifiers in CTL formulas are then restricted to fair paths. Since fairness is out of the scope of this paper we will only sketch our solution to models with fairness constraints without a formal account. All CTL* formulas are preserved by input elimination in the case the fairness constraints do not refer to inputs. However, if a fairness constraint is defined over one input, we can simply choose to avoid quantifying this input. On the other hand, in the case of a fairness constraint involving several inputs, a better idea is to create a new variable, representing the constraint.

For example, suppose we wish to assume $i_1 \vee i_2 \vee i_3 \vee i_4$ is infinitely often true. We can introduce a new variable to the model, defined as: $v_f := i_1 \vee i_2 \vee i_3 \vee i_4$, and use fairness constraint on v_f while quantifying out $i1,\ldots,i4$, and keeping v_f.

4. New greedy algorithm for early quantification

When introducing the existential quantification of free inputs, our motivation was to reduce the size of the transition relation. So far we assumed the quantification is done after the transition relation was built. Building the complete transition relation before quantification, using BDD representation, however, might not be possible, due to memory explosion.

As discussed in Section 0, the transition relation is a conjunction of next-state expressions and a combinational constraint. We can rewrite the definition of R' as:

$$R'(s,s_1) \Leftrightarrow \exists i\, R(i,s,s_1) \wedge C(i,s) \Leftrightarrow \exists i\, \wedge_{j=1..m} F_j(i,s,s_1) \wedge_{j=1..n} C_j(i,s)$$

In practice it turns out that only few of the next state functions and constraints depend on a certain input variable. Thus, existential quantification can be done on the minimal cluster of the transition relation containing the input i. For example, suppose that our transition relation is a conjunction of the following functions:

$$R = F_1(a,b,c) \wedge F_2(b,c,v_1) \wedge F_3(b,d,v_2,v_3) \wedge F_4(d,v_2)$$

where a,b,c and d are primary input signals, while v_1, v_2 and v_3 are state signals.

Quantification produces then the following result:

$$R' = \exists b.(\exists c.(F_2(b,c,v_1) \wedge \exists a.F_1(a,b,c)) \wedge \exists d.(F_3(b,d,v_2,v_3) \wedge F_4(d,v_2)))$$

As demonstrated in the above example, quantification is performed by creating different clusters for expressions involving different quantifiable inputs, and later combining clusters if they share the same input that needs to be quantified.

The heuristic for quantification is therefore based on choosing an input for quantification and then conjuncting all clusters involving that input.

It is obvious from our example that the order in which the various inputs are quantified has a major effect on the efficiency of the algorithm.

In the example above, we may choose an alternative ordering of quantification, as in:

$$R' = \exists a,c.(\exists d.(F_4(d,v_2) \wedge \exists b.(F_1(a,b,c) \wedge F_2(b,c,v_1) \wedge F_3(b,d,v_2,v_3))))$$

Assuming the size of BDD is respective to the number of variables in its support, this order is less efficient than the first order of quantification, since the number variables in the support peaks at six, while in the first suggested order, at all intermediate operations there are never more than three variables in the support after quantification.

The heuristic we use for choosing the quantification order is based on a greedy algorithm, trying to minimize the number of unquantified inputs in each cluster, using a tree structure to represent the successive quantification steps. In this respect, our quantification heuristic is similar to the one proposed by [HKB96], for efficiently computing either a monolithic or a partitioned transition relation [BCL91], with one essential difference. While [HKB96] suggested choosing pairs of clusters to conjunct, quantifying variables as soon as possible, our approach is based on choosing variable to quantify and then conjuncting all clusters that contain that variable.

The algorithm we propose for choosing the variables to quantify, and accordingly, the clusters to conjunct, is based on a binary quantification tree, where each node is associated with one cluster. For each node, n, the left branch leads to a quantification sub-tree where the cluster associated with the node n can be ignored during quantification defined by this sub-tree. The right branch leads to a quantification sub-tree in which the cluster associated with node n is required. Each leaf of the tree represents one or several variables to be quantified.

We build the tree recursively according to the following rules: (1) Given a list of clusters and a group of variables to quantify, if the cluster at the head of the list does not depend on any of the variables in the group, remove this cluster from the list. Else, build a new node in the tree, and associate the cluster at the head of the list ,C, to this node. (2) Build the left and right branches of this new node, by calling recursively the algorithm. (2a) The left sub-tree is constructed by calling the algorithm with the reminder of the cluster-list (removing C from the cluster list), and all quantification variables that do not affect the cluster C. (2b) The right sub-tree is constructed from the reminder of the cluster-list, and all quantification variables that affect the cluster C. (3) If the cluster-list is empty, create a leaf associated with the group of variables to be quantified. Note that each variable to be quantified can appear only in one leaf.

For each leaf we calculate a weight, by estimating the support of the resulting conjunction represented by all the right branches down to the leaf, after quantifying the variables in the leaf. Choosing the variables to quantify and the clusters to conjunct is done by choosing the leaf with the minimal weight.

We avoid re-creating the tree after each quantification, by modifying only those parts of the tree that are relevant to the conjuncted clusters, and adding the new cluster that results after quantification is done to the bottom of the tree.

This heuristic have several advantages over the heuristic in [HKB96]. One advantage is that the number of choices in each step is linear to the number of quantified variables, while in their heuristic, it is square in the number of clusters.

They overcome this difficulty by considering each time only the group of the smallest clusters.

Another important advantage of our heuristic, is that we keep the quantification local, by avoiding conjuncting clusters that do not result in quantification of variables, while considering conjunctions of more than two clusters at a single step. This way, even though the order of quantification is not necessarily optimal (It is shown in [HKB96] that finding the optimal ordering of quantification is an NP-complete problem), we ensure that for a given quantification order we do only the required conjunctions. It would be interesting to experimentally compare our algorithm to [HKB96].

5. Generating a counter-example

One drawback of input elimination is that as inputs are no longer part of the transition system, the values of the eliminated inputs for an error trace can no longer be derived directly from the transition relation.

Using the known algorithms for counter example generation [CGMZ95] will produce an accurate error trace for all non-eliminated signals, but no value for input signals. In case the user needs the values of the inputs for debugging the model, we need an additional algorithm that reconstructs the values of the input variables from an error trace that does not indicate the values of these variables.

We can reconstruct the input values of an error trace using satisfiability algorithm. Two consequent steps in a given error trace represents the current and next values of a transition. We can rebuild the transition relation for each pair of states in the trace, while restricting the current and next values of variables to their corresponding values in the error trace. The restriction operation can be done on each elementary partition, thus the full transition relation is not required. Finding the input values is done by looking for an assignment that satisfies the BDD.

The size of the BDDs required for reconstructing the input values can be further reduced in the case the user is interested only in the value of some of the inputs in a counter example. In such case we can both restrict the values of sequential elements, as well as quantify out unrequired input variables, while building the transition relation.

6. Data abstraction

The idea of data abstraction is to reduce the size of the model, by letting the user specify a mapping from the concrete variables of the model, to new, fewer, abstract variables [CGL92]. The mapping consists of a list of concrete variables to be eliminated, and a list of abstract variables together with their definition in terms of the concrete variables.

Using the mapping provided by the user, we derive a smaller, abstract model, which simulates the concrete model. This method enables the user to reduce the model by "throwing out" portions of the data, while preserving some properties of it. Since the abstract model simulates the concrete model, any proof of ACTL* formula on the abstract model is sound (no *false positive*), but inappropriate abstraction may result in a *false negative*, that is, a property that holds for the concrete model may fail in the abstract model. It is up to the user to choose an abstraction mapping that should preserve the required property.

To demonstrate this concept, consider the following example: A multi-stage pipeline has 8bit data path. Any number in the pipeline greater than 67 is considered as illegal data, and all other values are legal data. We want to prove that only legal data will be found at the pipeline's output. A possible abstraction is mapping each pipeline stage of the datapath to a 1 bit value, which is 0 for illegal data and 1 for legal data.

6.1 Definition of the abstract model

In this Section, we describe how an abstract model, as defined in [CGL92] can be derived from the concrete model and the abstraction mapping.

Definition 6. Given a labeled transition system $M = <S,S_0,R,L>$, and an abstraction mapping h: $S \rightarrow A$, we generate the *annotated* model M* by applying the abstraction mapping to the model: $M^* = <S^*,S_0^*,R^*,L^*>$, defined as:
1. $S^* = A \times S$
2. $(a_0,s_0) \in S_0^* \Leftrightarrow s_0 \in S_0 \wedge a_0 = h(s_0)$
3. $R^*((a,s),(a_1,s_1)) \Leftrightarrow R(s,s_1) \wedge a = h(s) \wedge a_1 = h(s_1)$
4. $L^*(a,s) = L(s)$

Furthermore, S can be described as a composition of two state spaces, $S = E \times Z$. While E is the state space of the concrete variables to be eliminated, Z represents the state space of all other concrete variables.

Theorem 3. The systems M and M* are *bisimular*.

Definition 7. The *abstract* model $M' = <S',S_0',R',L'>$ is obtained by quantifying out all variables in E:
1. $S' = A \times Z$
2. $(a_0,z_0) \in S_0' \Leftrightarrow \exists e_0 . a_0 = h(e_0,z_0) \wedge (e_0,z_0) \in S_0$
3. $R'((a,z),(a_1,z_1)) \Leftrightarrow \exists e,e_1 . a = h(e,z) \wedge a_1 = h(e_1,z_1) \wedge R((e,z),(e_1,z_1))$
4. $L'(a,z) = \cup_{\{e: h((e,z)) = a\}} L((e,z))$

Note that the definition of M' is more restrictive than definition 3 (abstract approximation), as we use "if and only if", and thus having only the initial states and transitions required. This system is known as *minimal transition system*.

Theorem 4. M' simulates M*

As a consequence of Theorem 3 and Theorem 4 we conclude, using the transitivity of the simulation relation, that M' simulates M.

6.2 Reducing data-abstraction problem to input elimination

The abstraction method described above requires to build the entire transition relation using a BDD before quantifying a part of the concrete variables. This method is impractical for the case where the transition relation of the concrete model is too big.

In Section 0 we introduced a mechanism that enables existentially quantifying inputs early during the construction of the transition system, and thus avoid building the complete transition system. In this section, we show that the problem of data abstraction can be reduced to input elimination, using simple transformations at the netlist level.

Using the translation of memory elements into next-state functions, as described in section 0, we note that all next-state functions are functions of current state variables, with the exception of clocks and asynchronous reset signals that can appear both in the current and the next state. In the following discussion, we assume that clocks and asynchronous reset signals are not going to be abstracted.

For the sake of simplicity assume we have only one abstracted variable ε which we want to replace with an abstract variable α. Assume the next-state function for ε is: $next(\varepsilon)=F_\varepsilon(\varepsilon,z)$ and the mapping function for α is: $\alpha=h(\varepsilon,z)$. According to definition 7, the new transition relation is defined as:

$$\exists\varepsilon, next(\varepsilon).\ \alpha = h(\varepsilon,z) \land next(\alpha)= h(next(\varepsilon),next(z)) \land R((\varepsilon,z),(next(\varepsilon),next(z)))$$

Next, we decompose the relation R into two conjuncts: one which defines the next state function of ε and the other, $R^{\cdot}$, which is the rest of R that does not refer to $next(\varepsilon)$. We know that $R^{\cdot}$ does not refer to $next(\varepsilon)$ according to our translation scheme of latches (see Section 0). Thus we get:

$$\exists\varepsilon, next(\varepsilon).\ \alpha = h(\varepsilon,z) \land next(\alpha)= h(next(\varepsilon),next(z)) \land next(\varepsilon)=F_\varepsilon(\varepsilon,z)) \land$$
$$R^{\cdot}((\varepsilon,z),(next(z)))$$

We now use the following relation between existential quantification and composition of functions: Let $f(x)$ and $g(v,y)$ be two Boolean functions. Then:

(**) $g(f(x),y) \equiv \exists v.\ g(v,y) \land (v = f(x))$

using formula (**) above:

$$\exists\varepsilon.\ \alpha = h(\varepsilon,z) \land next(\alpha)= h(F_\varepsilon(\varepsilon,z),next(z)) \land R^{\cdot}((\varepsilon,z),(next(z)))$$

Observe, that by eliminating $next(\varepsilon)$ from the transition relation ε becomes a free input in the above formula.

The above elimination of $next(\varepsilon)$ is performed via a manipulation of the circuit's netlist. The existential quantification of ε is then performed using the same algorithm described in section 0, by treating ε as free input.

The above algorithm is easily extended to deal with an arbitrary number of abstract and abstracted variables.

7. Experimental results

Input elimination has been tried on several actual verification problems of Intel's future micro-processors. Experiments show the following conclusions regarding input elimination:

1. Though it is theoretically possible that input elimination would increase the size of the BDD representing the transition relation, no such case has been identified.
2. Input elimination can significantly reduce the size of *transition relation* for models with 30% inputs and more. Reduction was ranged from 50% to 1% of the original size. It is important to note here that models with 30% inputs or more are the majority of verified models.
3. Input elimination results in significant reduction of both memory consumption and verification time, and enables several cases that could not complete using conventional model checking.

In one real life example, a model with 70% inputs that took over 48 hours to complete using standard model checking, took 5 minutes using input elimination. Another example, using 324 variables, of which 264 are free inputs, could not complete using standard model checking, but complete in 7 minutes using input elimination.

The following table contains results of some of the tests that were done on Intel future micro-processors, with and without input elimination:

Table 1.

Test	total variables	Input variables	Time w/o input elimination	Time with input elimination
real1	181	122	space out	40s
real2	107	26	space out	4100s
real3	204	88	space out	6221s
real4	138	109	185000s	300s
real5	117	44	1070s	130s
real6	324	264	space out	460s
real7	273	124	325s	22s
real8	297	70	space out	90s

All of the above examples are complex control-intensive circuits.

Experiments on data abstraction were done as well. In one experiment on real-life design, we reduced the number of variables in a model to one fifth of the original number, using data abstraction. Even greater reduction is possible for designs that contains more redundancy and wide data path.

8. Conclusions

Input elimination is a technique that has been proven to increase the capacity of model checking. This paper presents three aspects of input elimination in hardware context, which are re-encoding of the model, early quantification of primary inputs and generation of counter example.

We addressed in this paper several problems that need to be solved when applying input elimination to hardware models. In particular, hardware model usually contains many level sensitive latches and control signals of latches that refers to both the current and next value of the inputs, and therefore prevent the elimination of those inputs. We present a technique for re-encoding the model, and thus enable the elimination of most inputs.

We have presented a new greedy algorithm for deciding the order for early quantification of input variables. The suggested algorithm deals with the problem of possible explosion of intermediate results by choosing a quantification order that enables doing the quantification on clusters which are as small as possible.

We also have addressed the problem of reconstructing the values of input variables in an error trace, and introduced an algorithm that performs that task.

Finally, we have shown that a more general framework of *model abstraction* can be reduced to input elimination using simple manipulations of the model. By reducing model abstraction to input elimination we are able to produce an abstract model without building the concrete one, and without loss of any information in addition to the one implied directly from the abstraction mapping.

Acknowledgments: We thank Ranan Fraer, Pei-Hsin Ho and Moshe Vardi for their review and many useful comments.

9. References

[BCL91]   J.R. Burch, E. M. Clarke, D.E. Long. Representing circuits more efficiently in symbolic model checking. In Proceedings of the *Design Automation Conference*, pages 403-407, San Francisco, CA, June 1991.

[CCLQ97]   G. Cabodi, P. Camurati, L. Lavagno, S. Quer. Disjunctive Partitioning and Partial Iterative Squaring. *Design Automation Conference*, 1997.

[CGH94]   E.M. Clarke, O. Grumberg, H. Hamaguchi. Another look at LTL model checking. Formal Methods in System Design, Volume 10, Number 1, February 1997. Also in CAV'94.

[CGL92]   E.M. Clarke, O. Grumberg, D.E. Long. Model checking and Abstraction. In *Symposium on Principles of Programming Languages*, ACM, October, 1992.

[CGL96]   E.M. Clarke, O. Grumberg, D.E. Long. Model Checking. In Springer-Verlag Nato ASI Series F, Volume 152, 1996 (a survey on model checking, abstraction and composition).

[CGMZ95] E.M. Clarke, O. Grumberg, K.McMillen, X. Zhao. Efficient generation of counter examples and witnesses in symbolic model checking. In *DAC 95*.

[DeM94]   G. D. De Micheli. Synthesis and Optimization of digital circuits. McGraw Hill, 1994.

[GB94]   D. Geist, I. Beer, Efficient Model Checking by Automated Ordering of Transition Relation Partitions. In Proceedings of *Computer Aided Verification*, D.L. Dill Ed. LNCS 818, Springer-Verlag, 1994.

[GL94]   O. Grumberg, D. E. Long. Model checking and modular verification. ACM *Trans. Programming Languages and Systems*, 1994.

[HKB96]   R. Hojati, S. Krishnan, R. Brayton. Early Quantification and Partitioned Transition Relation. In Proceedings of International Conference on Computer Design, 1996.

[Lon93]   D. E. Long. Model Checking, Abstraction, and Compositional Verification. PhD thesis, Carnegie-Mellon University, July 1993.

[McM97]   K.L. McMillan. A compositional rule for hardware design refinement. In O. Grumberg editors, Computer Aided Verification, Haifa, Israel, 1997, Springer-Verlag.

[RABPP95] R. K. Ranjan, A. Aziz, R. K. Brayton, C. Pixley and B. Plessier. Efficient BDD Algorithms for Synthesizing and Verifying Finite State Machines. In Workshop Notes of Intl. Workshop on Logic Synthesis, Tahoe City, CA, May 1995.

Symbolic Simulation of the JEM1 Microprocessor

David A. Greve

Rockwell Collins Advanced Technology Center
Cedar Rapids, IA 52498 USA
dagreve@collins.rockwell.com

Abstract. Symbolic simulation is the simulation of the execution of a computer system on an incompletely defined, or symbolic, state. This process results in a set of expressions that define the final machine state symbolically in terms of the initial machine state. We describe our use of symbolic simulation in conjunction with the development of the JEM1[1], the world's first Java[2] processor. We demonstrate that symbolic simulation can be used to detect microcode design errors and that it can be integrated into our current design process.

1 Introduction

Traditional microcode verification techniques, which include extensive testing and demanding design reviews, have historically provided us with reasonable levels of assurance concerning the correctness of complex microcoded processors. However, the high cost of failure associated with devices used in ultra-critical applications or those undergoing mass production demands that the verification techniques employed in the design of that device provide extremely high confidence of correct design functionality, even in the face of ever increasing design complexity and reduced time to market.

Formal verification provides a high level of confidence in the functionality of a design by establishing that a formal model of an implementation satisfies a given formal specification. Figure 1 presents the standard formal verification commuting diagram. Typically, the formal verification of a microcoded microprocessor involves proving that the sequence of microinstructions f1,f2,...,fn result in a state change at the microarchitecture level that corresponds, via some abstraction function, to the state change resulting from the application of the machine instruction, F, at the macroarchitecture, or programmer's, level.

Although formal verification provides the highest degree of certainty that an implementation meets a given specification, our experience with programs employing formal verification is that it is time consuming, expensive, and not

---

[1] JEM and JEM1 are trademarks of Rockwell
[2] Java and all Java-based trademarks and logos are trademarks or registered trademarks of Sun Microsystems, Inc. in the United States and other countries.

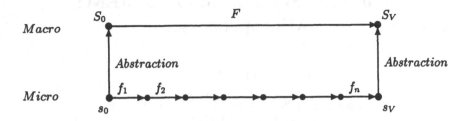

Fig. 1. Formal Verification Commuting Diagram

sufficiently mature to incorporate directly into our design process. We are exploring alternatives to formal verification that can provide high degrees of assurance: symbolic simulation is one such alternative.

Symbolic simulation is the simulation of the execution of a computer system on an incompletely defined, or symbolic, state. This process results in a set of expressions that define the final machine state symbolically in terms of the initial machine state. One uses an automated reasoning tool to derive a symbolic expression for the behavior of a computer system from a formal model of that system, a process corresponding to the bottom sequence of arrows in Figure 1. However, to establish the ultimate correctness of the derived behavior, corresponding to the upper segments in Figure 1, symbolic simulation relies on visual inspections of the results rather than mathematical proof.

Symbolic simulation is therefore a semi-formal technique which attempts to capitalize on the advantages of formal verification. By eliminating the need for formalizing and maintaining a high level specification as well as the need to derive formal proofs of correctness we reduce time and cost. By retaining the formal model of the implementation and using automated reasoning tools to derive expected behavior we retain much of the value of formal verification.

In this paper we review the goals and objectives of a recently completed symbolic simulation project, illustrate some of the techniques used in implementing the program, and discuss the outcome of the program. We also present several suggestions and observations that should be considered by future programs employing symbolic simulation.

2   Background

Our group has experimented with the application of formal methods to the verification of microcoded microprocessors [8, 7, 12]. The primary goal of these programs was to evaluate the viability of formal methods in industrial settings. As previously mentioned, although formal verification can be used to uncover design errors, our experience has indicated that it has a serious drawback: the methodology is still too immature to apply in an industrial setting.

We discovered in applying formal verification that many of the errors that we ultimately detected in our formal models were revealed during the symbolic simulation of the microcode, rather than during equivalence checking with an abstract specification. Individuals acquainted with the expected operation of a sequence of microcode can often identify errors in symbolic expressions easily because they manifest themselves as unusual looking results. The implication of this discovery is that the mere symbolic simulation of microcode is nearly as valuable as full formal verification in many situations. An early study of symbolic simulation can be found in [1]. The use of formal methods in the absence of an abstract specification was also explored in [5].

We also observed that, to a large extent, this type of symbolic analysis is amenable to automation. In the case of simple sequential microcode with the appropriate infrastructure in place, one need only indicate to the automated reasoning system the number of microcycles required to execute the sequence. Armed with this information the reasoning system can automatically expand definitions and rewrite terms to compute the symbolic result of that sequence of code. This simple technique does not necessarily apply to microcode loops, but it is a relatively simple matter to identify such sequences and deal with them on a case by case basis.

We concluded that symbolic simulation is the aspect of formal verification that is currently the most viable in our design environment. It retains much of the value of formal verification by retaining the formal model of the implementation and using automated reasoning tools to derive expected behavior. It also provides a simple, largely automatable methodology that makes it applicable in an industrial setting. Finally, by eliminating the need for developing and maintaining a high level specification and its associated, often fragile, formal proofs of correctness, symbolic simulation minimizes the time and cost of implementing the program.

This paper describes how symbolic simulation was applied in development of the JEM1, the world's first Java processor. We begin in Section 3 with a brief overview of the microarchitecture of the JEM1. Section 4 describes the elements of the symbolic simulator itself and Section 5 provides the results of the JEM1 symbolic simulator effort.

3 JEM1 Microarchitecture Overview

The JEM1 is the world's first direct execution Java processor [13, 4], a processor whose instruction set is a superset of the instructions specified for the Java Virtual Machine (JVM) [6, 2]. For a machine supporting an Instruction Set Architecture (ISA) as sophisticated as that of the JVM, the hardware implementation of the JEM1 is surprisingly simple. The JEM1 is a classic microprogrammed machine, including a control store ROM, microsequencer, and datapath, with nearly every processor operation controlled by microcode.

A crucial element of the microarchitecture is the control store ROM which, in a microprogrammed processor, contains the microcode that controls the machine. Each word in the control store ROM represents a single microinstruction which, in turn, corresponds to one line of microcode. Each microinstruction is logically partitioned into several fields, with each field responsible for controlling a specific portion of the machine. Because the JEM1 implements nearly all of the JVM instructions directly in microcode, the control store ROM is large and the resulting microcode is quite complex.

The microsequencer circuit is responsible for generating the microaddresses that are used to access the microcode ROM. The behavior of this circuit is largely under microcode control and allows the microprogrammer to sequence to the next sequential microcode location, to conditionally or unconditionally jump to a specific location, or to perform a single level of call and return. It also provides the ability to parse the next opcode and vector to its associated microcode entry point.

The datapath provides a multiport register file, a 32 bit ALU and barrel shifter, as well as status registers and shift linkages. This basic hardware is sufficient to support microcode implementations of all of the JVM defined arithmetic instructions over 32 and 64 bit integers and floating point numbers.

4 The JEM1 Symbolic Simulator

The JEM1 symbolic simulator is conceptually composed of several layers. The base of the simulator is the formal reasoning system which in our case is the PVS reasoning system. PVS is used to construct a formal specification of the JEM1 and its associated microcode. This model is ultimately controlled by a set of supporting software which assists in the automation of the simulation process. Finally, at the top level is the simulation environment itself. In the subsequent sections we describe each of these components in more detail.

4.1 PVS

PVS is the automatic reasoning system upon which we built the JEM1 symbolic simulator. Developed at SRI International's Computer Science Laboratory, PVS (Prototype Verification System) is an automated reasoning system for "specifying and verifying digital systems" [11, 9, 10]. The system consists of a specification language, a parser, a typechecker, and an interactive proof checker. It supports a specification language that is based on a simply typed higher-order logic, and provides a large number of prover commands that allow machine-checked reasoning about expressions in the logic. The primitive proof steps involve, among other things, the use of arithmetic and equality decision procedures, automatic rewriting, and BDD-based Boolean simplification. There is also support for the automation of reasoning in PVS via a facility for constructing new proof commands.

For the purposes of modeling simple digital logic, PVS provides the built-in Boolean type, bool, and most of the primitive Boolean operators including NOT, AND, and OR, IF-THEN-ELSE, CASES and equality. It is also possible to develop a more abstract model of digital logic using the enumerated type system provided by PVS. PVS provides direct support for reasoning about enumerated types and allows their use in CASES and equality expressions.

PVS allows the overloading of operators, including the built-in operators, to work with arbitrary types. This overloading capability allows us to use symbols that closely resemble those normally found in the digital logic domain when describing hardware functionality. PVS does not, however, provide a built-in type for modeling groups of bits, herein referred to as bitvectors. For this reason, a great deal of effort was required to develop and evaluate various techniques for representing and manipulating such constructs.

4.2 Formalization of the JEM1

Using the logic of PVS to describe the microarchitecture of the JEM1 is the first step. The microarchitecture specification is a formal description of the hardware which implements the JEM1 processor. Each major component of the microarchitecture, such as the ALU or the register file, is described in one or more PVS theories which defines its internal state, inputs, and outputs. These specifications, together with a variety of "glue" theories describing the data and control paths between them, define the microarchitecture over which the microcode executes.

The microarchitecture specification describes the processor from the perspective of a microcode programmer and abstracts away some of the details of the actual hardware implementation. Time, for example, is modeled in the microarchitecture using the natural numbers, where one unit of time corresponds to one *microcycle*, the time required for execution of a single microinstruction. Although a memory transaction may take an arbitrary amount of real time to complete, from the perspective of the microcoder each memory access takes a single microcycle. By abstracting away from the physical clock, it is possible to provide a more concise definition of the microarchitecture and automate much of the simulation process.

The microcode ROM is modeled as an uninterpreted function that accepts as input a microaddress and produces as output a microinstruction. Defining a line of microcode in the ROM involves the introduction of an axiom that states that the value of the ROM function evaluated at the microaddress of interest is the desired microinstruction. The process of constructing these axioms has been automated and uses the same files used to program the ROM in the actual device. We have chosen to use axioms rather than deriving similar results from a functional definition to minimize the time required for PVS to typecheck the theories associated with a particular microcode segment.

4.3 Supporting Software

Several aspects of symbolic simulation benefit from informal software analysis. For the JEM1 symbolic simulation program we crafted special purpose software to perform microcode translation, loop elimination and symbolic output reformulation.

As previously mentioned, the microcode translation software converts the output of the microassembler, including field names and enumeration values, into the format outlined by the formal PVS microcode representation used at the microarchitecture level. Figure 2 presents output of the microcode translation program. The **Label_IAND** lemma states that the value of the symbolic label **IAND** is microaddress 1918. The **uAX_IAND** axiom then defines the microinstruction located at location 1918 of the microcode ROM.

```
uCode_IAND : THEORY

BEGIN

   IMPORTING uCode_Labels,uCode_Definition

   Label_IAND : LEMMA IAND = 1918

   uAX_IAND : AXIOM uROM(1918) = (#
BS                            := FETCH_lt_1,
BS32                          := BS32eq11,
NM                            := MAP,
MODIFY                        := udef_MODIFY_31,
OVR_LK_CTRL                   := LOCK,
CNTR_CTRL                     := LD_CNTR,
SPARE_CTRL                    := udef_SPARE_CTRL_7,
TS                            := STATUS,
STATUS_REG                    := STgetsTFF2,
NIBL                          := NIBL_F,
TAU_CTRL                      := TAU_CTL,
CT                            := PCplus1,
PP                            := POP1,
RS                            := RisA,
SS                            := SisB,
FN                            := RandS,
FN3                           := FN3is1,
DN                            := BgetsFN,
CARRY_SELECT                  := CIisZ,
SIGN_REG                      := SIGNgetsSIGN,
IM_DATA                       := ZEX_IMx4,
AA                            := AisV,
BA                            := BisVminus1,
MIN_STACK                     := SVgt1,
MAX_STACK                     := NO_MAX,
JA                            := 4095,
uC                            := -1,
        { ... }
   #);

END uCode_IAND
```

Fig. 2. Partial PVS Specification of IAND Microinstruction

The process of loop elimination begins with the development of an informal software model of the JEM1 microsequencer. As mentioned in Section 3, the microsequencer for the JEM1 is quite simple and so it is easy to model its behavior in software. This informal model of the microsequencer enables us to perform software analysis of the control flow graph resulting from the evaluation of the JEM1 microcode using that model. This control flow analysis provides a convenient means of identifying and encapsulating microcode loops. It is also possible, using the same control flow analysis, to locate shared code segments. Once the software has identified such constructs, it partitions the microcode into segments containing only sequential, non-looping microcode sequences and generates a clock function for each of those segments. The clock function is a PVS function that calculates how many microcycles are required to execute the given microcode sequence based on the current state of the machine. These microcode segments can then be presented to PVS for symbolic simulation. Because the segments are sequential and contain no loops, using the supplied clock function we are able to direct PVS to automatically perform the simulation of the entire segment and produce the desired symbolic results.

Once the simulation is complete, PVS dumps symbolic results into a file. These results are run through a final conversion tool that reformulates the symbolic results into a representation that is easier to read and which can be fed back into PVS.

4.4 Simulation Environment

The symbolic simulation itself is performed completely within the confines of PVS. The top level symbolic simulation lemma for any given sequence of microcode claims that the state of the processor following the execution of that sequence of microcode is the same as it was initially. This conjecture is not expected to be a theorem; rather, the process of deriving a counterexample forms the basis of symbolic simulation.

A generic proof script is executed on the top level lemma which causes the final state of the processor to be derived via symbolic evaluation of the sequence of microcode under consideration. The evaluation of the microcode is performed with the theorem prover via expansion of function definitions and the rewriting and simplification of terms. The final steps in the proof script replace the equality over the state with a conjunction asserting that each element of the processor state has remained unchanged. When PVS fails to prove this assertion, what remains is a simplified conjunction containing only those assertions concerning the state elements that have in fact changed. PVS then dumps this unproven conjunction into a file. This unproven conjunction is the raw result of the symbolic simulation. A portion of the file resulting from the execution of the IAND microcode sequence is shown in Figure 3.

The file containing the raw symbolic simulation results is then run through the reformatting tool to generate a reformulation of the top level symbolic simulation theory. This reformulation results in a top level lemma that claims that the state of the machine following the execution of the microcode is equal to the symbolic simulation results just obtained. By incorporating this change and re-running the symbolic simulation proof script, it is possible to perform simple formal regression testing. In such a case, one expects PVS processing of the proof script to succeed. Figure 4 illustrates the reformulation of the results for the portion of the IAND microcode simulation presented in Figure 3. It is these

```
uSim_IAND-uSimulation.sequents

uSimulation :

  |-------
{1} ((pc |= PC(T0) + 1) = (pc |= PC(T0)))
  & ((vector |= V(T0) - 1) = (vector |= V(T0)))
  & ((skmt |= (V(T0) = 0)) = (skmt |= (1 + V(T0) = 0)))
  & ((Vm0 |= stack(1) AND stack(0)) = (Vm0 |= stack(1)))
  & ((carry |= unspecified_CARRY(T0) ^ (32))
    = (carry |= CARRY(T0)))
  & { ... }
```

Fig. 3. Partial sequent file for the IAND microcode sequence

reformulated results that we provide during the microcode design reviews to assist in establishing the correctness of the code.

We want to emphasize that the entire symbolic simulation process, including the microcode conversion, the running of the simulation, and the reformulation of the symbolic results, is completely automated. The only user interaction required is the initiation of each of these tasks.

5 Program Results

By the end of the symbolic simulation project, the JEM1 control store ROM contained 1689 lines of microcode. The process of eliminating loops and identifying common subroutines in this microcode ultimately produced 521 unique microcode segments, or approximately 3 lines of microcode per segment.

Each of these microcode segments results in five PVS files: one defining the microcode, one defining the clock function, one stating the clock function in terms of rewrites rules, a next state theory, and a top level simulation file. Each of these files also has an associated PVS proof script. The entire JEM1 symbolic simulator specification, including bitvector libraries, requires 5.7 megabytes of disk space.

The JEM1 symbolic simulation effort lasted for approximately 6 months and was manned at a level of approximately one half an engineer during that time. In the course of this program, we were able to perform symbolic simulation for all of the 1689 lines of microcode in the control store ROM. Symbolic simulation results, however, were available for only 3 of the early microcode walkthroughs, constituting coverage of 62 of the 518 microcode segments.

5.1 Issues

This program served to highlight three specific weaknesses in our symbolic simulation techniques: slow symbolic execution; the need for model validation; and the inability to represent complex symbolic results clearly.

```
Next_IAND[(IMPORTING uState_Definition) TO: time]: THEORY

BEGIN

  st: VAR micro_state

  i : VAR uState_Record

  IMPORTING uCode_Labels

  IMPORTING uSim_AbstractDefs[TO]

% Branch 0 :

  F_0(st)(i) : uState_Element[uState_Record, i] =
    CASES (i) OF
          pc      : pc      |= PC(TO) + 1,
          vector  : vector  |= V(TO) - 1,
          skmt    : skmt    |= (V(TO) = 0),
          Vm0     : Vm0     |= stack(1) AND stack(0),
          carry   : carry   |= unspecified_CARRY(TO) ^ (32),

      { ... }

          ELSE    st(i)
    ENDCASES;

  NextSt(st)(i) : uState_Element[uState_Record, i] = F_0(st)(i);

END Next_IAND
```

Fig. 4. Partial reformulated results for the IAND microcode sequence

Simulation Speed The slow symbolic simulation speed of PVS was perhaps the greatest detriment to this program. The fact that it took several minutes to execute a single line of microcode impacted the rate at which bitvector libraries could be evaluated, the turnaround time involved in finding and fixing model and microcode errors, and ultimately our ability to produce symbolic results in time for upcoming walkthroughs.

The CPU time required to perform the symbolic simulation for all 1689 lines of microcode was nearly 17 days, not including typechecking. This averages out to approximately 14 minutes per line of microcode. In general, it took around 6 minutes to execute a microcode segment containing only a single line of microcode. However, the time required to execute segments containing more than one line of microcode was apparently super-linear in the number of lines of code in the sequence. For example, from Table 1 we can see that approximately 20% of the total simulation time, corresponding to the top four entries in the table, was spend on just 7 particularly long and complex microcode segments. All times are for PVS 2.0 Alpha Plus (patch level 2.394) running on a Sparc 20 with 96 MB of main memory.

Microcode Segment(s)	time (sec)
BOOT_RET_4	18845
F2I	11752
DDIV+F2L+D2I+D2L	72603
BIST_CONT	190364
others	1143780
Total	1437344

Table 1. Simulation Times for Selected Microcode Segments

Future symbolic simulation efforts will require techniques for improving the throughput of the formal reasoning system used to perform the actual symbolic simulation.

Model Validation Another issue which exacerbated the problem of simulator performance involved errors in the formal model at the time of the design reviews. These errors existed because the processor formalization itself had never undergone any other form of model validation. An interesting impact of the decision to not validate the processor model up front was that it moved some of the cost of establishing model correctness from early in the program, during model development, to late in the program, during microcode inspections. The symbolic simulations ultimately exercised many of these errors which subsequently became apparent either during the simulation or the subsequent microcode walkthrough. Model errors required the regeneration and revalidation of simulation results. This task had to be performed in real time between design reviews. Unfortunately, due to the speed of the symbolic simulator, this was often not possible.

Efforts are already underway to explore the use of executable formal models as a part of the standard design process [3]. By using a single model as both a conventional microcode simulator for use in microcode development and as a basis for symbolic simulation, it is believed that the model validation issue can be resolved.

Result Presentation One daunting challenge of symbolic simulation is the concise representation of symbolic results. In order to maintain the high value of formal verification in the absence of formal equivalence checking, it is essential that the microarchitects be capable of reviewing the symbolic results with enough understanding to detect the sometimes subtle nuances which distinguish modeled behavior from desired behavior. In cases where the microarchitect cannot interpret the symbolic results, much of the value of symbolic simulation is lost.

Symbolic simulation results can often represent intermediate steps in a complicated numerical computation. For this reason, it is essential to have a bitvector library that is capable of reducing common bitvector operations, including arithmetic, logical operations, bit extraction, bit concatenation, and arbitrary combinations thereof, into the simplest possible form. Note that this requirement is not the same as requiring bitvector decision procedures. The objective of the bitvector library is to simplify the presentation of a bitvector operation, not necessarily to establish the equality of two bitvector expressions. The bitvector libraries used in this program were adapted from those used in our previous formal verification and were therefore relatively complete. However, there is still substantial room for improvement in this area.

The more difficult cases involved the longer, more complex sequences of microcode, especially those that performed multiple memory updates and dereferences. Such sequences resulted in symbolic expressions that were nearly impossible to read. This problem was addressed to some extent by providing the simulator with local knowledge of the memory mapped data structures being manipulated. Unfortunately, this is difficult to automate and doesn't work well in every case. Deriving a truly palatable representation for symbolic results in general is still, to a large extent, an open issue.

It should be noted that, even in the case where the symbolic results are difficult to read, they can still provide value if used as a regression test suite. By including symbolic simulation results as a part of this test suite, it is possible to load previous symbolic results back into the automated reasoning system and provide formal assurance that the behavior of the microcode has remained unchanged. As mentioned in Section 4.4, our current tool set already supports this approach.

5.2 Advances

Although the symbolic simulation program suffered from some significant setbacks, we believe that it was successful in demonstrating that it is possible to use symbolic simulation to find microcode errors and that the process can be automated and used in an industrial setting. These two aspects are discussed in greater detail in the following sections.

Incremental Improvement Symbolic simulation provides an intermediate step between our current verification approach, involving design reviews and functional testing, and full formal verification. The primary goal of symbolic simulation is to provide an incremental improvement over the current verification methodology. We believe that by demonstrating that microcode errors can be detected through the inspection of symbolic results, we have met this goal. Some examples of the problems identified in the course of the symbolic simulation program were:

An extraneous memory transaction. The incorrect specification of a particular microoperation resulted in a line of microcode that unexpectedly generated a write transaction to memory. The incorrect memory transaction was obvious in the symbolic results because no memory transactions were expected to take place during the microcode sequence.

Incorrectly specified shift value. A line of microcode erroneously employed a shift value of 15, rather than the desired value of 16. The fact that the symbolic result would not simplify and given that the task at hand was to create a 32-bit bit mask, the shift value of 15 appeared unusual.

Unintended alteration of a register value. A line of microcode employed a microoperation which, in conjunction with other microoperations in the microinstruction, resulted in the unintended side effect of zeroing a specific state register. This error was actually detected as part of a routine microcode testing procedure. The result, however, was verified by the symbolic simulation.

Use of an unspecified value. An error in the formal processor model resulted in a microcode branch based upon the value of an operation which, in the model, was unspecified. This particular error was easily identified because the term "unspecified" appeared in the formulation of the branch condition in the symbolic result.

A microassembler inconsistency. The tools employed to automate the formalization of the microcode revealed an inconsistency in the software used to assemble the microcode.

It is likely that all of these errors would have been discovered by traditional verification techniques, although some sooner than others. However, they are presented here to underscore the fact that many microcode errors are trivial to identify symbolically simply because they manifest themselves as unusual looking results.

It should be noted that, in addition to providing the basis for detecting errors via informal inspections, symbolic results also can act as the first step in the larger process of full formal verification. A program like the one we employed with the JEM1 is attractive because it allows one to partition the formal verification problem into two manageable portions, symbolic simulation and then formal equivalence checking with an abstract specification, with value being added in each step.

Automation As previously mentioned, one of the advantages of symbolic simulation over formal verification is that, because symbolic simulation is amenable to automation, it can be more easily integrated into the current design environment. We feel that the JEM1 symbolic simulation program was a significant demonstration of the ability to automate this process. By employing software tools to convert and analyze the microcode under consideration, the entire process, from translation to simulation to formulating the results, has been automated. In this particular program our ability to automate the process far outstripped our ability to produce and consume the symbolic results.

6 Conclusion

Symbolic simulation is an incremental improvement over the current verification process and a possible step towards inserting formal verification into the traditional design cycle. We have demonstrated this capability in the design cycle of the world's first Java processor, the JEM1. Although some deficiencies in our system hindered our work, we believe that we have demonstrated the validity of this approach as well as its potential. SRI is currently at work to improve the performance of PVS for applications such as ours and our future work in this area will study the impact of faster theorem provers, different formal representations, and extensions to the current analysis tool suite to provide even greater levels of automation.

References

1. Robert S. Boyer, Bernard Elspas, and Karl N. Levitt. SELECT – a formal system for testing and debugging programs by symbolic execution. Technical report, Stanford Research Institute, Menlo Park, CA, 1975. CSL-20.
2. James Gosling, Bill Joy, and Guy Steele. *The Java Language Specification*. Addison Wesley, Reading, Massachusetts, 1996.
3. David Greve, Matthew Wilding, and David Hardin. Efficient simulation using a simple formal processor model. Technical report, Rockwell Collins Advanced Technology Center, April 1998. (available at http://home.plutonium.net/ hokie/docs/efm.ps).
4. David A. Greve and Matthew M. Wilding. Stack-based Java a back-to-future step. *Electronic Engineering Times*, page 92, January 12, 1998.
5. Robert B. Jones, Carl-Johan H. Seger, and David L. Dill. Self-consistency checking. In Mandayam Srivas and Albert Camilleri, editors, *Formal Methods in Computer-Aided Design – FMCAD*, volume 1166 of *Lecture Notes in Computer Science*. Springer-Verlag, 1996.
6. Tim Lindholm and Frank Yellin. *The Java Virtual Machine Specification*. Addison Wesley, Reading, Massachusetts, 1996.
7. Steven P. Miller, David A. Greve, Matthew M. Wilding, and Mandayam Srivas. Formal verification of the AAMP-FV microcode. Technical report, Rockwell Collins, Inc., Cedar Rapids, IA, 1996.
8. Steven P. Miller and Mandayam Srivas. Formal verification of the AAMP5 microprocessor: A case study in the industrial use of formal methods. In *WIFT'95: Workshop on Industrial-Strength Formal Specification Techniques*, Boca Raton, FL, 1995. IEEE Computer Society.
9. S. Owre, N. Shankar, and J. M. Rushby. *The PVS Specification Language (Beta Release)*. Computer Science Laboratory, SRI International, Menlo Park, CA, February 1993.
10. S. Owre, N. Shankar, and J. M. Rushby. *User Guide for the PVS Specification and Verification System (Beta Release)*. Computer Science Laboratory, SRI International, Menlo Park, CA, February 1993.
11. N. Shankar, S. Owre, and J. M. Rushby. *The PVS Proof Checker: A Reference Manual (Beta Release)*. Computer Science Laboratory, SRI International, Menlo Park, CA, February 1993.
12. Matthew M. Wilding. Robust computer system proofs in PVS. In C. Michael Holloway and Kelly J. Hayhurst, editors, *LFM97: Fourth NASA Langley Formal Methods Workshop*. NASA Conference Publication no. 3356, 1997. (http://atb-www.larc.nasa.gov/Lfm97/).
13. Alexander Wolfe. First Java-specific MPU rolls. *Electronic Engineering Times*, page 1, September 22, 1997.

Symbolic Simulation:
An ACL2 Approach

J Strother Moore[1]

Department of Computer Sciences
The University of Texas at Austin
Austin, TX 78712-1188
moore@cs.utexas.edu

Abstract. Executable formal specification can allow engineers to test (or *simulate*) the specified system on concrete data before the system is implemented. This is beginning to gain acceptance and is just the formal analogue of the standard practice of building simulators in conventional programming languages such as C. A largely unexplored but potentially very useful next step is *symbolic simulation*, the "execution" of the formal specification on indeterminant data. With the right interface, this need not require much additional training of the engineers using the tool. It allows many tests to be collapsed into one. Furthermore, it familiarizes the working engineer with the abstractions and notation used in the design, thus allowing team members to speak clearly to one another. We illustrate these ideas with a formal specification of a simple computing machine in ACL2. We sketch some requirements on the interface, which we call a *symbolic spreadsheet*.

1 Introduction

The use of formal methods requires relatively high up-front costs to create a formal specification of the desired system or component. The cost of doing proofs, e.g., to relate a formal specification to an implementation or lower-level model, is even higher. Still more effort is required to construct mechanically checked proofs. These concerns are formidable barriers to the adoption of formal methods by industry.

Those of us in the formal methods community recognize that specification without proof is still valuable; we also recognize that careful mathematical argument without mechanically checked formal proof is valuable. The more carefully one records and analyzes design decisions, the more likely bugs will be found. The earlier the process starts, the earlier the bugs will be found. While valid, these arguments justify what might be called "informal formal methods," the use of conventional mathematical techniques in the service of hardware and software design, without mechanized support.

There is no doubt that such methods are effective and economical in the hands of experts – virtually all of mankind's deep mathematics has been done without mechanized support. But the promise of formal methods is to harness mathematical rigor, precision, and abstraction to help the man or woman in the cubicle. Thus, many of us have focused on tools.

We are working on making our tools "smarter" and more convenient to use. This work includes better proof procedures, better interfaces, better tutorials and introductory material, and more comprehensive libraries of previously-formalized results. With some moderate amount of training in the use of the tools, this will make it possible for lead engineers to create formal design specification documents. There is no need, in my opinion, for the average engineer to have the skills necessary to do this.

However, it is important that these formal design documents be useful to a much wider audience than those who write them. One mechanism is provided by the possibility that formal specifications can be executable. This idea is some-what contrarian in the theorem proving and formal methods communities, which tend to favor abstraction over other attributes of a logic. But in the hardware design community, executable specifications are almost standard practice, if one regards the ubiquitous simulators, written in C and other conventional program-ming languages, as specifications.

These simulators play a dual role. First, they allow engineers to run the design on sample input, for example to help debug the requirements. Second, the source code may be inspected to clarify or disambiguate the informal design documents produced by the lead engineers. Formal specifications, written in executable logics such as Pure Lisp or other functional programming languages, can serve the same purposes while encouraging a somewhat more abstract specification style and providing a migration path to proof. Because evaluation and simulation are already well-understood by engineers, the use of the formal specification requires little or no training beyond the logical notation used for constants.

A more radical suggestion is that the formal specifications be used to provide a "symbolic simulation" tool. Of course, symbolic simulation (or "symbolic eval-uation") is not new. Programmers have been using it since the earliest days — indeed, symbolic evaluation of a program is often easier than execution when one is limited to paper and pencil methods. Symbolic evaluation played a key role in the first version of the Boyer-Moore Pure Lisp theorem prover [15, 2] where it was called, simply, "evaluation." It was also used in the SELECT system [3], where it was combined with path assertions and counterexample generation for linear arithmetic constraints to produce an extended program testing environment.

With a symbolic simulation capability an engineer can "run" a design on certain kinds of indeterminant data, thereby covering more cases with one test. Because of its close connection with simulation, symbolic simulation is easy to grasp; indeed, it so naturally follows simulation that one may not notice its power at first.

With such tools, properly packaged, the formal specification could be in-spected and analyzed by many people with knowledge of the design issues and applications. This has two beneficial effects. First, "bugs" or other undesirable features of the design are more liable to be found early. Second, it will both educate the work force and raise expectations of clarity and abstraction. In particular, engineers will learn to read the specification notation used by the lead engineers. Language influences how we think. Seeing the abstract ideas of a design rendered into syntax is helpful. Given examples, people can generalize from them and will use the notation informally to communicate – and to *reason* – about the design. Expectations are raised in the sense that engineers will come to value the clarity and abstraction of formal specifications, once the language is familiar.

Symbolic simulation tools are thus an important bridge between current practice and the more wide-spread use of formal methods. I believe that when symbolic simulation tools are widely available, industry will find in its ranks a larger-than-expected number of engineers who are able to exploit the expressive power of formal notation to produce cleaner and more reliable designs. Furthermore, I believe this clarity, combined with the re-usability of previously formalized notions, will make it possible to create new designs faster than is currently done.

Greve makes many of these same points in [9], where he discusses a symbolic simulator for the JEM1 microprocessor. His paper gives specific examples of actual design bugs found by engineers using a symbolic simulator.

In the rest of this paper, I use one particular formal logic and theorem prover to illustrate the points just made. The system I use is ACL2. "ACL2" stands for "A Computational Logic for Applicative Common Lisp" [6, 14]. It was developed by Matt Kaufmann and me as a successor to the Boyer-Moore theorem prover, Nqthm [5]. The main idea was to replace the home-grown Pure Lisp of Nqthm with applicative Common Lisp so that formal models could be executed more efficiently and on a wider variety of platforms. However, symbolic simulation as a general technique can probably be provided by virtually any theorem proving system (see, for example, [9] which uses PVS [7]) that provides automated term rewriting, because at the logical level symbolic simulation is just "simplification."

2 Formalizing Computing Machines

Consider a simple computing machine whose state is given by a program counter, a control stack of suspended program counters, a memory, a status flag, and a program ROM. How might we formalize such a machine in ACL2? The most commonly used approach is described in [4]. We only sketch the formal model here. The ACL2 script corresponding to the results of this paper is available at http://www.cs.utexas.edu/moore/publications/symsim-script/index.html.

We represent the state of the machine as a 5-tuple. The five components of the state are accessed by functions named, respectively, pc, stk, mem, halt, and code. Each is defined in the obvious way to retrieve the appropriate element of a linear list. New states are constructed by making a list of the five components, e.g., (list *pc stk mem halt code*). Invariants ("guards") are maintained to insure that certain relationships hold among the components. For example, mem is always a list of integers. Because we generally construct new states by modifying a few fields in old states, we use a Common Lisp "macro" to write most of our states. For example, (modify *s* :pc x_1 :halt x_2) is the state whose components are the same as those of s except that the pc is x_1 and the halt flag is x_2. That is, the modify expression above denotes (list x_1 (stk s) (mem s) x_2 (code s)). Note that, despite the name, modify does not destructively change the state but constructs a new "copy." ACL2 is an applicative language. We omit the definition of modify.

Individual instructions at the ISA level are given semantics by defining functions that appropriately modify the current state of the machine. For example, (MOVE 2 0) is an instruction. At the level of abstraction used in this example, we represent instructions as lists, e.g., '(MOVE 2 0). Informally, the MOVE instruction takes two addresses and moves the contents of the second into the first. We formalize the semantics of MOVE by defining a function that takes one additional argument, the current state of the machine. The function returns the state

```
(defun move (a b s)
  (modify s
          :pc (pc+1 (pc s))
          :mem (put a (get b (mem s)) (mem s)))) ,
```

obtained by incrementing the program counter and changing memory as described.

Once such a function is defined for every instruction, the "execute" part of the machine's "fetch-execute" cycle is defined by case analysis on the opcode of the given instruction.

```
(defun execute (ins s)
  (let ((op (opcode ins))
        (a (a ins))
        (b (b ins)))
    (case op
          (MOVE  (move a b s))
          (MOVI  (movi a b s))
          (ADD   (add a b s))
          (SUBI  (subi a b s))
          (JUMPZ (jumpz a b s))
          (JUMP  (jump a s))
          (CALL  (call a s))
          (RET   (ret s))
          (otherwise s)))) .
```

The "fetch-execute" step is then defined by composition, with suitable handling of the halt status flag.

```
(defun step (s)
  (if (halt s)
      s
      (execute (current-instruction s) s))) .
```

Finally, the machine's basic cyclic behavior is then defined

```
(defun sm (s n)
  (if (zp n)
      s
      (sm (step s) (+ n -1)))))
```

as an "iterated step function". It steps the state s n times. The name "sm" stands for "small machine."

3 ACL2 as an Execution Engine

This model is easily programmed in applicative Common Lisp. One immediate consequence is that the model can be executed. That is, if you supply an explicit initial state and some number of instructions to execute, sm can be executed on any Common Lisp host to return the final state. Using evaluation you can test the system specification.

Below we show a particular program in the sm language. The program is named TIMES and it computes the product of two natural numbers by repeated addition. The comments explain how it works.

```
(TIMES (MOVI 2 0)    ; 0  mem[2] <- 0
       (JUMPZ 0 5)   ; 1  if mem[0]=0, go to 5
       (ADD 2 1)     ; 2  mem[2] <- mem[1] + mem[2]
       (SUBI 0 1)    ; 3  mem[0] <- mem[0] - 1
       (JUMP 1)      ; 4  go to 1
       (RET))        ; 5  return to caller
```

If called with two naturals i and j in memory locations 0 and 1, the program leaves $i \times j$ in memory location 2 and clears location 0 (by "counting i down"). The list constant shown above will be denoted by π. It represents a typical entry in the code component of a state.

Consider the following explicit state. Call this state α.

```
(st :pc   '(TIMES . 0)
    :stk  nil
    :mem  '(7 11 3 4 5)
    :halt nil
    :code '(π))
```

The program counter, pc, of α is a pair containing the symbol TIMES and a 0, indicating that the next instruction is the 0th instruction of the TIMES program in the code of the state. The stack component, stk, of α is empty. The α state has only five memory locations, containing, respectively, 7, 11, 3, 4, and 5. The halt flag is nil.

If we evaluate (step α) in ACL2 we get

```
(st :pc '(TIMES . 1)
    :stk NIL
    :mem '(7 11 0 4 5)
    :halt nil
    :code '(π)) .
```

The **MOVI** instruction at **pc** 0 of our **TIMES** program has been executed. The program counter has been incremented by one and memory location 2 has been cleared. "Single stepping" like this is often useful. Note that we could have used the expression (**sm** α 1) to run α one step.

To run α 31 steps, evaluate the ACL2 expression (**sm** α 31). This produces the following state:

```
(st :pc '(TIMES . 5)
    :stk NIL
    :mem '(0 11 77 4 5)
    :halt T
    :code '(π)).
```

The program counter points to the 5th instruction of **TIMES**. Observe that location 0 has been cleared, location 1 still contains 11, and location 2 contains 77. The halt flag has been set. The code still contains the list containing π.

This is an example of simple execution. The example illustrates looping but not subroutine **CALL**. A suitable interface would make it possible for an engineer not trained in formal methods – but familiar with the informal design documents for the **sm** machine – to use the formal specification to do tests of the design. This point was illustrated in the ACL2 demonstration accompanying Dave Hardin's talk at the 1998 Computer Aided Verification conference [12], in which an ACL2 model of the JEM1 ALU was integrated into a JEM1 simulator written in C.

The **sm** example is so small that it does not illustrate an important point: ACL2's execution capability can handle much larger system designs. Indeed, on "well-typed" (i.e., "gold" definitions [14]), ACL2's execution capability is just Common Lisp. We discuss performance measures of ACL2's execution and symbolic simulation of **sm** examples later in this paper.

In [6] we discuss a project in which Bishop Brock used ACL2 to formalize the Motorola CAP digital signal processor[8]. A model similar to the one described here was used, but it was orders of magnitude more complex. From [6]:

> The CAP design follows the 'Harvard architecture', i.e., there are separate program and data memories. The design includes 252 programmer-visible data and control registers. There are six independently addressable data and parameter memories. The data memories are logically partitioned into 'source' and 'destination' memories; the sense of the memories may be switched under program control. The arithmetic unit includes four multiplier-accumulators and a 6-adder array. The CAP executes a 64-bit instruction word, which in the arithmetic units is further decoded into a 317-bit, low-level control word. The instruction set includes no-overhead looping constructs and automatic data scaling. As many as 10 different registers are involved in the determination of the next program counter. A single instruction can simultaneously modify well over 100 registers. In practice, instructions found in typical applications simultaneously modify several dozen registers. Finally, the CAP has a three-stage instruction pipeline which contains many programmer-visible pipeline hazards.

The ACL2 specification of the CAP could be used as described above for simulation. In fact, the ACL2 model executed several times faster than the compiled SPW (Signal Processing Workbench) simulator and yet accurately modeled every bit in the processor, every cycle.

More recently, ACL2's execution capability was exploited at AMD. As part of a project to verify certain floating-point designs for the AMD-K7™, Art Flatau of AMD, wrote a mechanical translator from AMD's RTL language (essentially a subset of Verilog) to ACL2. This translator was used to produce ACL2 models of the floating-point circuits to be studied. However, before investing the time to try to prove the models correct, AMD managers insisted that the translator be "vetted" against the production RTL simulator. The ACL2 and RTL models were executed on some 80 million test vectors and found to return the same results. Only after this successful test was it deemed worthwhile to try to prove the ACL2 models correct. Such corroborative evidence would have been much harder to gather had the formal models not been executable. It is noteworthy that the 80 million test vectors failed to expose errors in the designs – errors later found by proof.

ACL2 is currently being used in an experiment at Rockwell-Collins to construct an executable specification of their JEM1, the world's first silicon Java Virtual Machine[17, 11].

4 ACL2 as a Theorem-Proving Engine

Our example makes it so clear that the definition of sm is "just" a Lisp program that it may be more appropriate to argue that it can be used as a specification! With ACL2 we can prove the following simple theorem about the specification.

```
(defthm sm-+
  (implies (and (natp i) (natp j))
           (equal (sm s (+ i j))
                  (sm (sm s i) j))))
```

This theorem shows that sm runs compose. The theorem is proved automatically by ACL2, by an induction on i, followed by simplification of both the base case and the induction step under the axioms and definitions involved. It takes ACL2 about 12 seconds to find the proof.

The user of ACL2 can help the theorem prover by giving it hints. For example, the proof above takes so long because, in the induction step, the system unnecessarily case splits on the instruction executed by step. This is obvious when one looks at ACL2's output during the proof: one sees a case for each instruction opcode. The proof would take even longer if our definition of step defined more opcodes. But the definition of step is actually irrelevant to this theorem! The system does not "know" that, but the user may – or may at least intuit it. If the user gives the system the hint to "disable step," which means to try to find a proof without using the definition of step, the system succeeds in finding a proof and only takes 0.12 seconds. This is just an example of the introduction of abstraction into the proof process. The details of step are irrelevant.

By exploiting such knowledge the user can dramatically speed up proofs; more importantly, the user can lead ACL2 to proofs that it would not find on its own.

The most common way for the user to give hints to the system is to build in rewrite rules about newly defined concepts. The user formulates these rules as theorems for the system to prove. Once they are proved the system interprets these theorems as rules and uses them automatically during simplification. For example, the sm-+ theorem, above, implicitly instructs the system to rewrite all expressions of the form (sm s (+ i j)) into the form (sm (sm s i) j). To be effective at extending the rule-base, the ACL2 user must understand how the system interprets previously proved theorems as rules.

The user can collect definitions, theorems and other forms of hints and advice into "books." Books can be "certified" once and then "included" into an ACL2 session. This has the effect of configuring the ACL2 simplifier (and all other proof techniques) as specified in the book. Multiple books can be included. The interaction of independently developed rules must be considered, but there are some hooks in the system to help authors codify their strategies.

It takes a lot of expertise to develop books. It is not unlike trying to teach a new class. A lot of material must be organized in ways that, when done, seem obvious; but many other, less-effective organizations are available and have to be considered. In [4] we describe such a book for sm. We show how to lead ACL2 to a proof of the following theorem about the TIMES program.

```
(defthm times-correct
  (implies (and (statep s0)
                (< 2 (len (mem s0)))
                (equal i (get 0 (mem s0)))
                (equal j (get 1 (mem s0)))
                (<= 0 i)
                (equal (current-instruction s0) '(CALL TIMES))
                (equal (assoc-eq 'TIMES (code s0)) 'π)
                (not (halt s0)))
           (equal (sm s0 (times-clock i))
                  (modify s0
                          :pc  (pc+1 (pc s0))
                          :mem (put 0 0
                                    (put 2 (* i j)
                                         (mem s0))))))))) .
```

This theorem can be read as follows. Consider a state with a memory containing at least three items (which, by definition of statep, must be integers). Let i and j be the 0th and 1st, respectively, and suppose $0 \le i$. Suppose the current instruction of the state points to the instruction (CALL TIMES) and that TIMES is defined by our previously exhibited π. We can paraphrase this rather long hypothesis by saying the state is poised to execute our TIMES on natural numbers i and j. The theorem tells us what the state will look like if we run it a certain number of steps. The number is not explicitly given, but is computed by times-clock as a function of i. The resulting state is a modification of the starting one obtained by incrementing the program counter by one, depositing a 0 into location 0, and depositing $i \times j$ into location 2.

It takes ACL2 less than 2 seconds to prove the theorem above. However, even with well-designed books, proving theorems like this requires a certain amount of training in how to use the book, how to approach the proof at a high level, and how to interact with ACL2. We explain some of the techniques used in [4].

What ACL2 can achieve in the hands of an expert is illustrated by David Russinoff's work in [16]. Russinoff used ACL2 to check proofs of the correctness of the AMD-K7 hardware for floating-point addition, subtraction, multiplication, division and square root. Using the translator mentioned above, Russinoff translated AMD's HDL descriptions (at the RTL level) into ACL2 functions. Russinoff then developed books containing thousands of lemmas about floating-point arithmetic. Using these books, he checked his proofs of the compliance of the hardware to the IEEE floating point standard. Bugs were found and corrected.

5   ACL2 as a Symbolic Simulator

Can the formal specification be made accessible to engineers not wishing to do formal proofs? The answer is yes: use it to drive a symbolic simulator for the design. We now illustrate that with our **sm** model.

Consider the following state:

```
(st :pc   '(TIMES . 0)
    :stk   nil
    :mem  (list i j x y z)
    :halt nil
    :code '(π))
```

This state is like α except that the five memory locations have unspecified content. We use the variables i, j, x, y and z to denote those contents and assume them to be integers. We use β to denote the state above.

Recall the TIMES program π:

```
(TIMES (MOVI 2 0)    ; 0  mem[2] <- 0
       (JUMPZ 0 5)   ; 1  if mem[0]=0, go to 5
       (ADD 2 1)     ; 2  mem[2] <- mem[1] + mem[2]
       (SUBI 0 1)    ; 3  mem[0] <- mem[0] - 1
       (JUMP 1)      ; 4  go to 1
       (RET))))      ; 5  return to caller
```

What is the result if we start a simulation on β and run for 4 steps? Assume that i and j are natural numbers and that i is positive. Then the answer is obvious: After 4 steps, location 0 contains $i-1$ and location 2 contains j. In addition, the program counter is (TIMES . 4), i.e., the next instruction is the JUMP back to 1.

Here is that problem, posed as a conjecture to ACL2:

```
(implies (and (ints i j x y z)
              (< 0 i))
         (equal (sm β 4) v))
```

Here v is a simple variable symbol. Note that its only occurrence in the conjecture is as the right-hand side of the conclusion. The conjecture could not possibly be a theorem under these circumstances (unless, of course, the hypotheses are contradictory). Nevertheless, the attempt to prove it with ACL2, using the above mentioned book, reduces to the goal of proving that v is

```
(st :pc   '(TIMES . 4)
    :stk  nil
    :mem  (list (- i 1) j j y z)
    :halt nil
    :code '(π)) .
```

That is, the rules in the book configure ACL2's simplifier into a symbolic simulator for the machine code in our specification.[1]

It is clear that an interface is required so as to hide the simplification process from the user. We have not constructed such an interface. But for the purposes of this paper we imagine one. We call it a *symbolic spreadsheet*. As its name suggests, we imagine a collection of "boxes" containing data. Boxes are linked via operations, with some boxes representing input and others representing output. However, unlike conventional spreadsheets, the data is symbolic, the links connecting boxes are formally defined logical functions, and the processing done by the spreadsheet is symbolic simplification. Familiar notation ought to be used where possible (e.g., in arithmetic expressions). We imagine being able to collect boxes together into larger structures, so that one of our states can be represented as a hierarchy of boxes on the spreadsheet. Obviously, it should be possible to hide data, i.e., to display the "memory" box by the contents of locations 0 and 2 only. Furthermore, it should be possible to have multiple states on the screen at once, so one can compare different states.

Such a spreadsheet should permit a rather simple configuration in which the user fills in a "form" to describe an initial symbolic state, such as β, and sees the result of stepping that state in another such form.

In our view, the difficulty is not so much the interface as the simplification. We argue in this paper that the ACL2 simplifier can be configured to do this job. We therefore continue to present our results as formal ACL2 terms, rather than as displayed in our imagined spreadsheet.

What if we start in β and run 4 steps, then 1 more (getting back to the top of the loop) and then 3 more? Then we should see i decremented twice and we should see the sum of two j's in location 2. Of course, this happens only if we know that i exceeds 1. Indeed, the proof attempt produces the goal to prove the unknown v equal to

---

[1] It is not necessary to phrase the problem as a bogus theorem-proving challenge. It is possible to invoke the ACL2 simplifier directly.

```
(st :pc   '(TIMES . 4)
    :stk  nil
    :mem  (list (- i 2) j (+ j j) y z)
    :halt nil
    :code '(π))
```

This illustrates another requirement on the spreadsheet. We need an "assumptions" box which contains assumptions about the variables. This can be menu-driven to limit the assumptions to those supported by the underlying rules.

What happens if we forget to say that i exceeds 1? That is, suppose we just have that i is positive? The result is a two-way case split. In one case, we have the additional hypothesis that (- i 1) exceeds 0 and the goal state shown above. In the other we have the additional hypothesis that (- i 1) is 0 and the goal state

```
(st :pc   '(TIMES . 5)
    :stk  nil
    :mem  (list 0 j j y z)
    :halt t
    :code '(π))
```

in which the program has halted.

In our imagined spreadsheet, the execution of this branching symbolic computation results in two copies of the output state being displayed, each with its own assumptions box. The two states might be "stacked", a visual arrangement that would immediately alert the user to the fact that the computation branched.

6   Extensibility of the Symbolic Simulator

So far we have used the symbolic simulator only to run primitive instructions. It is worthwhile to point out that it is extensible. Of course, it requires an "expert" to extend it because extension is done by adding new theorems to the database driving ACL2. But suppose that someone proves **times-correct** as stated above.

The symbolic simulator can then run calls of the TIMES code. For example, a run of length (+ (times-clock i) 2) starting in the symbolic state

```
(st :pc   '(MAIN . 0)
    :stk  nil
    :mem  (list i j x y z)
    :halt nil
    :code '(π
              (MAIN (CALL TIMES)
                    (ADD 4 2)
                    (SUBI 4 1)))))
```

produces the state

```
(st :pc   '(MAIN . 3)
    :stk  nil
    :mem  (list 0 j (* i j) y (+ (* i j) z -1))
    :halt nil
    :code '(π
             (MAIN (CALL TIMES)
                   (ADD 4 2)
                   (SUBI 4 1)))) .
```

Is this correct? The MAIN program calls TIMES, multiplying i times j and leaving the result in location 2. Then the MAIN program adds location 2 into location 4 and subtracts 1. The final value of location 4 ought to be $(i \times j) + z - 1$. So the simulator produced the expected results, regardless of the values of the variables. Note also that the simulator run shows that location 0 is cleared by this code sequence and that locations 1 and 3 are unchanged.

Actually, for this example to work the expression specifying the length of the run should be (cplus (times-clock i) 2). As noted in [4], it is convenient to maintain an isolation between arithmetic expressions denoting run-lengths and other expressions, so the former can be used by user to control proof decomposition. The interface to our symbolic simulator could mitigate this somewhat by translating arithmetic operators in the "run length" box to their "clock operator" counterparts. But the user would still have to understand how to formulate "clock expressions" so as to decompose the execution. For example, the equivalent expression (cplus 2 (times-clock i)) would have a very different effect on the simulator.

7 Performance

How fast is ACL2's symbolic simulation? That is, how fast is the ACL2 rewriter? To put it in perspective, we start by measuring the performance of ACL2 evaluation. All of our measurements were conducted on the small machine model sm and carried out on a 200 MHz Sun Microsystems Ultra 2 with 512 MB of memory, running ACL2 Version 2.2 built on Gnu Common Lisp.

ACL2 is applicative Common Lisp, provided the Common Lisp primitives are only applied in their intended domains. For example, the Common Lisp function car is intended to be applied to conses and to nil and the Common Lisp function + is intended to be applied to numbers. Common Lisp implementations are not required to check at runtime whether their arguments are suitable; that is the user's responsibility. Implementations are thus efficient but not "safe."

ACL2 functions, on the other hand, are axiomatized to be total. In our axioms, car returns nil if applied outside its intended domain and + treats non-numeric arguments as though they were 0. This notion of intended domain is formalized in ACL2 by the use of *guards*, arbitrary ACL2 formulas that specify the intended relationships between the input variables. By proving certain mechanically generated *guard conjectures*, ACL2 can guarantee that a given ACL2 function is Common Lisp compliant or "gold," which means that its execution on arguments satisfying its guard is "safe." See [14] for details.

If an ACL2 function is known to be Common Lisp compliant, it can be evaluated (on arguments satisfying its guard) via direct Common Lisp execution. In practice this means we execute binary code compiled from the function definition. If, on the other hand, an ACL2 function is not known to be compliant, or the actual arguments do not satisfy the guard, evaluation is performed by a purpose-built ACL2 interpreter that completes the Common Lisp primitives in accordance with the axioms. In practice, this means we run binary code compiled from a translation of the function definition in which function symbols have been mapped to completed counterparts which do runtime guard checks. When we talk of the execution speed of ACL2 functions we must specify whether we mean the speed of "possibly uncompliant" code or "compliant" code. Here we provide measures of both.

The definition of sm is Common Lisp compliant provided the guard on (sm s n) requires s to be a "well-formed state" and n to be a natural number. We do not exhibit the guards in this paper but they are given in the previously mentioned script available on the web. Since guards are optional, it is possible to strip them out to obtain "possibly non-compliant" code.

How much work is it to provide guards and prove compliance? Supplying guards for all of the functions in the sm system requires defining six predicates used nowhere but in guards (i.e., the notions of syntactically well-formed program counters, stacks, memories, instructions, programs, and systems of programs), as well as supplying a guard for each function in the sm system. In addition, about twenty additional lemmas have to be proved in order to lead ACL2 to the proof that sm and all of its subroutines are Common Lisp compliant. It took me several hours to invent appropriate guards.[2] Without guards, the small machine system can be admitted (syntax checking plus termination proofs) in less than a second. Verifying the guards requires about 5 seconds of additional proof.

How fast can ACL2 execute sm? We used the following expression:

```
(sm (st :pc    '(MAIN . 0)
        :stk   nil
        :mem   (list 0 0 0 0 0)
        :halt  nil
        :code  (list π
                    '(MAIN (MOVI 0 10000)
                           (MOVI 1 1000)
                           (CALL TIMES)
                           (RET))))
    40007)
```

---

[2] The guards on a function must imply the guards on all the subfunctions used in its definition, including recursive calls. In general this may be as hard as finding inductive invariants, but in practice it is not difficult. The difficulty in choosing guards is more stylistic: should one endeavor merely to insure that the Common Lisp primitives are used properly or should one strengthen the guard formulas so that they capture the correctness specification?

This requires **sm** to execute 40,007 instructions to multiply 10,000 times 1000 (by 10,000 repeated additions) leaving 10 million in memory address 2. The number 40,007 is just (+ 2 (times-clock i) 1), where i is 10,000.

If **sm** is regarded as non-compliant, the computation takes 7.39 seconds. If **sm** is regarded as compliant, it takes 0.53 seconds. This illustrates the value of guard verification if execution speed is of importance.

Since **sm** is an instruction interpreter, it is convenient to translate this performance into small machine instructions per second. Non-compliant execution proceeds at 40,007/7.39 or 5,414 small machine instructions per second in this example. Compliant execution proceeds at about 75,000 small machine instructions per second in this example. Because **sm** represents memory as a linear list of values, the speed degrades as memory size increases. Our particular experiment uses a very small memory and all the writes target the first three locations, reducing "copying" time. In more realistic tests of a comparable ACL2 model, Greve, Hardin and Wilding in [10] measured simulation speeds of about 19,000 instructions per second. A model written in C of the same processor provided 2.47 million instructions per second. The authors of [10] describe modifications to ACL2 that allowed them to achieve speeds of 1.85 million instructions per second. While some of the techniques used in [10] impose a burden on the user to insure fidelity with the axioms, I highly recommend the paper to those wishing to use ACL2 to simulate formal processor models.

Now we consider symbolic simulation. Here there is no difference between compliant and non-compliant models: the computation is done by ACL2's rewrite engine. The expression we have chosen to symbolically simulate is

```
(sm (st :pc    '(MAIN . 0)
        :stk   nil
        :mem   (list 1000 j x y z)
        :halt  nil
        :code  (list π
                     '(MAIN (CALL TIMES) (RET))))
    (+ (* (+ 1000 1) 4) 1))
```

in a context in which the variables are assumed integral. Obviously, this simplifies to

```
(st :pc    '(MAIN . 1)
    :stk   nil
    :mem   (list 0 j (* 1000 j) y z)
    :halt  T
    :code  (list π
                 '(MAIN (CALL TIMES) (RET))))
```

and requires the symbolic simulation of 4,005 small machine steps.

It takes ACL2 about 55 seconds to symbolically simulate this expression. This translates to about 72 symbolic instructions per second.

Since symbolic simulation is just simplification, an arbitrary amount of search (through the lemma data base) might be involved in a given symbolic simulation.

The simplication of the expression above produces, in addition to the final state shown, a list of all the rules used. If one poses the original symbolic simulation problem again, and this time gives ACL2 the hint to use only the rules listed, the time required drops to about 21 seconds.

Remarkably, if one discounts the time required to track the rules being used the time drops to about 46 seconds (with no hint) and to about 17 seconds (with the hint). The latter performance translates to 235 symbolic instructions per second.

The performance of a symbolic simulation engine is very dependent upon the data base of rules available. In addition, as the rule tracking observation above illustrates, "extraneous" aspects of a theorem prover may affect performance. We might therefore ask how many rewrites are involved in this symbolic simulation.

First, what is a "rewrite?" (a) Is it a call of a program in the simplifier which might replace a term with another term? (b) Is it the attempt to apply a conditional rewrite rule? (c) Is it the successful application of such a rule? Or (d) is it the application of such a rule on a path that actually leads to the final result? Interpretation (a) would let us count as a rewrite any call of ACL2's rewriter or type facility on a term, since any such call might return a changed term. Interpretation (b) excludes the use of built-in rules, such as the reduction of (equal x x) to t. The difference between (b) and (c) has to do with whether a rule is just "tried" (meaning we try to match the left-hand side of the rule and then try to relieve the hypotheses, etc.) or "used" (meaning the try was successful and the right-hand side of the rule was substituted for the target). Finally (d) brings to light the fact that the majority of rewrites usually happen on non-productive branches, i.e., branches in the proof search that do not lead to success and which are ultimately abandoned. The use of the hint, above, prunes out many (but not all) of these unsuccessful branches.

How many "rewrites" are involved in the symbolic evaluation of the term above? First consider the symbolic simulation without any hint. Approximately 1,000,000 calls of rewriting routines occur. Approximately 900,000 rules are tried and 425,000 are applied. However, only about 150,000 rule applications are actually involved in the final result. If we provide the hint, only about 400,000 calls occur. Approximately 160,000 rules are tried and virtually all of them are actually applied and used in the final result. These statistics are somewhat rough but give an idea of the amount of symbolic manipulation work involved in symbolic simulation.

It is useful also to map this to the number of rewrites per symbolic instruction simulated. Recall that 4,005 instructions are simulated in this experiment. So, without the hint, we try about 225 rules per instruction, actually apply about 100 and actually need about 40. In this example, the hint limits the search almost perfectly.

It should be noted that "long" symbolic simulation runs such as this one are generally impossible to do in the presence of indeterminant branching, since the answer state then grows exponentially. We timed a long run to amortize the cost through the general theorem prover entrance and to demonstrate that in appropriate contexts ACL2 can do such runs.

8 Conclusion

Our conclusions were drawn in the introduction, which might be appropriately re-read now. A symbolic simulator could be of great use to a design team, in part because it is accessible to many more people on the team than a verification tool would be. Furthermore, it leads naturally to verification and so represents a technology driver.

We have illustrated how such a simulator might be constructed with ACL2. The simple nature of the problem we tackled here may make some readers think this is an unrealistic proposal for designs of industrial scale. However, ACL2 has been used successfully to handle very large problems. Indeed, in the Motorola CAP work [6], Brock used ACL2 in exactly the fashion described producing states that sometimes required several megabytes of text to print fully. ACL2's simplifier is up to the task. One of Brock's problems was how to glean information from such large symbolic states. The spreadsheet would help render such states surveyable.

Two subtasks remain. The first is to construct an interface that invites the engineer to use it. This could become a lucrative product if successful. ACL2 is in the public domain and represents the heart of the tool.

The second subtask is to construct the books necessary to configure the simulator for a particular industrial model. The place to start is with current "heavy duty" ACL2 users who are proving theorems about large systems. It is likely that their existing books already contain the bulk of the required rules, simply because their books have been designed to do proofs and so often codify simplification of symbolic states.

9 Acknowledgments

These ideas have been kicking around in the Nqthm and ACL2 user communities for many years. A symbolic simulation capability is basically a first step in any ACL2 project aimed at code proofs and our ideas for how to harness the simplifier in this capacity have been developed by many, especially those who participated in the CLI "short stack" work[1] and its extensions, including Bill Bevier, Bob Boyer, Bishop Brock, Art Flatau, Warren Hunt, Matt Kaufmann, Matt Wilding, and Bill Young. I am especially indebted to Warren Hunt and Bishop Brock for the current view of the symbolic spreadsheet. I am also indebted to Dave Greve, Dave Hardin and Matt Wilding for their work in integrating ACL2 models into the JEM1 processor design process.

References

1. W. R. Bevier, W. A. Hunt, J S. Moore and W. D. Young. Special Issue on System Verification *Journal of Automated Reasoning* **5**(4), 1989.
2. R. S. Boyer and J S. Moore, Proving Theorems about Pure LISP Fucntions, *JACM*, **22**(1), pp. 129–144, 1975.

3. R. S. Boyer, K. N. Levitt and B. Elspas, SELECT–A Formal System for Testing and Debugging Programs, *Proceedings of the International Conference on Reliable Software*, IEEE Catalogue Number 75CHO940-7CSR, pp. 234–245, 1975.

4. R. S. Boyer and J S. Moore. Mechanized Formal Reasoning about Programs and Computing Machines. In R. Veroff (ed.), *Automated Reasoning and Its Applications: Essays in Honor of Larry Wos*, MIT Press, 1996.

5. R. S. Boyer and J S. Moore. *A Computational Logic Handbook, Second Edition*. Academic Press, London, 1997.

6. B. Brock, M. Kaufmann, and J S. Moore. ACL2 Theorems about Commercial Microprocessors. In *Proceedings of Formal Methods in Computer-Aided Design (FM-CAD'96)*, M. Srivas and A. Camilleri (eds.), Springer-Verlag, November, 1996, pp. 275–293.

7. J. Crow, S. Owre, J. Rushby, N. Shankar, and M. Srivas. A Tutorial Introduction to PVS, presented at *Workshop on Industrial-Strength Formal Specification Techniques*, Boca Raton, FL, April 1995 (see http://www.csl.sri.com/pvs.html).

8. S. Gilfeather, J. Gehman, and C. Harrison. Architecture of a Complex Arithmetic Processor for Communication Signal Processing in *SPIE Proceedings, International Symposium on Optics, Imaging, and Instrumentation,* **2296** *Advanced Signal Processing: Algorithms, Architectures, and Implementations V*, July, 1994, pp. 624–625.

9. D. A. Greve, Symbolic Simulation of the JEM1 Microprocessor, Technical Report, Advanced Technology Center, Rockwell Collins Avionics and Communications, Cedar Rapids, IA 52498, April, 1998 (also appearing in this volume, *The Proceedings of FMCAD '98*.

10. D. A. Greve, D. S. Hardin and M. M. Wilding, Efficient Simulation Using a Simple Formal Processor Model, Technical Report, Advanced Technology Center, Rockwell Collins Avionics and Communications, Cedar Rapids, IA 52498, April, 1998.

11. D. A. Greve and M. M. Wilding Stack-based Java a back-to-future step, Electronic Engineering Times, Jan. 12, 1998, pp. 92.

12. D. S. Hardin, M. M. Wilding, and D. A. Greve, Transforming the Theorem Prover into a Digital Design Tool: From Concept Car to Off-Road Vehicle, in A. J. Hu and M. Y. Vardi (eds.) *Computed Aided Verification: 10th International Conference, CAV '98*, Springer-Verlag LNCS 1427, pp. 39–44, 1998.

13. M. Kaufmann. ACL2 Support for Verification Projects. In *15th International Conference on Automated Deduction (CADE)* (to appear, 1998).

14. M. Kaufmann and J S. Moore. An Industrial Strength Theorem Prover for a Logic Based on Common Lisp. In *IEEE Transactions on Software Engineering* **23**(4), April, 1997, pp. 203–213.

15. J S. Moore, *Computational Logic: Structure Sharing and Proof of Program Properties*, Ph. D. dissertation, University of Edinburgh, Scotland, 1973.

16. D. M. Russinoff. A Mechanically Checked Proof of IEEE Compliance of the Floating Point Multiplication, Division, and Square Root Algorithms of the AMD-K7^{TM} Processor URL http://www.onr.com/user/russ/david/k7-div-sqrt.html.

17. A. Wolfe. First Java-specific MPU Rolls Electronic Engineering Times, Sept 22, 1997, pp. 1.

Verification of Data-Insensitive Circuits: An In-Order-Retirement Case Study*

A. Pnueli and T. Arons

Weizmann Institute of Science, Rehovot, Israel

Abstract. There is a large class of circuits (including pipeline and out-of-order execution components) which can be formally verified while completely ignoring the precise characteristics (e.g. word-size) of the data manipulated by the circuits. In the literature, this is often described as the use of *uninterpreted functions*, implying that the concrete operations applied to the data are abstracted into unknown and featureless functions. In this paper, we briefly introduce an abstract unifying model for such *data-insensitive* circuits, and claim that the development of such models, perhaps even a theory of *circuit schemas*, can significantly contribute to the development of efficient and comprehensive verification algorithms combining deductive as well as enumerative methods.

As a case study, we present in this paper an algorithm for out-of-order execution with in-order retirement and show it to be a refinement of the sequential instruction execution algorithm. Refinement is established by deductively proving (using PVS) that the register files of the out-of-order algorithm and the sequential algorithm agree at all times if the two systems are synchronized at instruction retirement time.

1 Introduction

As the complexity of hardware designs has grown, so has the need for advanced validation techniques. The need for formal verification tools to support industrial design processes is now recognized [22] and is apparent by the introduction of commercial verification tools. However, while circuit comparison and – to a lesser extent – property verification based on symbolic model-checking [5] have found their way into industrial applications (c.f. e.g. [10, 7]), coping with the complexity of industrial designs remains a key challenge, requiring complementary proof-methods to be combined in verification environments.

Obviously, no single verification method can tackle the entire design in one go, and it is necessary to reduce a large monolithic verification problem into an (often large) number of smaller problems. This reduction can be done in two complementary ways. A "bottom-up" approach partitions the design into sufficiently small units and attempts to verify each unit separately, while providing some approximate representation of the other units as a much simplified environment. This is the approach of *compositional verification* discussed in many

---

* This research was supported in part by a gift from Intel, a grant from the Minerva foundation, and an *Infrastructure* grant from the Israeli Ministry of Science.

articles such as [13], [1], [23], [27]. Traditionally, the separate verification of the individual units is performed by model checking, while using the locally established properties to infer global properties of the entire system is sometimes aided by deductive techniques.

A dual "top-down" approach handles a much larger part of the design (ideally the entire design) but reduces the complexity by abstracting away some of the detail, replacing some small circuits by an abstract operation, and bit-level data-structures by more abstract data types. Typically, we may need deductive techniques in order to justify and compute the abstractions, but often the abstracted system is small enough (and finite-state) so that it can be model-checked. Such use of abstraction is recommended and exemplified in [9], [20], [3], [12], [4], [11], and [15].

An important source of abstraction is that many components of a microprocessor design consist of an intricate controller managing the flow of data between various registers and arithmetical units, but their correctness is completely independent of the specific character of the flowing data. This gave rise to the idea that such circuits can be verified using uninterpreted (abstract) data domains and *uninterpreted functions* operating on them. When verifying such circuits, we usually *compare* the design against a *reference* machine and the only concern is that whenever the circuit applies an operation to (say) two arguments, then so does the reference machine, independently of the nature of the data or the operations. This idea has been successfully and effectively applied in [6], [14], [19], and the papers [8] and [31] in this volume.

The main principle in this particular abstraction is a clear separation between data and control, with the assumption that when the control part of the state is finite (and small or very regular) and the data remains uninterpreted, some verification and decision problems will become more tractable. A similar programme was undertaken in the early 70's in the study of programs in a sufficiently abstract way, leading to the theory of *program schemes* [26] which extensively studied an abstract model of (both imperative and declarative) programs assuming a simple (often finite) control, and uninterpreted data and operation. This theory proved very productive in recognizing some simple undecidable cases, as well as serving as an appropriate platform in which one can study program transformations, such as elimination of "go-to's" and mutual translations between recursion and iteration [18].

We believe that a similar study of abstract "circuit schemes" of particular classes can yield similar advantages and can serve as a useful platform for the development of various verification algorithms.

A first step in this direction has been taken in [17], by introducing the abstract model of a *Herbrand Automaton*. This model combines finite-state control with uninterpreted data and function registers, thus yielding a finite representation of infinite-state machines. Herbrand automata are used to model the proposed out-of-order (or pipeline) design as well as the reference machine which performs the same stream of instructions in a strictly sequential mode. The problem of verifying that a proposed out-of-order design correctly implements the reference model (which serves as a specification) is reduced to checking the equivalence of two Herbrand Automata.

To make this problem (which is undecidable in the general case) tractable, we restrict our attention to automata which apply each operation requested by an instruction only a bounded number of times. For such *Automata of Bounded Application* (HABA's), we show that the equivalence problem is decidable. In [17], we presented a decision algorithm for a general HABA, which is based on annotation of all reachable control states by invariants that hold on all visits to this location. For the special case of 1-bounded automata, we presented a more efficient decision algorithm, which is based on the comparison of the two automata over programs with operations taken from a limited set of four boolean operations.

It is interesting to note that the verification methods proposed by the papers [8] and [31] presented in this volume can, to some extent, be viewed as special cases of the general annotation method described in [17], where the additional abstractions explain their efficiency which is significantly higher than that of [17]. In retrospect, one can identify a narrower class of Herbrand Automata for which the more efficient approach can be justified in general. For example, one of the assumptions that simplifies the construction and management of annotations is that the reference machine is always "faster" than the out-of-order design, in the sense that every operation is performed by the reference machine *not later* than its execution by the OOO design.

In the rest of this paper, we present a case study of the verification of an out-of-order design with in-order retirement, which is verified using deductive techniques with the PVS tool.

2 Case Study: Out-of-Order Execution Designs

Modern out-of-order super-scalar microprocessors use dynamic scheduling to increase the number of instructions executed per cycle. These processors maintain a fixed-size window into the instruction stream, analyzing the instructions in the window to determine which can be executed out of order to improve performance. Branch prediction and register renaming are employed in order to keep the window full, while result-buffering techniques maintain the in-order-execution model required by the architecture.

In this paper we propose a proof-method for proving correctness of such processor designs based on *refinement*, and illustrate it by showing correctness of an algorithm for out-of-order execution with in-order-retirement. Our model is based on the Tomasulo algorithm in [34, 16] and [21], with modifications for in-order-retirement adapted from [25]. This paper is a continuation of the work on out-of-order execution presented in [16] and [2], extending the methodology to deal with in-order-retirement. As pointed out in [2] a correct choice of the stage in the instruction execution cycle at which to synchronize the out-of-order and sequential systems has a great impact on the complexity of the refinement relation and proof of correctness. Whereas synchronizing at instruction dispatch time proved efficient for Tomasulo's algorithm [2], in this paper we exploit the properties of in-order-retirement by synchronizing on instruction retirement. We share with [16] and [2] the ability to cope with *generic* designs, establishing correctness for an arbitrary number of functional units performing arbitrary arithmetic operations. Arithmetic operations are modeled as *uninterpreted functions*, allowing us to verify instruction scheduling without performing arithmetic verification.

Our methodology is intuitively simple: we force system IOR, exhibiting out-of-order execution with in-order-retirement, and the sequential system, SEQ, to retire instructions simultaneously. The in-order-retirement of system IOR implies that its register file is always compatible with the sequential execution of those instructions which have been retired. This fact is exploited by comparing only the systems' register files which, recording only retired instructions, are proved to be equal. In this way our refinement relation can largely ignore the complicated data structures in system IOR and any instructions which are in progress. We prove consistency of the two systems at every cycle during the execution, not only when execution has completed. Consistency between the data structures internal to system IOR is proved without reference to system SEQ. Verification is performed within the PVS [30] theorem prover.

Recent papers [28, 31–33] propose new techniques for hardware verification. In [28] Tomasulo's algorithm is verified using the SMV verifier, and an impressive level of automation is achieved. The proof, however, is dependent on the configuration and arithmetic operators and any modification requires that the new system be verified afresh.

In [31] symbolic model checking is used to verify Tomasulo's algorithm. A reference file containing symbolic values is used to implement uninterpreted functions, and abstraction techniques are developed to reduce the size of the model. The correctness of the method is proved using PVS. This approach is applicable only to designs of limited size, although this problem may be alleviated somewhat by combining it with other approaches.

A model of in-order-retirement is verified in [32] using the Stanford Validity Checker. A two-part approach is used: First it is shown that the implementation refines an intermediate in-order-execution abstraction. Functional equivalence between this abstraction and the specification system is then shown by incrementally flushing the pipeline and buffers in the intermediate abstraction.

In [33] a model including in-order-retirement, exceptions, and speculative instruction execution is verified. An intermediate model comprising a table of history variables is used to verify the system in ACL2. While the model verified is impressively detailed, the proof has the disadvantage of being specific to one configuration.

In comparison with the above works we believe our approach is both intuitive and straightforward, requiring neither specialized tools, intermediate models, nor flushing mechanisms. Our refinement relation is uncomplicated and its correctness is proved in a direct manner. The generality of the proof for all configurations and the use of uninterpreted functions are advantages shared with some of these papers.

We base this paper on the semantic model of *synchronous transition systems* ([24]), a variant of the clocked transition systems used in [29] in particular providing the concept of *steps*. Steps correspond to clock-cycles at the hardware-level. Transitions are expressed using first-order transition predicates.

The rest of the paper is structured as follows: The next section presents a short summary of the underlying mathematical model of synchronous transition systems and their refinement theory. Ultimately, the implementation has to be compatible with the sequential reference model presented in section 4. Section 5 introduces the out-of-order execution design. In section 6 we define the refinement relation, and in section 7 we prove the invariants needed by the refinement relation, thus completing the proof of correctness.

3 Synchronous Transition Systems and their Refinement

As our computational model we take synchronous transition systems ([24]). A *synchronous transition system* (STS) $S = \langle V, \Theta, \rho \rangle$, consists of the following components:

- V : A finite set of typed *system variables*. We define a *state* s to be a type-consistent interpretation of V, assigning to each variable $u \in V$ a value $s[u]$ over its domain. We denote by Σ the set of all states.
- Θ : The *initial condition*. A satisfiable assertion characterizing the initial states.
- ρ : A *transition relation*. This is an assertion $\rho(V, V')$, which relates a state $s \in \Sigma$ to its possible successors $s' \in \Sigma$ by referring to both unprimed and primed versions of the system variables. An unprimed version of a system variable refers to its value in s, while a primed version of the same variable refers to its value in s'.

Let $S = \langle V, \Theta, \rho \rangle$ be an STS. A *computation* of S is an infinite sequence of states σ: $s_0, s_1, s_2, \ldots$, satisfying the following requirements:

- *Initiation*: s_0 is initial, i.e., $s_0 \models \Theta$.
- *Consecution*: State s_{j+1} is a S-successor of s_j, for each $j = 0, 1, \ldots$.

3.1 Refinement between Systems

Refinement is the comparison of an *abstract system* $S_A = \langle V_A, \Theta_A, \rho_A \rangle$ and a *concrete system* $S_C = \langle V_C, \Theta_C, \rho_C \rangle$. The abstract system serves as a *specification* capturing all the acceptable correct computations of the system. Correctness of the concrete system is established by proving that S_C *refines* S_A. Refinement means that every computation of S_C corresponds to some computation of S_A.

Let Σ_A and Σ_C denote the sets of abstract and concrete states respectively. Let Ω, referred to as the *domain of observations*, denote a set of elements. Let $\mathcal{O}_A : S_A \mapsto \Omega$ and $\mathcal{O}_C : S_C \mapsto \Omega$ be two functions termed the *abstract* and *concrete observation functions*, respectively. These functions indicate the parts of the systems which are compared in the refinement relations.

Given a system S and an observation function $\mathcal{O} : \Sigma \mapsto \Omega$, we define an *observation of S* to be an infinite sequence of Ω-elements obtained by applying $\mathcal{O}$ to a computation of S. Let $Obs_{\mathcal{O}}(S)$ denote the set of all observations of S according to the observation function $\mathcal{O}$.

Given systems S_A and S_C with observation functions $\mathcal{O}_C$ and $\mathcal{O}_A$ respectively, S_C *refines S_A according to the observation pair* $(\mathcal{O}_C, \mathcal{O}_A)$, denoted

$$S_C \sqsubseteq_{(\mathcal{O}_C, \mathcal{O}_A)} S_A,$$

if $Obs._{\mathcal{O}_C}(S_C) \subseteq Obs._{\mathcal{O}_A}(S_A)$.

When the identity of the observation functions is obvious form the context we simply write $S_C \sqsubseteq S_A$ to indicate that S_C refines S_A.

3.2 Verifying refinement

Given a concrete system $S_C = \langle V_C, \Theta_C, \rho_C \rangle$ with observation function $\mathcal{O}_C$, and an abstract system $S_A = \langle V_A, \Theta_A, \rho_A \rangle$ with observation function $\mathcal{O}_A$, such that $V_C \cap V_A = \emptyset$, we define an *interpolating system*

$$S_I = \langle V_C \cup V_A, \Theta_C \wedge \Theta_A, \rho_C \wedge \rho_A^* \rangle$$

where $\rho_A^*(V_C, V_C', V_A, V_A')$ may refer to all variables in $V_C \cup V_A$ in their primed and unprimed versions.

The intention of the interpolating system S_I is that it emulates the joint behavior of S_C and S_A in a way that would allow any previously admissible step of S_C and immediately match it with an S_A-step. Thus, $\rho_C \wedge \rho_A^*$ should not exclude any possible S_C-step, but is allowed to select among the possible S_A-steps a single one that matches the S_C-step. Intuitively, ρ_A^* is a modification of ρ_A taking as parameters V_C and V_C' in order to choose from the successors defined by ρ_A the one which matches the S_C-step. We further require that the projection of an S_I-computation onto V_A is a legal computation of S_A.

In any interpolating system S_I satisfying the above requirements the problem of showing that $S_C \sqsubseteq S_A$ is reduced to the problem of showing that $\mathcal{O}_C = \mathcal{O}_A$ is an invariant of S_I.

We formalize this as refinement rule REF:

R1. $\exists V_A' : \rho_A^*$
R2. $\rho_A^* \longrightarrow \rho_A$
R3. $S_I \models \Box(\mathcal{O}_C = \mathcal{O}_A)$
$\overline{\qquad S_C \sqsubseteq S_A \qquad}$
Rule REF: Proving refinement

Premise R1 ensures that it is always possible to take a ρ_A^*-step, implying that the progress of system S_I is never blocked by ρ_A^*. Premise R2 requires that every

S_I-step is also an S_A-step. Premise R3 implies that the observation functions of S_C and S_A are equal for all states in all computations of S_I.

The three premises together ensure that, for every σ_C a computation of S_C, a computation of S_A can be found which completes σ_C into a computation σ_I of S_I. Since the equality of $\mathcal{O}_C$ and $\mathcal{O}_A$ is an invariant of the compound computation σ_I, the observation functions of S_A and S_C are equal at every point of σ_I, proving that S_C refines S_A.

In the proof of premise $R3$ one often uses auxiliary invariants of system S_I, most of which are also invariants of system S_C or of system S_A. To simplify the proofs of these invariants, we observe that every invariant of S_C or S_A is automatically an invariant of S_I.

4   The Reference Model: System SEQ

In this section we present system SEQ (Fig. 1) which is to serve as a reference model. System SEQ executes in a strictly sequential manner an input program consisting of non-branching register-to-register instructions. It accepts two parameters, N, the number of instructions, and R the maximum register index.

Instructions are stored in an array *prog* of length N. Each instruction has an *operation*, a *target* and two *source* operands. A program counter, *top*, points to the next instruction in *prog*. A register file *reg* records the current values of each register $0..R$.

At each step system SEQ either delays, in which case no change is made in the system, or executes the instruction pointed to by *top*. The value computed by the instruction is stored in $reg[prog[top].tgt]$ and *top* is incremented by one.

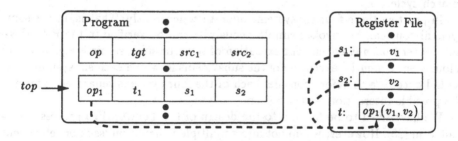

Fig. 1. Execution of one instruction in system SEQ

5   The Out-Of-Order Algorithm

In this section we provide a detailed description of our algorithm for out-of-order data-driven instruction execution with in-order-retirement. Our definitions are based on the descriptions in [21, 16] and [25].

Instructions flow from the instruction queue to the retirement buffer, where they assume their places in the queue for retirement, and the dispatch buffer, where they await availability of their source operands and a free execution unit. Once both operands are available execution of the instruction can be initiated by the appropriate functional unit. Results are written back to the retirement and dispatch buffers. Once an instruction reaches the head of the retirement queue it is retired, updating the register file with the instruction result.

In Fig. 2 we define system IOR, our implementation of the out-or-order execution, in-order-retirement algorithm. The data structures are illustrated in Fig. 3. We have added a number of auxiliary variables to our model in order to simplify the proofs. These variables are distinguished by underlining in Fig. 2. and by shading in Fig. 3. Auxiliary variables are only updated and copied from one record to another, so it is easy to prove that they do not affect the flow of control. The status field in the dispatch and retirement buffers is allowed to range over three values, **busy**, **write_b**, and **retire** instead of just busy and not-busy. However, the check in the algorithm is only on whether or not the value is busy, so this extended range does not influence the flow of control.

The functionality of system IOR can be divided into four subsystems:

- DISPATCH: This module dispatches instructions in program order.
- EXECUTE: This module executes instructions in the dispatch buffer.
- WRITE_BACK: This module writes back completed instructions.
- RETIRE: This module retires the buffer at the head of the retirement queue.

While only one instruction is dispatched or retired per cycle, modules EXECUTE and WRITE_BACK are parameterized by the number of functional units: when one of these modules in invoked, each functional unit in the system may execute or write-back results. Multiple instructions may be executed and written back in each cycle.

In practice these four subsystems operate concurrently. That is, in the same cycle all four can be invoked simultaneously. It can be verified in the PVS theorem prover that any concurrent execution of the four subsystems is equivalent to a four-step sequential execution of the subsystems in which each subsystem is executed once. We therefore consider each of the four systems separately, ignoring the possible interaction between them.

We have tried to be faithful to the design of the PentiumPro processor presented in [25]. However, we do not allow multiple instruction issue or retirement in a single cycle. We have also abstracted away the pipeline. However, the non-determinism in our model compensates for this – pipeline progress and delays are reflected by the non-deterministic rate of progress of instructions in our model. Our model admits all behaviors possible in a system in which the pipeline is explicitly modeled.

The data structures

In addition to the parameters N and R of system SEQ, system IOR expects the following parameters: U, the number of functional units; Z, the number of slots in the dispatch buffer; and B, the size of the retirement buffer.

system IOR($N, R, U, Z, B: \mathbb{N}^+$) **is**

types

REG_ID	$= [0..R]$;
FU_ID	$= [1..U]$;
$PRODUCER$	$= [0..B]$;
$VALUE$	$=$ floating-point-number ;
OP_TYPE	$= \{$fpadd, fpsub, fpmlt, fpdiv$\}$;
$STATUS$	$= \{$busy, write_b, retire$\}$;
$INST$	$= [op: OP\_TYPE,\ tgt: REG\_ID,\ src:$ **array**$[1..2]$ **of** $REG\_ID]$;
SRC_TYPE	$=$ **array**$[1..2]$ **of**
	$\qquad [st: STATUS,\ reg: REG\_ID,\ prd: PRODUCER,\ v: VALUE\ ]$;
DB_TYPE	$= [\ oc:$ **boolean**, $op: OP\_TYPE,\quad tgt: REG\_ID,$
	$\qquad\qquad prd: PRODUCER,\ arg: SRC\_TYPE\ ]$;
RB_TYPE	$= [\ st: STATUS,\ op: OP\_TYPE,\ tgt: REG\_ID,$
	$\qquad\qquad v: VALUE,\qquad arg: SRC\_TYPE\ ]$;
RES_TYPE	$= [active:$ **boolean**, $tgt: REG\_ID,\ prd: PRODUCER,\ v: VALUE\ ]$;

variables

$\qquad prog:$ **array**$[0..N]$ **of** $INST$;

$\qquad RF:$ **array**$[0..R]$ **of** $VALUE$ **init** 0;

$\qquad RTT:$ **array**$[0..R]$ **of** $[busy:$ **boolean**, $prd: PRODUCER\ ]$ **init** $\langle false, 0 \rangle$;

$\qquad DB:$ **array**$[1..Z]$ **of** DB_TYPE **init**

$\qquad\qquad\qquad \langle false,$ fpadd, 0, 0, $\langle$retire, 0, 0, 0$\rangle, \langle$retire, 0, 0, 0$\rangle \rangle$;

$\qquad RB:$ **array**$[1..B]$ **of** RB_TYPE **init**

$\qquad\qquad\qquad \langle$retire, fpadd, 0, 0, $\langle$retire, 0, 0, 0$\rangle, \langle$retire, 0, 0, 0$\rangle \rangle$;

$\qquad result:$ **array**$[1..U]$ **of** RES_TYPE **init** $\langle false, 0, 0, 0 \rangle$;

$\qquad top: [1..N{+}1]$ **init** 1;

$\qquad next, oldest: PRODUCER$ **init** 1;

functions

$\qquad fu\_table: OP\_TYPE \mapsto FU\_ID$;

$\qquad succ: PRODUCER \mapsto PRODUCER$;

$\qquad occ: PRODUCER \mapsto$ **boolean**;

behavior

$$\text{DISPATCH} \quad \Big\|\ \underset{FU=1}{\overset{U}{\big\|}}\ \text{WRITE\_BACK}[FU] \quad \Big\|\ \underset{FU=1}{\overset{U}{\big\|}}\ \text{EXECUTE}[FU] \quad \Big\|\ \text{RETIRE}$$

end system

Fig. 2. System IOR

The register file (denoted by RF) and register translation table, RTT, are both simple arrays indexed by the register index. The register file contains the current value for the register. The register translation table records whether any instruction targeting the register is in progress, and if so, which retirement buffer refers to the most recent such instruction.

The result array, $result$, stores the values of computations until these are written back.

The dispatch buffer, DB, is an unordered array of slots, each containing a flag indicating whether it is occupied. This buffer contains all the information needed to perform the operation, and is accessed by the functional units when executing instructions.

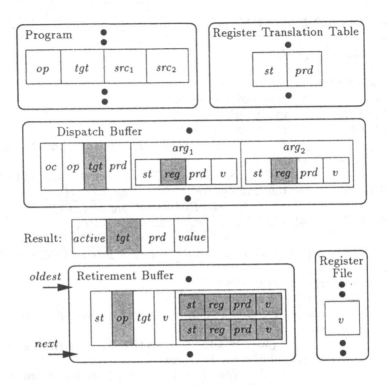

Fig. 3. Data structures for IOR

The retirement buffer, RB, is a circular array in which instructions are ordered according to their dispatch order. Its function is to ensure that instructions are retired in-order. The buffer is treated as a queue, with the the oldest buffer being "popped" off during retirement, while dispatched instructions are "pushed" onto the end of the queue. The pointers *oldest* and *next* point to the head of the queue and the next free buffer, respectively. The retirement buffer is empty when *next* and *oldest* point to the same slot. We never allow all the slots in the buffer to be occupied. The function *succ* defines a buffer's successor in the circular array, while *occ* determines (using *next* and *oldest*) whether a buffer is occupied.

The four subsystems

System DISPATCH: Instructions are *issued* from the instruction stream to buffers in both the dispatch and retirement buffers. In the dispatch buffer any unoccupied slot S_n may be used. In the retirement buffer the next slot in the queue, that pointed to by *next*, is used. Since we do not allow all the slots in the retirement buffer to be occupied, we dispatch only when the successor of *next* is not *oldest*. Operand registers are looked up in the register translation table which

```
system DISPATCH is
definitions
    ⟨o, t, s₁, s₂⟩ = prog[top]
    Sₙ          = choose S: [1..Z];
    delay       = choose d ∈ {true, false};
    ∀ j : {1,2}
        pⱼ  = RTT[sⱼ].prd;
        argⱼ = if    RTT[argⱼ].busy then ⟨ RB[pⱼ].st, sⱼ, pⱼ, RB[pⱼ].v ⟩;
                else ⟨ retire, sⱼ, 0, RF[sⱼ] ⟩;
behavior
    if top ≤ N ∧ ¬DB[Sₙ].oc ∧ succ(next) ≠ oldest ∧ ¬delay then
        DB[Sₙ] := ⟨ true, o, t, next, arg₁, arg₂ ⟩;
        RB[next] := ⟨ busy, o, t, 0, arg₁, arg₂ ⟩;
            top := top + 1;
            next := succ(next);
    end if
end system
```

records whether a value for the register is available in the register file (the status of the RTT entry is "not-busy") or, when the entry is busy, which buffer in the RB will produce the value for the register. If an operand's status is "not-busy", its value is simply copied from the register file to the dispatch buffer. Otherwise, data from the buffer producing the operand are copied from the operand's buffer in the RB to the operand field of the dispatch buffer. The operand also stores the index of the retirement buffer thus effecting register renaming – this field, rather than the index of the source register, is used to obtain the correct value for the operand. It uniquely defines the correct value for this operand, while the source register may be targeted by a number of other preceding and succeeding instructions. In our implementation operand information is also copied to auxiliary variables in the retirement buffer.

System EXECUTE: The dispatch buffer continuously snoops the bus, waiting for the values of any pending operands to be written back. Once both operands of an instruction are available the instruction may be *executed*. In our model the instruction operation determines which functional unit will eventually execute the instruction. The table *fu_table* maps operations to functional units. On each cycle each functional unit can execute an instruction if its result record is inactive. (An active result record indicates that a previously executed instruction is waiting to be written back.) On execution the result and retirement buffer index are passed to the *result* record of the functional unit and the dispatch buffer entry is freed. The active bit of the result record is set indicating that it is occupied and is waiting for access to the bus.

system EXECUTE(FU: $[1..U]$) **is**
definitions
   $enabled(S$: $[1..Z])$: **boolean**
       $= DB[S].oc \land DB[S].arg[1].st \neq$ **busy** $\land DB[S].arg[2].st \neq$ **busy**;
   e $=$ **choose** S: $[1..Z]$;
   $delay =$ **choose** $d \in \{true, false\}$;
behavior
   if $enabled(e) \land fu\_table[DB[e].op] = FU \land \neg result[FU].active \land \neg delay$ **then**
      $result[FU] := \langle true, DB[e].tgt, DB[e].prd,$
                    $do\_op(\ DB[e].op,\ DB[e].arg[1].v,\ DB[e].arg[2].v\ )\rangle$;
      $DB[e].oc\ \ := false$;
   end if
end system

system WRITE_BACK(FU: $[1..U]$) **is**
definitions
$\langle\ active_{FU},\ tgt_{FU},\ prd_{FU},\ val_{FU}\ \rangle = result[FU]$
$delay$ $=$ **choose** $d \in \{true, false\}$;
behavior
   if $result[FU].active \land \neg delay$ **then**
      $\langle\ RB[prd_{FU}].st,\ RB[prd_{FU}].v\ \rangle := \langle\ \textbf{write\_b},\ val_{FU}\ \rangle$
      $\forall\ b$: $[1..B], S$: $[1..Z], j$: $\{1..2\},$
      if $occ(b) \land RB[b].arg[j].st =$ **busy** $\land RB[b].arg[j].prd = prd_{FU}$
      then $\langle\ RB[b].arg[j].st,\ RB[b].arg[j].v\ \rangle\ := \langle\ \textbf{write\_b},\ val_{FU}\ \rangle$;
      if $DB[S].oc \land DB[S].arg[j].st =$ **busy** $\land DB[S].arg[j].prd = prd_{FU}$
      then $\langle\ DB[S].arg[j].st,\ DB[S].arg[j].v\ \rangle := \langle\ \textbf{write\_b},\ val_{FU}\ \rangle$;
   end if
end system

System WRITE_BACK: Results in the result record are *written back* by placing the result and its tag (producer field) on the bus. The retirement buffer matching the tag on the bus records the new value for the instruction and changes its status to **write_b**. Both the dispatch and retirement buffers use values on the bus to update their operand fields.

System RETIRE: Once it reaches the top of the retirement queue (it is pointed to by *oldest*) a buffer which has a valid result (its status is not **busy**) can be *retired*. During retirement the value stored in the buffer is copied to the register file. If the register translation table points to the retiring buffer the busy-flag of the relevant entry is reset, indicating that the most current value can be found in the register file. In our implementation the status of operands in the dispatch and retirement buffers is also updated to reflect retirement.

system RETIRE **is**
definitions
$delay =$ **choose** $d \in \{true, false\}$;
behavior
  if $occ(oldest) \land RB[oldest].st \neq$ **busy** $\land \neg delay$ **then**
    $RF[RB[oldest].tgt] := RB[oldest].v$;
    if $RTT[RB[oldest].tgt].prd = oldest$ **then** $RTT[RB[oldest].tgt].busy := false$;
    $\forall\, b : [1..B], S : [1..Z], j : \{1..2\}$,
      if $occ(b) \land RB[b].arg[j].st \neq$ **busy** $\land RB[b].arg[j].prd = oldest$
      then $RB[b].arg[j].st := $ **retire**;
      if $DB[S].oc \land DB[S].arg[j].st \neq$ **busy** $\land DB[S].arg[j].prd = oldest$
      then $DB[S].arg[j].st := $ **retire**;
  end if
end system

6   System IOR refines system SEQ

In this section we commence the proof that system IOR refines system SEQ. Our proof is dependent on invariants which we will prove in section 7.

    Both system SEQ and system IOR contain a program counter top, so we denote its SEQ-instance by top_A and its instance in IOR by top_C. We ensure that the two systems retire instructions simultaneously. System IOR may store in its retirement buffer instructions which have been issued but not yet retired and thus top_C will exceed top_A by the number of occupied slots in the retirement buffer. We define the function $adj\_top(top_C, next, oldest)$, which adjusts top_C by the number of outstanding instructions, returning the instruction index of the next buffer to be retired. We use adj_top and adj_top' as abbreviations for $adj\_top(top_C, next, oldest)$ and $adj\_top(top_C', next', oldest')$, respectively.

    The synchronization of the two systems at retirement time implies that adj_top always equals top_A. Correctness is shown by proving that the same value has been retired into the same register in the register files RF and reg of IOR and SEQ, respectively.

    Our observation functions are:

$$\mathcal{O}_A : (reg, top_A)$$
$$\mathcal{O}_C : (RF, adj\_top)$$

The initial conditions for the systems are:

$$\Theta_A : top_A = 1 \land \forall r : [0..R], reg[r] = 0$$

$$\Theta_C : \left(\begin{array}{l} \land \ top_C = 1 \land next = oldest \\ \land \ \forall r : [0..R], \ RF[r] = 0 \land RTT[r].busy = false \\ \land \ \forall S : [1..Z], \ DB[FU, S].oc = false \\ \land \ \forall FU : [1..U], \ result[FU].active = false \end{array} \right)$$

That is, we assume that both computations start at the first instruction with all registers storing the value zero, all dispatch and retirement buffers unoccupied.

For ρ_A^*, the restricted version of ρ_A, we take

if $\quad top_c \le N \,\wedge\, adj\_top' \ne adj\_top$
then $top_A' = top_A + 1 \,\wedge$
$\quad reg' = \lambda r \in [0..R]$ **if** $\quad r = prog[top_A].tgt$
$\qquad\qquad\qquad\qquad$ **then** $do\_op\,(prog[top_A].op,$
$\qquad\qquad\qquad\qquad\qquad\qquad reg[prog[top_A].src[1]], reg[prog[top_A].src[2]])$
$\qquad\qquad\qquad\qquad$ **else** $reg[r]$
else $\quad top_A' = top_A \,\wedge\, reg' = reg$

Thus ρ_A^*, the restricted transition relation for S_A, simply states that SEQ executes one instruction whenever an instruction is retired by system IOR.

To prove refinement we now prove that the three premises $R1, R2$ and $R3$ of rule REF hold in our system.

Premise $R1$ holds trivially by definition of ρ_A^*.

To prove $R2$ we note that the definition of ρ_A^* is identical to that of ρ_A except that the non-deterministic choice in ρ_A of whether to delay is decided deterministically.

Premise $R3$ asserts that $\alpha : \mathcal{O}_c = \mathcal{O}_A$ is an invariant of S_I. In order to prove the invariance of α we need to show that α holds at the initial state in S_I and that whenever α holds at the current S_I-state it will hold at the next S_I-state.

Showing that α holds at the initial state is trivial.

Assume that α holds in the current state of S_I. Thus, $adj\_top = top_A$ and the values in RF match the values in reg.

If system IOR does not retire an instruction in this step then SEQ delays. None of the registers in system IOR are changed and neither are the values in reg. Although top_C changes when an instruction is issued, adj_top does not. Thus, α holds at the next state in S_I.

When an instruction is retired by IOR adj_top is incremented by one. System SEQ will complete an instruction, incrementing its program counter, so $adj\_top' = top_A'$. We assume that before retirement occurred the register files of the two systems agreed. Thus we know that the two register files recorded the same values for the operands, and it suffices to prove that the correct value is written to register $prog[adj\_top].tgt$ without affecting other registers:

$$\varphi_1 : \left(\begin{array}{l} RF'[prog[adj\_top].tgt] \;=\; do\_op(\; prog[adj\_top].op, \\ \qquad\qquad\qquad\qquad\qquad RF[prog[adj\_top].src[1]], \\ \qquad\qquad\qquad\qquad\qquad RF[prog[adj\_top].src[2]]\;) \\ \wedge \qquad (\forall r : [0..R], r \ne prog[adj\_top].tgt \longrightarrow RF'[r] = RF[r]) \end{array} \right)$$

This property of system IOR is proved in section 7.

7 Some invariants of system IOR

In this section we outline the proofs of some of the invariants needed to prove the refinement relation.

We ultimately want to prove property φ_1 which asserts that the correct values are retired to the correct registers. Prior to its retirement the value in $RF'[prog[adj\_top].tgt]$ was stored in the retirement buffer pointed to by *oldest*. It is easy to prove that only register $RB[oldest].tgt$ is modified during retirement, so it suffices to prove that

$$\varphi_2 : \left(\begin{array}{l} RB[oldest].v = do\_op(\, prog[adj\_top].op, RF[prog[adj\_top].src[1]], \\ \qquad\qquad\qquad\qquad\qquad\qquad RF[prog[adj\_top].src[2]]\,) \\ \wedge \qquad\qquad\qquad\qquad\qquad RB[oldest].tgt = prog[adj\_top].tgt \end{array} \right)$$

It is not difficult to prove that $RB[oldest].tgt = prog[adj\_top].tgt$. We thus concentrate on proving that the value retired is correct.

We store as auxiliary variables the instruction operation and the source registers of the two operands. The invariants

$$RB[oldest].op = prog[adj\_top].op$$
$$\text{and} \quad RB[oldest].arg[j].reg = prog[adj\_top].src[j]$$

are easy to prove.

So, to prove φ_2 it suffices to prove that for any occupied buffer rb

$$\varphi_3 : \left(\begin{array}{l} RB[rb] = \langle\, \neg\textbf{busy}, tgt, v, op, \langle st_1, prd_1, reg_1, v_1, \rangle, \\ \qquad\qquad\qquad\qquad\qquad\quad \langle st_2, prd_2, reg_2, v_2, \rangle\,) \end{array} \right) \longrightarrow$$
$$st_1 \neq \textbf{busy} \wedge st_2 \neq \textbf{busy} \wedge v = do\_op(op, v_1, v_2)$$

providing that we also prove that the values stored in v_1 and v_2 are correct, i.e. these values will be found in registers reg_1 and reg_2 when buffer rb is retired.

We will now prove the correctness of the operands and then prove φ_3.

Tuples of the form $\langle st, prd, v \rangle$ occur repeatedly in the system and we call them records of type *TRIPLE_TYPE*. These records allow us to trace the progress of an instruction and match it to any operand fields which are waiting the completion of the instruction.

A *TRIPLE_TYPE* record is said to be *visible* if it is part of the main field of an occupied retirement buffer or one of its operands, or is part of one of the operand fields of an occupied dispatch buffer.

It can be shown that

$$\varphi_4 : visible(\langle st_1, prd_1, v_1\rangle) \wedge visible(\langle st_1, prd_1, v_2\rangle) \longrightarrow v_1 = v_2$$

That is, if two visible *TRIPLE_TYPE* records point to the same buffer and have the same status, they have the same value. The check on the status is necessary as when an entry in the retirement buffer is retired and a new instruction is placed in the vacated slot, operand fields in the system may use the same producer tag to refer to both the old and the new instructions.

We can prove that for any occupied buffer rb

$$\varphi_5 : \left(\begin{array}{l} RB[rb] = \langle\, \neg\mathbf{busy}, tgt, v, op, \langle st_1, prd_1, reg_1, v_1, \rangle, \\ \qquad\qquad\qquad\qquad\qquad\quad \langle st_2, prd_2, reg_2, v_2, \rangle\,\rangle \end{array} \right) \longrightarrow$$

$$\begin{array}{l} \mathbf{if} \quad st_j = \mathbf{retire} \\ \mathbf{then}\ \forall rb' : rb' \prec rb \longrightarrow RB[rb'].tgt \neq reg_j \\ \mathbf{else}\ \ prd_j \prec rb \wedge RB[prd_j].tgt = reg_j \wedge \\ \qquad\quad \forall rb' : prd_j \prec rb' \prec rb \longrightarrow RB[rb'].tgt \neq reg_j \end{array}$$

where $rb_1 \prec rb_2$ is true when both rb_1 and rb_2 are occupied and buffer rb_1 precedes buffer rb_2 in the queue defined by the *next* and *oldest* pointers.

Invariant φ_5 states that an operand field whose status is not "retired" points to the closest preceding retirement buffer with the correct target. If the operand is retired then there is no buffer preceding rb in the retirement buffer with the same target as the operand. This property allows us to assert that the value in the buffer prd_j will be retired to $RF[reg_j]$ before buffer rb is retired, and will not be overwritten before the retirement of rb.

Combining φ_4 and φ_5 it can easily be seen that for any occupied buffer rb

$$\varphi_6 : \left(\begin{array}{l} RB[rb] = \langle\, \neg\mathbf{busy}, tgt, v, op, \langle st_1, prd_1, reg_1, v_1, \rangle, \\ \qquad\qquad\qquad\qquad\qquad\quad \langle st_2, prd_2, reg_2, v_2, \rangle\,\rangle \end{array} \right) \longrightarrow$$

$$\begin{array}{l} \mathbf{if} \qquad st_j = \mathbf{retire} \quad \mathbf{then}\ v_j = RF[reg_j].v \\ \mathbf{else\ if}\ \ st_j = \mathbf{write\_b} \ \ \mathbf{then}\ v_j = RB[prd_j].v \end{array}$$

This, together with the precedence relation defined in φ_5, is the operand correctness property needed in φ_3. A simple corollary of φ_5 and φ_6 is that the values in the operand fields of the buffer pointed to by *oldest* are the values found in the register file.

We can now prove φ_3. Proving that the operand fields are not busy is easy. We need to show that the value in a buffer which is not busy can be obtained by applying the buffer operation to the operand values. The value v in $RB[rb]$ was obtained when an active result with target prd was written back. This result itself was calculated in a reservation station before being copied to the result register. By definition of our transition relation,

$$\left(\begin{array}{l} \wedge\ result'[FU] = \langle true, t, p, v \rangle \\ \wedge\ DB[S] = \langle true, t, p, op, \langle st_1, prd_1, reg_1, v_1 \rangle, \langle st_2, prd_2, reg_2, v_2 \rangle\,\rangle \end{array} \right) \rightarrow$$
$$v = do\_op(op, v_1, v_2)$$

That is, the value calculated by the dispatch buffer is copied to the result register together with the producer and target fields. Proving that the operation in $DB[S]$ is the same as that in $RB[tgt]$ is not difficult. Using φ_4 it can be shown that $\forall j : \{1,2\}, v_j = RB[tgt].args[j].v$. Thus, the value copied to $RB[tgt]$ equals the result of applying its operation to the values of its operands.

This completes the proof of φ_3. Invariants φ_3 and φ_6 allow φ_2 and φ_1 to be proved, completing the proof of the refinement relation.

8 Conclusion

We have presented a refinement-based proof-method for the verification of modern processor architectures, and demonstrated its applicability by showing the correctness of a data-path involving multiple functional units, register-renaming, dynamic scheduling, and out-of-order execution with in-order-retirement.

Our decision to synchronize the two systems at instruction retirement has obviated the need for an intermediate model or flushing, simplifying the refinement relation and its proof. The strength of our proof is its generality: we prove correctness for arbitrary configurations of unlimited size. The proof is also independent of the operations appearing in the instructions, and checks only the correctness of the out-of-order scheduling, not the calculation of mathematical expressions.

Still, this work needs to be extended in a number of ways:

- We would like to increase the degree of automation in the proofs. We are currently investigating the possibility of making better use of PVS strategies. Compilation of the employed notation for STS could ideally replace the current hand-translation.
- Our model is limited to non-branching programs in which no loads, stores or exceptions occur. Ongoing work considers extending the framework of this paper to incorporate some of these features.

References

1. M. Abadi and L. Lamport. Composing specifications. *Stepwise Refinement of Distributed Systems: Models, Formalism, Correctness*, LNCS-430:1–41, 1990.
2. T. Arons and A. Pnueli. Verifying tomasulo's algorithm by refinement. Technical report, Weizmann Institute, 1998.
3. S. Bensalem, A. Bouajjani, C. Loiseaux, and J. Sifakis. Properties preserving simulations. *CAV'92*:251–263, 1992.
4. N. Bjørner, I.A. Browne, and Z. Manna. Automatic generation of invariants and intermediate assertions. 1st *Intl. Conf. on Principles and Practice of Constraint Programming*, LNCS-976:589–623, 1995.
5. J.R. Burch, E.M. Clarke, K.L. McMillan, D.L. Dill, and J. Hwang. Symbolic model checking: 10^{20} states and beyond. *Inf. and Comp.*, 98(2):142–170, 1992.
6. J. R. Burch and D. L. Dill. Automatic verification of pipelined microprocessor control. *CAV'94*:68–80, 1994.
7. G. Barrett and A. McIsaac. Model-checking in a microprocessor design project. *CAV'97*, 1997.
8. R.E. Bryant and M. Velev. Deciding a theory of positive equality with uninterpreted functions. This volume.
9. P. Cousot and R. Cousot. Abstract interpretation: A unified lattice model for static analysis of programs by construction or approximation of fixpoints. *POPL'77*.
10. Y.A. Chen, E.M. Clarke, P.-H. Ho, Y. Hoskote, T. Kam, M. Khaira, J.OLeary, and X. Zhao. Verification of all circuits in a floating point unit using word-level modelchecking. *FMCAD'96*:1–18, 1996.

11. E.M. Clarke, O. Grumberg, and S. Jha. Verifying parametrized networks using abstraction and regular languages. *CONCUR'95*:395–407, 1995.
12. E.M. Clarke, O. Grumberg, and D.E. Long. Model checking and abstraction. *ACM Trans. Prog. Lang. Sys.*, 16(5):1512–1542, 1994.
13. E.M. Clarke, D.E. Long, and K.L. McMillan. Compositional model checking. *Proc. 4th IEEE Symp. Logic in Comp. Sci.*:353–362, 1989.
14. D. Cyrluk and P. Narendran. Ground temporal logic: A logic for hardware verification. *CAV'94*:247–259, 1994.
15. D. Dams, R. Gerth, and O. Grumberg. Abstract interpretation of reactive systems. *ACM Trans. Prog. Lang. Sys.*, 19(2), 1997.
16. W. Damm and A. Pnueli. Verifying out-of-order executions. *CHARME'97*:23–47, Montreal, 1997. Chapmann & Hall.
17. W. Damm, A. Pnueli, and S. Ruah. Herbrand automata for hardware verification. *CONCUR'98*, 1998.
18. S. Greibach. *Theory of program structures: schemes, semantics, verification*, volume 36 of *Lect. Notes in Comp. Sci.* Springer-Verlag, Heidelberg, 1975.
19. R. Hojati, A. Isles, D. Kirkpatrick, and R.K. Brayton. Verification using uninterpreted functions and finite instantiations. *FMCAD'96*:218 – 232, 1996.
20. N. Halbwachs, F. Lagnier, and C. Ratel. An experience in proving regular networks of processes by modular model checking. *Acta Informatica*, 29(6/7):523–543, 1992.
21. J.L. Hennessy and D.A. Patterson. *Computer Architecture: A Quantitative Approach*. Morgan Kaufmann Publishers Inc., 1996.
22. K. Keutzer. The need for formal methods for integrated circuit design. *FMCAD'96*:1–18, 1996.
23. R.P. Kurshan and K.L. McMillan. A structural induction theorem for processes. *Information and Computation*, 117:1–11, 1995.
24. Y. Kesten and A. Pnueli. An αSTS-based common semantics for SIGNAL, STATECHART, DC+, and C. Tech. report, Weizmann Institute, 1996.
25. Gwennap L. Intel's p6 uses decoupled superscalar design. *Microprocessor Report*, 9(2):9–15, 1995.
26. D.C. Luckham, D.M.R. Park, and M.S. Paterson. On formalized computer programs. *J. Comp. Sys. Sci.*, 4(3):220–249, 1970.
27. K.L. McMillan. A compositional rule for hardware design refinement. *CAV'97*.
28. K.L. McMillan. Verification of an implementation of Tomasulo's algorithm by compositional model checking. *CAV'98*:110–121, 1998.
29. Z. Manna and A. Pnueli. Clocked transition systems. *Logic and Software Engineering*:3 – 42. World Scientific, Singapore, 1996.
30. S. Owre, J.M. Rushby, N. Shankar, and M.K. Srivas. A tutorial on using PVS for hardware verification. *Proceedings of the Second Conference on Theorem Provers in Circuit Design*:167–188. FZI Publication, Universität Karlsruhe, 1994.
31. E. Clarke S. Berezin, A. Biere and Y. Zhu. Combining symbolic model checking with uninterpreted functions for out-or-order processor verification. This volume.
32. J.U. Skakkebaek, R.B. Jones, and D.L. Dill. Formal verification of out-of-order execution using incremental flushing. *CAV'98*:pp 98–110, 1998.
33. J. Sawada and Jr. W.A. Hunt. Processor verification with precise exceptions and speculative execution flushing. *CAV'98*:135–146, 1998.
34. R.M. Tomasulo. An efficient algorithm for exploiting multiple arithmetic units. *IBM J. of Research and Development*, 11(1):25–33, 1967.

Combining Symbolic Model Checking with Uninterpreted Functions for Out-of-Order Processor Verification *

Sergey Berezin, Armin Biere, Edmund Clarke, and Yunshan Zhu

Computer Science Department, Carnegie Mellon University
5000 Forbes Avenue, Pittsburgh, PA 15213, U.S.A.
{Sergey.Berezin,Armin.Biere,Edmund.Clarke,Yunshan.Zhu}@cs.cmu.edu

Abstract. We present a new approach to the verification of hardware systems with data dependencies using temporal logic symbolic model checking. As a benchmark we take Tomasulo's algorithm [10] for out-of-order instruction scheduling. Our approach is similar to the idea of uninterpreted function symbols [4]. We use symbolic values and instructions instead of concrete ones. This allows us to show the correctness of the machine independently of the actual instruction set architecture and the implementation of the functional units. Instead of using first order terms as in [4], we represent symbolic values with a new compact encoding. In addition, we apply some other reduction techniques to the model. This significantly reduces the state space and allows the use of highly efficient symbolic model checkers like SMV instead of special decision procedures. The correctness of the method has been proven formally with the PVS theorem prover.

1 Introduction

Modern microprocessors are becoming extremely complicated, involving superscalar pipelines and *out-of-order execution* (OOO). This complexity increases the demand for fast but reliable validation methods to ensure timely delivery of the product with as few errors as possible. Formal verification is the most precise technique that can guarantee the correctness of the design. However, the growing complexity of microprocessors makes formal verification increasingly difficult because data and control flow are tightly coupled. A formal model has to capture all of the data dependencies. As a result, the state space may become enormous. Straightforward model checking techniques [5, 6] can not handle this complexity because of the state explosion problem. Theorem proving [8, 12, 19] alone usually involves significant manual effort. Moreover, the proofs are too tedious to be easily manageable. Symbolic execution using uninterpreted function symbols [4] is based on extensive term rewriting and simple proof-theoretic reasoning,

---

* This research is sponsored by the Semiconductor Research Corporation (SRC) under Contract No. 97-DJ-294, the National Science Foundation (NSF) under Grant No. CCR-9505472, and the Defense Advanced Research Projects Agency (DARPA) under Contract No. DABT63-96-C-0071. Any opinions, findings and conclusions or recommendations expressed in this material are those of the authors and do not necessarily reflect the views of SRC, NSF, DARPA, or the United States Government.

and thus, can be easily automated. However, when the circuit becomes too large, each cycle in the symbolic execution produces large formulas. In addition, the number of cycles required to complete the verification grows as well.

Usually the terms that appear during symbolic execution are not arbitrary. Often, they share subterms and have similar structure. We introduce a special representation for such terms that reduces the number of copies of identical subterms. This representation is based on a data structure called the *reference file*. Terms that share common subexpressions simply have references to the same entries in the reference file. This greatly reduces the memory requirements and also simplifies the problem of checking equivalence between terms — we simply compare the references. We propose to use symbolic model checking techniques [3, 16] to perform the actual symbolic execution of the circuit. The correctness of the method has been proven in the PVS theorem prover [20].

Recently there has been a lot of work on the verification of superscalar microprocessors both with and without OOO execution. Burch and Dill [4] use the notion of *uninterpreted functions* to represent data and instructions symbolically. Interpreting these symbols results in a particular run of the concrete processor being verified. All of the formulas with uninterpreted function symbols that can be proven remain true under arbitrary interpretations. Thus, their approach can prove the correctness of a device regardless of the concrete implementation. This approach, however, requires special decision procedures for uninterpreted function symbols and does not use previously existing techniques like BDDs [1]. Skakkebæk et al. [21] propose an incremental flushing technique to verify an OOO design. Their approach is also based on uninterpreted function symbols and uses the SVC tool as a decision procedure. Our approach does not require any special decision procedures. We only need a powerful symbolic model checker.

Sajid et al. [18] have extended the decision procedures for uninterpreted function symbols to use BDDs. However, their work shares the disadvantage of [4] that their decision procedure can not be easily combined with symbolic model checking. Moreover, they do not have a notion of fairness nor can their method verify more involved temporal properties. They have also not investigated how their techniques can be applied to the verification of OOO.

Hojati and Brayton [11] have developed a formal description technique for integer combinational/sequential (ICS) systems. Their technique is even more general than uninterpreted functions ([4]). For a restricted class of models they show, by applying the notion of data independence [24], that 0-1 instantiation can reduce the size of the model to a finite number of states. This allows the usage of efficient symbolic model checkers. However, OOO is inherently data dependent and thus these abstraction techniques can not be applied. [11] also investigates how approximate reachability analysis can be performed for general ICS models. They give an algorithm that makes use of BDDs, but it is only used for reachability analysis and no limit on the number of terms occurring in the verification can be given. Our approach does not require a new algorithm, and the number of terms that need to be considered can be calculated a priori. Therefore, our model can be represented very efficiently in previously existing symbolic model checkers.

Velev and Bryant [22] use BDDs to verify pipelined microprocessor designs. It is not immediately clear if their approach can be extended to handle OOO. Instead of symbolic model checking their technique is based on symbolic trajectory evaluation.

McMillan in [17] has used model checking to verify a variant of Tomasulo's algorithm (see Sect. 2). However, his model has only one functional unit that computes only one concrete operation (integer addition). The main idea in his paper is to use compositional reasoning to reduce the complexity of the verification. This part is orthogonal to our approach and can be combined with our representation of uninterpreted functions. He also exploits symmetry to reduce the number of reservation stations, registers and their width. However, because the functional unit is modeled explicitly, his approach is unsuitable for verifying examples with a large number of instruction types or complex functional units (e.g. for operations like integer multiplication). He also does not consider liveness, which is also the case for [8]. Our representation makes it possible to abstract away the concrete instructions and functional units. Moreover, we have developed a number of new transformations in addition to the classical symmetry reduction [7] that significantly reduce the state space. In particular, these reductions enable us to verify liveness easily.

The approaches of Burch & Dill [4] and McMillan [17] are both valuable for verifying out-of-order execution algorithms. However, they both have major limitations. The decision procedures of Burch & Dill can not be used directly with symbolic model checking. McMillan's work, on the other hand, does use symbolic model checking techniques. But he does not use uninterpreted function symbols, and therefore, is unable to handle arbitrary instructions. Our approach does not suffer from these disadvantages. It includes the benefits of both together with a powerful new technique for symbolically representing the contents of registers. We believe that in order to verify realistic superscalar out-of-order designs automatically all of these techniques will be necessary.

Our paper is organized as follows: In Sect. 2 an overview of Tomasulo's algorithm is given. Section 3 presents the basic abstraction technique. Tomasulo's algorithm with reference file is presented in Sect. 4. Section 5 explains the overall strategy that we use for verifying the OOO algorithm. The next section describes other abstraction techniques that allow us to speed up the verification. Section 7 shows some experimental results. The paper concludes in Sect. 8 with a discussion of some directions for future research.

2 Out-of-Order Execution

As benchmarks for our method we have verified several configurations of our implementation of Tomasulo's algorithm [10]. This algorithm is the basic technique that is used to implement OOO in modern microprocessors. It also has been used in previous work on the verification of OOO [8, 17]. We will explain how OOO works and how Tomasulo's algorithm implements it. At the end of the section we briefly describe our model.

To achieve greater throughput of instructions, superscalar microprocessors use several functional units that can operate in parallel. However, if two instructions depend on each other (e.g. the later needs the result of the first instruction) one of them has to wait until the other has finished. In this case one functional unit is idle. But if a different instruction, potentially following the other two in the instruction sequence, does not depend on their results, then it can be executed on the free functional unit. This is the main idea behind OOO.

I0	R0 := R0 * R1
I1	R0 := R0 * R1
I2	R1 := R1 + R1

Fig. 1. An instruction sequence that allows OOO.

For instance, in Fig. 1 instruction I1 depends on the result computed by I0, and I2 does not have to wait for I0 or I1. This allows I2 to be executed in parallel to I0. Since a multiplication can take much longer than an addition, the execution of I2 could have finished before I0 has finished and I1 has started. In this case R1 is updated with the result of I2 before I1 is started. In this example the execution of I1 would read the wrong value from register R1. This is a data hazard (write after read) and has to be handled properly. Tomasulo's algorithm is designed to avoid such problems.

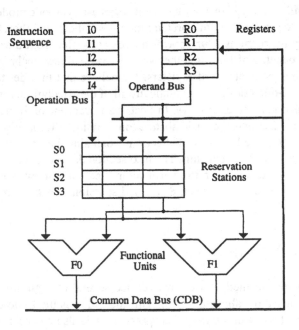

Fig. 2. A model of an implementation of Tomasulo's Algorithm.

The main idea behind Tomasulo's algorithm is to *dispatch* instructions from the instruction sequence into a pool of *reservation stations* (Fig. 2). From a reservation station it schedules an instruction for execution on an unoccupied functional unit as soon as all operands are available. To avoid data hazards two additional mechanisms are needed. First, in addition to the operator, the operands are also stored in the reservation station. This avoids write after read hazards that are explained in the previous paragraph. If the value of a register is not yet computed, the register contains a *tag* that points to the reservation station that will produce this value. Since the value of an operand may not be available while dispatching the instruction into the reservation station, the tag of the appropriate register is used instead of the actual value. Read after write hazards are now handled by updating the operands in the reservation stations together with the registers when results are written back.

In the actual hardware implementations of Tomasulo's algorithm tags and values are usually stored separately, and a special flag controls whether the value or the tag should be used. In our model tags and values share the same memory for efficiency. Figure 2 is a high level floorplan of our model. It is similar to the OOO unit described in [10]. It consists of a set of registers (the *register file*), a pool of reservation stations, and several functional units. The main difference is that reservation stations are not associated with specific functional units. In our paper we build a pool of reservation stations instead. The Pentium Pro[TM] [9] microprocessor also has a pool of reservation stations. A similar model was used in [17].

3 Basic Abstraction Techniques

Symbolic model checking techniques [3, 16] have proven to be of great value for the verification of reactive systems. They have made it possible to verify a large number of practical examples, often with enormous state spaces. The techniques are automatic and can generate counterexamples when a system fails to satisfy its specification. A variety of powerful model checkers are now available, and they are becoming widely used in industry. Because of the power of symbolic model checking techniques we decided to investigate whether they could be used in combination with uninterpreted function symbols to verify OOO designs.

The advantage of model checking over theorem proving [8, 19] is that it is much more automatic. But even with the use of symbolic methods like BDDs there is a limit on the size of the models that can be handled. We will show that the direct application of symbolic model checking to the verification of OOO designs is infeasible. To reduce the size of the model enough for symbolic techniques to be applicable, powerful abstraction techniques are needed. In this section we consider three different encodings of a microprocessor. The first two turn out to be impractical, since they do not provide enough reduction of the state space. We introduce the third representation that makes the use of model checking feasible and allows us to verify non-trivial OOO designs.

3.1 Unabstracted Representation

An OOO unit of a microprocessor can be described as a finite state machine. However, the number of states will be very large. Assume that the microprocessor has r registers of width w, s reservation stations, and f functional units. Each reservation station has to be able to save the contents of two registers (one for each operand of the instruction).

Each register contains w bits. This leads to a lower bound of $w \cdot (r + 2 \cdot s)$ on the number of bits n needed for encoding a state. For modern microprocessors we can assume $w \geq 32$, $r \geq 16$, $f \geq 6$, $s \geq 2 \cdot f = 12$ and derive $n \geq 32 \cdot (16 + 2 \cdot 12) = 960$. This number of state bits is clearly out of the scope of all currently available model checkers.

3.2 The Brute Force Approach

The most obvious abstraction is to use only a small number of bits for each register. This abstraction technique is described in [24, 11] and minimizes the size of registers (and reservation stations) containing data. While, in general, this is a conservative abstraction, it assumes *data independence*. Since the functional units operate on the data and the verification task is to compare the results of these operations, we can not make this assumption.

Burch & Dill ([4]) use *uninterpreted functions* to overcome this problem. In their approach the operations become uninterpreted function symbols and the data is represented by *terms* over these symbols. For their model of a microprocessor the verification task amounts to checking the equivalence of two deeply nested terms. A special decision procedure for quantifier-free first order logic with equality [4] is needed for this purpose. The original paper [4] only considered a single scalar pipeline. This work was extended to a superscalar in-order microprocessor in [2]. However, his project required a non-trivial amount of human interaction and manual modification of the model. The decision procedure of Burch & Dill can not make use of highly effective symbolic model checking techniques and is unable to handle OOO properly, as noted in [21].

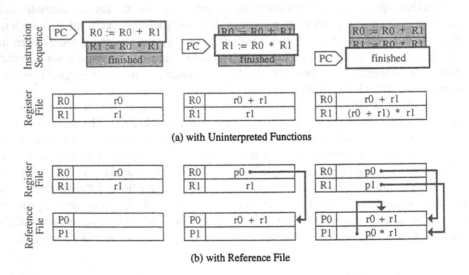

Fig. 3. Execution trace of a sequential machine.

We start with the same basic idea as [4]. The model of the microprocessor does not compute concrete values. It only manipulates symbolic terms made of constants and uninterpreted function symbols. This is explained in Fig. 3(a) where a symbolic execution trace of a sequential microprocessor is shown. The processor has two registers and the instruction sequence consists of two instructions. The registers R0 and R1 contain the initial symbolic values r0 and r1 respectively. The first instruction adds these two values and stores the symbolic result 'r0 + r1' into register R0. Note that '+' is treated as an uninterpreted function with no actual meaning. In particular, we can not assume commutativity or associativity of these functions. The second instruction computes the product of the result and the initial value r1 of register R1. The final symbolic result '(r0 + r1) * r1' is generated and stored in register R1.

In general, we associate a unique constant with the initial value of each register. Since the number of registers is finite, the number of constants is also finite. Thus, if we execute a finite number of instructions, the number of terms that occur during this symbolic execution will also be finite. Consequently, the processor model will be finite and can, in principle, be represented in a finite state model checker. A *direct encoding* of all possible terms would achieve our goal of combining uninterpreted function symbols with symbolic model checking. However, we will show that the number of bits needed to represent these terms grows exponentially with the number of instructions i. This makes this approach infeasible.

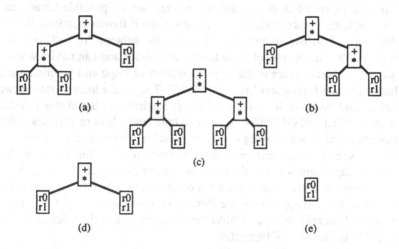

Fig. 4. All possible terms for 2 registers and 2 instructions. Every node in these graphs consists of two possible labellings. So each graph represents $2^{\#nodes}$ possible terms.

First, consider the example in Fig. 3(a). There are two constants and two uninterpreted function symbols of arity two ('+' and '*'). To count the number of possible terms consider the graphs (a) – (e) in Fig. 4. From (e) we get two terms of height 1 and 8 terms of height 2 from (d). The graphs (a) and (b) represent all $64 = 32 + 32$ terms of height 3 where one of the root children has a height of 1. The remaining graph (c) includes all the terms with the maximal number of nodes. These make up additional 128 terms. All together we have 202 different terms, and it takes 8 bits to encode one term.

In general, any term occurring in the symbolic execution will have a height no larger than $i + 1$. The height of a term is defined as the height of its syntax tree or, equivalently, the maximal number of nested function applications plus one. As a rough lower bound on the number of all possible terms we use a lower bound on the number of terms with maximal height and maximal number of subterms (e.g. Fig. 4(c)). There are at least as many terms as there are combinations of constants at the leaves. Because the tree has 2^i leaves and each leaf is one of r constants, a lower bound will be r^{2^i}.

Since this is doubly exponential, the number of bits needed in a binary encoding of that domain would grow exponentially with the number of instructions. As an example consider the case where $r = 4$, $s = 4$ and $i = 5$ (see Section 7). In a binary encoding we have to use at least $\log_2 4^{2^5} = 2^6 = 64$ bits for each register and other locations where a data value can be stored. With four registers and four reservation stations the number of state bits would be at least $64 \cdot (4 + 2 \cdot 4) = 768$. Note that 64 bits is often as big as the width of a register in a concrete model.

3.3 The Reference File

While the brute force approach of direct encoding of all possible terms is not feasible, it is important to note that in one execution trace not all possible terms can occur. Moreover, the same terms or subterms are *referenced* at different locations. For instance, in the final state of the execution trace in Fig. 3(a) the subterm 'r0 + r1' occurs both in register R0 and R1. In this model it has to be stored twice and can not be shared.

A similar problem occurs in the implementation of logic and functional programming languages like Prolog and Lisp [23, 14, 15]. They use a *heap* to store newly generated terms and thus enable *structure sharing*. Registers in the abstract machine for these languages (e.g. WAM [23]) only contain constant values or pointers to the heap. This prevents unnecessary copying and allows sharing of common subterms.

We use a special data structure, called *reference file*, similar to a heap. This is a much more compact encoding of the terms occurring during an execution. Each entry of the reference file contains an application of an uninterpreted function symbol. Each operand of the function application is either an initial value of a register or a pointer to another entry of the reference file. Unlike the heap, the size of the reference file is finite and is equal to the number of instructions i.

As an example, Figure 3(b) shows the execution of the same instruction sequence as in Fig. 3(a). Now all terms are represented with a reference file. After the first instruction the entry P0 of the reference file stores the corresponding function symbol ('+') together with its operands. In our case the operands are the constants 'r0' and 'r1'. Instead of the whole term ('r0 + r1') only the pointer ('p0') is written into the destination register R0. This result is further used by the second instruction. In the entry P1 of the reference file allocated for this instruction the first operand is again stored as a pointer ('p0') without being expanded. Finally, the pointer 'p1' is written to the destination register R1 of the second instruction.

Compared to the first execution trace in Fig. 3(a), the difference is that the registers do not contain terms anymore. Symbolic constants ('r0' and 'r1') or pointers to the reference file ('p0' and 'p1') are stored in the registers. When a functional unit finishes a computation, a new entry is allocated in the reference file for a newly generated term, and the registers are updated with pointers to it. Note that the terms occurring in the first run can easily be restored from the reference file by expanding the pointers.

Now, we can calculate an upper bound on the number of bits for the representation of the reference file. Each entry has to store a function symbol and two operands, where an operand is a constant or a pointer to the reference file. There are $i + r$ values for each operand and i function symbols. This requires $2 \cdot \log_2(i + r)$ bits to encode the two operands and $\log_2 i$ bits for a function symbol yielding the total of $2 \cdot \log_2(i + r) + \log_2 i$ bits per entry. The entire reference file consists of i entries, and thus, can be encoded with $O(i \cdot \log_2(i + r))$ bits.

We also need to encode the contents of the register file and the reservation stations. There, in addition to data values, we also need to store tags (see Section 2). This requires $\log_2(i + r + s)$ bits, since we have s different tag values. Each reservation station contains a busy bit, two operands and an opcode. This takes $s \cdot (1 + 2 \cdot \log_2(i + r + s) + \log_2 i)$ bits for all s reservation stations. We also need $r \cdot \log_2(i + r + s)$ bits for r registers. Putting it all together, our model of Tomasulo's algorithm can be encoded with $O((i + r + s) \cdot \log_2(i + r + s))$ state bits.

For an example with four registers ($r = 4$), four reservation stations ($s = 4$), and five instructions ($i = 5$) the exact number of bits is 16 for the register file, 48 bits for the reservation stations, and 55 bits for the reference file, giving the total of 119 bits.

In fact, the number of different instructions that have to be considered is $i = s + 1$, since only s instructions can be stored in the reservation stations at any time, and only one new instruction needs to be dispatched (see Section 5). With i instructions at most $3 \cdot i$ registers can be involved (2 operands and one destination register for each instruction). Thus, if the actual processor has more than $3 \cdot i$ registers, the unused registers can be eliminated. Hence, the only free parameter is the number of reservation stations s, and the upper bound simplifies to $O(s \cdot \log_2 s)$.

4 Tomasulo's Algorithm with Reference File

Now we can explain how our model of Tomasulo's algorithm actually works (Sect. 2) when combined with the reference file (Sect. 3). The configuration shown in Fig. 5 consists of three registers, three reservation stations, and two functional units. Since we only consider three instructions in this example, the number of entries in the reference file is also three (P0, P1, and P2).

The instruction sequence is the same as in Fig. 3(a) and (b) with one additional instruction. The first instruction places its result into the register R0. This register is then read by the two following instructions, which do not depend on each other. This allows the second and third instructions to be executed in parallel. The result of the first instruction will be written into the entry P0 of the reference file. Similarly, the results of the second and third instructions are written into P1 and P2 respectively. This establishes a one-to-one mapping between the instructions and the entries of the reference file.

When a reservation station becomes available, a new instruction can be dispatched into it. Dispatching means copying the value of the operands and the opcode into this

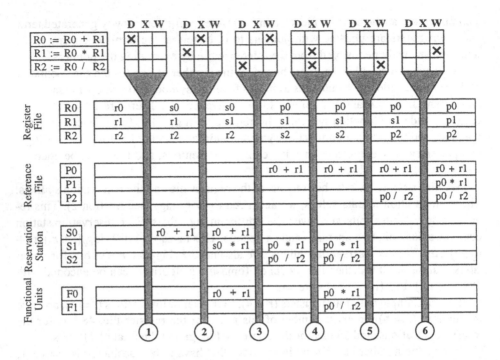

Fig. 5. Execution trace of an out-of-order machine using a Reference File (**D** = Dispatch, **X** = Execute, **W** = Write-Back).

free reservation station. In addition, the tag of the destination register is updated to point to this reservation station. For instance, when the first instruction 'R0 := R0 + R1' in Fig. 5 is dispatched, the value of the operands ('r0' and 'r1') together with the opcode ('+') are copied into S0. At the same time R0 is updated with 's0', a pointer to S0. At the next step the second instruction is dispatched. This is handled similarly to the first step. However, note that a dependency between S1 and S0 is generated by copying 's0' into the first operand field of S1.

Up to this point the reference file was not needed, since no actual computation took place. Now, in step three, the result of the execution of the first instruction is placed on the Common Data Bus (CDB) by the functional unit F0 and written back into register R0. In our abstract model the result of the computation is a new term. It consists of the uninterpreted function symbol '+' applied to the symbolic values 'r0' and 'r1'. The term is represented by a pointer ('p0') to the reference file entry P0 after P0 is updated appropriately. Thus, 'p0' is written back into register R0. At the same time the tag 's0' in S1 is updated with 'p0'.

Beside writing back, the third step involves dispatching the third instruction. This instruction also needs the result of the first instruction in its first operand. When copying the value of the first operand from register R0 to the reservation station, the new value 'p0' on the CDB has to be used instead of the old value 's0' in the register. Not forwarding the value from the CDB seems to be a frequent error while designing an OOO unit based on Tomasulo's algorithm. We made this mistake in a preliminary version of our model and during the verification the model checker produced a counterexample. A similar error is also reported in [17].

In the fourth step the second and third instructions are executed in parallel. After that, in the fifth step, the execution of both instructions is finished. However, only one functional unit at a time can use the CDB to write back. Our model nondeterministically chooses the second functional unit F1 to be the bus master. This illustrates that the reference file can be written out-of-order. Thus, the third entry P2 of the reference file is updated with the result of the third instruction while the second entry P1 remains empty. In the sixth step P1 is updated with the result of the second instruction. Intuitively, the reference file resembles a reorder buffer that never retires instructions. Therefore the final contents of the reference file will also be independent of the execution order.

5 Overall Verification Approach

The verification of an OOO design consists of proving partial and total correctness. The proof of the total correctness is accomplished by showing a liveness property, and in our model can be easily model checked. The partial correctness (safety) is usually stated in terms of sequential execution (Fig. 6). That is, executing an arbitrary sequence of instructions in both the OOO and the sequential machine produces the same result in the register file (for simplicity, we do not consider memory in our example here). A standard approach to deal with instruction sequences is to use induction on the length of the sequence [4]. Below we use s and t to denote states of the OOO machine, and p and q for the sequential machine.

In the induction step we have to compare states of the OOO and the sequential machines. We say that two states s and p are equivalent if *flushing* the OOO machine from s produces the same contents of the register file as in the state p. Flushing means executing all of the pending instructions in the reservation stations without dispatching new ones. Given two equivalent states s and p, the induction step consists of dispatching an arbitrary instruction I in the OOO machine and executing the same instruction in the sequential machine. Then we need to check that the resulting states t and q are again equivalent. This property is often described as a *commutative diagram* [4] (the rear side of the cube in Fig. 7). There are two paths from s to q in Fig. 7. The path where dispatching of I is followed by flushing is called the *OOO path*. The other path where flushing is followed by executing I is called the *sequential path*. Then the diagram commutes if and only if the two paths lead to the same state.

More formally, let $\text{Exec}(p, I)$ be a function that executes an instruction I sequentially in a state p. Similarly, $\text{Disp}(s, I)$ dispatches the instruction I in the state s yielding a new OOO state. And $\text{Flush}(s)$ flushes the OOO machine from the state s and returns the equivalent sequential state. Note that both Disp and Flush may span over several steps of the execution. Moreover, Flush may never terminate if the model has an error and there is a circular dependency among reservation stations. In other words, it may be a partial function. We verify that this function is always total, which implies the total correctness (liveness). We state the partial correctness of the processor as the following theorem:

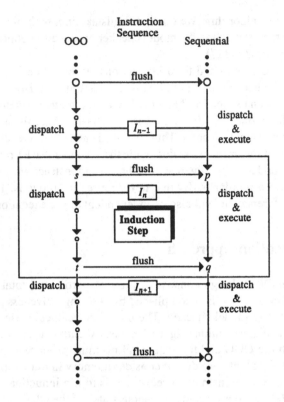

Fig. 6. Induction Principle

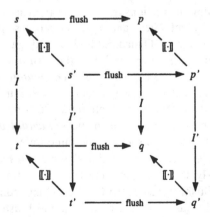

Fig. 7. The commutative diagrams of the concrete and abstract machines and their relationship.

Theorem 1 (Partial Correctness). *For any finite sequence of instructions $I_1, \ldots, I_n$ and two sequences of states $s_0, s_1, \ldots, s_n$ and $p_0, p_1, \ldots, p_n$ of the OOO and sequential machines respectively, such that*

$$\forall i < n. \ s_{i+1} = \text{Disp}(s_i, I_i) \quad and \quad p_{i+1} = \text{Exec}(p_i, I_i),$$

if $p_0 = \text{Flush}(s_0)$, then $p_n = \text{Flush}(s_n)$.

The proof is by induction over the length of the instruction sequence n. The base case of the induction ($n = 0$) is trivial. The inductive step relies on the following lemma:

Lemma 2 (Commutative Diagram). $\forall s. \forall I. \ \text{Exec}(\text{Flush}(s), I) = \text{Flush}(\text{Disp}(s, I))$.

If we prove Lemma 2, then the proof of Theorem 1 is easy. However, proving the lemma is hard and is the main bottleneck of the verification of OOO designs. In part, our method of proving this lemma is similar to the one of Burch & Dill [4]. As in their approach, we use an initial abstraction that replaces the concrete instructions and data values by uninterpreted constants and function symbols. But then we apply symbolic model checking techniques, including the reference file, as opposed to special decision procedures. In the rest of this section we describe the entire method more formally, and give an idea of how it can be encoded and justified formally in a theorem prover.

The first observation is that the number of instructions j stored in the processor at any given time is always finite and usually not very large. Thus, we only have to deal with at most $j + 1$ instructions, including the new instruction being dispatched. Since Tomasulo's algorithm is independent of the actual instructions, we can use $j + 1$ different instruction symbols instead. Registers initially contain distinct symbolic constants. Symbolic instructions are then used as function symbols to construct new terms over these constants. The correspondence between such a symbolic representation and the concrete microprocessor is given by the *semantic function* $[\![\cdot]\!]$. Proving the commutativity of the abstract diagram (the front side of the cube in Fig. 7) and providing an appropriate semantic function $[\![\cdot]\!]$ is sufficient for proving the commutativity of the concrete diagram. This, in turn, completes the proof of the main correctness theorem for arbitrary sequences of instructions. Formally, we need to prove the following two lemmas. We denote abstract (symbolic) values with primed letters (s', I', etc.), and concrete values by corresponding unprimed letters.

Lemma 3 (Abstract Diagram). $\forall s'. \forall I'. \ \text{Exec}(\text{Flush}(s'), I') = \text{Flush}(\text{Disp}(s', I'))$

Lemma 4 (Abstraction). *Let $[\![\cdot]\!]$ be a mapping of symbols from the abstract domain to the actual functions and constants in the implementation domain, such that*

$$\forall I'. \forall s'. \ \text{Disp}([\![s']\!], [\![I']\!]) = [\![\text{Disp}(s', I')]\!] \ and \ \forall I'. \forall p'. \ \text{Exec}([\![p']\!], [\![I']\!]) = [\![\text{Exec}(p', I')]\!].$$

Then, if $\forall s'. \forall I'. \text{Exec}(\text{Flush}(s'), I') = \text{Flush}(\text{Disp}(s', I'))$ in the abstract domain, then

$$\forall s'. \forall I'. \ \text{Exec}(\text{Flush}([\![s']\!]), [\![I']\!]) = \text{Flush}(\text{Disp}([\![s']\!], [\![I']\!])).$$

Figure 7 gives a graphical view of the approach: we prove the commutativity of the front (abstract) side of the cube (Lemma 3) using model checking and provide a semantic function $[\![\cdot]\!]$ that makes the left and right sides commute (conditions in Lemma 4). This implies the commutativity of the rear side, and that is exactly the inductive step for the concrete processor (Lemma 2).

We use the PVS theorem prover to prove Theorem 1 and Lemmas 2 and 4, providing Lemma 3 as an assumption. This is a relatively easy part of the verification. Lemma 2 immediately follows from Lemma 3 and Lemma 4. The proofs of Theorem 1 and Lemma 4 consist only of manipulating the next state functions and relations without expanding them, and thus, can be done only once for an arbitrary configuration of an OOO design. The constraints on the semantic function in Lemma 4 require some information of the actual implementation details and have to be checked for each concrete model. However, the proof of it only needs to establish that the transition relations of both concrete and abstract machines are structurally the same and differ only in the data types (bit vectors versus function and constant symbols). This lemma can be easily automated, or even avoided by creating a preprocessor that compiles the implementation model into the abstract domain.

Lemma 3 is proven automatically using our model checking technique. It is encoded in CTL as an AG formula of the form

$$AG(OOO.\text{finished} \wedge SEQ.\text{finished} \rightarrow$$
$$OOO.\text{regfile} = SEQ.\text{regfile} \wedge OOO.\text{reffile} = SEQ.\text{reffile}) .$$

To prove the liveness of the OOO unit we assume the following fairness constraints: every busy functional unit will eventually produce a result. Then we check the formula

$$AF \; OOO.\text{finished}$$

Note, that this task is much easier than proving the liveness property for the abstract commutative diagram since only one machine is involved.

6 Other Optimizations

With the abstraction techniques presented in previous sections we are able to verify several interesting configurations of OOO designs (Fig. 7). However, to verify designs that are closer to real microprocessors we need to apply other optimizations as well.

(Classical) Symmetry Reduction for Functional Units [7]. In our model the result of only one functional unit can be written back at each clock cycle. We make all the functional units completely general such that they can execute any type of instructions. Hence, by a symmetry argument we can assume that only one functional unit writes back on the CDB. This leaves all the other functional units idle, and they can be eliminated.

Execution of One Machine at a Time. To prove that the diagram in Fig. 7 commutes, we need to traverse both the sequential and the OOO paths (see page 11) and compare the results. In a model checker it amounts to running two independent machines from one and the same initial state. It turns out that starting one machine only

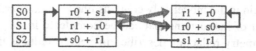

Fig. 8. Reducing the number of states by permuting reservation stations.

after the other one has finished greatly reduces the number of states and the complexity of the verification.

Partial Order on Reservation Stations. The next two reductions are very important and will be described in greater details in the full version of the paper. Here we only sketch them briefly. First, we notice that the OOO model is independent of the order of reservation stations. That is, applying an arbitrary permutation to the reservation stations preserves the transition relation (Fig. 8). Note that all the tags pointing to the permuted reservation stations have to be adjusted accordingly. This allows us to assume that all the tags in the reservation stations point "upwards", that is, to reservations stations with smaller indices. In addition, we can also require that all busy reservation stations have smaller indices than all free ones. This reduces the state space further.

Dynamic Permutation of Reservation Station. This is an extension of the previous reduction technique. Essentially, we construct a *quotient transition relation* under the permutations of reservation stations. In our implementation the quotient model applies an appropriate permutation to the reservation stations after every transition. This maintains the property that all busy RSs are followed by all free ones, as well as the tags are always pointing towards RSs with smaller indices.

AG Splitting and the Cone of Influence Reduction. Our specification of (partial) correctness consists of comparing many state variables of the OOO and the sequential machines. Instead of writing such a specification in one formula (see page 14), we split it into multiple formulas each comparing only a few state variables. This transformation is called *AG splitting*. Although ultimately we have to check all of the state variables, we can do it in multiple runs of the model checker. Each split formula depends on a small number of variables and another technique called the *cone of influence reduction* may eliminate some of the state variables from the model. In particular, notice that in our model we only write to the reference file and never read it. Thus, no state variable depends on any of the entries in the reference file. If we split our specification to compare one entry of the reference file at a time, then all the other entries can be removed from the model. A similar technique was used in [17].

7 Experimental Results

We conducted a set of experiments for several configurations of our model of Tomasulo's algorithm using an enhanced version of the SMV model checker ([16]). To make the description of our model more general the model was described in the M4 macro language. The M4 script can generate SMV code for an arbitrary configuration of the OOO unit. The parameters to the script are the number of functional units, reservation stations, registers and instruction symbols.

In all configurations the model only contains one Common Data Bus (CDB). This restricts the number of instructions that can be completed in one clock cycle to one. Therefore, the symmetry reduction described in Sect. 6 reduces the number of functional units to one. That is, proving the correctness of the system with only one functional unit implies the correctness of the same system with an arbitrary number of functional units. Nevertheless, we included several functional units in some of the configurations of our model to study how the complexity of the verification grows with the number of functional units. The table in Fig. 7 summarizes the results for a number of different configurations of the machine.

Fig. 9. Experimental results for different OOO machine configurations.

Regs	RSs	FUs	Instr.	Time (sec.)	States	BDD Nodes $\cdot 10^6$	Iter.
3	2	1	3	94	$5.4 \cdot 10^{14}$	0.7	7
3	3	1	4	240	$3.2 \cdot 10^{15}$	1.78	9
3	4	1	5	1480	$2.0 \cdot 10^{26}$	6.59	11
4	4	1	5	3362	$1.5 \cdot 10^{28}$	15.9	11
5	4	1	5	3941	$8.0 \cdot 10^{30}$	14.3	11
6	4	1	5	8427	$5.1 \cdot 10^{31}$	21.3	11
3	3	2	4	1116	$3.5 \cdot 10^{17}$	7.27	10
3	3	3	4	3380	$4.8 \cdot 10^{19}$	23.1	10
3	4	2	5	> 41505	$> 1 \cdot 10^{28}$	> 28.4	> 1

The actual implementation consists of an OOO machine and a sequential machine that start from one and the same state. In this state the dependencies between the reservation stations respect the order of their indices as described in Sect. 6. The sequential machine flushes all of the instructions from reservation stations and then executes a new instruction sequentially. This implements the sequential path of the commutative diagram (see page 11). The OOO machine dispatches the same new instruction as soon as one of the reservation stations becomes available, and then flushes the state. After both machines have finished their computation, the final values of the register and reference files are compared.

This finishes the main part of the partial correctness proof. In addition, to make sure that the application of the symmetry reductions from Sect. 6 is correct, we have to show that a certain invariant holds while executing both machines. The invariant states that the dependency graph for the reservation stations is acyclic. It is very important to check this invariant in order to prove the system correct. For example, one of our models had a subtle error that did not show up in the verification of the main correctness formula. The error happened when dispatching a new instruction in the OOO implementation occurred at the same cycle as writing back to a register, whose result was used by the instruction. The tag of that register had not been yet updated, and thus, was copied to the reservation station. At the next cycle, however, the register already had a new value, but the operands in the reservation station still maintained an old tag pointing to an empty RS. This tag is never replaced later, since the write back has already occurred, and this reservation station will never be ready to execute (see Sect. 4 for a more detailed description of this kind of error). This makes the safety property on page 14 vacuously true. We were able to catch this error only by checking that all tags point to busy reservation stations with smaller indices.

8 Conclusion

We have presented a set of new model checking techniques that combine symbolic model checking with the idea of uninterpreted function symbols. This allows us to prove more complex designs than with standard symbolic model checking or decision procedures for uninterpreted functions alone. Compared to straightforward symbolic model checking we are able to handle arbitrarily complex functional units since we can abstract from their actual implementation. On the other hand, we can use the broad spectrum of powerful symbolic model checking techniques, whereas decision procedures for uninterpreted functions can not be easily combined with symbolic model checking.

The key improvement is a new compact encoding of data terms that requires only $O(s \cdot \log_2 s)$ bits for an OOO design with s reservation stations as opposed to an exponential number of bits for a straightforward encoding.

In addition to that, we have developed a number of different transformations that dramatically reduce the size of the set of reachable states and increase the scale of the designs that we are able to handle. Most of the transformations can be formulated as symmetry reductions [7] and we want to investigate to what extent these reductions can be automated.

The technique includes a theorem proving part using PVS where we check that we did not overlook any important detail. First, we prove a general Theorem 1 together with Lemma 4. Both have to be proven only once, and can be applied directly to any configuration of an OOO design, for which Lemma 3 holds. This Lemma is actually proven by the SMV model checker.

We applied our technique to a set of non-trivial configurations of an OOO design based on Tomasulo's algorithm. We plan to extend it to handle additional hardware structures. In particular, we have already tried some preliminary experiments with memory operations (load and store instructions). In these experiments we assumed that the execution of load and store instructions is always in-order. A more realistic model would enable some degree of OOO execution. In this scenario an address might be the result of a previously executed instruction and must therefore be represented by a symbolic term. Since two addresses represented as terms might actually be the same even if the terms differ, a more involved analysis is required. We are also going to implement the reorder buffer and support for precise exception handling. Finally, it is very important to combine our method with compositional approaches like those of McMillan's [17] and incremental flushing of Skakkebæk et al. [21].

References

[1] R. E. Bryant. Graph-based algorithms for boolean function manipulation. *IEEE Transactions on Computers*, 35(8):677–691, 1986.

[2] J. R. Burch. Techniques for verifying superscalar microprocessors. In *33rd Design Automation Conference (DAC'96)*, pages 552–557, 1996.

[3] J. R. Burch, E. M. Clarke, and K. L. McMillan. Symbolic model checking: 10^{20} states and beyond. *Information and Computation*, 98:142–170, 1992.

[4] J. R. Burch and D. L. Dill. Automatic verification of pipelined microprocessor control. In D. L. Dill, editor, *CAV'94*, volume 818 of *LNCS*. Springer-Verlag, 1994.

[5] E. Clarke and E. A. Emerson. Design and synthesis of synchronization skeletons using branching time temporal logic. In *Proceedings of the IBM Workshop on Logics of Programs*, volume 131 of *LNCS*, pages 52–71. Springer-Verlag, 1981.

[6] E. M. Clarke, E. A. Emerson, and A. P. Sistla. Automatic verification of finite-state concurrent systems using temporal logic specifications. *ACM Transactions on Programming Languages and Systems*, 8(2):244–263, 1986.

[7] E. M. Clarke and S. Jha. Symmetry and induction in model checking. Number 1000 in LNCS. Springer-Verlag, 1995.

[8] W. Damm and A. Pnueli. Verifying out-of-order executions. In D. Probst, editor, *CHARME'97*. Chapman & Hall, 1997. To appear.

[9] L. Gwennap. Intel's P6 uses decoupled superscalar design. *Microprocessor Report*, 9(2):9–15, 1995.

[10] J. Hennessy and D. Patterson. *Computer Architecture: A Quantitative Approach*. Morgan Kaufmann Publishers, 1996.

[11] R. Hojati and R. K. Brayton. Automatic datapath abstraction of hardware systems. In *CAV'95*. Springer-Verlag, 1995.

[12] R. Hosabettu, M. Srivas, and G. Gopalakrishnan. Decomposing the proof of correctness of pipelined microprocessors. In Hu and Vardi [13], pages 122–134.

[13] Alan J. Hu and Moshe Y. Vardi, editors. *CAV'98*, number 1427 in LNCS, 1998.

[14] S. L. Peyton Jones. *The Implementation of Functional Programming Languages*. Prentice-Hall, 1987.

[15] Peter M. Kogge. *The Architecture of Symbolic Computers*. McGraw-Hill, 1991.

[16] K. L. McMillan. *Symbolic Model Checking: An Approach to the State Explosion Problem*. Kluwer Academic Publishers, 1993.

[17] K. L. McMillan. Verification of an implementation of tomasulo's algorithm by compositional model checking. In Hu and Vardi [13], pages 110–121.

[18] K. Sajid, A. Goel, H. Zhou, A. Aziz, and V. Singhal. BDD based procedures for a theory of equality with uninterpreted functions. In Hu and Vardi [13], pages 244–255.

[19] J. Sawada and W. A. Hunt. Processor verification with precise exceptions and speculative execution. In Hu and Vardi [13], pages 135–146.

[20] N. Shankar, S. Owre, and J. M. Rushby. *PVS Tutorial*. Computer Science Laboratory, SRI International, 1993.

[21] J. U. Skakkebæk, R. B. Jones, and D. L. Dill. Formal verification of out-of-order execution using incremental flushing. In Hu and Vardi [13], pages 98–109.

[22] M. N. Velev and R. E. Bryant. Bit-level abstraction in the verification of pipelined microprocessors by correspondence checking. 1998. Submitted for publication.

[23] D. H. D. Warren. An abstract prolog instruction set. Tech. Note 309, SRI, 1983.

[24] P. Wolper. Expressing interesting properties of programs in propositional temporal logic. In *Proceedings of the 13th annual ACM Symposium on Principles of Programming Languages (POPL'86)*, pages 184–193. ACM, 1986.

Formally Verifying Data and Control
with Weak Reachability Invariants

Jeffrey Su, David L. Dill, and Jens U. Skakkebæk

Computer Systems Laboratory,
Stanford University, Stanford, CA 94305, USA
Phone: (650) 725-9046, Fax: (650) 725-6949
E-mail: {xsu,dill,jus}@cs.stanford.edu

Abstract. Existing formal verification methods do not handle systems that combine state machines and data paths very well. Model checking deals with finite-
state machines efficiently, but model checking full designs is infeasible because
of the large amount of state in the data path. Theorem-proving methods may
be effective for verifying data path operations, but verifying the control requires
finding and proving inductive invariants that characterize the reachable states of
the system.

We present a new approach to verification of systems that combine control FSMs
and data path operations. Invariants are specified only for a small set of control
states, called *clean states*, where the invariants are especially simple. We avoid the
need to specify the invariants for the unclean states by symbolically simulating
over all paths to find the possible next clean states.

The set of all paths from one clean state to the next is represented by a regular
expression, which is extracted from the control FSMs. The number of paths is
infinite only if the regular expression contains stars. The method uses a heuristic
to generalize the symbolic state to cover all of the paths of the starred expression.
We have implemented a prototype tool for guiding an existing symbolic simulator
and verification tool and used it successfully to prove properties of the Instruction
Fetch Unit of TORCH, a superscalar microprocessor designed at Stanford. With
much less effort, we were able to find all the bugs in the unit that were found
earlier by manually strengthening the invariants.

1 Introduction

Existing formal verification methods do not handle systems that combine finite-state
machines (FSMs) and data paths very well. Model checking [6, 5, 4] the full design is
infeasible because of the large amount of state in the data path. Verifying the control
FSMs in isolation is difficult, because specifying them independently is difficult – the
design requirements are usually stated as properties of the data path, not the FSMs
themselves. The specification of the control is that it causes the data path property to be
satisfied. Abstracting the data path to reduce the amount of state is sometimes possible,

but it is subtle and may require changes in the control that introduce false errors or cause true errors to be missed.

Theorem-proving methods require finding and proving inductive invariants that characterize the reachable states of the system. This is not necessarily difficult for a system that is implementing an algorithm (e.g. a floating point unit). However, when a design has significant control complexity, finding invariants is primarily a tedious manual trial-and-error process.

One path to a solution to these problems would be to find ways to reduce the effort to find inductive invariants in these designs, through automation or methodology. [1] Although the problem of automatic invariant discovery has been studied over the years, there is not yet a complete solution to the problem [13, 11, 7, 3, 18, 2, 1]. In particular most of the work seems not to be applicable to register transfer level (RTL) hardware designs. Most current designs are described at RTL using a hardware description language (HDL) such as Verilog or VHDL, and are then manually or automatically synthesized.

Some of the invariants that are needed in a proof are *historyless* properties, by which we mean that they are provable with no assumptions about the previous state of the system. Equivalently, a historyless property is true of every state that has at least one predecessor in a state transition graph of the system behavior. The concept can be extended to include properties that hold for all states with at least one k-predecessor, where a k-predecessor is a state from which there is a path of length k to the state satisfying the invariants. Of course, for a historyless property to be an invariant, the initial state must satisfy the property. In RTL designs, historyless invariants are surprisingly useful, because they capture some important properties of data propagating through acyclic chains of registers. Also, multi-phase designs (where alternating layers of registers are clocked on different phases of a single clock) tend to lead to historyless invariants that relate the contents of consecutive latches which are clocked in different phases. The discovery and use of historyless invariants in RTL designs was explored in this conference in 1996 [24]. The discovery of historyless properties is also a component of the work cited above for finding invariants in software and protocol descriptions.

This paper attacks the invariant problem in another, complementary, way, by trying to simplify the problem. Examination of a number of designs has revealed a general tendency that can be exploited. Many systems can be thought of as processing a sequence of transactions, where processing a transaction involves a sequence of steps. When the system is not processing a transaction, we say it is in a *clean state*. This paper is based on the observation that *the invariants that needed for the clean states are much simpler than for the other states*. The reason for this is simple: much of the complexity of inductive invariants stems from capturing the bookkeeping that happens during the processing of a transaction.

The partial solution proposed here is to identify the clean states of the system and specify their invariants. These invariants are proved by symbolically simulating along every path from each clean state q to the next clean state q', and showing that if the invariant held in q, it will hold in each q' no matter what path was taken from q to q'.

---

[1] It is important to distinguish between the difficulties of *finding inductive invariants* vs. *proving inductive invariants*. In general, finding the invariants is much more difficult than proving them after they have been found.

The paths between the clean states are described using regular expressions.

The most serious technical difficulty is that there can be an infinite number of paths from q to q', because of cycles of unclean states along the path. However, in some systems at least, these cycles are simple wait loops, so it can be shown that paths that go around the cycle any number of times are equivalent to those that go around zero or one times (these ideas are made more precise below).

Viewed at the level of abstraction of the previous paragraph, there is little new about this approach. Indeed, it is very similar to very early work on program verification, especially the inductive assertions method of Floyd [9], which cuts all cycles in a program flow graph, then finds assertions that hold at the end of the cycle if they hold at the beginning. King specifically used symbolic simulation was to derive invariants [14]. Symbolic simulation along paths between major states has also applied to formal verification of microprograms [8, 15, 16]. The idea of using regular expressions to represent all possible execution paths comes directly from Tarjan [25], who suggested using regular algebra for program flow analysis.

However, RTL hardware design is quite different from sequential program and microprogram verification. To a programmer, RTL designs would appear to be very low-level. Control flow is encoded into one or several FSMs which are separated from the data path. Second, symbolic simulation of even one step results in a huge expression for the symbolic state, since hundreds of state variables may be updated simultaneously. In contrast, a single step in a sequential program or microprogram would typically be a small number of assignments to variables.

While the approach comes out of a tradition of program verification and analysis, these ideas have not previously been applied to RTL designs, however. The reason for this is probably that synthesizable HDL descriptions do not express control flow in the same way as sequential programs. Instead, FSM controllers are defined which are separate from the data path. The method proposed here extracts the regular expressions from the FSMs in the design, not the syntactic structure of the HDL. The other new insight is that, in many cases, finding an invariant around a loop *between clean states* is simple, because the loop often represents a wait state.

These results are preliminary. The proofs still require more effort than one would hope, the invariants are still large (but much smaller than without the method), and it has only been evaluated on one real design. However, it is a new approach that appears to have the potential to be a practical verification method for some designs that are difficult or impossible by other methods.

A simple example

A very simple example is depicted in Figure 1, which is used to make some of the above discussion more concrete. The example consists of three registers, controlled by a small state machine. Periodically, the *new_data* input to the state machine goes high, and, in the next cycle, a new value is loaded into *source*. The state machine then waits for a *ready* signal indicating that the new value can be transferred to *dest*. Then, half the data in *source* is copied to *middle* and the state machine enters state T_1; in the next cycle, that value is copied to *dest*; simultaneously, the remaining data is copied from

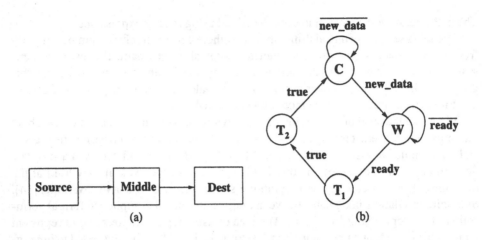

Fig. 1. (a) Data Path (b) Control FSM

source to *middle* and the state machine enters T_2. In the next cycle, the remaining data is copied from *middle* to *dest* and the state machine returns to state C.

Suppose the system is specified to have the property that, whenever it is in state C, the contents of *source* and *dest* registers are the same. Proving this by induction would require providing an inductive invariant (which is actually the induction hypothesis), which must be preserved by all state transitions. The inductive invariant would need to include properties in addition to the basic requirement: When in state T_1, the contents of *middle* equals the lower half of *source*, and when in T_2, the contents of *middle* equals the upper half of *source* and the contents of *dest* equals the lower half of *source*. Intuitively, the inductive invariant needs to track data movement step-by-step in its journey from *source* to *destination*.

This type of phenomenon occurs very frequently, but the systems are obviously more complex than this example, as are the additional properties required of the inductive invariants. In real memory systems, for instance, memory values may be loaded in many cycles and pass through several intermediate registers while being packed together in the right form. An invariant must relate all the intermediate registers with the memory and cache for all the different modes of execution. Hence, in practice, finding invariants is by far the most time-consuming task when verifying with a theorem prover.

The verification of a property only in clean states is illustrated in Figure 2.

Fig. 2. Verifying a property P only in the clean states.

The symbolic simulator operates on *symbolic states*, which map state variables to logical expressions. For each clean state, the simulator is presented with a symbolic state that initializes all storage elements to distinct symbolic constants. Also, a sequence of logical conditions on the data path and inputs must be satisfied for a particular path to be followed through the FSM. The conjunction of these conditions is called the *path constraint*. The result of symbolically simulating over a path is a new symbolic state and a path constraint.

To prove the invariant, it is then necessary to show that for each clean state that, if the path constraint is satisfied and the invariant holds on the initial symbolic state, it also holds for the symbolic states reached by symbolically simulating over all paths. The set of all paths from one clean state to the next is represented by a regular expression, which is extracted from the control FSMs. There number of paths is infinite only if the regular expression contains stars. The method uses a heuristic to generalize the symbolic state to cover all of the paths of the starred expression. From another viewpoint, the method compresses sequences of steps through the state graph so that there is only a single composite step from each clean state to the next. The composite step is computed by symbolically simulating along multi-step paths.

In the example of Figure 1, we would define one clean state, C. Only the initial invariant, that the *source* and *dest* registers are equal when in state C, would need to be proved, but it would need to be proved for the symbolic states yielded by simulating over all paths from C back to C.

This method has been used to prove invariants for the Instruction Fetch Unit of TORCH [22, 21], a superscalar microprocessor designed at Stanford. The same bugs were found as in an earlier effort [23], but with a major reduction in effort.

2 The Verification Method

Extracting regular expressions for control paths

Regular expressions are used because they make it easy to identify and handle cycles in the FSMs. Hence, the first step of the method is to obtain regular expressions describing all paths between the clean states. This requires extracting the state machine controllers in the design, constructing a single product machine (called "the FSM," below), locating the clean states within the FSM, and, for each clean state, deriving a regular expression describing the control paths to the next clean states.

Currently, all of these steps are manual, although everything can be done by well-known algorithms, except extraction of the FSMs from the HDL description. FSM extraction can probably be done automatically in many cases; however, for the designs we are considering, manual extraction by the designer is not difficult.

The FSM is a high-level finite state machine. The outputs are not modelled, since they are not relevant for invariant checking. The transitions are labelled with logical formulas, which are written in a quantifier-free fragment of first-order logic which includes Boolean signals and operators, uninterpreted functions, equality, bitvectors, arithmetic, and arrays. The logical formulas can include individual Boolean signals (which appear as propositional symbols), or predicates on the data path signals (e.g., "$r_1 = r_2$" to represent the output of a comparator between two registers).

The alphabet of the regular language consists of the set of input values to the FSM. Sets of inputs are represented as logical formulas (taken directly from the FSM).

More formally, a regular expression consists of

- The empty string, ϵ;
- A Boolean formula over the input signals to the FSM;
- A concatenation of two regular expressions, $(\alpha \cdot \beta)$;
- A union of two regular expressions, $(\alpha + \beta)$;
- Kleene closure of a regular expression, α^*.

In our examples, we assume that $*$ has higher precedence than $\cdot$, which has higher precedence than $+$, and drop parenthesis accordingly. In the Lisp implementation, the regular expressions are actually represented using Lisp syntax.

For every clean state, we can construct from the original FSM a finite automaton describing the set of all paths of non-zero length from that clean state to another clean state that do not have clean states except at the beginning and end. The regular expression of all input sequences accepted by this finite automaton can be computed by standard algorithms from finite automaton theory (see [10], for example).

In the example of Figure 1, the desired regular expression is:

$$\overline{new\_data} + new\_data \cdot \overline{ready}^* \cdot ready \cdot True \cdot True$$

In this case, the Boolean combinations are all single signals or their complements (indicated by the overlining). In general, of course, a single "symbol" in the regular expression may be a more complex Boolean expression.

Guiding symbolic simulation

The symbolic simulator simulates a high-level netlist (HLN), which is a graph structure representing a digital circuit. The vertices of the graph are circuit elements, such as adders, Boolean gates, registers, and memories. The edges in the graph represent arrays of wires. A *symbolic state* is a map from state variables (registers and memories) to logical expressions representing the symbolic values of the state variables. Given an HLN, a symbolic state S, and logical formulas for the circuit inputs, the symbolic simulator returns a symbolic state representing the updated values of the state variables after one clock cycle of execution of the circuit.

Symbolically simulating along the paths of a regular expression is called *path simulation*. Path simulation operates on pairs $\langle S, P \rangle$, where S is a symbolic state and P is a conjunction of Boolean formulas, called a *path constraint*. Given an HLN, a pair $\langle S, P \rangle$, path simulation produces a finite set of pairs $\{\langle S', P' \rangle\}$, where P' is the conjunction of P with additional path constraints.

For each clean state, we start with a symbolic state S that assigns the correct constants to the FSM state variables and distinct symbolic constants to all other state variables. Path simulation along the regular expression from the clean state to every next clean state yields a set of $\langle S', P' \rangle$ pairs representing the symbolic states and path constraints when the next clean states are reached.

If I is a Boolean formula on the state variables, $I(S)$ represents the formula obtained by substituting for each state variable in I the corresponding logical formula from S. To prove that I is an invariant, we must show that $P' \wedge I(S) \Rightarrow I(S')$ for each pair produced by the path simulation. The assumption that P' holds is justified, since the FSM could only have followed the path if all the conditions in P' were satisfied.

At times, it is necessary or desirable to approximate a set of pairs with a single pair. We say $\langle S', P' \rangle$ *approximates* $\langle S'', P'' \rangle$ if the validity of $P' \wedge I(S) \Rightarrow I(S')$ is a sufficient condition for the validity of $P'' \wedge I(S) \Rightarrow I(S'')$. The approximation is *conservative*: The approximation may cause the proof of a valid invariant to fail, but will never allow an invalid invariant to be proved.

A simple approximation is used, called the *merge* of a set of pairs into a single pair. If both states in the pair map a state variable to the same logical formula, the merged state maps it to the same formula; otherwise, the merged state maps the state variable to a *fresh symbolic constant*, which is a named constant that has not previously appeared in a symbolic state or path predicate. The merge of two pairs $\langle S, P \rangle$ and $\langle S', P' \rangle$ is a pair $\langle S'', P'' \rangle$, where S'' is the merge of S and S', and P'' is the disjunction of P and P'. Since validity is, by definition, truth in all interpretations, the fresh variables are implicitly universally quantified, so this approximation satisfies the definition above.

Path simulation is guided by the recursive structure of the regular expression. Henceforth, "simulate means "symbolic simulate."

Concatenation Concatenation is handled very simply: to simulate the paths in $\alpha \cdot \beta$, starting with a simulation state $\langle S, P \rangle$, first simulate α from $\langle S, P \rangle$ to obtain a simulation state $\langle S', P \wedge P_\alpha \rangle$, then simulate β from $(S', P \wedge P_\alpha)$ to obtain a symbolic state $\langle S'', P \wedge P_\alpha \wedge P_\beta \rangle$.

Union There are two approaches used to simulate $\alpha + \beta$ from pair $\langle S, P \rangle$. The most obvious approach is to simulate α and β separately, yielding two symbolic states. Further simulation would be performed from these states separately. An invariant could be proved by collecting the *set* of symbolic states after an entire simulation, and checking the invariant for each state in the set. The problem with this approach is that it may redo the same work many times, because the symbolic states will be very similar.

The second approach is to compute the *merge* of the two end states. This approximation seems crude, but (surprisingly), works very well for verifying the TORCH Instruction Fetch Unit, described in Section 4, and greatly reduces the complexity of verification. However, note that this may lead to false errors in other designs.

Currently, the choice of which method to use for unions is manual. The regular expression is split into separate expressions which are simulated separately to give several pairs. Pairs are merged for union operations within the individual expressions.

Repetition Obviously, one of the major problems is that there are an infinite number of paths when the regular expression contains stars. The star operator is handled by merging the results of simulating the starred expression a small number of times, to find a symbolic state that subsumes the symbolic states that would be computed by exactly simulating all possible numbers of iterations. This approach is similar to that used with MDGs by Zhou et al. [26].

The basic method is to repeatedly simulate the expression inside the loop, merging the result with the previous result until the symbolic state is identical to the result from the previous iteration, modulo renaming of the fresh variables. The pair $\langle S^*, P^* \rangle$ resulting from this process approximates all the pairs that would have resulted from simulating each of the paths represented by the cycle.

In some cases, this loop generalization is too conservative, resulting in false negatives. For instance, in the TORCH memory system design some pipelined registers take several cycles to reach a stable state when the FSM traverses the loop. If these registers are generalized early, the method will propagate the fresh variables to other state variables, causing them to be generalized unnecessarily. To get a better loop approximation, the user may direct the simulator to traverse the loop several times before the simulator performs the generalization.

3 Verification Process

The the verification process is illustrated in Figure 3. Given a description of the implementation in synthesizable Verilog, a translator converts the Verilog description into an HLN description. A HLN machine corresponds to a Verilog module and consists of a set of input names, output names, a set of type declarations which defines all input, output, and local variables, and a set of behavior descriptions.

The HLN description is fed to an interactive guidance tool. The guidance tool takes the HLN and the regular expression extracted from the FSM. It then simulates along the paths described by the regular expression, with user interaction to determine how to merge states in if-then-else constructs and around loops. Before the simulation, a symbolic state and a symbolic input set are created. The symbolic simulator then uses this state, input set, and the next state transition function to repeatedly execute the implementation and produce the next symbolic state. The simulator in turn calls a decision procedure which is used to simplify the state and used to check equality between states of separate execution paths. The end result is a set of simulation states, one for each simulation path.

A proof obligation is then created for each simulation state. Let $\langle S_i', P_i' \rangle$ be the simulation state resulting from simulation path number i, starting from $\langle S, True \rangle$. The proof obligation to prove correctness invariant I is then:

$$(P_i' \wedge I(S)) \Rightarrow I(S_i')$$

If the proof obligations generated for all simulation paths are valid and I is also true in the initial state, then I is an invariant in all reachable clean states.

The logical formula is then fed to SVC (the Stanford Validity Checker) for checking. SVC is a decision procedure for quantifier-free first-order logic and uses an algorithm similar to the algorithms by Shostak [20, 19] and Nelson-Oppen [17]. The input Boolean formula to SVC can contain Boolean operators, uninterpreted functions and interpreted functions, and distinct constants such as the Boolean truth and bit constants.

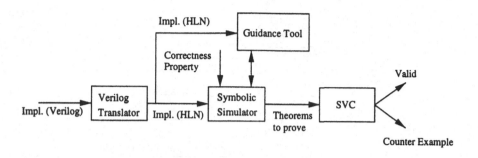

Fig. 3. The verification process.

It may also include the finite bitvectors and records used to model the state of the hardware. SVC will either return with "Valid", or a counter-example. As SVC is used here, the former indicates that the invariant holds, and the latter indicates that the invariant is wrong, or one of the approximation steps has lost a critical constraint.

4 TORCH

We have applied the approach to verifying the correctness of the Instruction Fetch Unit in the TORCH microprocessor. The TORCH design was created by Horowitz's group at Stanford University from 1991-1992 and later optimized. It was constructed for research into microprocessor architectures and has not been fabricated. TORCH is an extension of the MIPS R2000/3000 design [12], which is a 32 bit instruction architecture with a five stage pipeline. It has been simulated (nonsymbolically) extensively, although not to the same degree as in an industrial setting where the resources available for simulation are far greater. The Verilog source code of the TORCH is publicly available at http://www-flash.stanford.edu/torch/.

The TORCH architecture is sketched in Figure 4. TORCH extends the MIPS architecture with some extra optimizing features. It includes *two* asymmetric pipelines with dual issue and dual retirement. To hold various status bits introduced by compiler optimizations, the 32 bit instructions are extended with an extra byte, making the instructions 40 bit wide. For debugging purposes, TORCH can run in a special MIPS compatible mode, where the optimizations are turned off (and the extra byte ignored).

We have previously verified properties of the RTL description of the Instruction Fetch Unit (IFU) [23]. The IFU consists of four modules: an instruction cache (ICache), the IFetch Data Path, the IFetch Control, and the PC Unit Data Path. The description of the IFU consists of 1700 lines of Verilog and uses various common library routines that total approximately 300 additional lines. There are 60 bits of control state (including state machines and random registers with control information); 566 bits of the PC and saved PCs/next PCs, etc.; 1200 bits of explicit data registers. The cache and memory are modelled as unbounded arrays, but each memory word is 64 bits wide and each cache line is 320 bits wide. In addition, each cache line has a 24 bit tag. A block diagram appears in Figure 5.

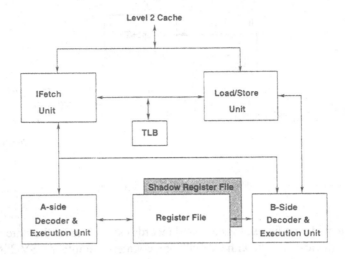

Fig. 4. A diagrammatic overview of the TORCH architecture.

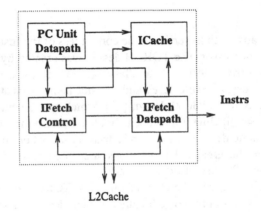

Fig. 5. The Instruction Fetch Unit (IFU).

The PC Unit Data Path maintains a program counter (PC) and calculates the next PC based on input from the surrounding modules, in particular the decode/execute module. The IFetch Data Path and Control return the instruction corresponding to a given PC. The PC is looked up in the ICache by matching the PC with the ICache tags. If there is a match, called a *hit*, the instruction is returned. Otherwise, there is a *miss* and the IFetch Control will output a stall signal and initiate communication with the main memory (in actuality, the ICache communicates with a level 2 cache, but the model here merges the level 2 cache and memory into a single unit) to fill the 8 instruction cache lines (8 * (32 MIPS bits + 8 bits status information) = 320 bits). Once its request is being serviced, the IFU receives two 32 bit words per cycle in 5 cycles, in total 10 words. The first two words contain the eight status bytes, and the last 8 words are the MIPS instructions. Each instruction is matched with its status byte and stored in the cache line. The cache has 1,024 cache lines. Following this, a *refetch* occurs, the stall signal is lowered, and the instruction is provided on the interface.

The unpacking and matching of MIPS instructions with status bytes is carried out to provide compatibility with the MIPS architecture. Information is stored in 32 bit words in memory. The unpacking is of course turned off when TORCH runs in MIPS mode and only 8 words are loaded from memory during a cache miss.

5 Verifying TORCH

The correctness invariant is:

> *For every instruction location, if the location is registered in a valid cache line, the contents of the cache line are the same as the contents of the line in memory.*

The wording of this invariant is intentionally vague. "Contents of" hides the fact that the true invariant is almost a page long, because the format of the data in the cache is different from the format in memory. As the data is transferred, an extra byte of status information is appended to every 32 bit word in the cache. So, comparing the contents of the cache and memory requires extracting the appropriate 32 bit fields before comparing them with the memory. Writing this expression is a bit tedious, but not intellectually difficult.

There is an FSM in the IFetch control logic and also one in the memory module. The two state machines together implement the data transfer protocol on each side of the memory bus. Each controls the data transfer on its side and keeps track of the progress of the current transfer the memory line, i.e., how many pairs of instructions that have been transferred. We form the product of the two FSMs to form a single FSM representing the data transfer. The TORCH mode FSM is illustrated in Figure 6. The FSM for the MIPS mode is similar, but slightly simpler.

The input signals of the FSM are: (1) cache miss (M) which triggers the cache miss process, (2) MIPS mode (P) or TORCH mode $\overline{P}$, (3) the ITLB miss (T) which signals a miss in the TLB, (4) a nondeterministic delay (D) modeling the delay in the main memory or level two cache, and (5) level two cache miss (L). The cache miss M is a predicate on several state variables, and all other signals are inputs to the memory system. The set of input symbols of the FSM is the set of predicates which consists of terms from the following set:

$$\{M, \overline{M}, \overline{M}, P, \overline{P}, T, \overline{T}, L, \overline{L}, D, \overline{D}\}$$

The regular expression corresponding to the FSM is a union of three major paths (see Figure 7). The first path in the expression represents a cache hit. The other two represent a cache miss with and without TLB miss, respectively.

The memory is an array and the cache is an array of records containing three fields: the valid bit (*valid*), the cacheline data (*data*), and the memory address (*addr*) of this

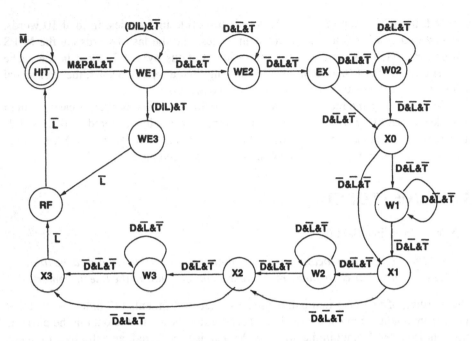

Fig. 6. The product of the two memory system FSMs for the TORCH mode. For clarity we have left out $\overline{P}$ (indicating TORCH mode) from every transition.

$$
\begin{aligned}
&[\ [\overline{P} \wedge \overline{M}] \\
&+ [(\overline{P} \wedge M \wedge \overline{T}) \cdot (\overline{P} \wedge (D \vee L) \wedge \overline{T})^* \cdot (\overline{P} \wedge (D \vee L) \wedge T) \cdot \overline{P} \cdot (\overline{P} \wedge \overline{L})] \\
&+ [(\overline{P} \wedge M \wedge \overline{T}) \cdot (\overline{P} \wedge (D \vee L) \wedge \overline{T})^* \cdot (\overline{P} \wedge \overline{D} \wedge \overline{L} \wedge \overline{T}) \cdot \\
&\quad (\overline{P} \wedge D \wedge \overline{L} \wedge \overline{T})^* \cdot (\overline{P} \wedge \overline{D} \wedge \overline{L} \wedge \overline{T}) \cdot \\
&\quad ((\overline{P} \wedge \overline{D} \wedge \overline{L} \wedge \overline{T}) + (\overline{P} \wedge D \wedge \overline{L} \wedge \overline{T}) \cdot (\overline{P} \wedge D \wedge \overline{L} \wedge \overline{T})^* \cdot (\overline{P} \wedge \overline{D} \wedge \overline{L} \wedge \overline{T})) \cdot \\
&\quad ((\overline{P} \wedge \overline{D} \wedge \overline{L} \wedge \overline{T}) + (\overline{P} \wedge D \wedge \overline{L} \wedge \overline{T}) \cdot (\overline{P} \wedge D \wedge \overline{L} \wedge \overline{T})^* \cdot (\overline{P} \wedge \overline{D} \wedge \overline{L} \wedge \overline{T})) \cdot \\
&\quad ((\overline{P} \wedge \overline{D} \wedge \overline{L} \wedge \overline{T}) + (\overline{P} \wedge D \wedge \overline{L} \wedge \overline{T}) \cdot (\overline{P} \wedge D \wedge \overline{L} \wedge \overline{T})^* \cdot (\overline{P} \wedge \overline{D} \wedge \overline{L} \wedge \overline{T})) \cdot \\
&\quad ((\overline{P} \wedge \overline{D} \wedge \overline{L} \wedge \overline{T}) + (\overline{P} \wedge D \wedge \overline{L} \wedge \overline{T}) \cdot (\overline{P} \wedge D \wedge \overline{L} \wedge \overline{T})^* \cdot (\overline{P} \wedge \overline{D} \wedge \overline{L} \wedge \overline{T})) \cdot \\
&\quad (\overline{P} \wedge \overline{L} \wedge \overline{T}) \cdot (\overline{P} \wedge \overline{L})\ ]\ ]
\end{aligned}
$$

Fig. 7. The regular expression corresponding to the TORCH mode FSM.

cacheline. The correctness invariant above (that the contents of valid cache lines are the same as the corresponding location in memory) must be proved. The base case of the induction proof is trivial, since no cache lines are valid (the initial state is the state immediately after initialization, and initialization flushes the cache). For the inductive step, is to prove that if the property holds in an clean state S, then it is also valid in the next clean state S'. A small trick is required to deal with the quantifier, since SVC doesn't support quantification: we substitute a fresh variable for the quantifier in the consequent (this is called *Skolemizing*) and manually instantiate the quantifier in the antecedent as necessary.

A prototype guidance tool has been implemented in Lisp and applied it in the verification of the IFU. In the proof, each of the three paths in the regular expression was handled separately. The proof for MIPS mode was similar. The run-times for simulation and subsequent verification are shown Figure 8. The last column in the table lists the possible number of paths after loop generalization. However, unions of paths were collapsed into a single path using the generalization method described above.

Regular Expr.	Mode	Run-time	Memory usage	Possible Number of paths
1	MIPS	00:17	16 MBytes	1
2	MIPS	03:43	52 MBytes	2
3	MIPS	30:22	72 MBytes	108
1	TORCH	00:17	16MBytes	1
2	TORCH	04:05	53 MBytes	2
3	TORCH	84:40	81 MBytes	324

Fig. 8. Combined run-times of the simulation and verification required for each path.

From the table of Figure 8, paths specified by the third and the sixth regular expressions take much more time than the rest of the paths do. This is due to the cost of simulation. Most of the simulation time is spent on finding the loop approximation and merging paths from union or loops, which needs to compare variables in two states. Since the memory system has 86 state variables, the comparison is quite expensive. However, without merging parallel paths, we would have 108 ($108=2^2 \times 3^3$) and 324 ($324=2^2 \times 3^4$) paths (each of the first two stars generating two paths and each union generating three paths) for MIPS 3 and TORCH 3, respectively. If the simulation had not merged these parallel paths, the time for simulating these paths would have been about 108 and 324 times of the listed time for these two regular expression. The proofs were carried out with many fewer strengthening invariants than previously needed. Apart from the property, 12 extra conjuncts were needed to strengthen the invariant.

Two of these simply list the reachable states of the FSM for the phase one and phase two latches of the state bits of the FSM. These invariants are evident from inspection of the HDL source, or could be computed automatically by well-known state enumeration algorithms. There is another invariant that says that the phase one and phase two latches are the next state and current state of the FSM. This invariant is also apparent from the HDL, or could be computed by extracting the next state function. There is also a simple invariant that says that the internal reset low when the input reset is low.

Four of these were simple historyless invariants and were found by manually applying and existing method [24]. Here is a typical one: $\overline{MemStallS1} \Rightarrow (FSMS1 = FSMS2)$. This invariant says that whenever $MemStallS1$ is false, the phase two variable ($FSMS2$) has been overwritten by the phase one variable ($FSMS1$).

Four of the conjuncts (see Figure 9) had to be found by trial and error. Some of these may be historyless properties that could not be found by the (incomplete) technique that was used. There was one unexpected issue in the design that may contribute to the need for these invariants. The FSM does not discover a read miss until the following cycle, when it transitions to reflect the cache miss. The *Hit* state in the FSM thus corresponds both to the hit states in the data path and the first miss state. Since the clean states are

$$TagS2 = PCChainS2r[29:3]$$
$$(FSMS2 = HIT) \wedge IStallS2 \Rightarrow (PC = PCChainS1r)$$
$$(FSMS2 = HIT) \wedge IStallS2 \wedge TORCHMODE \Rightarrow$$
$$(MemAddress = 5 \times PCChaincs2r) \wedge (MemAddress = 5 \times PCChainS1r)$$
$$(((FSMS2 = HIT) \wedge ICacheMiss) \vee FSMS2 \neq HIT) \wedge TORCHMODE$$
$$\Rightarrow (MemAddress = 5 \times PCChainS2r)$$

Fig. 9. Four conjuncts of invariants found by inspection and trial and error. These are logically accurate, but the notation and variable names have been modified for readability.

defined as the *hit* states, this requires the reachability invariant to be inductive on both kinds of states.

This effort uncovered the same bugs as found by the previous method [23]. The first bug is that it writes the tag of a noncacheable instruction into the ICache tag register file. As in the MIPS architecture, TORCH provides for both cacheable and noncacheable memory accesses. When the IFU fetches a noncacheable instruction, it causes a cache miss and sends the request to main memory. When the noncacheable data arrives at the IFU, the requested instruction is passed on to the decode/execute unit. Since the instruction is noncacheable, neither the data nor the tag should be written to the ICache. However, a bug in the implementation of the IFU causes the address tag of noncacheable instructions to be written into the ICache tag, while the noncacheable data is not.

The other two bugs were found from the control logic of the PC tag. The first bug causes the PC tag latch to not keep the current PC tag during an ICache miss. Instead, it incorrectly stores the tag of the immediately following instruction. At the end of the ICache miss, the tag corresponding to the next instruction is written into the ICache tag register file while the ICache data corresponding to the current PC is correctly written as ICache data.

We discovered a second bug immediately after the first one was corrected. Consider two branches B_1 and B_2 with target addresses that have different tags T_1 and T_2 but identical cache line indices. The bug manifests itself at the end of the ICache miss caused by the target T_1 of B_1: After the ICache line has been updated, the IFU issues an internal refetch command to the ICache. This refetch gets the designated instructions and its tag from the ICache, and the ICache tag is compared with the PC tag as in a normal fetch. However, the bug causes the next PC tag to be loaded into the PC tag latch before this comparison is done. Thus, if the new PC tag is different from the old PC tag, this refetch will generate another ICache miss. This second cache miss will cause the same data to be fetched, but will wrongly store target address T_2 of B_2 in the tag register file. Following this, the instruction for the target of B_1 is correctly returned. However, when the instruction for the target T_2 of B_2 is requested, T_2 is already in the target register file, which wrongly causes a hit. As a result, an instruction from the B_1 cacheline is incorrectly returned. As before, this causes TORCH to following a wrong path of execution.

The original effort took 2 person-months. It is hard to estimate the effort required by the new method, since the prototype tools were being developed while doing the verification. Furthermore, the user was familiar with the design of TORCH as well as the invariants. We guess that it would have taken 1-2 person-weeks without previous knowledge of TORCH and the invariants.

6 Discussion

The technique described here is not universal. There are several restrictions and assumptions about the design style:

- There is a simple property to be proved on some subset of the states (which we will designate as the clean states).
- The number of clean states is small.
- The product of the control FSMs is small (no more than hundreds of states).
- Cycles are wait loops, in which state variables do not change.
- The control is not pipelined.

Interestingly, at least some designs fall within these guidelines, yet state enumeration approaches are not easy to apply because of the difficulty of verifying the control FSMs independently of the data path. We are investigating ways to remove and generalize these restrictions.

The method needs to be more automated. The current implementation still involves some manual work, such as extracting the finite state machine and finding the regular expression. From the design in the HDL description, it should be possible to generate the transition function for the control state machine automatically and then generate the regular expressions using the existing algorithm. In addition, there needs to be a tool that finds the historyless invariants automatically.

References

1. S. Bensalem, Y. Lakhnech, and S. Owre. Computing abstractions of infinite state systems compositionally and automatically. In Alan J. Hu and Moshe Y. Vardi, editors, *Computer Aided Verification (CAV)98*, volume 1427 of *Lecture Notes in Computer Science*, pages 319–331, Vancouver, BC, Canada, June/July 1998. Springer-Verlag.
2. Saddek Bensalem, Yassine Lakhnech, and Hassen Saïdi. Powerful techniques for the automatic generation of invariants. In Rajeev Alur and Thomas A. Henzinger, editors, *Computer Aided Verification (CAV)96*, volume 1102 of *Lecture Notes in Computer Science*, pages 323–335, New Brunswick, NJ, July/August 1996. Springer-Verlag.
3. N. S. Bjørner, A. Browne, and Z. Manna. Automatic generation of invariants and intermediate assertions. *Theoretical Computer Science*, 173(1):49–87, February 1997.
4. R. E. Bryant, D. L. Beatty, and C.-J. H. Seger. Formal hardware verification by symbolic ternary trajectory evaluation. In *28th ACM/IEEE Design Automation Conference*, 1991.
5. J. R. Burch, E. M. Clarke, and D. E. Long. Representing circuits more efficiently in symbolic model checking. In *28th ACM/IEEE Design Automation Conference*, 1991.
6. J. R. Burch, E. M. Clarke, K. L. McMillan, and D. L. Dill. Sequential circuit verification using symbolic model checking. In *27th ACM/IEEE Design Automation Conference*, 1990.
7. M. Caplain. Finding invariant assertions for proving programs. In *International Conference on Reliable Software*, pages 165–171, 1975.

8. W. Carter, W. Joyner, and D. Brand. Symbolic simulation for correct machine design. In *16th Design Automation Conference Proceedings (1979)*, pages 280–286, June 1979.

9. R. Floyd. Assigning meaning to programs. In *Proc. Symposium in Applied Mathematics*, volume 19, pages 19–32, 1967.

10. R. W. Floyd. *The Language of machines: an introduction to computability and formal languages*. New York : Computer Science Press, 1994.

11. S. German and B. Wegbreit. A synthesizer of inductive assertions. *IEEE Transactions on Software Enginnering*, 1(1):68–75, March 1975.

12. G. Kane. *MIPS RISC Architecture*. Prentice Hall, 1988.

13. S. Katz and Z. Manna. A heuristic approach to program verification. In *Proceedings: 3rd International Joint Conference on Artificial Intelligence*, pages 500–512, 1976.

14. J King. A program verifier. In *Information Processing 71 Proceedings of the IFIP Congress*, volume 1, pages 234–249, 1972.

15. B. Levy. Microcode verification using sdvs-the method and a case study. In *17th MICRO (1984)*, pages 234–245, 1984.

16. R. Mueller and M. Ruda. Formal methods of microcode verification and synthesis. *IEEE Software*, 3(4):38–48, July 1986.

17. G.E. Nelson and D.C. Oppen. Simplification by cooperating decision procedures. *ACM Transactions on Programming Languages and Systems*, 1(2):245–257, October 1979.

18. Hassen Saïdi and Susanne Graf. Construction of abstract state graphs with PVS. In Orna Grumberg, editor, *Computer Aided Verification (CAV)97*, volume 1254 of *Lecture Notes in Computer Science*, pages 72–83, Haifa, Israel, June 1997. Springer-Verlag.

19. R.E. Shostak. A practical decision procedure for arithmetic with function symbols. *Journal of the ACM*, 26(2):351–360, April 1979.

20. R.E. Shostak. Deciding combinations of theories. Technical Report SRI-CSL-132, Computer Science Laboratory, SRI International, February 1982.

21. M. Smith, M. Horowitz, and M. Lam. Efficient superscalar performance through boosting. In *5th International Conference on Architectural Support for Programming languages and Operating Systems*, pages 248–259, Boston, MA, 1992. IEEE/ACM.

22. M. Smith, M. Lam, and M. Horowitz. Boosting beyond static scheduling in a superscalar processor. In *17th International Symposium on Computer Architecture*, volume 18-2, pages 344–354, Seattle, WA, May 1990. IEEE/ACM.

23. J. Su, L. Arditi, S. Das, J. U. Skakkebæk, and D. L. Dill. Formal verification of the TORCH microprocessor RTL design. Unpublished, 1998.

24. Jeffrey X. Su, David L. Dill, and Clark W. Barrett. Automatic generation of invariants in processor verification. In M. Srivas and A. Camilleri, editors, *Formal Methods in Computer Aided Design (FMCAD)*, volume 1166 of *Lecture Notes in Computer Science*, pages 197–201. Springer-Verlag, November 1996.

25. R. Tarjan. A unify approach to path problems. *Journal of the ACM*, 28(3):577–593, July 1981.

26. Z. Zhou, X. Song, S. Tahar, E. Cerny, F. Corella, and M. Langevin. Formal verification of the island tunnel controller using multiway decision graphs. In M. Srivas and A. Camilleri, editors, *Formal Methods in Computer Aided Design (FMCAD)*, volume 1166, pages 233–247. Springer-Verlag, November 1996.

Generalized Reversible Rules

C. Norris Ip

Cadence Berkeley Laboratories,
Cadence Design Systems, Inc., U.S.A.
ip@cadence.com

Abstract. A generalized notion of reversible rules is presented in this paper to perform state reduction in automatic formal verification. The key idea is that some of the transition rules in a design may be invertible, and therefore, they can be used to collapse subgraphs into abstract states, thereby reducing the state explosion problem.

This paper improves upon previous work to achieve the following goals: 1) the definition of reversible rules is simplified so that it is easy to apply the reduction method in practice; 2) the definition is generalized to allow more reduction in the size of the state graph.

The reduction algorithm can be combined with symmetry reduction techniques, for verification of invariants, deadlock-freedom, and stuttering-invariant temporal properties.

1 Introduction

Formal verification methods that rely on state space exploration (e.g. reachability analysis and model checking) are very effective in catching errors in designs [ZWR+80,BWHB86]. However, such methods suffer from the state explosion problem: the vast number of possibilities cannot be explored within reasonable time and memory.

Many techniques have been developed to tackle this problem. For example, symbolic techniques using BDDs can store a large set of states compactly and manipulate them efficiently, reducing memory usage and verification time to an acceptable level for many (but not all) designs [BCM+90,CBM89]. State reduction techniques, such as partial order reduction [Pel96,Val93,GW94] and symmetry reduction [Eme96], can be used to reduce the number of states that a verification algorithm needs to store and examine. The resulting memory usage and verification time were reduced by more than a few order of magnitudes.

In [ID96], a state reduction method was introduced to further reduce the size of a state graph, using the notion of reversible rules to collapse subgraphs of the state space into abstract states. It is motivated by the following observation:

> If the execution of a transition rule r generates the state q' from another state q, it is often possible to reconstruct q automatically from q'.

A rule with this property is called a *reversible rule*, and the corresponding reconstruction procedure is called its *reversed rule*. This property implies that

q and q' can often be generated from each other, and in such cases, only one of them needs to be stored in memory for verification purposes.

This paper describes a state reduction method based on the same observation. Compared to [ID96], the actual definition of reversible rules in this paper is simplified and generalized. First of all, it has been simplified so that it is easier to apply the state reduction method in practice. On one hand, the detection of many reversible rules and the construction of their corresponding reversed rules can easily be automated. On the other hand, the correctness of user-specified reversible rules and reversed rules can be guaranteed by a simple run-time check.

Furthermore, the definition in [ID96] requires the reversed rules to be able to reconstruct the original states all the times. The new definition allows them to fail sometimes, and therefore, effectively every transition rule in a design may be regarded as a generalized reversible rule, and the usefulness of a reversed rule depends on how often it succeeds in generating the original state. By removing this and other restrictions, more reversible rules can be used simultaneously to generate a smaller reduced state graph. Compared to the algorithm in [ID96], an extra reduction of more than 25% is possible, as shown in Section 7.

The new reduction algorithm has been evaluated using several designs described in the Murφ description language [DDHY92]. It has been combined with symmetry reduction to perform explicit state enumeration, obtaining extra reductions of 40% to 90% in the number of states in the reduced state graphs. Because the algorithm explicitly examines every state in the original state graph at least once, it takes longer to generate the reduced state graph than the time to generate the original state graph. Optimization to reduce the time requirement is being investigated.

The current implementation is sound and complete for verifying invariant and deadlock-freedom. Although not implemented in the current implementation, the resulting reduced state graph is sound and complete for linear time temporal logic, such as LTL formulae. In the domain of branching time temporal logic, it is sound (but not complete) for $\forall$CTL* formulae[1].

2   Background

This section summarizes the background for automatic formal verification using state space exploration. In reachability analysis and model checking, the behavior of a design is captured in a state graph (a Kripke structure), which encodes all possible executions of the design:

Definition 1 (state graph). *A state graph is a quadruple* $\langle Q, Q_0, \Delta, \text{error} \rangle$, *where Q is a set of states, $Q_0 \subset Q$ is a set of initial states, $\text{error} \in Q$ is a unique error state, and $\Delta \subseteq Q \times Q$ is a transition relation such that $q = \text{error}$ whenever $(\text{error}, q) \in \Delta$.*

---

[1] The claim in [ID96] about stuttering-invariant CTL model checking was a mistake, and all instances of "CTL" in the paper should be replaced by "LTL".

The special error state, **error**, is used as the successor whenever an user-specified invariant or an assertion in the description of a design is violated.

For protocols, concurrent programs, and many other application domains[2], a state graph is usually defined implicitly by a set of transition rules, T. Each transition rule is a function to generate a successor from a given state, encoding a specific atomic action in a design. Formally, for all $q_1, q_2 \in Q, (q_1, q_2) \in \Delta$ if and only if there exists $t \in T$ such that $q_2 = t(q_1)$.

In subsequent sections, $G = \langle Q, Q_0, \Delta, \mathbf{error} \rangle$ represents the original state graph of a design; T represents the corresponding set of transition rules; q (with/without subscripts) represents a state in Q; t (with/without subscripts) represents a transition rule in T.

The following definitions define the usual terminologies:

Definition 2 (successor/predecessor). *If $(q, q') \in \Delta$, then q' is a* successor *of q, and q is a* predecessor *of q'.*

Definition 3 (finite path). *A finite sequence of states $q_0, \ldots, q_n$ is called a* finite path *if q_i is a successor of q_{i-1} for all $1 \leq i \leq n$.*

Definition 4 (reachability). *A state q' is* reachable *from another state q if there exists a path $q, \ldots, q'$. A state q' is* reachable *in G if it is reachable from an initial state in Q_0.*

Definition 5 (deadlock state). *A state is a* deadlock *state if it has no successor other than itself.*

We usually denote $(q_1, q_2) \in \Delta$ as $q_1 \longrightarrow q_2$, denote $q_2 = t(q_1)$ as $q_1 \xrightarrow{t} q_2$, and denote $q_n = t_n(\ldots t_1(q_0)\ldots)$ as $q_0 \xrightarrow{t_1 \ldots t_n} q_n$.

Using a simple search algorithm, we can verify whether a deadlock state or error is reachable in a state graph. An on-the-fly algorithm is shown in Figure 1. This algorithm generates and explores new states only when all previous states are known to be error-free. The states are usually stored in a hash table, called *Reached* in the algorithm, so that it can be decided efficiently whether or not a newly-reached state is *old* (has been examined already) or *new* (has not been examined already). New states are stored in a queue of *active states*, called *Unexpanded* (whose successors still need to be generated and examined).

Model checking techniques can be used to verify the correctness of temporal properties, such as CTL* and LTL formulae [Wol87]. These temporal properties usually involves path formulae, which are evaluated over infinite path in the state graph:

Definition 6 (infinite path). *A infinite sequence of states $q_0, \ldots$ is called an* infinite path *if q_i is a successor of q_{i-1} for all $1 \leq i$.*

Temporal property checking is discussed in Section 6.

---

[2] For hardware implementations, each transition rule can be regarded as a state transition function for a particular value assignment of the primary inputs. However, the large number of possible value assignments usually leads to an explosion in the number of transition rules, so a set of symbolic transition relations is typically used in place of the transition rules in these cases.

```
procedure explicit_algorithm()
    Reached = Unexpanded = Q₀;
    while Unexpanded ≠ ∅ do
        remove a state q from Unexpanded;
        generate_next_states(q);
procedure generate_next_states(state q)
    deadlock = TRUE;
    for each transition rule t ∈ T Do
        q' = t(q);
        if q' = error then stop and report error;
        if q' ≠ q then deadlock = FALSE;
        if q' is not in Reached then
            put q' in Reached and Unexpanded;
        if deadlock then stop and report deadlock;
```

Fig. 1. An on-the-fly algorithm for simple error checking: two data structures are maintained, one for storing previously examined states (*Reached*), and one for storing states whose successors have not be generated (*Unexpanded*). A large state graph means that a large memory is needed to store these two data structures.

3 Motivating Examples

This section presents two motivating examples for the reduction technique using reversible rules. They illustrate what a reversible rule is and how it may be used to reduce a state graph.

3.1 Message-Passing Cache Coherence Protocols

Cache coherence is a way of implementing a shared-memory abstraction in a multiprocessor with caches (e.g. [LLG+90]). Using a message-passing communication network, whenever a processor wants to load a cache entry, it sends a request to the memory. The memory handles requests from the processors, and keeps track of which processors have read-only copies or writable copies of the cache entry.

In these protocols, whenever a processor has an invalid cache entry and no outstanding requests (processor state I), and it has two options: to issue a request for a read-only copy (processor state a) or a writable copy (processor state b). These events typically generate subgraphs similar to the ones shown in Figure 2a, contributing to state explosion: in subgraph B, 9 states are generated from two processors in state I; in general, 3^k states are generated from a state with k processors in the processor state I. If we collapse these subgraphs into abstract states, we can obtain a smaller state graph for verification.

The reduction algorithm using reversible rules was developed to do exactly that. The operation in which a processor issues a request and changes its internal state to 'pending' can be regarded as a reversible rule, because the original state

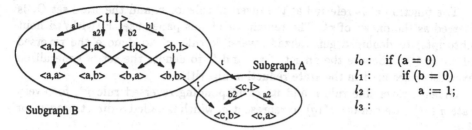

Fig. 2. Motivating examples: a) subgraphs in a typical cache coherence protocol, b) a program to be executed concurrently.

before such an operation can be reconstructed by removing the request and changing the internal state from 'pending' to 'normal'.

3.2 Concurrent Programs

Reversible rules are also found in concurrent programs. For example, consider two processors running the program in Figure 2b concurrently. The transition from program location l_0 to l_1, the transition from program location l_1 to l_2, and the assignment are three atomic actions. If both programs start at local l_0 and both a and b are 0 initially, either processor may changes its program location to l_1 and then to l_2. Therefore, a subgraph with 9 states will be generated[3]. These transitions can be regarded as reversible rules, because the original state can be reconstructed by changing l_2 back to l_1 and changing l_1 back to l_0. Using the reduction algorithm with generalized reversible rules, this subgraph can be collapsed into one abstract state.

4   Generalized Reversible Rules

This section defines reversible rules, and discusses how they can be detected and how the corresponding reversed rules may be constructed.

4.1   Definition

A generalized reversible rule has a very simple definition (cf. [ID96]):

Definition 7 (generalized reversible rule). *A transition rule $r \in T$ is a generalized reversible rule if there exists a function $r^* : Q \to Q$ and a state set $Q_r \subseteq Q$ such that for every state $q \in Q_r$, $r^*(q) \neq q \wedge r(r^*(q)) = q$.*

---

[3] For those readers familiar with other state reduction techniques, note that symmetry reduction may reduce it to 6 states, and partial order reduction may reduce it to 5 states.

The function r^* is referred as the reversed rule for r, and the state set Q_r is referred as the on-set of r^*. The remainder of this paper uses r (with/without subscripts) to denote a generalized reversible rule, r^* to denote the reversed rule of r, Q_r to denote the on-set of r^*, and U to denote the set of generalized reversible rules used in the state reduction algorithm.

Given a reversible rule r and its corresponding reversed rule r^*, for every state $q \in Q_r$, we can use $r^*(q)$ to represent q, which is called a direct progenitor of q:

Definition 8 (direct progenitor). *A state q' is a direct progenitor of another state q if there exists a reversible rule r such that $q \in Q_r$ and $r^*(q) = q'$.*

Definition 9 (progenitor). *A state q' is a progenitor of another state q if $q' = q$ or there exists states $q_0, ..., q_n$ such that $q = q_0$, $q' = q_n$, and q_i is a direct progenitor of q_{i-1} for all $1 \leq i \leq n$.*

In most cases, there are many possible choices for the reversed rule. The usefulness of a generalized reversible rule depends on the choice of the reversed rule and the corresponding on-set. For example, for every rule r, we can arbitrarily choose a state q such that $r(q) \neq q$, and define the reversed rule as $r^*(r(q)) = q$, with Q_r being the set $\{r(q)\}$. However, it would not be very useful for state reduction because it can only be used to convert $r(q)$ to q. In order to achieve a significant state reduction, r^* should be chosen so that Q_r is as large as possible.

The definition of reversible rules implies that a state q can be generated from any of its progenitors by executing a sequence of reversible rules:

Lemma 1 (basic lemma). *If q' is a progenitor of q and $q' \neq q$, there exists $r_0, ..., r_k$ such that $q = r_k(...(r_0(q'))...)$.*

Corollary 1. *If q' is a progenitor of q, q is reachable from q'.*

4.2 Detection

There are two ways of supplying reversible rules for the state reduction algorithm:

- Many reversible rules can be detected by a simple analysis of the description of a design, and their reversed rules can be synthesized from the description.
- The user can provide reversed rules to some transition rules, and the algorithm can automatically check whether a state belongs to the on-set of such rules or not.

This subsection discusses automatic detection of reversible rules in two application domains: message-passing protocols and concurrent programs.

Message-Passing Protocols. In these protocols, components are separated by a message-passing communication network, and they interact by issuing requests and waiting for replies. The *request* phase is typically modeled by reversible rules. For example, in Murφ [DDHY92], a request may be initiated by the following rule:

```
Ruleset i: ProcessIDs Do
  Rule ''request''
    state[i] = normal ==> state[i] := pending; send_request(i);
  End;
End;
```

This segment of a Murφ program defines a set of rules, one for each process. The rule for process i checks whether process i has pending requests. If not, it sets its state to 'pending', and issues a request on the network. By a simple transformation, a reversed rule can be synthesized: the original state can be reconstructed by setting the process state back to 'normal', and remove the request from the network. The on-set of this reversed rule consists of the states with process i in the 'pending' state and a request from process i in the network.

Concurrent Program. In concurrent programs, many reversible rules are usually found in *control flow statements*, such as **if**, **while**, **for**, and **switch**. For example, consider the following **if** statement in C syntax:

$$l_0: \quad \text{if (a == 1000)}$$
$$l_1: \qquad \text{a = 0;}$$
$$\qquad \text{else}$$
$$l_2: \qquad \text{a = a + 1;}$$
$$l_4:$$

Depending on the value of a, a program may change the program location from l_0 to either l_1 or l_2. This action describes a reversible rule, because we can obtain the original state by changing the program location back to l_0. The corresponding on-set is the states in which either the program is at l_1 and a is 1000, or the program is at l_2 and a is not 1000.

On the other hand, an increment operation and other similar operations are also reversible. For example, if the only way to get to the program location l' is by executing $a = a + 1$ at location l, we can reconstruct the original state by changing the location l' back to l and decrementing a by 1.

For these reversible rules, a reversed rule with a large on-set can be automatically constructed, and therefore, the whole process of finding reversible rules and applying them to perform state reduction can be automated.

Other reversible rules may also be specified; however, it is important to understand when such rules would be useful. A reversed rule with a small on-set or a reversed rule that generates unreachable states may not be useful. In these situations, the user needs to decide whether to use these rules to perform state reduction, and if so, to select reversed rules that are effective in reducing the state graph.

5 Reduction Algorithms for Invariants and Deadlock Checking

This section describes the basic reduction algorithm for checking invariants and deadlock-freedom, as shown in Figure 3. The goal of this reduction algorithm is to represent a set of states Q' with a common progenitor q, so that only q needs to be stored in memory. Each iteration of the algorithm expands one progenitor in three steps:

1. **local search:** (procedure "local_search") A localized state enumeration algorithm is used to expand a progenitor q into a set of states Q', by recursively executing the reversible rules.
2. **successor generation:** (procedure "generate_next_state") Successors of the states in Q' are generated by executing the non-reversible transition rules.
3. **progenitor generation:** (procedure θ) Progenitors of these successors are generated by executing the reversed rules.

```
procedure reduction_algorithm()
    Reached = Unexpanded = {  θ(q)  | q ∈ Q₀};
    while Unexpanded ≠ ∅ do
        remove a state s from Unexpanded;
        local_search(s);

procedure local_search(state s)
    Local_Reached = Local_Unexpanded = {s};
    while Local_Unexpanded ≠ ∅ do
        remove a state s from Local_Unexpanded;
        generate_next_states(s);
        for each transition rule r ∈ U do
            s' = r(s);
            if s' = error then stop and report error;
            if s' ≠ s then deadlock = FALSE;
            if s' is not in Local_Reached then
                put s' in Local_Reached and Local_Unexpanded;
        if deadlock then stop and report deadlock;

procedure generate_next_states(state s)
    deadlock = TRUE;
    for each transition rule t ∈ T \ U do
        s' = t(s);
        if s' = error then stop and report error;
        if s' ≠ s then deadlock = FALSE;
        if θ(s') is not in Reached then
            put θ(s') in Reached and Unexpanded;
```

Fig. 3. The basic state reduction algorithm using reversible rules: θ is a procedure to generate a progenitor for the argument, and some of the changes from the on-the-fly algorithm shown in Figure 1 are highlighted an underline.

Consider the state graph in Figure 4. In the first iteration, progenitor a is expanded to the set $\{\,a, b_1, b_2, c\,\}$ in step 1; the successors d and e are generated in step 2; and the state e is converted to its progenitor d in step 3. Therefore, the algorithm only put state d into the set *Reached* and the queue *Unexpanded*. In the second iteration, d is expanded into the set { d, e } in step 1, and the state c is generated step 2. Because a is a progenitor of c and a is already in *Reached*, no new progenitor is generated, and the algorithm terminates with only two states in the reduced state graph.

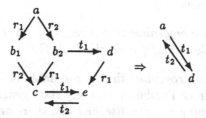

Fig. 4. An example to illustrate the reduction algorithm: a state graph of 6 states is reduced to a state graph of 2 states.

In this reduction algorithm, every edge and every state of the original state graph are explored. Therefore, no run-time reduction is obtained, but the memory usage for the reduced state graph is reduced to 2 states. During the local search, 4 states are temporarily stored in *Local_Reached*, so there is really no memory reduction for this example. However, as shown in Section 9, for larger examples, the memory needed during local search is usually negligible compared to the reduction in the size of the state graph.

On the other hand, a bad implementation of θ may fail to find any progenitor for c in the second iteration (or it may find a progenitor different from a). Therefore, it is possible for the reduced state graph to include c and the successors of c would be examined again in another iteration. In this case, the total run-time would be longer, and the reduction would be smaller.

The remainder of this section presents the proof for the correctness of the algorithm, and discusses how to implement θ.

5.1 Correctness of the Reduction Algorithm

Lemma 2 forms the basis for the soundness of the reduction algorithm:

Lemma 2. *Assuming the reduction algorithm continues the search after detecting an error, if a state is reachable in G, it is examined by the reduction algorithm.*

Proof. If a state q' is reachable from an initial state q in G, there exists a path $q_0, ..., q_n$ such that $q = q_0$ and $q' = q_n$. In the very first line of the state reduction algorithm, $\theta(q_0)$ is generated and stored in *Reached*; therefore, q_0 is generated in step 1 of the iteration expanding $\theta(q_0)$.

Consider a state q_{i-1} that is generated in step 1 of an iteration. If $q_{i-1} \xrightarrow{r} q_i$ for some reversible rule r, unless a deadlock state or an error state has been examined subsequently, q_i is also generated in the same step of the same iteration.

Otherwise, if $q_{i-1} \xrightarrow{t} q_i$, for some non-reversible rule t, q_i is generated in step 2 of the same iteration. $\theta(q_i)$ is then generated in step 3 and q_i will be re-generated in step 1 of the iteration expanding $\theta(q_i)$.

By induction on the path, q_n is examined in step 1 of an iteration.

The completeness of the reduction algorithm depends on whether the progenitors are reachable in G or not:

Lemma 3. *If θ always returns a progenitor that is reachable in G, and a state q is examined by the reduction algorithm, q is reachable in G.*

Proof. Since θ always returns a progenitor that is reachable in G, every state stored in *Reached* is reachable in G. Furthermore, every state generated in step 1 and step 2 is expanded from a progenitor in *Reached*, and therefore, is reachable in G.

The reduction algorithm shown in Figure 3 explores a reduced state graph $\langle Q', Q'_0, \Delta', \textbf{error} \rangle$, such that:

- Q' are the progenitors returned by θ;
- Q'_0 are the progenitors of Q_0 in G.
- $(q, q') \in \Delta'$ if and only if one of the following is true:
  - $q' = \textbf{error}$, and a deadlock state or **error** is generated in step 1 or step 2 when expanding q.
  - q' is a progenitor generated in step 3 when expanding q.

The remainder of this paper refers to such reduced state graphs as $G_{U,\theta}$, where U and θ are, respectively, the set of reversible rules and the progenitor generator in the algorithm.

The correctness of the reduction algorithm for checking invariants and deadlock-freedom is a direct consequence of lemmas 2 and 3:

Theorem 1 (soundness). *If a deadlock state or **error** is reachable in G, **error** is reachable in $G_{U,\theta}$.*

Theorem 2 (completeness). *If θ always return a progenitor that is reachable in G, and **error** is reachable[4] in $G_{U,\theta}$, a deadlock state or **error** is also reachable in G.*

5.2 Implementing a Progenitor Generator

This subsection discusses how to implement the progenitor generator θ.

Constructing a direct progenitor. The following procedure can be used to generate a direct progenitor:

---

[4] A deadlock state may be found in $G_{U,\theta}$, even when a deadlock state is not reachable in G, if a subgraph generated by the reversible rules contains a loop.

```
direct_progenitor(state q)
    for each reversible rule r ∈ U do
        q' = r*(q); if (q' ≠ q ∧ r(q') = q) then return q';
    return q;
```

The guard in the **if** statement checks whether q is in the on-set of r. If so, return q' as a direct progenitor of q. If no direct progenitor exists, it returns q. When a reversed rule r^* is automatically constructed, we can enforce the property that if $r^*(q) \neq q$, we have $r(r^*(q)) = q$. If this is the case, a more efficient procedure can be used to generate a direct progenitor:

```
direct_progenitor(state q)
    for each reversible rule r ∈ U do
        q' = r*(q); if (q' ≠ q) then return q';
    return q;
```

Finding a good progenitor. The amount of reduction we get from the algorithm depends on how a progenitor is selected from the set of progenitors for a given state. In order to obtain maximum reduction in size of the state graph, we need to choose a progenitor that represents as many reachable states as possible. Because a direct progenitor q' of a state q can be used to represent q and all the other states that q represents, q' is at least as good as q. Using this observation, a procedure to find a good progenitor can be defined as:

```
progenitor(state s)
    i = 0; q = s; q' = direct_progenitor(q);
    while (q' ≠ q ∧ i++ < BOUND) { q = q'; q' = direct_progenitor(q); }
    return q';
```

Because the search for a better progenitor is not guaranteed to terminate, a pre-defined bound, BOUND, has to be used to stop the procedure from going into an infinite loop. In order to avoid generating different progenitors for states in a loop, other heuristics may be used: for example, each iteration can check whether the current progenitor has been stored in the reachable state set already. If so, the algorithm can stop the search and return the current progenitor.

Finding a reachable progenitor. In order to have a sound and complete reduction algorithm, the progenitor generator θ must construct a progenitor that is reachable from an initial state in G. A simple heuristic could use the current set of reached states as reference, using the procedure shown as follows:

```
reachable_progenitor(state q)
    q' = progenitor(q);
    if (q' is in Reached) then return q'; else return q;
```

One limitation about this procedure is that if one of the reversed rules always generates non-reachable states, no reduction would be obtained. To eliminate this limitation, a search algorithm could be used to examine more than one progenitor to find one that is reachable. In practice, an exhaustive search is probably too expensive. A compromise between the two extremes could be used.

6 Temporal Property Checking

The reduced state graph obtained by reduction using reversible rules can also be used for temporal property checking. Two variants of temporal logics are linear time temporal logic and branching time temporal logic [Wol87]. As described in this section, for linear time temporal logic, the reduction algorithm can be adapted to generate a reduced state graph that is both sound and complete for verifying stuttering-invariant linear time temporal properties. For branching time temporal logic, the reduction algorithm can be adapted to generate a reduced state graph that is sound for verifying stuttering-invariant ∀CTL* properties. The definition of ∀CTL* formulae can be found in [GL94].

A stuttering-invariant formula is a formula without the next-time operator (c.f. [Lam83]). Informally, such a formula is called stuttering-invariant because it cannot distinguish two similar paths if one of them can be obtained from another via the replacement of a state with a series of states that satisfy the same subset of atomic propositions in the formula.

In order to use the reduction algorithm for verification of stuttering invariant properties, the reversible rules need to satisfy the following restriction:

1. Given a temporal property f with atomic propositions P, for every $p \in P$, q satisfies p if and only if $r(q)$ satisfies p.
2. The subgraphs generated by the reversible rules are acyclic.

The first restriction guarantees that every state represented by the same progenitor satisfies the same subset of atomic propositions in a formula. The second restriction avoids truncating infinite paths in G into finite paths in $G_{U,\theta}$. With these restrictions, the reduced state graph is sound for verification of stuttering-invariant temporal properties:

Theorem 3 (soundness – linear time). *If U satisfies restrictions 1 and 2, and $G_{U,\theta}$ satisfies a stuttering-invariant linear time temporal formula f, G also satisfies f.*

Theorem 4 (soundness – branching time). *If U satisfies restrictions 1 and 2, and $G_{U,\theta}$ satisfies a stuttering-invariant ∀CTL* formula f, G also satisfies f.*

Proof. For any infinite path π in G, we can always find an infinite path π' in $G_{U,\theta}$ that satisfies the same stuttering-invariant path formula:

Because of restriction 2, π cannot have a suffix without non-reversible rules, i.e., it can be written as $q_0 \xrightarrow{r_{(1,1)},\ldots,r_{(1,k_1)},t_1} q_1 \longrightarrow \ldots \xrightarrow{r_{(n,1)},\ldots,r_{(n,k_n)},t_n} q_n \longrightarrow \ldots$

Consider the path segment $q_{i-1} \xrightarrow{r_{(i,1)},\ldots,r_{(i,k_i)},t_i} q_i$ in G. During the construction of $G_{U,\theta}$, step 2 and 3 of the iteration expanding $\theta(q_{i-1})$ generates q_i and $\theta(q_i)$ respectively. Therefore, $\theta(q_{i-1}) \xrightarrow{t_i} \theta(q_i)$ is a path in $G_{U,\theta}$. By induction on the infinite path π, we have $\theta(\pi) = \theta(q_0) \xrightarrow{t_1} \theta(q_1) \longrightarrow \ldots \xrightarrow{t_n} \theta(q_n) \longrightarrow \ldots$ as an infinite path in $G_{U,\theta}$. Because restriction 1 states that the reversible rules preserve the atomic propositions in f, and f is stuttering-invariant, π satisfies f if and only if $\theta(\pi)$ satisfies f. Therefore, if $G_{U,\theta}$ satisfies a linear time temporal formula f, G also satisfies f.

The proof for branching time temporal formula can be completed by a case analysis on the state formula, similar to the one in [GL94].

In both cases, false error reports may be generated because the reduced state graph may contain false paths that do not correspond to actual paths in the original state graph. Another restriction may be used to remove false error reports for linear time temporal properties:

3. Step 3 always uses a progenitor that has been generated in step 2 in the same iteration.

Theorem 5 (completeness — linear time). *If θ always return a progenitor that is reachable in G, U satisfies restrictions 1,2,and 3, and G satisfies a stuttering-invariant linear time temporal formula f, then $G_{U,\theta}$ also satisfies f.*

Proof. Consider an infinite path $\pi = q_0 \xrightarrow{t_1} q_1 \longrightarrow \ldots \xrightarrow{t_n} q_n \longrightarrow \ldots$ in $G_{U,\theta}$. Because of restriction 3, there exists $r_{i,1}, \ldots, r_{i,k_i}, t_k$ such that $q_{i-1} \xrightarrow{r_{i,1},\ldots,r_{i,k_i},t_k} q_i$ in G. Therefore, there exists a path $\pi' = q_0 \xrightarrow{r_{(1,1)},\ldots,r_{(1,k_1)},t_1} q_1 \longrightarrow \ldots \xrightarrow{r_{(n,1)},\ldots,r_{(n,k_n)},t_n} q_n \longrightarrow \ldots$ in G. Because restriction 1 states that the reversible rules preserve the atomic propositions in f, and f is stuttering-invariant, π satisfies f if and only if π' satisfies f. Therefore, if G satisfies f, $G_{U,\theta}$ also satisfies f.

7 Comparison to Related Work

The technique described in this paper is a generalization of the technique presented in [ID96]. The same intuitive notion of reversible rules is used:

If execution of a transition rule r generates the state q' from another state q, it is often possible to reconstruct q automatically from q'.

However, the actual definition and usage of reversible rules in [ID96] requires a few extra properties, such as:

1. the reversed rules always reconstruct the original state correctly;
2. the reversed rules are independent and they eventually generate a unique progenitor;
3. the reversed rules always generate a reachable state.

These properties pose a severe restriction on the use of reversible rules. For example, a nest if statement has two reversible rules that are not independent of each other. In this case, restriction 2 require dropping one of them from the set of reversible rules, whereas the new algorithm can use both and further reduce the size of the state graph.

Instead of imposing these restrictions on the reversed rules, the new algorithm executes reversed rules only on the states in the corresponding on-set, allows a flexible implementation of the progenitor generator θ, and uses a progenitor only when it can be determined to be reachable. Therefore, the new algorithm is easier to use in practice, and it is applicable to a wider class of reversible rules, therefore, a wider class of applications.

[ID96] also describes an optimization that takes advantage of a special property (called the singular property) in message passing protocols, which leads to a significant speed up in the reduction process. Although it has not been implemented with the new algorithm, the same optimization can be adapted for verification of message passing protocol. Optimization to reduce the time requirement in a more generic setting is being investigated.

As shown in Table 1, the new algorithm leads to an extra reduction of about 25% for MCSLOCK1, a distributed lock protocol described in Section 9.

Table 1. Comparison to the algorithm in [ID96]: an extra memory reduction of about 25% is obtained.

MCSLOCK1 with 5 processes	without rev. reduction	old alg. (15 rev. rules)	optimized old alg. (15 rev. rules)	new alg. (40 rev. rules)
Size	508,785	315,679 + 32	315,679 + 4	231,987 + 64
Time	517s	862s	460s	1605s

$\langle n1 \rangle + \langle n2 \rangle$: size of state graph $(n1)$ and maximum size of local searches $(n2)$

There are other works that also remove some of the states from the state set that are kept in the memory. [MK96] describes a technique to remove invisible transitions from a state graph, where a transition is invisible if it preserves the atomic propositions in the formula to be verified. It relies on the bit-state hashing method to preprocess the reachable portion of the state graph, and requires exploration of the reachable portion at least twice. [JJ91] describes a state space caching method to arbitrary remove states from the state set, relying on a depth-first search to guarantee termination. Both of them explore the state graph multiple times to compensate the removal of states, whereas the reduction method using reversible rules uses other reachable states to represent the states that were removed.

8 Reversible Rules and Symmetry

Reduction using reversible rules is orthogonal to the reduction using symmetry. Symmetry can be defined as an automorphism on the state graph and the transition rules [ID93]: States p and q are symmetric if there is an automorphism h such that $h(p) = q$. In order to combine symmetry and reversible rules to obtain an even smaller state graph, the set of reversible rules must also be symmetric:

Definition 10 (symmetric rule set). *Given an automorphism h on G, a set of rules $U \subseteq T$ is symmetric if and only if for every $r \in U$, we have $h(r) \in U$.*

Theorem 6. *If U is both symmetric and reversible in a state graph G, U is also reversible in the quotient graph obtained by symmetry reduction.*

9 Practical Results

The verification results for the following protocols are presented in Table 2:

- *an industrial directory-based cache coherence protocol* (ICCP) [DDHY92]: Five kinds of requests may be initiated by a processor: read, write, promote, uncache, and replacement. They were used as reversible rules to obtain about 90% reduction.
- *Stanford DASH multiprocessor's cache coherence protocol* (DASHC) [LLG$^+$90]: Three kinds of requests may be initiated by a processor: read, write, and write-back. They were used as reversible rules to obtain about 50% reduction.
- *Stanford DASH multiprocessor's lock protocol* (DASHL) [LLG$^+$92]: Two kinds of requests may be initiated by a processor: lock acquire and lock release. They were used as reversible rules to obtain about 40% reduction.
- *two distributed linked-list protocols* (LIST1,LIST2) [Dil95]: In both protocols, two kinds of requests may be initiated by a processor: add-to-list and remove-from-list. They were used as reversible rules to obtain about 50% to 65% reduction.
- *a concurrent program implementing the Peterson mutual-exclusion algorithm for n-processors* (PETERSON) [Pet81]: The algorithm contains one **while** statement and one increment operation. They were used as reversible rules to obtain about 80% reduction.
- *two concurrent programs implementing two distributed lock algorithms* (MCSLOCK1, MCSLOCK2) [MCS91]: MCSLOCK1 contains three **if** statements and two **while** statements, and MCSLOCK2 contains four **if** statements and two **while** statements. They were used as reversible rules to obtain about 50% reduction.

All these reversible rules belong to the category that could be automated (as mentioned in Section 4.2), although current implementation does require the user to specify the reversed rules.

Table 2. Practical results: memory reductions of about 40% to 90% is obtained, with increases in run-time of about 25% to 220%.

	ICCP (p5,v2) (145/35)	LIST1 (p5) (55/10)	LIST2 (p5) (34/10)	DASHC (p4,v2) (335/25)	DASHL (p4) (345/10)
Original size	$\approx 3 \times 10^7$	$\approx 1.3 \times 10^7$	$\approx 3.4 \times 10^7$	$\approx 5 \times 10^6$	$\approx 5.8 \times 10^6$
Size (r)	$\approx 3 \times 10^6$ +44,640	$\approx 3.6 \times 10^6$ +3,840	$\approx 1.6 \times 10^7$ +3,840	$\approx 2.7 \times 10^6$ +2,656	$\approx 3.7 \times 10^6$ +480
Size (s)	211,269	197,547	447,583	133,426	254,790
Size (s/r)	24,269 + 327	69,892 + 32	231,875 + 32	74,687 + 40	159,949+24
Time (s)	1148s	416s	1,470s	2,424s	3,249s
Time (s/r)	2652s	624s	2,359s	3,078s	4,282s

	PETERSON (p6) (54/18)	MCSLOCK1 (p5) (65/40)	MCSLOCK2 (p3) (48/21)
Original size	$\approx 2 \times 10^8$	$\approx 5.5 \times 10^7$	$\approx 3.2 \times 10^6$
Size (r)	$\approx 3.8 \times 10^7$ +23,040	$\approx 2.4 \times 10^7$ +6,780	$\approx 1.5 \times 10^6$ +72
Size (s)	351,682	508,785	544,617
Size (s/r)	72,859 + 32	231,987 + 64	258,757 + 12
Time (s)	314s	517s	288s
Time (s/r)	451s	1,605s	654s

s :   symmetry reduction
r :   reversible rules reduction
$p\langle n \rangle$:   system n processors
$v\langle n \rangle$:   system with n possible values in the cache line
$\langle n1 \rangle / \langle n2 \rangle$: total number of rules ($n1$) and number of reverisble rules ($n2$)
$\langle n1 \rangle + \langle n2 \rangle$: size of state graph ($n1$) and maximum size of local searches ($n2$)

Each system in the list is verified twice; the first time with symmetry reduction, and the second time with both symmetry reduction and reversible rule reduction. Furthermore, the implementation includes an estimator[5] to estimate the size of the original state graph without these reductions. Those estimations are included as references.

In order to get a fair comparison of the actual memory usages, we need to measure any auxiliary data structure that the reduction algorithm uses, i.e.,

---

[5] For symmetry reduction, the use of graph heuristics allows us to estimate efficiently the number of states that are equivalent to a representative. Therefore, the total number of states before reduction can be estimated from the states in the reduced state graph. For reduction using reversible rules, every state in the original state graph is at least examined once in a local search. Therefore, the total number of states examined in all local searches gives us an upper bound of the total number of states before reduction. From experiment of the same protocols with fewer components, the estimates are typically within 10% of the actual sizes.

the memory needed for the local search. Therefore, the maximum number of states examined in a single local search is included in the table to provide an accurate measurement of memory reduction. For example, the memory reduction for ICCP is $1 - (24269 + 327)/211269$, i.e., about 88%. In general, about 40% to 90% memory reduction is obtained, with increases in run-time of about 25% to 220%. The author is currently investigating optimization to reduce the run-time overhead.

The reversible rule reductions in these examples were performed using a progenitor generator that explicitly checks the soundness of the reversed rules, and generates reachable progenitors. Therefore, the results are sound and complete for checking invariants and deadlock-freedom.

10 Conclusion and Future Work

A generalized framework for state reduction using reversible rules is presented in this paper. Using a simplified and generalized definition of reversible rules, it is easy to detect reversible rules and to construct the corresponding reversed rules automatically. The reduction algorithm also includes a run-time check to ensure the correctness of user-specified reversible rules.

The resulting reduction algorithm generates a reduced state graph that is sound and complete for checking invariants and deadlock-freedom. It can also be used for checking stuttering-invariant temporal properties. Reductions from 40% to 90% have been obtained from several message-passing protocols and concurrent programs.

The reduction method using reversible rules are applicable to any design that has reversible rules. The author has found that message-passing protocols and concurrent programs are two application domains where designs usually have a lot of reversible rules. In other domains, reversible rules may be rare to find; it would be interesting to identify other domains that also have designs with a lot of reversible rules.

The author is currently investigating how this technique can be optimized to reduce verification time, and how it can be used in combination with partial order and symbolic model checking.

References

[BCM+90] J.R. Burch, E.M. Clarke, K.L. McMillan, D.L. Dill, and L.J. Hwang. Symbolic model checking: 10^{20} states and beyond. *5th IEEE Symposium on Logic in Computer Science*, 1990.

[BWHB86] J. Billing, M.C. Wilbur-Ham, and M.Y. Bearman. Automated protocol verification. *Protocol Specification, Testing, and Verification, V*, 1986.

[CBM89] Olivier Coudert, Christian Berthet, and Jean Christophe Madre. Verification of synchronous sequential machines based on symbolic execution. *Automatic Verification Methods for Finite State Systems*, 1989.

[DDHY92] David L. Dill, Andreas J. Drexler, Alan J. Hu, and C. Han Yang. Protocol verification as a hardware design aid. *IEEE International Conference on Computer Design: VLSI in Computers and Processors*, pages 522–525, 1992.

[Dil95] David L. Dill. Protocols used in class CS355. *Stanford University*, Spring 1994-1995.

[Eme96] E. Allen Emerson, editor. *Formal Methods in System Design, Special Issue on Symmetry in Automatic Verification*, volume 9(1/2). Kluwer Academic Publishers, August 1996.

[GL94] Orna Grumberg and David E. Long. Model checking and modular verification. *ACM Transaction on Programming Languages and Systems*, 16(3):843–871, May 1994.

[GW94] P. Godefroid and P. Wolper. A partial approach to model checking. *Information and Computation*, 110(2):305–326, May 1994.

[ID93] C. Norris Ip and David L. Dill. Better verification through symmetry. *11th International Symposium on Computer Hardware Description Languages and Their Applications*, pages 87–100, April 1993.

[ID96] C. Norris Ip and David L. Dill. State reduction using reversible rules. *33rd Design Automation Conference*, pages 564–567, June 1996.

[JJ91] C. Jard and Th. Jeron. Bounded-memory algorithms for verification on-the-fly. *3rd Workshop on Computer-Aided Verification*, July 1991.

[Lam83] L. Lamport. What good is temporal logic. *Information Processing 83*, pages 657–668, 1983.

[LLG+90] Daniel Lenoski, James Laudon, Kourosh Gharachorloo, Anoop Gupta, and John Hennessy. The directory-based cache coherence protocol for the DASH multiprocessor. *17th International Symposium on Computer Architecture*, 1990.

[LLG+92] Daniel Lenoski, James Laudon, Kourosh Gharachorloo, Wolf-Dietrich Weber, Anoop Gupta, John Hennessy, Mark Horowitz, and Monica Lam. The Stanford DASH multiprocessor. *Computer*, 25(3), 1992.

[MCS91] John M. Mellor-Crummey and Michael L. Scott. Algorithms for scalable synchronization on shared-memory multiprocessors. *ACM Transactions on Computer Systems*, 9(1), 1991.

[MK96] Hillel Miller and Shmuel Katz. Saving space by fully exploiting invisible transitions. *8th International Conference on Computer-Aided Verification*, pages 336–347, 1996.

[Pel96] D. Peled. Partial order reduction: Model-checking using representatives. *21st International Symposium on Mathematical Foundations of Computer Science*, 1996.

[Pet81] G.L. Peterson. Myths about the mutual exclusion problem. *Information Processing Letters*, 12(3), 1981.

[Val93] A. Valmari. On-the-fly verification with stubborn sets. *5th International Conference on Computer Aided Verification*, pages 397–408, June 1993.

[Wol87] Pierre Wolper. On the relation of programs and computations to models of temporal logic. *Colloquium on Temporal Logic in Specification*, 1987.

[ZWR+80] Pitro Zafiropulo, Colin H. West, Harry Rudin, D.D. Cowan, and Daniel Brand. Towards analyzing and synthesizing protocols. *IEEE Transactions on Communications*, 28(4), April 1980.

An Assume-Guarantee Rule for Checking Simulation*

Thomas A. Henzinger   Shaz Qadeer   Sriram K. Rajamani   Serdar Taşıran

EECS Department, University of California at Berkeley, CA 94720-1770, USA
Email:{tah,shaz,sriramr,serdar}@eecs.berkeley.edu

Abstract. The simulation preorder on state transition systems is widely accepted as a useful notion of refinement, both in its own right and as an efficiently checkable sufficient condition for trace containment. For composite systems, due to the exponential explosion of the state space, there is a need for decomposing a simulation check of the form $P \preceq_s Q$ into simpler simulation checks on the components of P and Q. We present an assume-guarantee rule that enables such a decomposition. To the best of our knowledge, this is the first assume-guarantee rule that applies to a refinement relation different from trace containment. Our rule is circular, and its soundness proof requires induction on trace trees. The proof is constructive: given simulation relations that witness the simulation preorder between corresponding components of P and Q, we provide a procedure for constructing a witness relation for $P \preceq_s Q$. We also extend our assume-guarantee rule to account for fairness assumptions on transition systems.

1 Introduction

In hierarchical verification, we need to check proof obligations of the form $P \preceq Q$, where P and Q are system descriptions and $\preceq$ is a preorder on system descriptions. The assertion $P \preceq Q$ holds if P describes the same system as Q, but possibly on a finer level of detail (or equivalently, Q describes the same system as P but possibly on a coarser level of abstraction). For example, P may be an RTL-level description of a pipelined processor, and Q may be an ISA description of the same processor. The assertion $P \preceq Q$ is therefore variously pronounced as "P implements Q," or "P refines Q," or "Q specifies P," or "Q abstracts P."

Mathematically, a popular choice for the preorder $\preceq$ is *trace containment*. In this case, $P \preceq Q$ asserts that every sequence of inputs and outputs that is possible for P is also possible for Q (at the same time, the more abstract specification Q may allow some traces that are not realized by the more concrete implementation P). While simple and intuitive, trace containment has several

* This research was supported in part by the Office of Naval Research Young Investigator award N00014-95-1-0520, by the National Science Foundation CAREER award CCR-9501708, by the National Science Foundation grant CCR-9504469, by the Defense Advanced Research Projects Agency grant NAG2-1214, by the Army Research Office MURI grant DAAH-04-96-1-0341, and by the Semiconductor Research Corporation contract 97-DC-324.041.

shortcomings. First, it is a practical impossibility to check trace containment automatically for all but the smallest examples, because the check is exponential in the number of states of Q if Q is nondeterministic(as specifications often are). Second, in top-down design, if system description P fleshes out detail that is left open in system description Q, there is a much tighter relation between P and Q than trace containment would indicate; namely, each implementation state of P corresponds to a specification state of Q. This tighter relation is captured mathematically by the notion of a *simulation relation*. Intuitively, Q simulates P iff, starting from the initial states and continuing ad infinitum, every input-output pair of P can be matched by the same input-output pair in Q [Mil71]. Clearly, if Q simulates P, then every trace of P is also a trace of Q. The converse is not true; that is, simulation is a stronger requirement than trace containment. However, it has been said that trace containment without simulation is more often than not due to coincidence rather than systematic design [Kur94].

While trace containment is defined *globally*, for input-output sequences of arbitrary length, simulation is defined *locally*, by considering individual input-output pairs for all states. It is this locality in the definition of simulation that leads to significant advantages. First, if Q is claimed to simulate P, then a witness to this claim can be produced in the form of a relation between states of P and states of Q, and the witness can be efficiently checked for correctness (the check is linear in the number of states of P and Q). Such witness relations are widely used in verification methods and tools, under various names like homomorphisms [Kur94] and refinement mappings [AL91,Lyn96]. Second, even if no witness is available, the existence of a simulation can be checked in polynomial time (the check is quadratic in the number of states of P and Q). The number of states of a system, however, depends exponentially on the size of the system description (note that n boolean variables give rise to 2^n states—this is the *state-explosion problem*). Thus, even algorithms that are linear in the number of states are often infeasible in practice, and techniques have been studied for dividing a given verification task into simpler subtasks.

Compositional techniques for dividing the verification task $P \preceq Q$ into simpler subtasks are guided by the structures of P and Q. If the refinement relation $\preceq$ is interpreted as trace containment, a number of compositional techniques are known. Specifically, if $P = P_1 \| P_2$ and $Q = Q_1 \| Q_2$, then in order to check $P \preceq Q$, it suffices to check both $P_1 \preceq Q_1$ and $P_2 \preceq Q_2$. This *compositional principle* for trace containment is propositionally valid whenever parallel composition corresponds to trace intersection (replace $\|$ by conjunction, and $\preceq$ by implication). Unfortunately, the compositional principle is often not helpful, because P_1 typically refines Q_1 only when constrained by an environment that behaves like P_2, and similarly, P_2 may refine Q_2 only when constrained by an environment that behaves like P_1. Under certain modeling assumptions (namely, nonblocking and finite nondeterminism), the compositional principle can be strengthened to an *assume-guarantee principle* [Sta85,CLM89,GL94,AL95,AH96,McM97]: in order to check $P \preceq Q$, it suffices to check both $P_1 \| Q_2 \preceq Q_1$ and $Q_1 \| P_2 \preceq Q_2$. Three observations about this proof rule are important. First, the rule addresses the issue that the environment of P_1 may have to be suitably constrained in order to implement Q_1, and similarly for P_2. Second, the rule avoids reasoning about the compound implementation $P_1\|P_2$, which typically has the largest of the involved state spaces. Third, unlike the compositional principle, the assume-guarantee principle is circular and therefore not propositionally valid—its proof requires induction on the length of traces.

By contrast to the case of trace containment, if the refinement relation $\preceq$ is interpreted as "is simulated by," then little is known about compositional techniques other than the fact that the compositional principle remains valid whenever parallel composition corresponds to the intersection of trees whose branches are traces (this is because Q simulates P iff every trace tree of P is also a trace tree of Q). In particular, it would be useful to have an assume-guarantee principle for simulation, which, given witnesses for the two subtasks $P_1\|Q_2 \preceq Q_1$ and $Q_1\|P_2 \preceq Q_2$, lets us construct a witness for $P \preceq Q$. In this paper, we show that under the same modeling assumptions under which the assume-guarantee principle is sound for trace containment, it is also sound for simulation. Second, we show how the compound witness can be constructed from the witnesses for the subtasks. Third, we show that in analogy to the case of trace containment, the assume-guarantee principle for simulation can be extended to account for fairness assumptions in system descriptions. As in the case of trace containment [AL95,AH96], the proof of soundness requires the modeling assumption of receptiveness [Dil89].

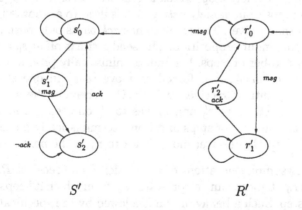

Fig. 1. Specifications of sender and receiver

We illustrate the assume-guarantee rule for simulation using an example. Figure 1 shows Moore-machine specifications for a sender S' and a receiver R' in a communication protocol. Each state is labeled with outputs that are true in that state, and each arc is labeled with conditions on inputs that need to be satisfied for the arc to be taken. A state with no label means that no output propositions are true in that state, and an arc with no label means that there

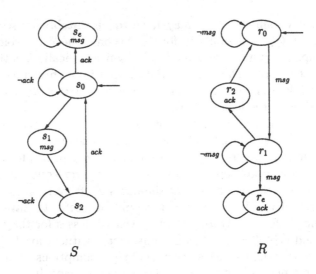

Fig. 2. Implemenmtations of sender and receiver

is no condition on input propositions to take that arc. The initial states s_0' of the sender and r_0' of the receiver are marked using arrows. The sender has one output proposition, namely msg, which is true whenever a message is produced, and one input proposition, namely ack, which is used to acknowledge the receipt of a message by the receiver. The receiver has ack as its only output proposition and msg as its only input proposition. The sender starts off at s_0' and stays there for an arbitrary number of steps. It nondeterministically produces a message by moving to s_1'. Once in s_1', it is forced to move to s_2' in one step. Then, the sender waits in s_2' until it receives an ack. On receiving the ack it goes back to s_0'. The receiver starts at r_0' and moves to r_1' on seeing a msg. It stays in r_1' for an arbitrary number of steps and nondeterministically moves to r_2'. After acknowledging at r_2', the receiver moves back to r_0' in the next step.

Figure 2 shows implementations of the sender S and receiver R. If S receives an ack while at s_0, it goes to an "error-state" s_e, from which it keeps sending messages in every step. Such a behavior is not allowed by the specification S'. Thus, S' does not simulate S. However, after composing with the specification of the receiver, it is seen that $S \parallel R' \preceq_s S'$, because no acknowledgments can be received by S while at s_0. The relation $\Theta_S = \{((s_0, r_0'), s_0'), ((s_1, r_0'), s_1'), ((s_2, r_1'), s_2'),$ $((s_2, r_2'), s_2')\}$ is a witness to this simulation. Similarly, R' does not simulate R, but $S' \parallel R \preceq_s R'$ with the relation $\Theta_R = \{((s_0', r_0), r_0'), ((s_1', r_0), r_0'), ((s_2', r_1), r_1'),$ $((s_2', r_2), r_2')\}$ as witness. In Section 3, we prove that the existence of simulation relations from $S \parallel R'$ to S' and from $S' \parallel R$ to R' is sufficient to conclude the existence of a simulation relation Ω from $S \parallel R$ to $S' \parallel R'$, and give a procedure to construct Ω, from Θ_S and Θ_R. For our example, the witness simulation relation Ω is $\{((s_0, r_0), \langle s_0', r_0' \rangle), ((s_1, r_0), \langle s_1', r_0' \rangle), ((s_2, r_1), \langle s_2', r_1' \rangle), ((s_2, r_2), \langle s_2', r_2' \rangle)\}$.

For simplicity we use Moore machines (with local Streett conditions) to model systems. Our results apply to other nonblocking, finitely nondeterministic, receptive models such as (Fair) Reactive Modules [AH96]. Section 2 defines Moore machines and establishes the connection between trace-tree containment and simulation. In Section 3 we prove the validity of the assume-guarantee rule for the simulation preorder and describe how the compound witness simulation relation is constructed from the witnesses for the components. Section 4 defines simulation on fair Moore machines. The assume-guarantee rule for fair simulation is proven in Section 5.

2 Simulation Relations on Moore Machines

Moore machines. A Moore machine is a tuple $P = \langle S^P, s^P, I^P, O^P, L^P, R^P \rangle$ where

- S^P is the set of states,
- $s^P \in S^P$ is the initial state,
- I^P is the set of input propositions,
- O^P is the set of output propositions disjoint from I^P,
- $L^P : S^P \to \mathcal{P}(O^P)$ is a function that labels each state with the subset of output propositions true in that state, and
- $R^P \subseteq S^P \times \mathcal{P}(I^P) \times S^P$ is the transition relation. We write $R^P(s, i, s')$ as shorthand for $(s, i, s') \in R^P$.

We restrict our attention to Moore machines P satisfying the following two properties:

1. *Nonblocking*: For all $s \in S^P$ and $i \subseteq I^P$, there exists a state τ such that $R^P(s, i, t)$.
2. *Finite nondeterminism*: For all $s \in S^P$, $i \subseteq I^P$, and $o \subseteq O^P$, there are at most a finite number of states τ such that $R^P(s, i, t)$ and $L^P(t) = o$.[1]

Run trees and trace trees. A (finite or infinite) *tree* is a set $\tau \subseteq \mathbb{N}^*$ such that if $xn \in \tau$, for $x \in \mathbb{N}^*$ and $n \in \mathbb{N}$, then $x \in \tau$ and $xm \in \tau$ for all $0 \le m < n$. The elements of τ represent nodes: the empty word ϵ is the root of τ, and for each node x, the nodes of the form xn, for $n \in \mathbb{N}$, are the children of x. The number of children of node x is denoted by $deg(x)$. A tree τ is finite if τ is a finite set. The *depth* of a node $\tau \in x$ is defined inductively as follows: (1) the depth of ϵ is 0, and (2) if the depth of $x \in \tau$ is d, then the depth of xn is $d + 1$. If τ is finite, then the depth of τ is defined as the maximum of depths over all nodes of τ. The nodes of τ with no children are called *leaves* of τ. A *path* ρ of τ is a finite or infinite set $\rho \subseteq \tau$ of nodes that satisfies the following three conditions: (1) $\epsilon \in \rho$, (2) for each

---

[1] For simplicity, we consider Moore machines with single initial states. Our results apply if there are multiple initial states, as long as the initial states satisfy finite nondeterminism: for all $o \subseteq O^P$, there are only a finite number of initial states s such that $L^P(s) = o$.

node $x \in \rho$, there exists at most one $n \in \mathbb{N}$ with $xn \in \rho$, and (3) if $xn \in \rho$, then $x \in \rho$. Given a pair of sets A and B, an $\langle A, B \rangle$-*labeled tree* is a triple $\langle \tau, \lambda, \delta \rangle$, where τ is a tree, $\lambda : \tau \to A$ is a node labeling function that maps each node of τ to an element in A, and $\delta : \tau \times \tau \to B$ is an edge labeling function that maps each edge $\langle x, xn \rangle$ of τ to an element in B. Then, every path $\rho = \{\epsilon, n_0, n_0 n_1, \ldots\}$ of τ generates a sequence $\Gamma(\rho) = \lambda(\epsilon) \cdot \langle \delta(\epsilon, n_0), \lambda(n_0) \rangle \cdot \langle \delta(n_0, n_0 n_1), \lambda(n_0 n_1) \rangle \cdots$ in $A \times (B \times A)^* \cup A \times (B \times A)^\omega$.

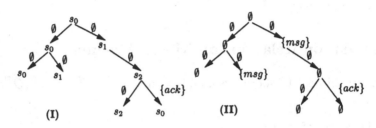

Fig. 3. (I) is an run tree for the Moore machine S in Figure 2, and (II) is the corresponding trace tree

An *run tree* T of P is a $\langle S^P, \mathcal{P}(I^P) \rangle$-labeled tree $\langle \tau, \lambda, \delta \rangle$ such that $\lambda(\epsilon) = s^P$, and for all edges $\langle x, xn \rangle$ we have $R^P(\lambda(x), \delta(x, xn), \lambda(xn))$. Note that for any depth, P has at least one run tree, and possibly many. A *trace tree* $T' = \langle \tau, \lambda', \delta \rangle$ of P is a $\langle \mathcal{P}(O^P), \mathcal{P}(I^P) \rangle$-labeled tree such that there is a run tree of $T = \langle \tau, \lambda, \delta \rangle$ of P, and for every $x \in \tau$ we have $L^P(\lambda(x)) = \lambda'(x)$. For brevity we say $T' = L^P(T)$, and call T a *witness* to T'. See Figure 3 for an example.

Tree containment. Consider two Moore machines $P = \langle S^P, s^P, I^P, O^P, L^P, R^P \rangle$ and $Q = \langle S^Q, s^Q, I^Q, O^Q, L^Q, R^Q \rangle$. We say that Q is *refinable* by P if (1) $O^Q \subseteq O^P$, and (2) $I^Q \subseteq I^P \cup O^P$.

Suppose Q is refinable by P. Let $T = \langle \tau, \lambda, \delta \rangle$ be a trace tree of P. We say that the *projection* of T on Q is a trace tree $[T]_Q = \langle \tau, \lambda', \delta' \rangle$ of Q, such that for all $x \in \tau$ we have $\lambda'(x) = \lambda(x) \cap O^Q$ and for all $x, xn \in \tau$, we have $\delta'(x, xn) = (\delta(x, xn) \cup \lambda(x)) \cap I^Q$. We say that Q *tree contains* P if (1) Q is refinable by P, and (2) for every trace tree T of P, the projection $[T]_Q$ is a trace tree of Q.

Composition. The *composition* of P and Q, denoted by $P \| Q$, exists if $O^P \cap O^Q = \emptyset$, and is defined to be the Moore machine $K = \langle S^K, s^K, I^K, O^K, L^K, R^K \rangle$ where

- $S^K = S^P \times S^Q$,
- $s^K = \langle s^P, s^Q \rangle$,
- $I^K = (I^P \cup I^Q) \setminus (O^P \cup O^Q)$,
- $O^K = O^P \cup O^Q$,
- $L^K(\langle p, q \rangle) = L^P(p) \cup L^Q(q)$ for all $\langle p, q \rangle \in S^K$,

- $R^K(\langle p_1, q_1 \rangle, i, \langle p_2, q_2 \rangle)$ iff $R^P(p_1, (i \cup L^Q(q_1)) \cap I^P, p_2)$ and $R^Q(q_1, (i \cup L^P(p_1)) \cap I^Q, q_2)$ for all $\langle p_1, p_2 \rangle, \langle q_1, q_2 \rangle \in S^K$ and $i \subseteq I^K$.

The branching behavior of a Moore machine is characterized by its set of trace trees. Composition of two Moore machines results in the intersection of the sets of trace trees for the component machines.

Proposition 1. *Consider two Moore machines P and Q such that the composition $P \parallel Q$ exists. Then T is a trace tree of $P \parallel Q$ iff (1) $[T]_P$ is a trace tree of P, and (2) $[T]_Q$ is a trace tree of Q.*

Simulation. Consider two Moore machines P and Q such that Q is refinable by P. A binary relation $\Theta \subseteq S^P \times S^Q$ is said to be a *simulation relation* from P to Q if and only if the following three conditions hold:

1. $(s^P, s^Q) \in \Theta$.
2. For all $(p, q) \in \Theta$, we have $L^P(p) \cap O^Q = L^Q(q)$.
3. For all $(p, q) \in \Theta$ and for all $i \subseteq I^P$ and $\tilde{p} \in S^P$ such that $R^P(p, i, \tilde{p})$, there exists $\tilde{q} \in S^Q$ such that $R^Q(q, (i \cup L^P(p)) \cap I^Q, \tilde{q})$ and $(\tilde{p}, \tilde{q}) \in \Theta$.

If such a relation Θ exists, Q is said to *simulate* P (written as $P \preceq_s Q$) with Θ as the *witnessing* simulation relation. Further, if such Θ exists and $(p, q) \in \Theta$, we say that state q of Q *simulates* state p of P. It is well known that Q simulates P iff in a game of a protagonist playing in Q against an adversary playing in P, the protagonist can match every move of the adversary by moving to a state with the same observation ad infinitum. It is also known that each strategy in such a game corresponds to a trace tree, and consequently, simulation is equivalent to tree containment.

Proposition 2. *Consider two Moore machines P and Q. Then $P \preceq_s Q$ iff Q tree contains P.*

3 Assume-Guarantee Rule for Simulation

Suppose we are given a specification $P' \parallel Q'$ and an implementation $P \parallel Q$, where $P \parallel Q' \preceq_s P'$ and $P' \parallel Q \preceq_s Q'$. Consider a specific state $\langle p, q' \rangle$ of $P \parallel Q'$. Suppose that this state is simulated by state p' of P'. Further, suppose that there exists q such that state $\langle p', q \rangle$ of $P' \parallel Q$ is simulated by state q' of P'. Then, it seems plausible (and indeed it is true, as we show below) that state $\langle p, q \rangle$ of $P \parallel Q$ is simulated by state $\langle p', q' \rangle$ of $P' \parallel Q'$. A difficulty arises when state $\langle p', q \rangle$ of $P' \parallel Q$ is simulated by state q'' of P' that is different from q'. We then examine the state $\langle p, q'' \rangle$ of $P \parallel Q'$ and find a state p'' of P' that simulates it. We continue by finding a state of Q' that simulates state $\langle p'', q \rangle$ of $P' \parallel Q$, etc. In this way, if we reach a cycle that includes p' and q', we are still able to show that state $\langle p, q \rangle$ of $P \parallel Q$ is simulated by state $\langle p', q' \rangle$ of $P' \parallel Q'$. Since our Moore machines have only single initial states, such a cycle should exist for the initial states, satisfying condition 1 for simulation relations. Finite nondeterminism ensures that condition 3 for simulation relations is satisfied as well.

Theorem 1. *Let P, Q, P', Q' be Moore machines such that $P \parallel Q$ and $P' \parallel Q'$ exist. Suppose that $P \parallel Q' \preceq_s P'$ and $P' \parallel Q \preceq_s Q'$ with witnessing simulation relations Θ_P and Θ_Q respectively, and every input of $P' \parallel Q'$ is either an input or output of $P \parallel Q$. Then, we can construct a simulation relation Ω from $P \parallel Q$ to $P' \parallel Q'$.*

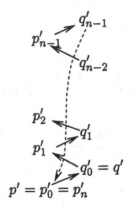

Fig. 4. Figure demonstrating the definition of Ω

Proof. Let $\Omega \subseteq (S^P \times S^Q) \times (S^{P'} \times S^{Q'})$ be defined as follows: $(\langle p, q \rangle, \langle p', q' \rangle) \in \Omega$ iff there exist $p'_0, p'_1, \ldots, p'_n \in S^{P'}$ and $q'_0, q'_1, \ldots, q'_{n-1} \in S^{Q'}$ such that (see Figure 4)

- $p'_0 = p'_n = p'$ and $q'_0 = q'$, and
- for all $0 \le i < n$, we have (1) $(\langle p'_i, q \rangle, q'_i) \in \Theta_Q$, and (2) $(\langle p, q'_i \rangle, p'_{i+1}) \in \Theta_P$.

We need to show that the conditions 1-3 of the definition of a simulation relation are satisfied by Ω. In the following (for 2 and 3), assume that $(\langle p, q \rangle, \langle p', q' \rangle) \in \Omega$, i.e., there exist $p'_0, p'_1, \ldots, p'_n \in S^{P'}$ and $q'_0, q'_1, \ldots, q'_{n-1} \in S^{Q'}$ satisfying the conditions mentioned above.

1. Since Θ_P and Θ_Q are simulation relations, we have $(\langle s^P, s^{Q'} \rangle, s^{P'}) \in \Theta_P$ and $(\langle s^{P'}, s^Q \rangle, s^{Q'}) \in \Theta_Q$. Let $n = 1$, $p'_0 = p'_1 = s^{P'}$, and $q'_0 = s^{Q'}$. Then, $(\langle s^P, s^Q \rangle, \langle s^{P'}, s^{Q'} \rangle) \in \Omega$.

2. Since $(\langle p'_0, q \rangle, q'_0) \in \Theta_Q$ and Θ_Q is a simulation relation, $L^Q(q) \cap O^{Q'} = L^{Q'}(q'_0)$. Since $q'_0 = q'$, we have $L^Q(q) \cap O^{Q'} = L^{Q'}(q')$. Also, $(\langle p, q'_i \rangle, p'_{i+1}) \in \Theta_P$ for all i from 0 to $n-1$. Therefore, $L^P(p) \cap O^{P'} = L^{P'}(p'_{i+1})$ for all i from 0 to n. Since, $p'_n = p'_0 = p'$, we have $L^P(p) \cap O^{P'} = L^{P'}(p')$. Hence, we get $L^{P \parallel Q}(\langle p, q \rangle) \cap O^{P' \parallel Q'} = L^{P' \parallel Q'}(\langle p', q' \rangle)$.

3. Let $R^{P \parallel Q}(\langle p, q \rangle, i, \langle \tilde{p}, \tilde{q} \rangle)$ for some $\langle \tilde{p}, \tilde{q} \rangle$ and $i \in \mathcal{P}(I^{P \parallel Q})$. Define $i' = i \cup L^P(p) \cup L^Q(q)$. Clearly, i' includes all inputs of the machines P, Q, P' and Q'. In the following, whenever we use i' as the input for some Moore machine

K, the intention is to take the projection $i' \cap I^K$ onto the set of inputs of K. We want to show that there exists $(\langle \tilde{p}, \tilde{q} \rangle, \langle \tilde{p}', \tilde{q}' \rangle) \in \Omega$ such that $R^{P' \parallel Q'}(\langle p', q' \rangle, i' \cap I^{P' \parallel Q'}, \langle \tilde{p}', \tilde{q}' \rangle)$.

- Since all machines are nonblocking, we have $R^{P'}(p'_0, i', \tilde{p}'_0)$ for some $\tilde{p}'_0$. Hence, $R^{P' \parallel Q}(\langle p'_0, q \rangle, i', \langle \tilde{p}'_0, \tilde{q} \rangle)$.
- The fact that $(\langle p'_0, q \rangle, q'_0) \in \Theta_Q$ implies that there exists some $\tilde{q}'_0$ such that $R^{Q'}(q'_0, i', \tilde{q}'_0)$ and $(\langle \tilde{p}'_0, \tilde{q} \rangle, \tilde{q}'_0) \in \Theta_Q$.
- $R^{Q'}(q'_0, i', \tilde{q}'_0)$ implies that $R^{P \parallel Q'}(\langle p, q'_0 \rangle, i', \langle \tilde{p}, \tilde{q}'_0 \rangle)$.
- $(\langle p, q'_0 \rangle, p'_1) \in \Theta_P$, therefore, there exists $\tilde{p}'_1$ such that $R^{P'}(p'_1, i', \tilde{p}'_1)$ and $(\langle \tilde{p}, \tilde{q}'_0 \rangle, \tilde{p}'_1) \in \Theta_P$

Repeating this process, we obtain $\tilde{p}'_0, \tilde{p}'_1, \tilde{p}'_2, \ldots$, and $\tilde{q}'_0, \tilde{q}'_1, \tilde{q}'_2, \ldots$ such that

- for all $k \in \mathbb{N}$, we have $R^{P'}(p'_0, i', \tilde{p}'_{kn})$ and $R^{Q'}(q'_0, i', \tilde{q}'_{kn})$, and
- for all $j \geq 0$, we have that $(\langle \tilde{p}'_j, \tilde{q} \rangle, \tilde{q}'_j) \in \Theta_Q$ and $(\langle \tilde{p}, \tilde{q}'_j \rangle, \tilde{p}'_{j+1}) \in \Theta_P$.

Since P' has the finite nondeterminism property, there exist $a, b \in \mathbb{N}$ with $b > a$ such that $\tilde{p}'_{an} = \tilde{p}'_{bn}$. Consider the states $\tilde{p}'_{an}, \tilde{p}'_{an+1}, \ldots, \tilde{p}'_{bn} \in S^{P'}$ and $\tilde{q}'_{an}, \tilde{q}'_{an+1}, \ldots, \tilde{q}'_{bn-1} \in S^{Q'}$. We know that for all $an \leq i < bn$, $(\langle \tilde{p}'_i, \tilde{q} \rangle, \tilde{q}'_i) \in \Theta_Q$ and $(\langle \tilde{p}, \tilde{q}'_i \rangle, \tilde{p}'_{i+1}) \in \Theta_P$. It follows from the definition of Ω that $(\langle \tilde{p}, \tilde{q} \rangle, \langle \tilde{p}'_{an}, \tilde{q}'_{an} \rangle) \in \Omega$. We also have $R^{P' \parallel Q'}(\langle p', q' \rangle, i' \cap I^{P' \parallel Q'}, \langle \tilde{p}'_{an}, \tilde{q}'_{an} \rangle)$. $\square$

4 Simulation Relations on Fair Moore Machines

Fair Moore machines. Consider a Moore machine $P = \langle S^P, s^P, I^P, O^P, L^P, R^P \rangle$. An *run* of P is a finite or infinite sequence $\bar{s} = s_0 \xrightarrow{i_1} s_1 \xrightarrow{i_2} s_2 \xrightarrow{i_3} s_3 \cdots$ such that $s_0 = s^P$ and $R^P(s_k, i_{k+1}, s_{k+1})$ for all $k \geq 0$. The set of all finite runs of P is denoted by Σ^P. Let $\bar{s} = s_0 \xrightarrow{i_1} s_1 \xrightarrow{i_2} s_2 \xrightarrow{i_3} \ldots \xrightarrow{i_{n-1}} s_{n-1}$. Then the length of $\bar{s}$, denoted by $|\bar{s}|$, is n, and the kth prefix for $0 \leq k \leq n$, denoted by $\bar{s}_k$, is $s_0 \xrightarrow{i_1} s_1 \xrightarrow{i_2} s_2 \xrightarrow{i_3} \ldots \xrightarrow{i_{k-1}} s_{k-1}$. If $\bar{s}$ is infinite, then $|\bar{s}|$ is defined to be ω. A *fairness constraint* F^P for P is a function that maps every infinite run of P to the binary set $\{fair, unfair\}$. A *fair Moore machine* $\mathcal{P} = \langle P, F^P \rangle$ consists of a Moore machine P and a fairness constraint F^P for P. A *fair* run of $\mathcal{P}$ is either a finite run of P or an infinite run $\bar{s}$ of P such that $F^P(\bar{s}) = fair$. A *fair run tree* of $\mathcal{P}$ is a run tree $\langle \tau, \lambda, \delta \rangle$ of P such that for every path ρ of τ, the run $\Gamma(\rho)$ of P is a fair run of $\mathcal{P}$. A *fair trace tree* of $\mathcal{P}$ is a trace tree of P that is witnessed by a fair run tree of $\mathcal{P}$.

We restrict ourselves to fair Moore machines $\mathcal{P}$ that are receptive [Dil89], i.e., in an infinite game of the system $\mathcal{P}$ against the environment, no matter what the environment does the system has a strategy to produce a fair run as the outcome of the game. Formally, a *receptiveness strategy* of $\mathcal{P}$ is a function from $\Sigma^P \times I^P$ to S^P. An infinite run $\bar{s} = s_0 \xrightarrow{i_1} s_1 \xrightarrow{i_2} s_2 \xrightarrow{i_3} \ldots$ of P is an *outcome* of the receptiveness strategy σ if for all $k \geq 1$, we have $s_k = \sigma(\bar{s}_k, i_k)$. The fair Moore machine $\mathcal{P}$ is *receptive* if there is a receptiveness strategy σ such that every outcome $\bar{s}$ of σ is a fair run of $\mathcal{P}$.

In the following, we consider two fair Moore machines, $\mathcal{P} = \langle P, F^P \rangle$ and $\mathcal{Q} = \langle Q, F^Q \rangle$.

Fair tree containment and composition. We say that Q *fair-tree contains* P if (1) Q is refinable by P, and (2) for every fair trace tree T of P the projection $[T]_Q$ is a fair trace tree of Q. The *composition* of P and Q, denoted by $P \parallel Q$, exists if $P \parallel Q$ exists, and is defined to be the fair Moore machine $\mathcal{K} = \langle K, F^K \rangle$, where

- $K = P \parallel Q$,
- $F^K(\bar{s}) = fair$ iff both $F^P([\bar{s}]_P) = fair$ and $F^Q([\bar{s}]_Q) = fair$, where $[\bar{s}]_P$ is the projection of $\bar{s}$ on P and $[\bar{s}]_Q$ is the projection of $\bar{s}$ on Q.

Proposition 3. *Consider two fair Moore machines P and Q such that the composition $P \parallel Q$ exists. Then T is a fair trace tree of $P \parallel Q$ iff (1) $[T]_P$ is a fair trace tree of P, and (2) $[T]_Q$ is a fair trace tree of Q.*

Fair simulation. Suppose that Q is refinable by P. Intuitively, Q fairly simulates P [HKR97] if there is a strategy in the simulation game that matches every fair run of P with a fair run of Q. Formally, a *simulation strategy* of Q with respect to P is a partial function from $\Sigma^P \times \Sigma^Q$ to S^Q. If $\bar{s} = s_0 \xrightarrow{i_1} s_1 \xrightarrow{i_2} s_2 \xrightarrow{i_3}$ $\ldots \xrightarrow{i_{n-1}} s_n \in \Sigma^P$, and $\bar{s}' = s_0' \xrightarrow{i_1'} s_1' \xrightarrow{i_2'} s_2' \xrightarrow{i_3'} \ldots \xrightarrow{i_{m-1}'} s_m' \in \Sigma^Q$, then the following three conditions are necessary for $\sigma(\langle \bar{s}, \bar{s}' \rangle)$ to be defined: (1) $n = m + 1$, (2) for all $0 \leq k < n$, we have $L^P(s_k) \cap O^Q = L^Q(s_k')$, and (3) for all $1 \leq k < m$, we have $(L^P(s_{k-1}) \cup i_k) \cap I^Q = i_k'$. It is required that if $\sigma(\langle \bar{s}, \bar{s}' \rangle) = s_n'$, then $L^P(s_n) \cap O^Q = L^Q(s_n')$. Given a finite or infinite run $\bar{s} = s_0 \xrightarrow{i_1} s_1 \xrightarrow{i_2} s_2 \xrightarrow{i_3} \ldots$ of P, the *outcome* $\kappa[\bar{s}]$ of the simulation strategy κ is the finite or infinite run $\bar{s}' = s_0' \xrightarrow{i_1'} s_1' \xrightarrow{i_2'} s_2' \xrightarrow{i_3'} \ldots$ of Q such that (1) $|\kappa[\bar{s}]| = |\bar{s}|$, and (2) for all $k \geq 1$, we have $s_k' = \kappa(\bar{s}_{k+1}, \bar{s}_k')$. A binary relation $\Theta \subseteq S^P \times S^Q$ is a *fair simulation relation* of P by Q if the following three conditions hold:

1. $(s^P, s^Q) \in \Theta$.
2. For all $(p, q) \in \Theta$, we have $L^P(p) \cap O^Q = L^Q(q)$.
3. There exists a simulation strategy κ of Q with respect to P such that, if $(p, q) \in \Theta$ and $\bar{s} = s_0 \xrightarrow{i_1} s_1 \xrightarrow{i_2} s_2 \xrightarrow{i_3} \ldots$ is a fair run of P, then the outcome $\kappa[\bar{s}] = s_0' \xrightarrow{i_1'} s_1' \xrightarrow{i_2'} s_2' \xrightarrow{i_3'} \ldots$ is a fair run of Q and $\kappa[\bar{s}]$ Θ-matches $\bar{s}$ (that is, $(s_k, s_k') \in \Theta$ for all $0 \leq k < |\bar{s}|$). We say that κ is a *witness* to the fair simulation Θ.

If such a relation Θ exists, Q is said to *fairly simulate* P, written $P \preceq_s^F Q$. We will state some properties of fair simulation. First, since all finite runs are fair by definition, we have the following proposition.

Proposition 4. *Consider two fair Moore machines $P = \langle P, F^P \rangle$ and $Q = \langle Q, F^Q \rangle$. If $P \preceq_s^F Q$, then $P \preceq_s Q$.*

The proof of Proposition 4 depends on the receptivenss of Q. Analogous to the equivalence between simulation and tree containment, fair simulation can be proved to be equivalent to fair tree containment [HKR97].

Proposition 5. *Consider two fair Moore machines P and Q. Then $P \preceq_s^F Q$ iff Q fair-tree contains P.*

5 Assume-Guarantee Rule for Fair Simulation

Let $Safe(\mathcal{P})$ be the fair Moore machine obtained by replacing the fairness constraint of $\mathcal{P}$ with the trivial fairness constraint that maps every infinite run to *fair*. We now present the assume-guarantee proof rule for fair simulation. The same rule was proved for fair trace containment in [AH96], with an essentially similar proof (we just use trace trees instead of traces to get the proof below).

Theorem 2. *Let* $\mathcal{P}, \mathcal{Q}, \mathcal{P}'$ *and* $\mathcal{Q}'$ *be Moore machines such that* $\mathcal{P} \parallel \mathcal{Q}$ *and* $\mathcal{P}' \parallel \mathcal{Q}'$ *exist. Suppose that* $\mathcal{P} \parallel Safe(\mathcal{Q}') \preceq_s^F \mathcal{P}'$, $\mathcal{P}' \parallel \mathcal{Q} \preceq_s^F \mathcal{Q}'$, *and every input of* $\mathcal{P}' \parallel \mathcal{Q}'$ *is either an input or output of* $\mathcal{P} \parallel \mathcal{Q}$*. Then,* $\mathcal{P} \parallel \mathcal{Q} \preceq_s^F \mathcal{P}' \parallel \mathcal{Q}'$*.*

Proof. Let $\mathcal{P} = \langle P, F^P \rangle, \mathcal{Q} = \langle Q, F^Q \rangle, \mathcal{P}' = \langle P', F^{P'} \rangle$, and $\mathcal{Q}' = \langle Q', F^{Q'} \rangle$. From Proposition 4, we know that $P \parallel Q' \preceq_s P'$ and $P' \parallel Q \preceq_s Q'$. Consequently, by Theorem 1 we know that $P \parallel Q \preceq_s P' \parallel Q'$. Therefore, from Proposition 2, we get that $P' \parallel Q'$ tree contains $P \parallel Q$. Hence, if T is a trace tree of $P \parallel Q$, then $[T]_{P' \parallel Q'}$ is a trace tree of $P' \parallel Q'$. It remains to be proved that if T is a fair trace tree of $\mathcal{P} \parallel \mathcal{Q}$, then $[T]_{P' \parallel Q'}$ is a fair trace tree of $\mathcal{P}' \parallel \mathcal{Q}'$.

For notational simplicity, we omit explicit projections in the following. Suppose T is a fair trace tree of $\mathcal{P} \parallel \mathcal{Q}$. From Proposition 3, T is a fair trace tree of both $\mathcal{P}$ and $\mathcal{Q}$. Further, we know that T is a trace tree of $P' \parallel Q'$. Thus, T is a fair trace tree of $Safe(\mathcal{Q}')$. Again, by Proposition 3, T is a fair trace tree of $\mathcal{P} \parallel Safe(\mathcal{Q}')$. Since $\mathcal{P} \parallel Safe(\mathcal{Q}') \preceq_s^F \mathcal{P}'$, we conclude (from Proposition 5) that T is a fair trace tree of $\mathcal{P}'$. Therefore, from Proposition 3, since T us a fair trace tree of both $\mathcal{P}'$ and $\mathcal{Q}$, it is a fair trace tree of $\mathcal{P}' \parallel \mathcal{Q}$. Since $\mathcal{P}' \parallel \mathcal{Q} \preceq_s^F \mathcal{Q}'$, T is a fair trace tree of $\mathcal{Q}'$. Finally from Proposition 3, since T is a fair trace tree of both $\mathcal{P}'$ and $\mathcal{Q}'$, it is a fair trace tree of $\mathcal{P}' \parallel \mathcal{Q}'$. □

References

[AH96]   R. Alur and T.A. Henzinger. Reactive modules. In *Proceedings of the 11th Annual Symposium on Logic in Computer Science*, pages 207–218. IEEE Computer Society Press, 1996.

[AL91]   M. Abadi and L. Lamport. The existence of refinement mappings. *Theoretical Computer Science*, 82(2):253–284, 1991.

[AL95]   M. Abadi and L. Lamport. Conjoining specifications. *ACM Transactions on Programming Languages and Systems*, 17(3):507–534, 1995.

[CLM89] E.M. Clarke, D.E. Long, and K.L. McMillan. Compositional model checking. In *Proceedings of the 4th Annual Symposium on Logic in Computer Science*, pages 353–362. IEEE Computer Society Press, 1989.

[Dil89] D.L. Dill. *Trace Theory for Automatic Hierarchical Verification of Speed-independent Circuits*. The MIT Press, 1989.

[GL94]   O. Grumberg and D.E. Long. Model checking and modular verification. *ACM Transactions on Programming Languages and Systems*, 16(3):843–871, 1994.

[HKR97] T.A. Henzinger, O. Kupferman, and S. K. Rajamani. Fair simulation. In *CONCUR 97: Theories of Concurrency*, Lecture Notes in Computer Science 1243, pages 273–287. Springer-Verlag, July 1997.

[Kur94] R.P. Kurshan. *Computer-aided Verification of Coordinating Processes.*
         Princeton University Press, 1994.

[Lyn96] N.A. Lynch. *Distributed Algorithms.* Morgan-Kaufmann, 1996.

[McM97] K.L. McMillan. A compositional rule for hardware design refinement. In
         CAV 97: Computer-Aided Verification, Lecture Notes in Computer Science
         1254, pages 24–35. Springer-Verlag, 1997.

[Mil71] R. Milner. An algebraic definition of simulation between programs. In *Proceedings of the 2nd International Joint Conference on Artificial Intelligence,*
         pages 481–489. The British Computer Society, 1971.

[Sta85] E.W. Stark. A proof technique for rely/guarantee properties. In *Proceedings
         of the 5th Conference on Foundations of Software Technology and Theoretical
         Computer Science,* Lecture Notes in Computer Science 206, pages 369–391.
         Springer-Verlag, 1985.

Three Approaches to Hardware Verification: HOL, MDG and VIS Compared

Sofiène Tahar[1], Paul Curzon[2], and Jianping Lu[1]

[1] ECE Department, Concordia University, Montreal, Canada.
{tahar, jianping}@ece.concordia.ca
[2] School of Computing Science, Middlesex University, London, UK
p.curzon@mdx.ac.uk

Abstract. There exist a wide range of hardware verification tools, some based on interactive theorem proving and other more automated tools based on decision diagrams. In this paper, we compare three different verification systems covering the spectrum of today's verification technology. In particular, we consider HOL, MDG and VIS. HOL is an interactive theorem proving system based on higher-order logic. VIS is an automatic system based on ROBDDs and integrating verification with simulation and synthesis. The MDG system is an intermediate approach based on Multiway Decision Graphs providing automation while accommodating abstract data sorts, uninterpreted functions and rewriting. As the basis for our comparison we used all three systems to independently verify a fabricated ATM communications chip: the Fairisle 4×4 switch fabric.

1 Introduction

Formal hardware verification techniques have established themselves as a complementary means to simulation for the validation of digital systems due to their potential to give very strong results about the correctness of designs. Many academic and commercial verification tools have emerged in recent years, which can be broadly classified into two contrasting formal verification techniques: interactive formal proof and automated decision graph based verification. This paper compares and contrast such tools using an Asynchronous Transfer Mode (ATM) switch fabric as a case study.

In the interactive proof approach, the circuit and its behavioral specification are represented in the logic of a general purpose theorem prover. The user interactively constructs a formal proof to prove a theorem stating the correctness of the circuit. Many different proof systems with a variety of interaction approaches have been used. In this paper we consider one such system: HOL [7], an LCF style proof system based on higher-order logic.

In the automated decision diagram approach the circuit is represented as a state machine. Techniques such as reachability analysis are used to automatically verify given properties of the circuit or verify machine equivalence. We consider the MDG [3] and VIS [2] tools. The VIS tool is based on a multi-valued extension of pure ROBDDs (Reduced Ordered Binary Decision Diagrams [1]).

The MDG system uses Multiway Decision Graphs [3] which subsume ROBDDs while accommodating abstract sorts and uninterpreted function symbols.

As the basis of our comparison, we used HOL, MDG and VIS to independently verify the Fairisle 4×4 switch fabric [9]. This is a fabricated chip which forms the heart of an ATM communication switch. The device, designed at the University of Cambridge is used for real applications in the Cambridge Fairisle network. It switches data cells from input ports to output ports within the ATM switch, arbitrating clashes and sending acknowledgments. It was not designed for the verification case study. Indeed, it was already fabricated and in use, carrying real user data, prior to any formal verification attempt.

Other groups have also used the 4×4 fabric as a case study. Schneider *et. al* [12] used a verification system based on the HOL theorem prover, MEPHISTO, to automate the verification of lower-level hardware modules against top-level block units of the fabric. Jakubiec and Coupet-Grimal are using the fabric in their work based on the Coq proof system [8]. Garcez also verified some properties of the 4×4 fabric using the HSIS model checking tool [6].

2   The Fairisle 4 by 4 Switch Fabric

The Fairisle switch forms the heart of the Fairisle network. It consists of a series of port controllers connected to a central switch fabric. In this paper, we are concerned with the verification of the switch fabric which is the core of the Fairisle ATM switch. The port controllers provide the interface between the transmission lines and the switch fabric, and synchronize incoming and outgoing data cells, appending control information to the front of the cells in a routing byte.

A cell consists of a fixed number of data bytes which arrive one at a time. The fabric switches cells from the input ports to the output ports according to the routing byte (header) which is stripped off before the cell reaches the output stage of the fabric. If different port controllers inject cells destined for the same output port controller (indicated by *route* bits in the routing byte) into the fabric at the same time, only one will succeed—the others must retry later. The routing byte also includes a priority bit (*priority*) that is used by the fabric during round-robin arbitration which gives preference to cells with the priority bit set. The fabric sends a negative acknowledgment to the unsuccessful input ports, and passes the acknowledgment from the requested output port to the successful one.

The port controllers and switch fabric all use the same clock, hence bytes are received synchronously on all links. They also use a higher-level cell frame clock—the *frame start* signal, which ensures the port controllers inject data cells into the fabric so routing bytes arrive together. The fabric does not know when this will happen. Instead, it monitors the *active* bit of the routing bytes: when any goes high the cells have arrived. If no input port raises the active bit throughout the frame then the frame is inactive—no cells are processed; otherwise it is active.

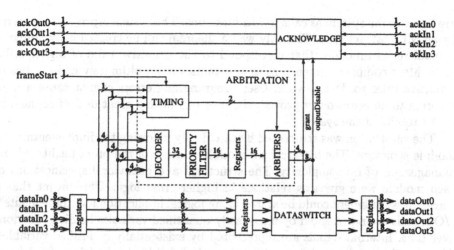

Fig. 1. The Fairisle ATM Switch Fabric

Figure 1 shows a block diagram of the 4×4 switch fabric. It is composed of an arbitration unit (timing, decode, priority filter and arbiters), an acknowledgment unit and a dataswitch unit. The timing block controls the timing of the arbitration decision based on the frame start signal and the time the routing bytes arrive. The decoder reads the routing bytes of the cells and decodes the port requests and priorities. The priority filter discards requests with low priority which are competing with high priority requests. It then passes the resulting request situation for each output port to the arbiters. The arbiters (in total four—one for each port) make arbitration decisions for each output port and pass the result to the other units with the grant signal. The arbiters indicate to the other units when a new arbitration decision has been made using the output disable signals. The dataswitch unit performs the switching of data from input port to output port according to the latest arbitration decision. The acknowledgment unit passes acknowledgment signals to the input ports. Negative acknowledgments are sent until a decision is made.

Each unit is repeatedly subdivided down to the logic gate level, providing a hierarchy of modules. The design has a total of 441 basic components including 162 1-bit flip flops. It is built on a 4200 gate equivalent Xilinx programmable gate array. The switching element can be clocked at 20 MHz and frame start pulses occur every 64 clock cycles.

3 The HOL Verification

In the first study, the Fabric was verified using the HOL90 Theorem Proving System [7]. This is a general purpose interactive proof system. It provides a range of proof commands of varying sophistication, including decision procedures. It is also fully user programmable, allowing user-defined, application-specific proof tools to be developed. The interface to the system is an ML interpreter. Proofs are input to the system as calls to ML functions. The system represents theorems by an ML abstract type. The only way a theorem can be created is by applying a small set of functions that correspond to the primitive rules of higher-order logic. More complex inference rules and tactics must ultimately call a series of primitive rules to do the work. User programming errors cannot cause a non-theorem to be erroneously proved: the user can have a great deal of confidence in the results of the system.

The verification was structured hierarchically following the implementation's module structure. The hierarchical, modular nature of the proof facilitated the management of its complexity. The structural and behavioral specifications of each module were given as relations in higher-order logic. This meant that a correctness statement could be stated using logical implication for "implements". I/O signals are represented by universally quantified variables holding functions over time. Internal signals are represented by existentially quantified variables. The overall correctness statement for the switch fabric has the (simplified) form:

```
∀aIn aOut dOut d frameStart.
  ENVIRONMENT frameStart d ⊃
  FABRIC4B4 ((d,frameStart,aIn), (dOut, aOut)) ⊃
    ∃last. FABRIC4B4_SPEC last ((d,frameStart,aIn), (dOut, aOut))
```

The correct operation of the fabric relies on an assumption about the environment. Cells must not arrive at certain times around a frame start. The relation ENVIRONMENT above, specifies this condition in a general way.

A correctness theorem of the above form was proved for each module stating that its implementation down to the logic gate level satisfied the specification.

In conducting the overall proof, the human verifier needs a very clear understanding of why the design is correct, since a proof is essentially a statement of this. Thus performing a formal proof involves a deep investigation of the design. It also provides a means to help achieve that understanding. Having to write formal specifications for each module helps in this way. Having to formulate the reasons why the implementation has that behavior gives much greater insight. In addition to uncovering errors, this can serve to highlight anomalies in the design and suggest improvements, simplifications or alternatives [4].

3.1 The Specifications

The structural specification of a design describes its implementation: the components it consists of and how they are wired together. The original designers of the fabric used the relatively simple Qudos HDL [5], to give structural descriptions of the hardware. This description was used to simulate the design prior to fabrication. The Xilinx netlist was also generated from this description. The descriptions used in the HOL verification were hand-derived from the Qudos descriptions. Qudos structural descriptions can be mimicked very closely in HOL up to surface syntax. However, the extra expressibility of HOL was used to simplify and generalize the description.

In HOL words of words are supported. Therefore, a signal carrying 4 bytes can be represented as a word of 4 8-bit words, rather than as one 32-bit signal. This allows more flexible indexing of bits, so that the module duplication operator FOR can be used. Arithmetic can also be used to specify which bit of a word is connected to an input or output of a component. For example, we can specify that for all i, the $2i$-th bit of an output is connected to the i-th bit of a subcomponent. This, again, meant that we could avoid writing essentially identical pieces of code several times, as was necessary in the Qudos specifications. When an additional module, used in several places is introduced, the verification task is reduced as that module need only be verified once.

It should be stressed that while the descriptions of the implementation were modified in the ways outlined above, no simplification was made to the implementation itself to facilitate the verification. The netlists of the structural specifications used corresponds to that actually implemented.

The behavioral specification against which the structural specification was verified describes the actual un-simplified behavior of the switch fabric. It is presented at a similar level of abstraction to that used by the designers, describing the behavior over a frame in terms of interval temporal operators (i.e. timing diagrams). The behavior of each output is specified as a series of interval specifications. The start and end times are specified in terms of the frame times given in an assumption. The values output are functions of the inputs and state at earlier times.

3.2 Time Taken

The module specifications (both behavioral and structural) were written prior to any proof. This took between one and two person-months. No breakdown of this time has been kept. Much of the time was spent in understanding the design. The behavioral specifications were more difficult. The specifier had no previous knowledge of the design. There was a good English overview of the intended function of the switch fabric. This also outlined the function of the major components. However, it was not sufficient to construct an unambiguous behavioral specification of all the modules. The behavioral specifications were instead constructed by analyzing the HDL. This was very time-consuming.

Approximately two person-months were spent performing the verification. Of this, one week was spent proving theorems of general use. Approximately three weeks were spent verifying the upper modules of the arbitration unit, and a further week was spent on the top two modules of the switch. 3–4 days were spent combining the correctness theorems of the 43 modules to give a single correctness theorem for the whole circuit. The remaining time of just over two weeks was spent proving the correctness theorems for the 36 lower level units. The proofs of the upper-level modules were generally more time-consuming for several reasons: there were more intervals to consider; they gave the behavior of several outputs; and those behaviors were defined in terms of more complex notions. They also contained more errors which severely hampered progress. Apart from standard libraries, the work did not build on previous theories.

It takes several hours of machine time on a Sparc 10 to completely rebuild the proofs from scratch by re-running the scripts in batch mode. Single theories representing individual modules generally take minutes to rebuild. A large proportion of the time is actually spent restarting HOL and loading in appropriate parent theories and libraries for each theory. In the initial development of the proof the machine time is generally not critical, as the human time is so much greater. However, since the proof process consists of a certain amount of replay of old proofs (e.g. when mistakes are made), a speed up would be desirable.

If changes are made to the design, it is important that the new verification can be done quickly. Since theorem proofs are very time consuming, this is especially important. This problem is attacked in several ways in the HOL approach: the proofs can be made generic; their modular nature means that only affected modules need to be reverified; and proofs of modules which have changed can often be replayed with only minor changes. After the original verification had been completed, several real variations on the design were also verified. Each took only a matter of hours or days.

One of the biggest disadvantages of the HOL system is that its learning curve is very steep. Furthermore, interactive proof is a time-consuming activity even for an expert. Much time is spent dealing with trivial details of a proof. Recent advances in the system such as new simplifiers and decision procedures (not used in this study) may alleviate these problems.

3.3 Errors

No errors were discovered in the fabricated hardware. Errors that had inadvertently been introduced in the structural specifications (and could just as easily have been in the implementation) were discovered. The original versions of the behavioral specifications of many modules contained errors.

A strong indication of the source of detected errors was obtained. Because each module was verified independently, the source of an error was immediately narrowed down to being in the current module, or in the specification of one of its submodules. Furthermore, because performing the proof involves understanding why the design is correct, the exact location of the error was normally obvious from the way the proof failed. For example, in one of the modules, two wires were inadvertently swapped. This was discovered because the subgoal ([T, F] = [F, T]) was generated in the proof attempt. One side of this equality originated from the behavioral specification and one from the structural specification. It was clear from the context of the subgoal in the proof attempt that two wires were crossed. It was also clear which signals were involved. It was not immediately clear, however, which specification (structural or behavioral) was wrong.

A further example of a discovered error concerned the time the grant signal (Figure 1) was read by the dataswitch. It was specified that the two bits of each grant signal were read on a single cycle. However, the implementation read them on consecutive cycles. This resulted in a subgoal of the form $grant\ t = grant\ (t + 1)$. No information was available in the goal to allow this to be proven, suggesting an error. On this occasion it was in the specification.

3.4 Scalability

In theory, the HOL proof approach is scalable to large designs. Because the approach is modular and hierarchical, increasing the size of the design does not necessarily increase the complexity of the proof. However, in practice the modules higher in the hierarchy generally (though not always) take longer to verify. This is demonstrated by the fact that two of the upper most modules took approximately half of the total verification time—a matter of weeks.

The extra time arises in part because there are more cases to consider. The situation is made worse if the interfaces between modules are left containing lots of low level detail. For example, for the switch fabric, low level modules required assumptions to be made about their inputs. These assumptions had to be dealt with in the proofs of higher level modules adding extra proof work manipulating and discharging them. If the proof is to be tractable for large designs, it is important that the interfaces between modules are as clean as possible. The interfaces of the Fairisle fabric could have been much simpler. We demonstrated this by redesigning the fabric with cleaner interfaces [4].

4 The MDG Verification

In the second study, the same circuit was verified using a decision graph approach. A new class of decision diagrams called *multiway decision graphs* (MDGs) was used to represent sets of states and the transition and output relations [3]. Based on a technique called *abstract implicit enumeration* [3], hardware verification tools have been developed which perform combinational circuit verification, safety property checking and equivalence checking of two sequential machines [3].

The formal system underlying MDGs is many-sorted first-order logic augmented with a distinction between abstract and concrete sorts. Concrete sorts have enumerations, while abstract sorts do not. A data value can be represented by a single variable of abstract sort, rather than by concrete boolean variables, and a data operation can be represented by an uninterpreted function (cross-operator) symbol. MDGs permit the description of the output and next state relations of a state machine in a similar way to the way ROBDDs do for FSMs. We call the model an *abstract state machine* (ASM) since it may represent an unbounded class of FSMs, depending on the interpretation of the abstract sorts and operators. For circuits with large datapaths, MDGs are thus much more compact than ROBDDs. As the verification is independent of the width of the datapath, the range of circuits that can be verified is greatly increased. Because of the use of uninterpreted functions, reachability analysis on MDGs may not terminate in some cases when circuits include some specific cyclic behavior [3]. We did not encounter this problem in the current study.

The MDG operators and verification procedures are packaged as MDG tools implemented in Prolog [3]. The ATM circuit we investigate here is an order of magnitude larger than any other circuit verified using MDGs.

We described the actual hardware implementation of the switch fabric at two levels of abstraction. We gave a description of the original Qudos gate-level implementation and a more abstract RTL description which holds for an arbitrary word width n. Using the MDG equivalence checking, we verified the gate-level implementation against the abstract (RTL) hardware model. Here the n-bit words of abstract sort of the latter were instantiated to 8 bits using uninterpreted functions which encode and decode abstract data to boolean data and vice-versa [13]. Besides, we used a few rewriting rules to map 8-bit constants of concrete sort to generic ones of abstract sort.

Starting from timing-diagrams describing the expected behavior of the switch fabric, we derived a complete high-level behavioral specification in the form of a state machine. This specification was developed independently of the actual hardware design and includes no restrictions with respect to the frame size, cell length and word width. Using implicit reachability analysis, we checked its equivalence against the RTL hardware model when both seen as abstract state machines. That is, we ensured that the two machines produce the same observable behavior by feeding them with the same inputs and checking that an invariant stating the equivalence of their outputs holds in all reachable states.

By combining the above two verification steps, we hierarchically obtain a complete verification of the switch fabric from a high-level behavior down to the gate-level implementation. Prior to the full verification, we also checked both behavioral and RTL structural specifications against several specific safety properties of the switch. Here, we combined an environment state machine with each switch fabric specification yielding a composed machine which represented the required platform for checking if the invariant properties hold in all reachable states of the specification [13]. Although the properties we verified do not represent the complete behavior of the switch fabric, we were able to detect several injected design errors in the structural model.

4.1 The Specifications

As with the HOL study, we translated the Qudos HDL gate-level description into a suitable HDL description; here a Prolog-style HDL, called MDG-HDL. As in the HOL study, extra modularity was added over the Qudos descriptions, while leaving the underlying implementation unchanged. A structural description is usually a (hierarchical) network of components (modules) connected by signals. The MDG-HDL comes with a large library of predefined, commonly used, basic components (such as logic gates, multiplexors, registers, bus drivers, ROMs, etc.). Multiplexors, registers and drivers can be modeled at the Boolean or the abstract level using abstract terms as inputs and outputs.

Hardware descriptions in MDG are very similar up to syntax to HOL. The data sorts of the interface and internal signals must always be specified. MDG does not provide a module replication facility, so repeated elements must be explicitly written out multiple times, nor an ability to structure words, so this description cannot be abstracted as in HOL.

Besides the gate-level description, we also provided a more abstract (RTL) description of the implementation which holds for arbitrary word width n. Here, the data-in and data-out lines are modeled using an abstract sort *wordn*. The ac-

tive, priority and *route* fields are accessed through corresponding cross-operators (functions). In addition to the generic words and functions, the RTL specification also abstracts the behavior of the dataswitch unit by modeling it using abstract data multiplexors instead of logic gates. We thus obtain a simpler implementation model of the dataswitch which reflects the switching behavior in a more natural way and is implemented with fewer components and signals. For more details about the abstraction techniques used, refer to [13].

MDG-HDL is also used for behavioral descriptions. A behavioral description is given by high-level constructs as ITE (If-Then-Else) formulas, CASE formulas or tabular representations. The tabular constructor is similar to a truth table but allows first-order terms in rows. It can be used to define arbitrary logic relations. In the MDG study, we gave the behavioral specification of the switch fabric in two different forms: 1) as a complete high-level behavioral state machine and 2) as a set of *safety* properties which reflect the essential behavior of the switch fabric as it is used in its environment.

The main behavioral description of the switch fabric was as an abstract state machine (ASM) which reflects its complete behavior under the assumption that the environment maintains certain timing constraints on the arrival of the frame start signal and headers. This ASM reproduces the exact behavior of the switch fabric during the initialization phase, the arrival of a frame start, the arrival of the routing bytes, and the end of the frame. The generation of the acknowledgment and data output signals is described by case analysis on the result of the round-robin arbitration. This is done in MDG-HDL using ITE and tabular constructs.

Although this ASM specification describes the complete behavior of the switch fabric, we also validated (in an early stage of the project) the fabric implementation by property checking. This is useful as it gives a quick verification result at low cost. We verified that the structural specification satisfies its requirements when the ATM switch fabric works under the control of its operating environment, i.e., the port controllers. We provided for this purpose a set of safety properties which reflect the essential behavior of the switch fabric, e.g., for checking of correct priority computation, circuit reset or data routing. We first modeled the environment as a state machine with one state variable s of enumerated (concrete) sort [1..68]. This allowed us to map the time points for initialization, frame start, header arrival and frame end to specific states. We then described the properties as invariants which should hold in all reachable states of the fabric model. Examples of properties are described in [13].

4.2 Time Taken

The translation of the Qudos design description to the MDG-HDL gate-level structural model was straightforward and took about one person-week. The description of the RTL structural specification including modeling required about one person-week. The time spent for understanding the expected behavior and writing the behavioral specification was about one person-week. The time taken for the verification of the gate-level description against the RTL model, including the adoption of abstraction mechanisms and correction of description errors, was about two person-weeks. The verification of the RTL structural specification against the behavioral model required about one person-week of work. The user time required to set up four properties, build the environment state machine, conduct the property checking on the structural specification and interpret the results was about one person-week. Checking of these same properties on the behavioral specification took about one hour. The average time for the introduction and verification of a design error was less than an hour. The experimental results are shown in Table 1. The CPU time given is for a SPARC station 10.

Like ROBDDs, the MDGs require a fixed node ordering. The variable ordering plays an important role as it determines the canonical attribute of the graphs and the size of the graphs which greatly affects its efficiency. A bad ordering easily leads to a state space explosion as occurred after an early ordering attempt. In contrast to VIS which provides heuristics for several node ordering techniques including dynamic ordering, node ordering in MDG has to be given by the user explicitly. This takes much of the verification time. On the other hand, unlike ROBDDs where all variables are Boolean, time must be spent assigning to every variable used an appropriate sort and type definitions must be provided for all functions. In some cases, rewrite rules may need to be provided to partially interpret the otherwise uninterpreted function symbols.

Because the verification is essentially automatic, the work re-running a verification for a new design is minimal compared to the initial effort since the latter includes all the modeling aspects. Much of the effort is spent on re-determining a suitable variable ordering. Depending on the kind of design changes adopted, the original variable ordering may need major changes for a modified design.

The MDG gate-level specification is a concrete description of the implementation. In contrast, the RTL structural and ASM behavioral specifications are generic. They abstract away from frame, cell and word sizes, provided the environment timing assumptions are kept. Design changes at the gate-level that still satisfy the RTL model behavior would hence not affect the verification against the ASM specification. For property checking, specific assumptions about the operating environment were made, (e.g. that the frame interval is 64 cycles). This is sound since the switch fabric will be used under the behest of its operating environment (the port controllers) which ensure this. While this reduces the verification cost, a disadvantage is that the verification must be completely redone if the operating environment changes, though only a few parameters have to be changed in the description of the (simple) environment state machine [13].

4.3 Errors

As with the HOL study, no errors were discovered in the implementation. For experimental purposes, we injected several errors into the structural specifications and checked them using either the set of properties or the behavioral model. Errors were automatically detected and automatically generated counter-examples were used to identify them. The injected errors included the main errors introduced accidently in the HOL study and in addition following three examples:

Table 1. Experimental Results for the MDG and VIS (Model checking) Verifications

Verification	MDG			VIS		
	CPU Time (s)	Memory (MB)	MDG Nodes	CPU Time (s)	Memory (MB)	BDD Nodes
Gate-Level to RTL	183	22	183,300			
RTL to Beh. Model	2920	150	320,556			
P1: Data Output Reset	202	15	30,295	3593	32	93,073,140
P2: Ack. Output Reset	183	15	30,356	833	5	28,560,871
P3: Data Routing	143	14	27,995	3680	41	79,687,78
P4: Ack. Output	201	15	33,001	415	5	4,180,124
Error (i)	20	1	2,462	83	4	1,408,477
Error (ii)	1300	120	150,904	49	2	250,666
Error (iii)	1000	105	147,339	15	1	85,238

(i) We exchanged the JK flip-flop inputs that produce the output disable signal. This prevented the circuit from resetting. (ii) We used, at one point, the priority bit of input port 0 instead of input port 2. (iii) We used an AND gate instead of an OR gate. Experimental results for these errors, when checked by verifying the RTL model against the behavioral specification, are given in Table 1.

While checking properties on the hardware structural description, we also discovered errors that we mistakenly introduced in the structural specifications. However, we were able to easily identify and correct them using the counter-example facility of the MDG tools. Also, during the verification of the gate-level model, we found a few errors in the description that were introduced during the translation from Qudos HDL to MDG-HDL. These were easily removed by comparing both descriptions, as they included the same collection of gates. Finally, many trivial typing errors were found at an early stage by the error messages output after each compilation of the specification's components.

4.4 Scalability

Like all FSM-based verification, MDG proof is not directly scalable to large designs due to the state space explosion. Unlike other approaches, MDGs can cope with datapath complexity as they use data of abstract sort and uninterpreted functions. Still, a direct verification of the gate-level model against the behavioral model or even against the set of properties was practically impossible. We overcame this by providing an abstract RTL structural specification which we instantiated for the verification against the gate-level model. To handle large designs, major effort is required to set up the model abstraction levels.

5 The VIS Verification

We also verified the fabric using VIS [2], another decision diagram based tool. It integrates the verification, simulation and synthesis of finite-state hardware systems. It uses a Verilog front-end and supports fair CTL model checking, language emptiness checking, combinational and sequential equivalence checking, cycle-based simulation, and hierarchical synthesis. Its fundamental data structure is a multi-level network of latches and combinational gates. The variables of a network are multi-valued, and logic functions over these variables are represented by an extension of BDDs: multi-valued decision diagrams.

VIS operates on the intermediate format BLIF-MV. It includes a compiler from Verilog to BLIF-MV. It extracts a set of interacting FSMs that preserves the behavior of the Verilog program defined in terms of simulated results. Through the interacting FSMs, VIS performs fair CTL model checking under Buchi fairness constraints. The language of a design is given by sequences over the set of reachable states that do not violate the fairness constraint. Also VIS can check the combinational and sequential equivalence of two designs. Sequential verification involves building the product FSM, and checking whether a state where the values of corresponding outputs differ can be reached from the set of initial states of the product machine. If model checking or equivalence checking fails, VIS reports the failure with a counter-example.

We translated the original Qudos HDL gate-level description of the switch fabric into Verilog HDL. We also derived a complete high-level behavioral specification in the form of a finite state machine according to the timing diagrams describing the expected behavior of the switch fabric. This specification was developed independently of the actual hardware design and uses a different design hierarchy to the structural one. Using these Verilog specifications, we attempted to obtain a complete verification of the switch fabric from a high-level behavioral specification down to the gate-level implementation through equivalence checking. This verification was similar to that in the MDG case. However, it did not succeed in VIS due to state space explosion. We therefore attempted to separately verify the submodules of the fabric based on the same design hierarchy as the structural one. This is similar to the HOL study, and involved writing separate Verilog RTL behavioral specifications for each submodule. We succeeded in verifying the equivalence of the behavioral specification of each submodule and its corresponding structural specification by VIS sequential equivalence checking. Through this verification, we checked that the implementation of each submodule satisfies its specification. Unlike the HOL verification, we could not verify the correctness of the connections among the submodules of the switch fabric. For real designs, this step would be useful to verify if the logic synthesis is correct.

As an alternative to equivalence checking, we attempted model checking of the switch fabric. Unlike MDG, model checking is the main verification approach in VIS. As for the MDG approach an environment state machine was needed [11]. To ease the model checking we compressed the 68 states into 7 states. Again we failed to verify the whole switch fabric due to the state space explosion. We succeeded in model checking a simplified fabric with its datapath and control path reduced from 8 bits to the minimum 1 bit and 4 bits, respectively.

Both behavioral and structural specification were written in Verilog, so we were able to perform their simulation in Verilog-XL directly. It was very useful for detecting some syntax and semantic errors of the descriptions before performing equivalence or model checking. In addition, we extracted some safety properties from the generalization of simulation vectors. These safety properties were further used in model checking, enabling the detection of design errors that were omitted by simulation. The Verilog-XL graphical interface also eased the analysis of counter-examples which were generated by model and equivalence checking. Furthermore, as the RTL behavioral specification was written in Verilog, we were able to synthesize the structural specification with some timing constraints directly using the Synopsys Design Compiler. We performed equivalence checking between the submodules of the RTL behavioral specification and the submodules of the synthesized structural one to ensure the correctness of the synthesis.

5.1 The Specifications

The Verilog structural specification of the fabric is very similar to the other descriptions. A big advantage of the VIS Verilog front-end is the ease of importing existing (industrial) designs with no extra overhead of manual translation. Moreover, it allows the direct interaction of VIS with other commercial tools for simulation and synthesis. However, the fabric structure had to be reduced to 4 bits and the datapath further to one bit to enable the model checking procedure to terminate. The control path could not be reduced below 4 bits as the data includes the header control information. For more details about the abstraction and reduction techniques adopted refer to [11].

We gave the behavioral specification of the fabric in two forms: an RTL description as a state machine of the whole fabric and a set of liveness and safety properties covering its essential behavior. In addition, as with HOL, behavioral specifications of submodules of the design hierarchy were developed. VIS-Verilog HDL is used for behavioral specification. It contains two new features over standard Verilog: a nondeterministic construct, $ND, to specify non-determinism on wire variables; and symbolic variables which use an enumerated type mechanism similar to the one available in the MDG system.

Unlike in MDG, we extensively used property checking to verify the fabric in VIS as it is optimized for model checking. Moreover, thanks to the expressiveness of CTL, properties can be defined more easily in VIS. With MDG a property (invariant) is described in MDG-HDL using ITE and tabular constructs. Before using model checking to verify the overall behavior of the switch fabric, we set up an environment state machine and developed a set of properties. The nondeterministic construct ($ND) of VIS-Verilog HDL eases the establishment of an environment state machine. We used it to express the inputs of the switch fabric. CTL can represent both safety and liveness properties. The latter can be used to detect deadlock or livelock which is difficult using simulation. 58 CTL properties were verified. We first verified a number of safety properties including all those used in the MDG study. In addition, we verified many CTL liveness properties. Example properties that we checked on the fabric model can be found in [11].

Due to the state space explosion, we succeeded in checking only a few properties on the abstracted fabric directly. Instead we adopted several techniques that divide a property into sequentially or parallelly related sub-properties in a similar manner to the compositional reasoning proposed in [10]. Details about the specific property division techniques we used are reported in [11].

5.2 Time Taken

The translation of the Qudos structural description to Verilog was straightforward taking about three person-days. The time spent for understanding the expected behavior and writing the behavioral specification was around ten person-days. The time taken for the simulation of both RTL behavioral and structural specification in Verilog-XL, including the development of test-bench files, was about three person-days. The verification of the RTL behavioral specification against the structural specification was done automatically, and took around one person-day. The user time required to set up 58 CTL properties, build the related environment state machine, construct the appropriate property division and conduct the model checking took approximately three weeks. The injection and verification of an error took less than one hour.

The experimental results of model checking, which were obtained on a SPARC station 20, are shown in Table 1. VIS generates comparatively more BDD nodes than the MDG system does. This is due to the data abstraction within MDG that is absent in VIS. The equivalence checking of the whole switch fabric ran for three days before running out of memory. The same problem occurred with the dataswitch module. Equivalence checking of the arbitration module was successful but it took two days of machine time. The lower level modules such as the timing unit were verified in seconds. We also failed to verify the properties on the original switch fabric after two days of machine time. Finally we reduced the datapath of the switch fabric from 8 to 1 bit. The successful model checking results in Table 1 are based on this reduced model.

Since VIS is based on ROBDDs, the node ordering has a dramatic influence on the speed of both equivalence checking and model checking. Unlike MDG, VIS provides dynamic ordering facilities to reduce the cost of manual variable ordering.

The experimental results given in Table 1 were obtained using VIS dynamic ordering. It is to be noted that in some cases a manually optimized ordering, e.g., an interleaved order of the bits of the data words, would have enhanced the VIS verification.

We also applied cascade and parallel property divisions (practical approaches to compositional reasoning). Using these techniques, we enhanced the model checking by up to 200 times. However, we had to establish environment state machines and abstract the circuit first. The derivation of reduced models from the original structure and the division of properties was very time consuming. For a cascade property division, we built a new partial environment state machine for each target sub-circuit. For parallel property division, we disassembled a circuit at different symmetric locations and later composed the properties.

5.3 Errors

As in the HOL and MDG studies, no errors were discovered in the switch fabric implementation. We injected the same errors as for MDG into the implementation and checked them using either model checking or equivalence checking. Experimental results are reported in Table 1. Like MDG, VIS provides a counter-example generation facility to help identify the source of design errors. Injected errors were hence automatically detected and further viewed graphically with Verilog-XL. Through checking the equivalence between the RTL behavioral and the structural specifications of the submodules, we discovered errors that we mistakenly introduced in the structural specification. Also, during model checking, we found connection errors that were mistakenly introduced in the RTL behavioral and structural specifications. We easily identified and corrected these errors from the counter-examples.

5.4 Scalability

The VIS proof approach is not directly scalable to large designs due to state space explosion. To solve this problem the datapath complexity must be decreased by abstraction and reduction. In a large design like the switch fabric, we also had to apply compositional reasoning [10]. The environment state machine must imitate the behavior of the models which are associated with the target model. It must also have fewer components than the original models. Consequently, the environment state machine is especially hard to develop when the concurrent interaction between the target model and its associated models is complex.

6 Conclusions

The structural descriptions are very similar. HOL provides significantly more expressibility allowing more natural specifications. Some generic features were included in the MDG description that were not in the HOL description. This could have been done with minimal effort, however. Due to its Verilog front-end, (commercial) designs can be imported into VIS with no extra overhead of a manual translation, which is one reason for its popularity. This also allows direct interaction with commercial tools for simulation and synthesis.

The behavioral descriptions are totally different. The MDG and VIS specifications are based on a state machine model while HOL's is based on interval operators explicitly describing the timing behavior in terms of frames corresponding to whole ATM cells arriving. In the MDG and VIS specifications the frame abstraction is not used: the description is firmly at the byte level. Verilog allows direct testing of the specifications using commercial simulation facilities, however. Unlike Verilog descriptions, HOL's higher-order logic and MDG-HDL descriptions are not directly executable. All describe the behavior in a clear and comprehensive form. Writing the behavioral specifications took longer in HOL and VIS, as separate specifications were needed for each module. In MDG this was not necessary as the whole design was verified in one go.

An advantage of MDG and VIS is that a property specification is easy to set up and verify. For both systems it was necessary to introduce an environment state machine in order to restrict the possible inputs to the switch fabric. It is verified that the specification satisfies its requirements under specific working conditions. It can greatly reduce the full verification cost by catching errors at an early stage. In this respect VIS, with its very efficient CTL based model checking, outperforms its MDG counterpart. Properties are easier to describe in CTL than are invariants in the MDG system. Currently, MDG tools do not provide CTL property specification or liveness property verification. Work on the integration of a recently developed MDG model checking algorithm based on a restricted first-order temporal logic (Abstract CTL–ACTL) is ongoing.

The HOL verification was much slower, taking several months. This time includes the verification of each of the modules and of their combination. Much of the time was spent on the connection of the highest level modules (which VIS failed on). Using HOL, many lemmas had to be proved and much effort was required to interactively create the proof scripts. For example, the time spent verifying the dataswitch was about three days. The proof script was over 500 lines long (17 KB). The MDG and VIS verifications were achieved automatically without the need of a proof script. For MDG, however, careful management of the MDG node ordering was needed (which currently has to be done manually). This could take hours or a few days of work. In contrast, VIS provides several options for variable ordering heuristics which eliminate the ordering overhead. However, major effort was spent here developing abstract models of the switch fabric units to manage the state explosion of the boolean representation. Furthermore, the HOL and MDG verifications succeeded in verifying the whole switch fabric but VIS failed to verify even the smallest 1-bit datapath version of the fabric using equivalence checking. Additional time was spent hierarchically verifying submodules as with HOL but their combination could not be verified.

In all the approaches, the work needed to verify a modified design is greatly reduced once the original has been verified. MDG and HOL allow generic verification to be performed (e.g. word sizes are unspecified), though HOL is more flexible. No generic verification is possible in VIS. Because MDG and VIS are automated and fast, re-verification times are largely the time taken to modify the specifications and, for MDG, to find a new variable ordering. With HOL the behavioral specifications of many modules and the proof scripts themselves may need to be modified. For model checking in VIS, new environment machines, and model abstraction and reduction techniques may be required.

An advantage of the HOL approach over the others is the confidence in the tool the LCF approach offers. Although the VIS (and to a certain extent the MDG) software package has been successfully tested on several benchmarks and has been considerably improved, they cannot guarantee the same level of proof security as HOL. Compared to MDG, VIS is a more mature tool. It is implemented in a well-engineered fashion in C as compared to the prototype implementation of MDG in Prolog. Moreover, VIS is very widely used in both academia and industry, giving confidence in its correctness.

Table 2. Summary of the Comparison

Area	Feature	HOL	MDG	VIS
Specification	Behavioral Specification - Time Taken		++	+
	Structural Specification - Time Taken	+	+	++
	Behavioral Specification - Expressibility	++	+	
	Structural Specification - Expressibility	++		+
Verification	Full Verification Completed	++	+	
	Machine Time Taken		+	+
	Total Verification Time Taken		++	+
	Verification Time for Design Modifications	+	+	+
	Property Checking		+	++
	Equivalence Checking		++	+
	Scalability	++		
	Confidence in Tool	++		+
Errors	Detect Errors	++	++	++
	Locating Errors	+	++	++
	Avoid Error Introduction	+	+	+
	False Error Reports		++	+
Design	Impart Understanding of Design	++	+	+
	Suggesting Design Improvements	++	+	+
	Commercial Front End			++

All the approaches highlight errors, and help determine their location. However, the way this information manifests itself differs. VIS and MDG are more straightforward, outputting a trace of the input sequence that leads to the erroneous behavior. Errors are detected automatically and can be diagnosed with the help of the counter-example facility. In addition, due to its front-end, VIS counter-examples can be analyzed using commercial tools such as XL-Verilog. In HOL, errors manifest themselves as unprovable goals. The form of the goal, the context of the proof and the verifier's understanding of the proof are combined to track down the location, and understand its cause.

With the MDG (and to a certain extent VIS) verification approach the verifier does not need be concerned with the internal structure of the design being verified. This means that no understanding of the internals is obtained by doing the verification. In contrast, with HOL, a very detailed understanding of the internal structure is needed. The verifier must know why the design works the way it does. The process of doing the verification helps the verifier achieve this understanding. This means that internal idiosyncrasies in the implementation are likely to be spotted, as are other potential improvements.

A summary of the main comparison points is given in Table 2. Each system is given a rough rating of either "++", "+" or nothing to indicate how favorably the system comes out with respect to that feature. In conclusion, the major advantages of HOL are: the expressibility of the specification language; the confidence afforded in its results; the potential for scalability and the insight into the design that is obtained. The strength of MDG and VIS is in their speed; their relative ease of use and their error detection capabilities. MDG has the advantage of using abstract data types and uninterpreted functions with a rewriting facility, hence allowing larger circuits to be verified—but with the drawback that an MDG verification may not terminate in some cases. VIS is a very efficient model checker supporting the CTL expressiveness for both liveness and safety properties. Moreover, VIS outperforms MDG due to its maturity in the use of efficient graph manipulation techniques. The VIS Verilog front-end and mature C implementation make VIS very attractive to industry.

Acknowledgments

We are grateful for the help of X. Song and E. Cerny at the Univ. of Montreal, I. Leslie and M. Gordon at Cambridge, R. Brayton at Berkeley, F. Somenzi at Colorado, H. Thimbleby at Middlesex, Z. Zhou at Texas Instruments and M. Langevin at Nortel.

References

1. R. Bryant. Graph-based Algorithms for Boolean Function Manipulation. *IEEE Trans. on Computers*, C-35(8):677–691, 1986.
2. R. Brayton et. al. VIS: A System for Verification and Synthesis In R. Alur and T. Henzinger, eds, *Computer Aided Verification*, LNCS 1102, 428–432, Springer-Verlag, 1996.
3. F. Corella, Z. Zhou, X. Song, M. Langevin, and E. Cerny. Multiway Decision Graphs for Automated Hardware Verification. *Formal Methods in System Design*, 10(1):7–46, 1997.
4. P. Curzon and I.M. Leslie. Improving Hardware Designs whilst Simplifying their Proof. *Designing Correct Circuits*, Workshops in Comp., Springer-Verlag, 1996.
5. K. Edgcombe. *The Qudos Quick Chip User Guide*. Qudos Limited.
6. E. Garcez and W. Rosenstiel. The Verification of an ATM Switching Fabric using the HSIS Tool. In *IX Brazilian Symp. on the Design of Integrated Circuits*, 1996.
7. M.J.C. Gordon and T.F. Melham. *Introduction to HOL: A Theorem Proving Environment for Higher-order Logic*. Cambridge University Press, 1993.
8. L. Jakubiec, S. Coupet-Grimal, and P. Curzon. A Comparison of the Coq and HOL Proof Systems for Specifying Hardware. In E. Gunter and A. Felty, eds, *Theorem Proving in Higher Order Logics: Short Presentations*, 63–78, 1997.
9. I.M. Leslie and D.R. McAuley. Fairisle: An ATM Network for the Local Area. *ACM Communication Review*, 19(4):327–336, 1991.
10. D.E. Long. Model Checking, Abstraction and Compositional Verification. *Ph.D thesis*, Carnegie Mellon University, July 1993.
11. J. Lu and S. Tahar. Practical Approaches to the Automatic Verification of an ATM Switch Fabric using VIS. In *Proc. IEEE Great Lakes Symp. on VLSI*, 368–373, 1998.
12. K. Schneider and T. Kropf. Verifying Hardware Correctness by Combining Theorem Proving and Model Checking. In J. Alves-Foss, editor, *Higher Order Logic Theorem Proving and Its Applications: Short Presentations*, 89–104, 1995.
13. S. Tahar, Z. Zhou, X. Song, E. Cerny, and M. Langevin. Formal Verification of an ATM Switch Fabric using Multiway Decision Graphs. In *Proc. IEEE Great Lakes Symp. on VLSI*, 106–111, 1996.

An Instruction Set Process Calculus

Shiu-Kai Chin[1] and Jang Dae Kim[2]

[1] EE/CS Dept., Syracuse University, Syracuse, NY 13244
skchin@syr.edu
[2] National Semiconductor, 2900 Semiconductor Dr., Santa Clara, CA 95052
jdkim@galaxy.nsc.com

Abstract. We have created a calculus for reasoning about hardware and firmware at the algorithmic state machine (ASM) and instruction-set processor (ISP) levels of description. The calculus is a value-passing process algebra that extends the Mealy machine model to include parallel composition. It supports reasoning about the composed behavior of synchronous ASM and ISP components and microcode. We present an overview of the calculus and its application including an example showing the equivalence of a microcoded machine to its target instruction set specified by both ASM and ISP descriptions. The calculus, its properties, and the examples have been deeply embedded, proved, and verified as conservative extensions to the logic of the Higher Order Logic (HOL90) theorem prover.

1 Introduction

The *Instruction-set Process Calculus* (IspCal) is a calculus for describing processes at the algorithmic and instruction-set levels. Its purpose is to support compositional reasoning about the equivalence between *structures* of processes and *behaviors*.

The focus of IspCal on synchronous composition is similar to Milner's SCCS [11], Milne's Circuit Calculus (CIRCAL) [10], and Esterel [3]. IspCal differs from previous work in that it targets **instruction set processor** (ISP) [1] descriptions by having value-passing, delayed assignment of register values, and a semantics tuned to support the Mealy model of computation. This view is consistent with synchronous hardware design. The processes that are described are **algorithmic state machines** (ASM) [4][7] (e.g., hardware flowcharts), or ISP descriptions stating how instructions change registers and outputs.

Our objectives for IspCal are threefold: (1) to make connections to existing computer engineering design models by extending the Mealy machine model of hardware to reason about composed ISP and ASM descriptions; (2) to reason about correctness of microcoded architectures; and (3) to show equivalence between structural hardware and microcode descriptions and ASM or ISP specifications. IspCal describes behavior at a *cycle simulation* level, (i.e., changes occur on a clock-cycle basis).

IspCal consists of a *behavioral* language and a *structural* language. The behavioral language typically is used to describe ASM or ISP-level behavior. It

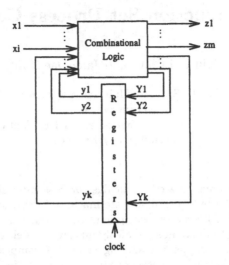

Fig. 1. Finite-State Machine

includes operators such as prefix, summation, guarded expressions, and mutual recursion. The structural language typically is used to describe implementations consisting of composed ASM or ISP processes. It includes operators for parallel composition and hiding.

The syntax and semantics of IspCal, all the equational laws, examples, and proofs have been done as conservative extensions to the higher-order logic of the HOL [6] theorem-prover. The techniques of defining IspCal within HOL are similar to the work done by Nesi [13]. Details of IspCal in HOL and the HOL proofs of the examples are found in http://www.cat.syr.edu/~chin/

The rest of this paper is organized as follows. Section 2 describes the Mealy machine model of hardware. Section 3 describes the syntax and semantics of behavioral IspCal expressions while Section 4 describes the syntax and semantics of structural IspCal expressions. A microprogrammed controller is described in Section 5, and a datapath for a CPU is described in Section 6. The controller and datapath are combined in Section 7 to form a CPU. The conclusions are in Section 8.

2 The Mealy Machine Model

The behaviors described by IspCal are an extension to the sequential Mealy machine model [8]. The model sketched in Figure 1 is used widely by computer engineers and circuit designers. Sequential machine behavior is specified as a sequence of events occurring at discrete points in time corresponding to clock cycles.

The machine has inputs $x_1 \ldots x_i$ and outputs $z_1 \ldots z_m$. The *present state* of the machine is given by state variables $y_1 \ldots y_k$. The *next state* of the machine is given by next-state variables $Y_1 \ldots Y_k$. The state register holding $y_1 \ldots y_k$

updates its contents with $Y_1 \ldots Y_k$ when *clock* has an *active* edge. The outputs $z_1 \ldots z_m$ and the next-state variables $Y_1 \ldots Y_k$ are functions of inputs $x_1 \ldots x_i$ and present-state variables $y_1 \ldots y_k$.

More formally, a *synchronous sequential machine* M is described by a 5-tuple

$$M = (I, O, S, \delta, \lambda),$$

where I, O, and S are finite, non-empty sets of *inputs, outputs,* and *states,* $\delta : (I \times S) \to S$ is a state-transition function taking elements of I and S and returning a state, and $\lambda : (I \times S) \to O$ is the output function taking elements of I and S as inputs and returning an element of O.

3 Behavioral IspCal

IspCal extends the Mealy machine model by adding several CCS-like [12] [11] operators such as parallel composition, choice, hiding, and value-passing. *Actions* are sets of input, output, and state assignments occurring *simultaneously* within a single clock cycle.

3.1 Syntax

The abstract syntax of behavioral IspCal uses expressions from the following syntactic domains: **BExp**, the set of boolean expressions be; **DExp**, the set of data expressions d; **Dvar**, the set of data identifiers x; **Port**, the set of port names p; **Event**, the set of events e; **BIsp**, the set of behavioral IspCal terms P; and **Var**, the set of process identifiers X.

We take for granted the syntax of data expressions, port names, and process and data identifiers. They are formally defined in our HOL implementation and are omitted here for reasons of space. **Data expressions** include: *Boolean expressions* with operators such as not, and, or, etc.; *indexed arrays*, (e.g. if x is an array then $x[i]$ is the ith element of array x); *arithmetic expressions* with arithmetic operators such as addition, subtraction, multiplication, and division; and *command expressions* corresponding to (micro) instructions. We use the HOL notion for **conditional expressions**: *if A then B else C* is written as $A \to B \mid C$ where $(true \to B \mid C) = B$ and $(false \to B \mid C) = C$.

Event expressions have three forms: $p?x$ (inputs to variables x through ports p); $p!d$ (values d assigned to ports p); and $x := d$ (assignment of data values d to state variables x). Thus their abstract syntax can be given as follows:

$e ::= p?x \mid p!d \mid x := d$

Prefix action expressions are finite sets of event expressions.

$pre ::= \{e_0; e_1; e_2; \ldots; e_k\}$

The abstract syntax of **process expressions** is as follows:

$P ::= pre.P \mid P_1 + P_2 \mid be \to P \mid X \mid Proc\, X\, Pl$

Process list expressions appear within the definition of recursive process expressions. They are finite sets of expressions $X_i \Leftarrow P_i$

$$Pl ::= \{X_0 \Leftarrow P_0; X_1 \Leftarrow P_1; X_2 \Leftarrow P_2; \ldots; X_n \Leftarrow P_n\}$$

We implicitly assume that there is only one definition for each process identifier X_i in the set. Process terms P_i cannot be recursive terms of the form $Proc\ X\ Pl$. As mutual recursion is supported, there is no need for nested recursion.

3.2 Substitution of Process Identifiers

$P[P'/X']$ denotes the term obtained by substituting process term P' for every occurrence of X' in process term P. The substitution rules are defined recursively as follows:

$$(pre.P)[P'/X'] = pre.(P[P'/X]')$$
$$(P1 + P2)[P'/X'] = (P1[P'/X']) + (P2[P'/X'])$$
$$(b \rightarrow P)[P'/X'] = b \rightarrow (P[P'/X'])$$
$$X[P'/X'] = \begin{cases} P' \text{ if } X = X' \\ X \text{ otherwise} \end{cases}$$

Expanding recursive process terms $Proc\ X\ Pl$ is done by substitution of process expressions P_i into free process identifiers X_i. As nested recursion is prohibited in the syntax, we don't have to specify the case for $Proc\ X\ Pl$ in the definition of process substitution. Multiple parallel (simultaneous) substitution of free process identifiers $(X'_0, X'_1, X'_2, \ldots, X'_m)$ in a behavioral IspCal term P with new process terms $(P'_0, P'_1, P'_2, \ldots, P'_m)$, respectively, is denoted by $P[P'_0/X'_0, P'_1/X'_1, P'_2/X'_2, \ldots, P'_m/X'_m]$.

3.3 Semantics

An inductively defined labeled-transition relation defines the semantics of behavioral IspCal. Inductively defined relations are defined by *rules*. The labeled-transition relations are the least relations satisfying the rules. This corresponds to the way relations are defined inductively in the HOL theorem-prover [6] by Melham [9].

Let V be the set of data values, v. A *store* σ is a mapping from data identifiers x to data values v. That is,

$$\sigma : Var \rightarrow V$$

Each store σ specifies a value, written $\sigma(x)$, for data identifier x. For any store σ, $\sigma[x \mapsto v]$ is the store that agrees with σ except it maps the identifier x to value v.

We assume the evaluation semantics are given for data and Boolean expressions. (This is described in detail as part of our HOL implementation but is omitted here for space reasons). We write $\langle d, \sigma \rangle \longrightarrow^* v$ to indicate that data expression d given store σ evaluates to v. We use similar notation for the evaluation of Boolean expressions. We let $[exp]\sigma$ denote the value of exp when evaluated with store σ.

Execution of an assignment statement $x := d$ given store σ results in an update of store σ to $\sigma[x \mapsto v]$ where $v = [d]\sigma$. Execution of multiple assignment statements results in multiple updates $\sigma[\ldots, x_i \mapsto v_i, \ldots]$ where $v_i = [d_i]\sigma$.

The operational semantics describe how a configuration — a process paired with a store — evolves to another configuration. A store keeps track of bindings of data identifiers of values.

The set of configurations Cfg is $Cfg = \{\langle P, \sigma \rangle | P \in BIsp \wedge \sigma \in store\}$ where σ is a store containing free data or state variables in P. The operational semantics is defined by a labeled transition system over process configurations.

An **event** is a member of the set $\{p.v | p \in Port \wedge v \in V\}$ where V is the set of data values. Intuitively, the event $p.v$ represents the presence of the value v on port p.

An **action** is simply a finite set of events $\{p_1.v_1; \ldots; p_n.v_n\}$.

A transition relation $\alpha \overset{act}{\Longrightarrow} \beta$ means the configuration α may perform the action act and become β. The transition label act is a subset of events $Event = \{p.v \mid p \in Port \wedge v \in V\}$. The labeled transition relation $\overset{act}{\Longrightarrow}$ is a subset of $Cfg \times Cfg$, where act is the transition label.

Four rules define $\Longrightarrow$, a family of relations indexed by actions. They are defined as the least relations satisfying the rules.

1. The prefix rule describes the behavior of a Mealy machine.

$$\frac{\forall i : k < i \leq n. \langle d_i, \sigma[x_1 \mapsto v_1, \ldots, x_k \mapsto v_k]\rangle \longrightarrow^* v_i \quad \begin{matrix} \forall i\, j.\, p_i = p_j \supset v_i = v_j \\ \forall i\, j.\, x_i = x_j \supset v_i = v_j \end{matrix}}{\langle \{p_1?x_1; \ldots; p_k?x_k; p_{k+1}!d_{k+1}; \ldots; p_m!d_m; x_{m+1} := d_{m+1}; \ldots; x_n := d_n\}.P, \sigma\rangle}$$

$$\overset{\{p_1 . v_1; \ldots; p_m . v_m\}}{\Longrightarrow}$$

$$\langle P, \sigma[x_{m+1} \mapsto v_{m+1}, \ldots, x_n \mapsto v_n]\rangle$$

Inputs $x_1, \ldots, x_k$ appear on ports $p_1, \ldots, p_k$. Ports $p_{k+1}, \ldots, p_m$ have output values. The expressions determining the output values are $d_{k+1}, \ldots, d_m$. The state variables are $x_{m+1}, \ldots, x_n$. The updated values of $x_{m+1}, \ldots, x_n$ are determined by expressions $d_{m+1}, \ldots, d_n$. Just as in the Mealy machine model, the values of the output and state expressions depend on the input values assigned to $x_1, \ldots, x_k$ and on the values of the state variables $x_{m+1}, \ldots, x_n$ in store σ. Each output and state expression d_i is evaluated given the input values $v_1, \ldots, v_k$ and the values of free state variables in d_i stored in σ. This is represented by $\forall i : k < i \leq n. \langle d_i, \sigma[x_1 \mapsto v_1, \ldots, x_k \mapsto v_k]\rangle \longrightarrow^* v_i$.

The conditions when $p_i = p_j$ and $x_i = x_j$ mean ports and state variables have only one value at a time. Notice that instantiation of input events does not change the store; only assignment statements can update stores and updates are in effect only after labeled transitions.

2. The second rule defines summation similar to CCS [12]. Intuitively, the process $P_1 + P_2$ can choose to behave like either P_1 or P_2.

$$\frac{\langle P_1, \sigma \rangle \xrightarrow{act} \langle P_1', \sigma' \rangle}{\langle P_1 + P_2, \sigma \rangle \xrightarrow{act} \langle P_1', \sigma' \rangle} \qquad \frac{\langle P_2, \sigma \rangle \xrightarrow{act} \langle P_2', \sigma' \rangle}{\langle P_1 + P_2, \sigma \rangle \xrightarrow{act} \langle P_2', \sigma' \rangle}$$

3. The third rule defines the behavior of *guarded* expressions.

$$\frac{\langle P, \sigma \rangle \xrightarrow{act} \langle P', \sigma' \rangle \quad \langle be, \sigma \rangle \longrightarrow^* true}{\langle be \rightarrow P, \sigma \rangle \xrightarrow{act} \langle P', \sigma' \rangle}$$

4. The final rule defines *expansion* of a recursive process *Proc X Pl*, as similar to the **fix** operator of SCCS [11], when there is exactly one instance of $X \Leftarrow P_X(X_1, \ldots, X_n)$ in *Pl* defining X. As free occurrences of process identifiers X_i in the body P_X represent recursion, they are replaced by recursive process terms *ProcX_i Pl* in which X_i are the initial process identifiers of the new recursion terms.

$$\frac{\langle P_X[ProcX_1Pl/X_1, \ldots, ProcX_nPl/X_n], \sigma \rangle \xrightarrow{act} \langle P', \sigma' \rangle}{\langle Proc\, X\, Pl, \sigma \rangle \xrightarrow{act} \langle P', \sigma' \rangle} \quad \begin{array}{l} X \text{ is uniquely} \\ \text{defined in } Pl \text{ by} \\ X \Leftarrow P_X(X_1, \ldots, X_n) \end{array}$$

3.4 Bisimulation Equivalence

Two behavioral IspCal configurations $\langle P_1, \sigma_1 \rangle$ and $\langle P_2, \sigma_2 \rangle$ are equivalent if they are **bisimilar**, written $\langle P_1, \sigma_1 \rangle \sim \langle P_2, \sigma_2 \rangle$. Our definition of bisimulation corresponds to Milner's [12].

$$\langle P_1, \sigma_1 \rangle \sim \langle P_2, \sigma_2 \rangle =_{def} \exists \mathcal{R}. \, Bisim\, \mathcal{R} \wedge (\langle P_1, \sigma_1 \rangle, \langle P_2, \sigma_2 \rangle) \in \mathcal{R}$$

where $\mathcal{R} \subseteq Cfg \times Cfg$ identifying bisimilar states or processes. $\mathcal{R}$ is a **bisimulation**, written *Bisim* $\mathcal{R}$, when for all P_1, P_2, σ_1, σ_2, if $(\langle P_1, \sigma_1 \rangle, \langle P_2, \sigma_2 \rangle) \in \mathcal{R}$ then for all actions a:

(a) $\langle P_1, \sigma_1 \rangle \xrightarrow{a} \langle P_1', \sigma_1' \rangle$ implies $\exists P_2' \sigma_2'. \langle P_2, \sigma_2 \rangle \xrightarrow{a} \langle P_2', \sigma_2' \rangle \wedge (\langle P_1', \sigma_1' \rangle, \langle P_2', \sigma_2' \rangle) \in \mathcal{R}$.

(b) $\langle P_2, \sigma_2 \rangle \xrightarrow{a} \langle P_2', \sigma_2' \rangle$ implies $\exists P_1' \sigma_1'. \langle P_1, \sigma_1 \rangle \xrightarrow{a} \langle P_1', \sigma_1' \rangle \wedge (\langle P_1', \sigma_1' \rangle, \langle P_2', \sigma_2' \rangle) \in \mathcal{R}$.

Bisimilar configurations make transitions that preserve bisimilarity. Given two configurations, if $\langle P_1, \sigma_1 \rangle \sim \langle P_2, \sigma_2 \rangle$, then for any action a,

(a) $\langle P_1, \sigma_1 \rangle \xrightarrow{a} \langle P_1', \sigma_1' \rangle$ implies $\exists P_2' \sigma_2'. \langle P_2, \sigma_2 \rangle \xrightarrow{a} \langle P_2', \sigma_2' \rangle \wedge \langle P_1', \sigma_1' \rangle \sim \langle P_2', \sigma_2' \rangle$

(b) $\langle P_2, \sigma_2 \rangle \xrightarrow{a} \langle P_2', \sigma_2' \rangle$ implies $\exists P_1' \sigma_1'. \langle P_1, \sigma_1 \rangle \xrightarrow{a} \langle P_1', \sigma_1' \rangle \wedge \langle P_1', \sigma_1' \rangle \sim \langle P_2', \sigma_2' \rangle$.

Bisimilarity is an equivalence relation.

(Commutivity and Associativity) For all behavioral IspCal terms P, Q, R, and store σ,

1. $\langle P + Q, \sigma \rangle \sim \langle Q + P, \sigma \rangle$
2. $\langle P + (Q + R), \sigma \rangle \sim \langle (P + Q) + R, \sigma \rangle$

(Substitutivity) For all behavioral IspCal terms P, Q, R and store σ,

1. $\langle P, \sigma \rangle \sim \langle Q, \sigma \rangle \supset \langle P + R, \sigma \rangle \sim \langle Q + R, \sigma \rangle$
2. $\langle P, \sigma \rangle \sim \langle Q, \sigma \rangle \supset \langle b \rightarrow P, \sigma \rangle \sim \langle b \rightarrow Q, \sigma \rangle$
3. $X \Leftarrow P_X(X_1, \ldots, X_n) \in Pl \wedge (\forall X \Leftarrow P_i, X \Leftarrow P_j \in Pl. \, P_i = P_j) \supset$
   $\langle Proc\, X\, Pl, \sigma \rangle \sim \langle P_X[Proc\, X_1\, Pl/X_1, \ldots, Proc\, X_n\, Pl/X_n], \sigma \rangle$

4 Structural IspCal

4.1 Syntax

Structural IspCal (SIsp) terms are constructed from behavioral IspCal terms. A basic SIsp term ($Cell\ P\ \sigma\ s$) consists of a behavioral IspCal term P, its (local) store σ, and a port set s declaring all the ports process P may use. Structural IspCal terms, C_1 and C_2, are composed synchronously by $C_1\ \|\ C_2$. SIsp term $C \setminus s$ hides ports s of the component C. The abstract syntax for SIsp terms is:

$$C ::= Cell\ P\ \sigma\ s\ |\ C_1 \| C_2\ |\ C \setminus s$$
$$s ::= \{p_0; p_1; p_2; \ldots; p_n\}$$

where $P \in BIsp$, σ is a local store for P, $p \in Port$, and $s \subseteq Port$.

4.2 Sort

The *sort* of a structural IspCal term C is the set of all possible real events $p.v$ that the term C may perform. We define a recursive function $SORT$ that computes the sort of structural IspCal terms as follows:

$$SORT(Cell\ P\ \sigma\ s) = \{p.v \mid p \in s \wedge v \in V\}$$
$$SORT(C_1 \| C_2) = SORT(C_1) \cup SORT(C_2)$$
$$SORT(C \setminus s) = SORT(C) - \{p.v \mid p \in s \wedge v \in V\}$$

4.3 Operational Semantics of Structural IspCal Terms

The semantics of structural IspCal terms is defined by a labeled transition system $\overset{act}{\Longrightarrow} \subseteq SIsp \times SIsp$. As in the case of behavioral IspCal, $C_1 \overset{act}{\Longrightarrow} C_2$ means that the component C_1 may perform the action act and become C_2. The rules for the labeled transition system for structural IspCal are as follows:

1. $$\frac{\langle P, \sigma \rangle \overset{act}{\Longrightarrow} \langle P', \sigma' \rangle}{Cell\ P\ \sigma\ s \overset{act}{\Longrightarrow} Cell\ P'\ \sigma'\ s}\ act \subseteq \{p.v \mid p \in s \wedge v \in V\}$$
   Basic cells behave according to their process definitions.

2. $$\frac{C \overset{act}{\Longrightarrow} C'}{(C \setminus s) \overset{act - \{p.v \mid p \in s \wedge v \in V\}}{\Longrightarrow} (C' \setminus s)}$$
   Cells with hidden ports make the same transitions as the same cells without hidden ports except the events corresponding to the hidden ports are missing.

3. $$\frac{C_1 \overset{a_1}{\Longrightarrow} C_1'\quad C_2 \overset{a_2}{\Longrightarrow} C_2'}{(C_1 \| C_2) \overset{a_1 \cup a_2}{\Longrightarrow} (C_1' \| C_2')}\ a_1 \cap SORT(C_2) = a_2 \cap SORT(C_1)$$
   Synchronous composition has the union of its components' actions as its transition label. The components make transitions in lock-step synchronism.

The side condition for the composition rule is the *interacting condition*. It requires events on shared ports of C_1 and C_2 to be the same. Intuitively, when two components are composed with $\|$, similarly named ports of C_1 and C_2 are connected together and both components share the same clock.

4.4 Bisimulation Equivalence

Bisimulation relations for SIsp terms are defined as they are for BIsp configurations. $\mathcal{R}$ is a bisimulation when for all SIsp terms C_1, C_2, if $(C_1, C_2) \in \mathcal{R}$ then for all actions a:

(a) $C_1 \overset{a}{\Rightarrow} C_1'$ implies $\exists C_2'. C_2 \overset{a}{\Rightarrow} C_2' \wedge (C_1', C_2') \in \mathcal{R}$.

(b) $C_2 \overset{a}{\Rightarrow} C_2'$ implies $\exists C_1'. C_1 \overset{a}{\Rightarrow} C_1' \wedge (C_1', C_2') \in \mathcal{R}$.

As before, bisimilarity for SIsp processes is defined as

$$C_1 \sim C_2 =_{def} \exists \mathcal{R}.\ Bisim\ \mathcal{R} \wedge (C_1, C_2) \in \mathcal{R}$$

The properties of bisimilarity for SIsp processes are the same as the properties for bisimilarity for BIsp configurations. If $C_1 \sim C_2$ then for any action a,

(a) $C_1 \overset{a}{\Rightarrow} C_1'$ implies $\exists C_2'. C_2 \overset{a}{\Rightarrow} C_2' \wedge C_1' \sim C2'$

(b) $C_2 \overset{a}{\Rightarrow} C_2'$ implies $\exists C_1'. C_1 \overset{a}{\Rightarrow} C_1' \wedge C_1' \sim C_2'$.

The following properties are essential for hardware composition and hold for SIsp bisimulation relations.

(**Commutivity and Associativity**) For all structural IspCal agents C, C_1, C_2, and C_3,

1. $(C_1 \parallel C_2) \sim (C_2 \parallel C_1)$
2. $((C_1 \parallel C_2) \parallel C_3) \sim (C_1 \parallel (C_2 \parallel C_3))$

(**Substitutivity**) For all structural IspCal agents C_1, C_2, C_3 and stores σ_1, σ_2,

1. $\langle P1, \sigma_1 \rangle \sim \langle P2, \sigma_2 \rangle \supset (Cell\ P_1\ \sigma_1\ s) \sim (Cell\ P_2\ \sigma_2\ s)$
2. $C_1 \sim C_2 \wedge (SORT(C_1) = SORT(C_2)) \supset (C_1 \parallel C_3) \sim (C_2 \parallel C_3)$
3. $C_1 \sim C_2 \supset (C_1 \setminus s) \sim (C_2 \setminus s)$

5 Microprogrammed Controller

We describe a microprogrammed controller using both an ASM model and an ISP model. The two models are proved equivalent, provided they start from similar states. We implement the controller using a sequencer and a ROM. The implementation is proved equivalent to both the ASM and ISP models of the controller. This controller is then used in Section 7 to implement a PDP-8 instruction set machine [2].

5.1 User's Model

The controller has an input port op and an output port *control*. The ASM description of the microprogrammed controller uses symbolic control states in place of registers in the control part. It has 27 such control states in the following ASM description of the controller. Figure 2 shows the state diagram of the controller. The IspCal description is below. Note, op is used both as a port name and as a variable name. When op appears in data expressions op is a variable.

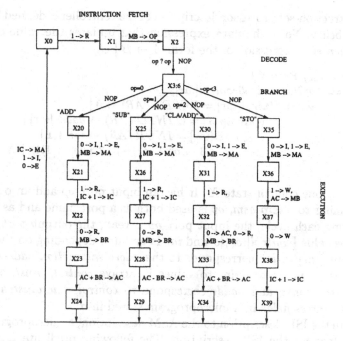

Fig. 2. Flow chart for the microprogrammed controller

Input and output events are in every prefix in *micro_asm* (even if the input values are not used) as ports on physical devices always have some value.

```
micro_asm =def {
  X0  ⟸ {op?op; control!SET R}.X1;
  X1  ⟸ {op?op; control!MB INTO OP}.X2;
  X2  ⟸ {op?op; op := op; control!NOP}.
        ((op = 0 → X3) + (op = 1 → X4) + (op = 2 → X5) + (¬op < 3 → X6));
  X3  ⟸ {op?op; control!NOP}.X20;
  X4  ⟸ {op?op; control!NOP}.X25;
  X5  ⟸ {op?op; control!NOP}.X30;
  X6  ⟸ {op?op; control!NOP}.X35;
  X20 ⟸ {op?op; control!RESET I, SET E, MB TO MA}.X21;
  X21 ⟸ {op?op; control!SET R, INCR IC}.X22;
  X22 ⟸ {op?op; control!RESET R, MB INTO BR}.X23;
  X23 ⟸ {op?op; control!ADD}.X24;
  X24 ⟸ {op?op; control!IC INTO MA, SET I, RESET E}.X0;
  X25 ⟸ {op?op; control!MB TO MA, RESET I, SET E}.X26;
  X26 ⟸ {op?op; control!SET R, INCR IC}.X27;
  X27 ⟸ {op?op; control!MB INTO BR, RESET R}.X28;
  X28 ⟸ {op?op; control!SUBTRACT}.X29;
  X29 ⟸ {op?op; control!IC INTO MA, SET I, RESET E}.X0;
  X30 ⟸ {op?op; control!MB TO MA, RESET I, SET E}.X31;
  X31 ⟸ {op?op; control!SET R}.X32;
  X32 ⟸ {op?op; control!MB INTO BR, RESET AC, RESET R}.X33;
  X33 ⟸ {op?op; control!ADD}.X34;
  X34 ⟸ {op?op; control!IC INTO MA, SET I, RESET E}.X0;
  X35 ⟸ {op?op; control!RESET I, SET E, MB TO MA}.X36;
  X36 ⟸ {op?op; control!SET W, AC INTO MB}.X37;
  X37 ⟸ {op?op; control!RESET W}.X38;
  X38 ⟸ {op?op; control!INCR IC}.X39;
  X39 ⟸ {op?op; control!IC INTO MA, SET I, RESET E}.X0;
}
```

The successor of $\langle Proc\, X_2\ micro\_asm, \sigma \rangle$, denoted as $X_{3:6}$ in Figure 2, is $\langle ((op = 0 \rightarrow Proc\, X_3\ micro\_asm) + (op = 1 \rightarrow Proc\, X_4\ micro\_asm) + (op = 2 \rightarrow Proc\, X_5\ micro\_asm) + (\neg(op < 3) \rightarrow Proc\, X_6\ micro\_asm), \sigma[op \mapsto v]\rangle$ for some value v, which behaves similar to $Proc\, X_3\ micro\_asm$, $Proc\, X_4\ micro\_asm$, $Proc\, X_5\ micro\_asm$, or $Proc\, X_6\ micro\_asm$, depending on the value v received in state X_2.

The instruction-set processor description of the controller is defined by $micro\_isp$ shown below. Note, the data expression that updates the value of iar is a nested *if-then-else* expression of the form $A \rightarrow B \mid C$.

$$micro\_isp =_{def} Proc\ X\ \{$$
$$X \Leftarrow \{op?op; control!control[iar];$$
$$iar := ((uinst[iar] = RESET\ IAR) \rightarrow 0 \mid$$
$$((uinst[iar] = C\ INTO\ IAR) \rightarrow uaddr[iar] \mid$$
$$((uinst[iar] = INCR\ IAR) \rightarrow iar + 1 \mid$$
$$(op < 3 \rightarrow op + iar + 1 \mid 3 + iar + 1))))\}.X$$
$$\}$$

$micro\_isp$ has one control state X. It has an input port op and an output port $control$. Similar to $micro\_asm$, op is used both as a port name and as a variable name. During each cycle, the input port op is read, the output port $control$ is assigned the value $[control[iar]]\sigma$ and iar is updated depending on the values of op and $[uinst[iar]]\sigma$. iar corresponds to the *micro-instruction address register*. It is used as an index to point to values in arrays $control$, $uinst$, and $uaddr$. Arrays $control$, $uinst$, and $uaddr$ correspond to *control, microinstruction,* and *next micro-address fields* in a microprogram stored in σ.

To relate the ISP description to the ASM description, a microprogram should be in the store for the ISP description. The following predicate $is\_micropgm$ specifies the stored microprogram by explicitly stating the values of the control, micro-instruction, and next micro-address fields. Notice the correspondence between the values of $control$ in the microprogram with the values assigned to the $control$ port in $micro\_asm$. The predicate $is\_micropgm$ is an abstract description of the microprogram in σ.

$\forall \sigma.\ is\_micropgm\ \sigma =_{def}$
$([uinst[0]]\sigma = INCR\ IAR) \wedge ([control[0]]\sigma = SET\ R) \wedge$
$([uinst[1]]\sigma = INCR\ IAR) \wedge ([control[1]]\sigma = MB\ INTO\ OP) \wedge$
$([uinst[2]]\sigma = ADD\ OP\ TO\ IAR) \wedge ([control[2]]\sigma = NOP) \wedge$
$([uinst[3]]\sigma = C\ INTO\ IAR) \wedge (uaddr[3]]\sigma = 20) \wedge ([control[3]]\sigma = NOP) \wedge$
$([uinst[4]]\sigma = C\ INTO\ IAR) \wedge (uaddr[4]]\sigma = 25) \wedge ([control[4]]\sigma = NOP) \wedge$
$([uinst[5]]\sigma = C\ INTO\ IAR) \wedge (uaddr[5]]\sigma = 30) \wedge ([control[5]]\sigma = NOP) \wedge$
$([uinst[6]]\sigma = C\ INTO\ IAR) \wedge (uaddr[6]]\sigma = 35) \wedge ([control[6]]\sigma = NOP) \wedge$
$([uinst[20]]\sigma = INCR\ IAR) \wedge ([control[20]]\sigma = RESET\ I, SET\ E, MB\ TO\ MA) \wedge$
$([uinst[21]]\sigma = INCR\ IAR) \wedge ([control[21]]\sigma = SET\ R, INCR\ IC) \wedge$
$([uinst[22]]\sigma = INCR\ IAR) \wedge ([control[22]]\sigma = RESET\ R, MB\ INTO\ BR) \wedge$
$([uinst[23]]\sigma = INCR\ IAR) \wedge ([control[23]]\sigma = ADD) \wedge$
$([uinst[24]]\sigma = RESET\ IAR) \wedge ([control[24]]\sigma = IC\ INTO\ MA, SET\ I, RESET\ E) \wedge$
$([uinst[25]]\sigma = INCR\ IAR) \wedge ([control[25]]\sigma = MB\ INTO\ MA, RESET\ I, SET\ E) \wedge$
$([uinst[26]]\sigma = INCR\ IAR) \wedge ([control[26]]\sigma = SET\ R, INCR\ IC) \wedge$
$([uinst[27]]\sigma = INCR\ IAR) \wedge ([control[27]]\sigma = MB\ INTO\ BR, RESET\ R) \wedge$
$([uinst[28]]\sigma = INCR\ IAR) \wedge ([control[28]]\sigma = SUBTRACT) \wedge$
$([uinst[29]]\sigma = RESET\ IAR) \wedge ([control[29]]\sigma = IC\ INTO\ MA, SET\ I, RESET\ E) \wedge$
$([uinst[30]]\sigma = INCR\ IAR) \wedge ([control[30]]\sigma = MB\ INTO\ MA, RESET\ I, SET\ E) \wedge$
$([uinst[31]]\sigma = INCR\ IAR) \wedge ([control[31]]\sigma = SET\ R) \wedge$
$([uinst[32]]\sigma = INCR\ IAR) \wedge ([control[32]]\sigma = MB\ INTO\ BR, RESET\ AC, RESET\ R) \wedge$
$([uinst[33]]\sigma = INCR\ IAR) \wedge ([control[33]]\sigma = ADD) \wedge$
$([uinst[34]]\sigma = RESET\ IAR) \wedge ([control[34]]\sigma = IC\ INTO\ MA, SET\ I, RESET\ E) \wedge$
$([uinst[35]]\sigma = INCR\ IAR) \wedge ([control[35]]\sigma = RESET\ I, SET\ E, MB\ INTO\ MA) \wedge$
$([uinst[36]]\sigma = INCR\ IAR) \wedge ([control[36]]\sigma = SET\ W, AC\ INTO\ MB) \wedge$
$([uinst[37]]\sigma = INCR\ IAR) \wedge ([control[37]]\sigma = RESET\ W) \wedge$
$([uinst[38]]\sigma = INCR\ IAR) \wedge ([control[38]]\sigma = INCR\ IC) \wedge$
$([uinst[39]]\sigma = RESET\ IAR) \wedge ([control[39]]\sigma = IC\ INTO\ MA, SET\ I, RESET\ E)$

With the microprogram in the store of the ISP description, we can prove that $micro\_isp$ is equivalent to $Proc\ X_0\ micro\_asm$ if the ISP model starts at instruction address 0.

$$\vdash \forall \sigma_1 \sigma_2.([iar]\sigma_1 = 0) \wedge is\_micropgm\sigma_1 \supset \langle micro\_isp, \sigma_1 \rangle \sim \langle ProcX_0 micro\_asm, \sigma_2 \rangle$$

We proved the following relation $\mathcal{R}$ is a bisimulation, i.e. $\vdash Bisim\ \mathcal{R}$, for any σ_1' and σ_2'. $\mathcal{R}$ must be given by the designer.

$$
\begin{aligned}
\mathcal{R} = &\{(\langle micro\_isp, \sigma_1' \rangle, \langle Proc\,X_0\ micro\_asm, \sigma_2' \rangle) \mid ([\![IAR]\!]\sigma_1' = 0) \wedge is\_micropgm(\sigma_1')\} \cup \\
&\{(\langle micro\_isp, \sigma_1' \rangle, \langle Proc\,X_1\ micro\_asm, \sigma_2' \rangle) \mid ([\![IAR]\!]\sigma_1' = 1) \wedge is\_micropgm(\sigma_1')\} \cup \\
&\{(\langle micro\_isp, \sigma_1' \rangle, \langle Proc\,X_2\ micro\_asm, \sigma_2' \rangle) \mid ([\![IAR]\!]\sigma_1' = 2) \wedge is\_micropgm(\sigma_1')\} \cup \\
&\{(\langle micro\_isp, \sigma_1' \rangle, \langle (op = 0) \rightarrow Proc\,X_3\ micro\_asm + (op = 1) \rightarrow Proc\,X_4\ micro\_asm + \\
&\quad (op = 2) \rightarrow Proc\,X_5\ micro\_asm + \neg(op < 3) \rightarrow Proc\,X_6\ micro\_asm, \sigma_2' \rangle) \mid \\
&\quad ([\![IAR]\!]\sigma_1' = 3) \wedge ([\![op]\!]\sigma_1' = 0) \wedge is\_micropgm(\sigma_1')\} \cup \\
&\{(\langle micro\_isp, \sigma_1' \rangle, \langle (op = 0) \rightarrow Proc\,X_3\ micro\_asm + (op = 1) \rightarrow Proc\,X_4\ micro\_asm + \\
&\quad (op = 2) \rightarrow Proc\,X_5\ micro\_asm + \neg(op < 3) \rightarrow Proc\,X_6\ micro\_asm, \sigma_2' \rangle) \mid \\
&\quad ([\![IAR]\!]\sigma_1' = 4) \wedge ([\![op]\!]\sigma_1' = 1) \wedge is\_micropgm(\sigma_1')\} \cup \\
&\{(\langle micro\_isp, \sigma_1' \rangle, \langle (op = 0) \rightarrow Proc\,X_3\ micro\_asm + (op = 1) \rightarrow Proc\,X_4\ micro\_asm + \\
&\quad (op = 2) \rightarrow Proc\,X_5\ micro\_asm + \neg(op < 3) \rightarrow Proc\,X_6\ micro\_asm, \sigma_2' \rangle) \mid \\
&\quad ([\![IAR]\!]\sigma_1' = 5) \wedge ([\![op]\!]\sigma_1' = 2) \wedge is\_micropgm(\sigma_1')\} \cup \\
&\{(\langle micro\_isp, \sigma_1' \rangle, \langle (op = 0) \rightarrow Proc\,X_3\ micro\_asm + (op = 1) \rightarrow Proc\,X_4\ micro\_asm + \\
&\quad (op = 2) \rightarrow Proc\,X_5\ micro\_asm + \neg(op < 3) \rightarrow Proc\,X_6\ micro\_asm, \sigma_2' \rangle) \mid \\
&\quad ([\![IAR]\!]\sigma_1' = 6) \wedge \neg([\![op]\!]\sigma_1') < 3) \wedge is\_micropgm(\sigma_1')\} \cup \\
&\{(\langle micro\_isp, \sigma_1' \rangle, \langle Proc\,X_i\ micro\_asm, \sigma_2' \rangle) \mid is\_micropgm(\sigma_1') \wedge \\
&\hspace{6cm} ([\![IAR]\!]\sigma_1' = i)\ for\ i = 20, \ldots, 39\}
\end{aligned}
$$

5.2 Controller Implementation

The controller is implemented using a sequencer and a read-only memory (ROM). The sequencer and ROM are composed to implement the controller as shown in Figure 3. The difference between this description and $micro\_isp$ is that $micro\_isp$ did not describe any implementation details of how the ROM was controlled by the sequencer. In particular, the details of how values for the $addr$, $uinst$, and $uaddr$ ports $internal$ to the implementation of $micro\_imp$ being computed were not stated.

The behavior of the sequencer and ROM are described as ISP models. The sequencer has an instruction address register iar. It has input ports op, $uinst$, and $uaddr$, and an output port $addr$. The sequencer in IspCal is defined as:

```
microseq =def Proc X {
    X ⟸ {op?op; uinst?uinst; uaddr?uaddr; addr!iar;
        iar := ((uinst = RESET IAR) → 0 |
                ((uinst = C INTO IAR) → uaddr |
                ((uinst = INCR IAR) → iar + 1 |
                (op < 3 → op + iar + 1 | 3 + iar + 1))))}.X
}
```

$uinst$ is the micro-instruction from the ROM. $uaddr$ is the next micro-address from the ROM. The ROM is addressed by $addr$.

The microprogram ROM, $microrom$, has three output ports — $uinst$ for current microinstruction code, $uaddr$ for the next address, and $control$ for the micro-control output.

```
microrom =def Proc X {
    X ⟸ {addr?addr; uinst!uinst[addr]; uaddr!uaddr[addr]; control!control[addr]}.X
}
```

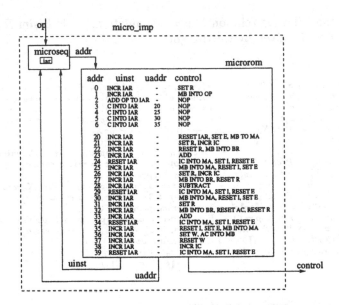

Fig. 3. Structural implementation of the controller

The term *microrom* only defines how memory is read. The actual memory contents are in store σ. *micro_imp* is defined to be the composition of the sequencer and the ROM with the internal ports *uinst*, *uaddr*, and *addr* hidden.

$$micro\_imp\ \sigma_1\ \sigma_2 =_{def}$$
$$(Cell\ microseq\ \sigma_1\ \{op; uinst; uaddr; addr\}\ \|$$
$$Cell\ microrom\ \sigma_2\ \{addr; uinst; uaddr; control\}) \backslash \{uinst; uaddr; addr\}$$

The following theorem asserts the equivalence between the structural implementation *micro_imp* and the ISP description of the controller *micro_isp*, if they start from similar memory states.

$$\vdash \forall \sigma\ \sigma_1\ \sigma_2.$$
$$([iar]\sigma = [iar]\sigma_1) \wedge (\forall i.[uinst[i]]\sigma = [uinst[i]]\sigma_2) \wedge$$
$$(\forall i.[uaddr[i]]\sigma = [uaddr[i]]\sigma_2) \wedge (\forall i.[control[i]]\sigma = [control[i]]\sigma_2) \supset$$
$$(micro\_imp\ \sigma_1\ \sigma_2) \sim (Cell\ micro\_isp\ \sigma\ \{op; control\})$$

The bisimulation relation $\mathcal{R}$ between *micro_imp* and *micro_isp* is given below. Note that any pair $(micro\_imp\ \sigma_1\sigma_2, Cell\ micro\_isp\ \sigma\{op; control\})$ is contained in the relation $\mathcal{R}$ for any σ, σ_1, and σ_2 as long as the stores σ, σ_1, and σ_2 satisfy the constraints in the set specification of $\mathcal{R}$ shown below.

$$\mathcal{R} = \{(micro\_imp\ \sigma_1\sigma_2, Cell\ micro\_isp\ \sigma\{op; control\})\ |$$
$$([iar]\sigma = [iar]\sigma_1) \wedge (\forall i.\ [uinst[i]]\sigma = [uinst[i]]\sigma_2) \wedge$$
$$(\forall i.\ [uaddr[i]]\sigma = [uaddr[i]]\sigma_2) \wedge (\forall i.\ [control[i]]\sigma = [control[i]]\sigma_2)\}$$

By transitivity of the bisimulation relation, the implementation is also equivalent with the ASM description of the controller provided the microprogram is

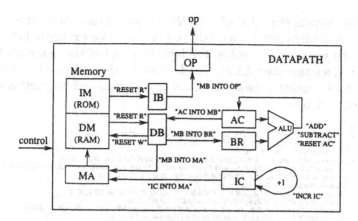

Fig. 4. Datapath Registers and their Connection

in the store for the implementation. The above theorem does not constrain σ in any way as the behavior of *micro_asm* is unaffected by σ.

$\vdash \forall \sigma \, \sigma_1 \, \sigma_2.$

$\qquad ([iar]\sigma_1 = 0) \wedge is\_micropgm \, \sigma_2 \supset$

$\qquad (micro\_imp \, \sigma_1 \, \sigma_2) \sim (Cell \, (Proc \, X0 \, micro\_asm) \, \sigma \, \{op; control\})$

6 A Datapath Example

Datapath modules are simple to describe in IspCal using value-passing. The example in this section is modeled after Harris 6100 microprocessor in [2]. Our simplified model is shown in Figure 4.

Main memory is divided into instruction memory (IM) and data memory (DM). The read-only instruction memory holds the operation (OP code) portion of instructions. The data memory holds instruction addresses and data. Data are read from or written to the data memory. Registers IB and DB are buffer registers for IM and DM, respectively. The remaining registers are the instruction register (OP), memory address register (MA), instruction counter (IC), accumulator (AC) and another register (BR).

The instruction counter (IC) keeps track of program instructions. Normally IC is incremented with each instruction execution. The contents of the counter are transferred into MA at the beginning of each instruction cycle. The MA register provides memory address to both IM and DM.

When an instruction word is read from the memory, the OP-code portion of the word is stored in the OP register later to determine what instruction is to be performed. The memory address register points to the memory location to be read from or written to.

We modeled only 13 micro-control instructions for the datapath in this example. All the registers in the datapath store numerical values in this model. The OP register also stores numbers because the micro-controller performs arithmetic operations on the value received from the OP register. Note that the *op* port always outputs the value of the *OP* register. This reflects the fact that a physical port always carries some value on it unless we explicitly hide the port.

As all the registers are visible, the IspCal model of the datapath is an ISP machine,(i.e., a single-state ASM). The ISP model shown below has a single control state X. All registers are state variables. The data and instruction memories are indexed variables.

```
datapath =def Proc X {
    X ⟸ {control!SET R; op!OP; DB := DM[MA]; IB := IM[MA]}.X +
        {control!MB INTO OP; op!OP; OP := IB}.X +
        {control!NOP; op!OP}.X +
        {control!RESET I, SET E, MB INTO MA; op!OP; MA := DB}.X +
        {control!SET R, INCR IC; op!OP; IC := IC + 1; DB := DM[MA]; IB := IM[MA]}.X +
        {control!RESET R, MB INTO BR; op!OP; BR := DB}.X +
        {control!ADD; op!OP; AC := AC + BR}.X +
        {control!IC INTO MA, SET I, RESET E; op!OP; MA := IC}.X +
        {control!SUBTRACT; op!OP; AC := AC - BR}.X +
        {control!MB INTO BR, RESET AC, RESET R; op!OP; AC := 0; BR := DB}.X +
        {control!SET W, AC INTO MB; op!OP; DB := AC}.X +
        {control!RESET W; op!OP; DM[MA] := DB}.X +
        {control!INCR IC; op!OP; IC := IC + 1}.X
}
```

There are 13 micro-control instructions that can appear on the *control* port. All instructions except NOP involve register updates. The microinstruction command placed on the *control* port (from the controller) determines which instruction to perform.

Two types of lemmas together characterize the behavior of *datapath*. Lemmas of the first type describe the state transition for each instruction. An example of the first lemma type is shown below for the microinstruction *SET R*. (The other microinstructions have similar lemmas).

$$\vdash \forall \sigma.\ \langle datapath, \sigma \rangle \overset{\{control.SETR; op.[OP]\sigma\}}{\Longrightarrow}$$
$$\langle datapath, \sigma[DB \mapsto [DM[MA]]\sigma; IB \mapsto [IM[MA]]\sigma]\rangle$$

This lemma says the configuration $\langle datapath, \sigma \rangle$ can make a $\{control.SET\ R; op.[OP]\sigma\}$ move to a new configuration $\langle datapath, \sigma[DB \mapsto [DM[MA]]\sigma; IB \mapsto [IM[MA]]\sigma]\rangle$.

The second type of lemma lists all possible states reachable by a single transition. In this case, if a transition was made, the transition corresponded to one of the 13 microinstructions.

$$\vdash \forall \sigma a P \sigma'. \langle datapath, \sigma \rangle \overset{a}{\Longrightarrow} \langle P, \sigma' \rangle \supset$$
$$(P = datapath) \wedge$$
$$((a = \{control.SETR; op.[OP]\sigma\}) \wedge$$
$$\quad (\sigma' = \sigma[DB \mapsto [DM[MA]]\sigma; IB \mapsto [IM[MA]]\sigma]) \vee$$
$$(a = \{control.MBINTOOP; op.[OP]\sigma\}) \wedge (\sigma' = \sigma[OP \mapsto [IB]\sigma]) \vee$$
$$(a = \{control.NOP; op.[OP]\sigma\}) \wedge (\sigma' = \sigma) \vee$$
$$(a = \{control.RESETI, SETE, MBINTOMA; op.[OP]\sigma\}) \wedge$$
$$\quad (\sigma' = \sigma[MA \mapsto [DB]\sigma]) \vee$$
$$(a = \{control.SETR, INCRIC; op.[OP]\sigma\}) \wedge$$
$$\quad (\sigma' = \sigma[IC \mapsto [IC]\sigma + 1; DB \mapsto [DM[MA]]\sigma; IB \mapsto [IM[MA]]\sigma]) \vee$$
$$(a = \{control.RESETR, MBINTOBR; op.[OP]\sigma\}) \wedge (\sigma' = \sigma[BR \mapsto [DB]\sigma]) \vee$$
$$(a = \{control.ADD; op.[OP]\sigma\}) \wedge (\sigma' = \sigma[AC \mapsto [AC]\sigma + [BR]\sigma]) \vee$$
$$(a = \{control.ICINTOMA, SETI, RESETE; op.[OP]\sigma\}) \wedge$$
$$\quad (\sigma' = \sigma[MA \mapsto [IC]\sigma]) \vee$$
$$(a = \{control.SUBTRACT; op.[OP]\sigma\}) \wedge$$
$$\quad (\sigma' = \sigma[AC \mapsto [AC]\sigma - [BR]\sigma]) \vee$$
$$(a = \{control.MBINTOBR, RESETAC, RESETR; op.[OP]\sigma\}) \wedge$$
$$\quad (\sigma' = \sigma[AC \mapsto 0; BR \mapsto [DB]\sigma]) \vee$$
$$(a = \{control.SETW, ACINTOMB; op.[OP]\sigma\}) \wedge (\sigma' = \sigma[DB \mapsto [AC]\sigma]) \vee$$
$$(a = \{control.RESETW; op.[OP]\sigma\}) \wedge (\sigma' = \sigma[DM[[MA]\sigma] \mapsto [DB]\sigma]) \vee$$
$$(a = \{control.INCRIC; op.[OP]\sigma\}) \wedge (\sigma' = \sigma[IC \mapsto [IC]\sigma + 1]))$$

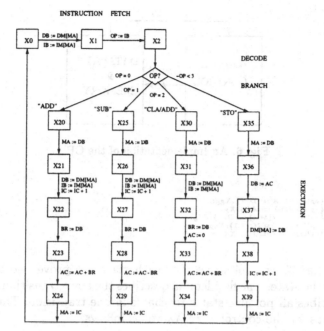

Fig. 5. Behavioral ASM description of the CPU

The two lemma types are used when reasoning about bisimulation.

7 A CPU Example

Composing the controller and datapath implements a PDP-8 instruction set machine [2]. The instruction set implemented here includes *ADD*, *SUBTRACT*, *CLEAR AND ADD*, and *STORE*. We show that the composition of the controller and datapath is bisimilar to the instruction set behavior as specified by an ASM. The behavioral ASM model is shown in Figure 5. Its representation as an IspCal term *top_asm* is shown below. The structural model is shown in Figure 6.

```
top_asm =def {
    X0 ⇐ {op!OP; DB := DM[MA]; IB := IM[MA]}.X1;
    X1 ⇐ {op!OP; OP := IB}.X2;
    X2 ⇐ {op!OP}.((OP = 0 → X3) + (OP = 1 → X4) + (OP = 2 → X5) + (¬ OP < 3 → X6));
    X3 ⇐ {op!OP}.X20;
    X4 ⇐ {op!OP}.X25;
    X5 ⇐ {op!OP}.X30;
    X6 ⇐ {op!OP}.X35;
    X20 ⇐ {op!OP; MA := DM}.X21;
    X21 ⇐ {op!OP; IC := IC + 1; DB := DM[MA]; IB := IM[MA]}.X22;
    X22 ⇐ {op!OP; BR := DM}.X23;
    X23 ⇐ {op!OP; AC := AC + BR}.X24;
    X24 ⇐ {op!OP; MA := IC}.X0;
    X25 ⇐ {op!OP; MA := DM}.X26;
    X26 ⇐ {op!OP; IC := IC + 1; DB := DM[MA]; IB := IM[MA]}.X27;
    X27 ⇐ {op!OP; BR := DM}.X28;
    X28 ⇐ {op!OP; AC := AC − BR}.X29;
    X29 ⇐ {op!OP; MA := IC}.X0;
    X30 ⇐ {op!OP; MA := DM}.X31;
    X31 ⇐ {op!OP; DB := DM[MA]; IB := IM[MA]}.X32;
    X32 ⇐ {op!OP; AC := 0; BR := DM}.X33;
    X33 ⇐ {op!OP; AC := AC + BR}.X34;
    X34 ⇐ {op!OP; MA := IC}.X0;
```

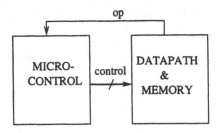

Fig. 6. An Implementation of the CPU

$$X_{35} \Longleftarrow \{op!OP; MA := DM\}.X_{36};$$
$$X_{36} \Longleftarrow \{op!OP; DB := AC\}.X_{37};$$
$$X_{37} \Longleftarrow \{op!OP; DM[MA] := DM\}.X_{38};$$
$$X_{38} \Longleftarrow \{op!OP; IC := IC + 1\}.X_{39};$$
$$X_{39} \Longleftarrow \{op!OP; MA := IC\}.X_0$$
$$\}$$

top_asm has 27 control states. For each state, we prove two lemmas that characterize the state. The first lemma describes the state transition. The second lemma describes all possible states reachable in one transition. The characterization lemmas for the control state X_0 are as follows.

$$\vdash \forall \sigma.\, Cell(Proc\, X_0\, top\_asm)\, \sigma\, \{op\} \overset{\{op.[OP]\sigma\}}{\Longrightarrow}$$
$$Cell(Proc\, X_1\, top\_asm)\sigma[DB \mapsto [DM[MA]]\sigma; IB \mapsto [IM[MA]]\sigma]\, \{op\}$$
$$\vdash \forall \sigma\, a\, C.\, Cell(Proc\, X_0\, top\_asm)\, \sigma\, \{op\} \overset{a}{\Longrightarrow} C \supset$$
$$(a = \{op.[OP]\sigma\}) \wedge$$
$$(C = Cell(ProcX_1 top\_asm)\sigma[DB \mapsto [DM[MA]]\sigma; IB \mapsto [IM[MA]]\sigma]\{op\})$$

Similar characterization lemmas are proved for the rest of the control states. The branching operation needs some explanation. The characterization lemmas for the state X_2 are shown below.

$$\vdash \forall \sigma.\, Cell(Proc\, X_2\, top\_asm)\, \sigma\, \{op\} \overset{\{op.[OP]\sigma\}}{\Longrightarrow}$$
$$Cell((OP = 0 \to Proc\, X_3\, top\_asm) + (OP = 1 \to Proc\, X_4\, top\_asm) +$$
$$(OP = 2 \to Proc\, X_5\, top\_asm) + (\neg OP < 3 \to Proc\, X_6\, top\_asm))$$
$$\sigma[DB \mapsto [DM[MA]]\sigma; IB \mapsto [IM[MA]]\sigma]\, \{op\}$$
$$\vdash \forall \sigma\, a\, C.\, Cell(Proc\, X_2\, top\_asm)\, \sigma\, \{op\} \overset{a}{\Longrightarrow} C \supset$$
$$(a = \{op.[OP]\sigma\}) \wedge$$
$$(C = Cell((OP=0 \to ProcX_3 top\_asm) + (OP=1 \to ProcX_4 top\_asm) +$$
$$(OP=2 \to ProcX_5 top\_asm) + (\neg OP < 3 \to ProcX_6 top\_asm))\sigma\{op\})$$

The successor of the process $Cell(ProcX_2 top\_asm)\sigma\{op\}$ is a conditional process

$$Cell((OP=0 \to Proc\, X_3\, top\_asm) + (OP=1 \to Proc\, X_4\, top\_asm) +$$
$$(OP=2 \to Proc\, X_5\, top\_asm) + (\neg OP < 3 \to Proc\, X_6\, top\_asm))\, \sigma\, \{op\},$$

rather than process $ProcX_3 top\_asm$, $ProcX_4 top\_asm$, $ProcX_5 top\_asm$, or $Proc\, X_6\, top\_asm$. The conditional process, however, behaves like them depending on the stored value for the state variable OP. If OP is 0, for example, it behaves like $Proc\, X_3\, top\_asm$ as shown in the following lemmas.

$\vdash \forall \sigma. \, ([OP]\sigma = 0) \supset$
$\quad Cell(((OP = 0) \to ProcX_3 top\_asm) + ((OP = 1) \to ProcX_4 top\_asm) +$
$\quad\quad ((OP = 2) \to ProcX_5 top\_asm) + \neg(OP < 3) \to ProcX_6 top\_asm)\sigma\{op\}$
$\quad \overset{\{op.[OP]\sigma\}}{\Longrightarrow} Cell(Proc\,X_{20}\,top\_asm)\,\sigma\,\{op\}$

$\vdash \forall \sigma\,a\,C. \, ([OP]\sigma = 0) \supset$
$\quad Cell(((OP = 0) \to ProcX_3 top\_asm) + ((OP = 1) \to ProcX_4 top\_asm) +$
$\quad\quad ((OP = 2) \to ProcX_5 top\_asm) + \neg(OP < 3) \to ProcX_6 top\_asm)\sigma\{op\}$
$\quad \overset{a}{\Longrightarrow} C \supset$
$\quad (a = \{op.[OP]\sigma\}) \wedge (C = Cell(Proc\,X_{20}\,top\_asm)\,\sigma\,\{op\})$

The ASM description of the microprogrammed controller and the datapath of the previous section are composed together with their (local) stores σ_1 and σ_2. The *control* port is abstracted away as the *control* signal is invisible at the top-level.

$micro\_asm\_datapath\,X\,\sigma_1\,\sigma_2 =_{def}$
$\quad (Cell(ProcXmicro\_asm)\sigma_1\{op; control\} \| Celldatapath\sigma_2\{op; control\}) \backslash \{control\}$

We prove the bisimulation equivalence between the specification and implementation using all the characterization lemmas. The correctness theorem is shown below.

$\vdash \forall \sigma\,\sigma_1. \, Cell\,(Proc\,X_0\,top\_asm)\,\sigma\,\{op\} \sim micro\_asm\_datapath\,X_0\,\sigma_1\,\sigma$

The following relation $\mathcal{R}$ is proved to be a bisimulation. The above pair is contained in $\mathcal{R}$ where $\sigma_1, \sigma \in Store$.

$\mathcal{R} = \{(Cell\,(Proc\,X_0\,top\_asm)\,\sigma\{op\}, micro\_asm\_datapath\,X_0\sigma_1\sigma) \mid for\,any\,\sigma,\,\sigma_1\} \cup$
$\quad \{(Cell\,(Proc\,X_1\,top\_asm)\,\sigma\{op\}, micro\_asm\_datapath\,X_1\sigma_1\sigma) \mid for\,any\,\sigma,\,\sigma_1\} \cup$
$\quad \{(Cell\,(Proc\,X_2\,top\_asm)\,\sigma\,\{op\}, micro\_asm\_datapath\,X_2\,\sigma_1\,\sigma) \mid [OP]\sigma$
$\quad = [IB]\sigma\,and\,any\,\sigma_1\} \cup$
$\quad \{(Cell\,((OP = 0 \to Proc\,X_3\,top\_asm) + (OP = 1 \to Proc\,X_4\,top\_asm) +$
$\quad\quad (OP = 2 \to Proc\,X_5\,top\_asm) + (NegOP < 3 \to Proc\,X_6\,top\_asm))$
$\quad \sigma\,\{op\},$
$\quad Cell\,((op = 0 \to Proc\,X_3\,micro\_asm) + (op = 1 \to Proc\,X_4\,micro\_asm) +$
$\quad\quad (op = 2 \to Proc\,X_5\,micro\_asm) + (Negop < 3 \to Proc\,X_6\,micro\_asm))$
$\quad \sigma_1\{op; control\} \|$
$\quad datapath\,\sigma\{op; control\} \backslash \{control\}) \mid [op]\sigma_1 = [OP]\sigma\,and$
$\quad [OP]\sigma = [IB]\sigma\} \cup$
$\quad \{(Cell\,(Proc\,X_i\,top\_asm)\,\sigma\{op\}, micro\_asm\_datapath\,X_i\,\sigma_1\,\sigma) \mid for\,i =$
$\quad 20,\dots,39\,and\,any\,\sigma\,and\,\sigma_1\}$

8 Conclusions

Siewiorek, Bell, and Newell [5] identified several levels of design abstraction including the instruction-set processor (ISP) and register-transfer levels (RTL). Previous work at the ISP and RTL levels [1][7] is missing a means to reason *compositionally* about *structures* of ISP and RTL components in an algebraic fashion.

The contribution of IspCal is the capability to reason compositionally at the RTL and ISP levels. It can show equivalence between *structures* and *behavior* specified by ASM or ISP descriptions. As IspCal is a process algebra, it is straightforward to define a modal logic for it using the methods outlined by Stirling [14].

Acknowledgments Many thanks go to Susan Older who helped us with the high-level description of the semantics of IspCal. This work was partially supported by the New York State Center for Advanced Technology in Computer Applications and Software Engineering (CASE).

References

1. M. R. Barbacci. Instruction Set Processor Specifications ISPS: The Notation and its Applications. *IEEE Trans. Comp.*, C-30(7), January 1981.
2. Thomas C. Bartee. *Digital Computer Fundamentals — Sixth Edition*. McGraw-Hill, New York, 1984.
3. Gerard Berry. The Foundations of Esterel. In G. Plotkin, C. Stirling, and M. Tofte, editors, *Proof, Language, and Interaction: Essays in Honour of Robin Milner*. MIT Press, 1998.
4. Christopher R. Clare. *Designing Logic Systems Using State Machines*. McGraw-Hill, 1973.
5. D. Siewiorek, C. Bell, and A. Newell. *Computer Structures — Principles and Examples*. McGraw-Hill, New York, 1982.
6. M.J.C. Gordon. A proof generating system for higher-order logic. In G. Birtwistle and P. A. Subramanyam, editors, *VLSI specification, verification and synthesis*. Kluwer, 1987.
7. Gordon L. Smith, Ralph Bahnsen, and Harry Halliwell. Boolean Comparison of Hardware and Flowcharts. *IBM J. Res. Develop.*, 26(1):106 – 116, January 1982.
8. Z. Kohavi. *Switching and Finite Automata Theory*. McGraw-Hill, New York, 1982.
9. T. F. Melham. A Package for Inductive Relation Definitions in HOL. In *Proceedings of the 1991 International Tutorial and Workshop on the HOL Theorem Proving system*. IEEE Computer Society Press, Davis, California, August 1991.
10. George Milne. Circal and the representation of communication, concurrency and time. In *ACM Transactions on Programming Languages and Systems*, April 1985.
11. Robin Milner. Calculi for synchrony and asynchrony. *Theoretical Computer Science*, 25:267 – 310, 1983.
12. Robin Milner. *Communication and Concurrency*. Prentice Hall, New York, 1989.
13. Monica Nesi. A Formalization of the Process Algebra CCS in Higher Order Logic. Technical Report 278, University of Cambridge, December 1992.
14. Colin Stirling. Modal and Temporal Logics for Processes. In *Logics for Concurrency*, number 1043 in Lecture Notes in Computer Science, pages 149 – 237. Springer-Verlag, 1996.

Techniques for Implicit State Enumeration of EFSMs

James H. Kukula[1], Thomas R. Shiple[2], and Adnan Aziz[3]

[1] Synopsys, Inc., Beaverton, OR. kukula@synopsys.com
[2] Synopsys, Inc., Mountain View, CA. shiple@synopsys.com
[3] University of Texas, Austin, TX. adnan@ece.utexas.edu

Abstract. BDD-based implicit state enumeration usually fails for systems with wide numeric datapaths. It may be possible to take advantage of higher level structure in designs to improve efficiency. By treating the integer variables as atomic types, rather than breaking them into individual bits, one can perform implicit state enumeration using Presburger arithmetic decision procedures; the complexity of this approach is independent of the width of the datapath. Since BDDs grow with the width of the datapath, we know that at some width BDDs will become less efficient than Presburger techniques. However, we establish that for widths of practical interest, the BDD approach is still more efficient.

1 Introduction

The BDD-based approach to functional verification of finite state systems has enjoyed some success. This approach proceeds by first building BDDs for the next state functions and then using implicit state enumeration to explore the state transition graph [8]. For systems with up to a few hundred flip-flops this technique can be vastly more efficient than techniques that manipulate states explicitly. However, this approach usually fails for systems with wide numeric datapaths because the BDDs become too large. One way to attack this state explosion problem is to analyse a simpler abstract model of the system [19, 7, 9, 13, 25]. We investigate techniques for exact analysis of the original system, exploring alternative representations for state sets and relations.

We focus on systems whose datapaths consist of addition and comparison operations on integer data. The control component of such systems determines what operations to perform on the data, and in turn the results of data comparisons influence the control component. Our aim is to discover whether the system can reach any undesirable states; we do this by performing state reachability analysis. A control component operating alone in an unconstrained environment can in general reach more states than when constrained by interaction with a datapath. Thus it is important to analyze the system as a whole, to avoid erroneously reporting states being reachable when in fact they are not.

Our goal is to verify real digital systems, whose datapaths have fixed finite widths. One approach to verifying such systems is to generalize them to systems having datapaths with unbounded width. At first glance this generalization from finite width datapaths to unbounded width datapaths would only seem to make the verification problem more difficult. However, since the analysis of the general case is independent of the datapath width, beyond a certain point this approach will be more efficient. The more general case may also better model the user's actual verification problem. For example, a designer might construct a parameterized library component that can be instantiated with arbitrary width. One would like to verify all possible instantiations.

Specifically, we model digital systems as extended finite state machines (EFSMs) [6], which are FSMs interacting with datapaths. In EFSMs, the state variables are partitioned into control variables (variables over finite domains) and data variables (unbounded integers). The behavior of an EFSM is defined by a set of transitions, each of which updates the values of both control and data variables. A transition has individual control states as origin and destination, but can specify the machine's behavior for whole sets of data states. Each transition has a gating predicate over the data variables, which must be satisfied for the transition to fire, and an update function that defines how the data variables change when the transition fires. The EFSM model naturally captures a frequently used machine structure, where a small, irregularly structured control component interacts with a large but regular datapath component. Note that since data variables are unbounded integers, the state space of an EFSM is infinite.

Figure 1 depicts a simple EFSM with five control states, seven input variables $(r, i_{a_x}, i_{a_y}, i_{b_x}, i_{b_y}, i_{d_x}, i_{d_y})$, seven data variables $(a_x, a_y, b_x, b_y, d_x, d_y, i)$, and ten transitions. This machine reads data and then checks a series of inequalities that determines whether the variable i should be assigned a 0 or 1 value.

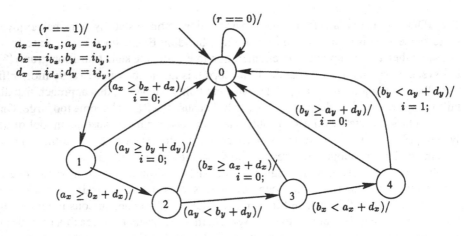

Fig. 1. An EFSM.

BDDs are commonly used to represent sets of states of finite state machines. A different mechanism is needed to handle the possibly infinite sets of states of EFSMs. The theory of formal languages deals with infinite sets, in particular with infinite sets of strings. With this in mind, we represent the values of the integer components of EFSM states as bit strings, and view sets of states as languages. Furthermore, it is well known that regular languages can be represented by finite state automata; since the sets of EFSM states we consider form regular languages, we can represent them by automata. Also, since a relation over states can be viewed as a set of pairs of states, the transition relation of an EFSM can also be viewed as a set of strings, and represented by an automaton.

Note that, while EFSMs and language-recognizing automata are both state transition systems, we use them in very different ways. EFSMs are the machines we wish to verify, for example a signal processing chip that reads in raw data and writes out filtered data. In order to verify the signal processing chip, we generalize its datapath to unbounded width. We represent a single state of the chip by a string, and a set of states by a set of strings, *i.e.*, by a language. To compactly represent this language we use the automaton that recognizes it. This automaton reads in strings that represent states of the signal processing chip. If the automaton accepts a string, then the state is in the set we are considering, otherwise not. So in summary, EFSMs are the systems we are verifying, and automata are just a data structure used to represent sets of EFSM states and transitions.

We use implicit state enumeration to check whether an EFSM can reach any undesirable states. For finite state systems, implicit state enumeration can be performed by means of various BDD operations such as intersection and existential quantification. In a completely parallel manner, implicit state enumeration for EFSMs can be performed by means of similar operations on automata. In fact, this parallel is strong since BDDs can be thought of simply as the special case of acyclic automata.

The purpose of this paper is to compare the BDD and Presburger approaches to implicit state enumeration of EFSMs. One could argue that this comparison is ill-founded. First, reachability using BDDs is theoretically guaranteed to converge because the underlying model of the designs is finite state. On the other hand, since the Presburger approach treats the data variables as unbounded, the state space is infinite and convergence is not guaranteed. Nonetheless, in some practical cases the Presburger approach does converge and invariants can be proven; and in cases when it does not, often the sequential depth of the finite state model is exponential so the BDD approach cannot practically achieve convergence anyway. A second reason to question the comparison is that the Presburger approach performs implicit state enumeration for *all* datapath widths, and hence provides a much stronger result than the BDD approach. This is relevant if one wants to verify a parameterized library component that can be instantiated with arbitrary width. In spite of these objections, we still believe that the Presburger approach is, *a priori*, a reasonable alternative to the bit-level BDD approach, and hence comparing their performance is a valuable endeavor.

The main contribution of this paper is a comparison of the automata-based approach for verification of EFSMs with the BDD-based approach for verification of finite state systems. The automata-based approach provides a compact representation of the reached state set which is independent of the width of the datapath, whereas the time and space expenses of BDDs grow as the datapath gets wider. Our experiments show, however, that for typical datapath widths, BDD computations remain less expensive than automata-based computations.

In Section 2 we discuss related work on the analysis of EFSMs. In Section 3 we introduce the EFSM model, and discuss the use of automata to represent the states and transitions of EFSMs. In Section 4 we provide experimental results comparing the BDD approach to the automata approach for state reachability. We conclude with a summary, evaluation, and future plans in Section 5.

2 Related Work

The use of BDDs for FSM analysis is well known [8]. The other half of the present comparison is on the use of automata-based Presburger decision procedures to attack EFSM reachability. Automata-based decision procedures for Presburger arithmetic were first outlined by Büchi [4]. Boudet and Comon [2] gave a more concrete procedure. Wolper and Boigelot have aims similar to Boudet and Comon, but use a representation of interacting automata, rather than a single automaton [24]. Henriksen *et al.* [12] implemented practical procedures for automata-based decisions in the tool Mona. Our automata representation and procedures are essentially the same as those described by Henriksen.

There has been much activity over the last decade in the analysis of EFSMs. Cheng and Krishnakumar formally defined EFSMs with data variables that constitute an n-dimensional linear space [6, 18]. Their first paper addresses functional test generation while the second discusses the computation of reachable states using finite unions of convex polyhedra as the underlying representation. Bultan *et al.* [5] used the matrix-based Presburger engine Omega [15, 22] to compute the reachable states for EFSMs. Shiple *et al.* [23] compare Omega to their automata-based Presburger engine Shasta, applied to the problem of EFSM reachability. This comparison found that while the raw speed of either engine can be superior to the other by a factor of 50 or more, the asymptotic performance of Shasta is equal or superior to that of Omega. They also compared Shasta to Mona, and found Shasta to always be within a factor of 2 in runtime. Thus, we feel that Shasta, the Presburger engine used in this paper, is representative of the capability of Presburger approaches for implicit state enumeration of EFSMs.

Automata-based and similar decision procedures have been used for various related verification problems. Godefroid and Long use an extension of BDDs, called QBDDs, to analyze systems containing unbounded queues [10]. The Mona tool translates formulas in the monadic second-order logic over strings into automata [1]. The authors show how this machinery can be used in a deductive framework to perform equivalence checking on iterated systems. The tool MOSEL follows the same approach as Mona [14]. Closely related is the work by Gupta [11]; she introduces linear inductive functions, which are sequential counterparts to BDDs, to represent iterated systems. She uses LIFs to perform implicit state enumeration on these systems by successively unrolling the systems in time.

3 Implicit State Enumeration for EFSMs

The basic approach of using BDDs to analyze designs with fixed width datapaths is well known. In this section, we focus on the technique of analyzing generalizations of these designs within a Presburger framework, using automata as the underlying data structure.

3.1 EFSM Model

As mentioned in the introduction, conceptually the variables of an EFSM are partitioned into control and data. However, we can treat finite domain control variables as special cases of infinite domain data variables. This allows us to treat all variables uniformly, simplifying the formalism.

Thus, we formally define an EFSM to be a tuple (k_i, k_o, k_s, I, T), where[1]

- $k_i \in N$ gives the number of input variables,
- $k_o \in N$ gives the number of output variables,
- $k_s \in N$ gives the number of state variables,
- $I : N^{k_s} \rightarrow B$ is the characteristic function of the set of valid initial states, and
- $T : N^{k_i} \times N^{k_s} \times N^{k_s} \times N^{k_o} \rightarrow B$ is the characteristic function of the transition relation for the machine. $T(x_i, x_s, x'_s, x_o)$ is true iff the EFSM, upon receiving input x_i, can make a transition from state x_s to state x'_s producing output x_o.

The transition relation T encodes both the transition gating predicates over the data variables and the update functions for the data variables. For example, consider the transition from control state 1 to 0 in the EFSM shown in Figure 1, and let s be the control state variable. This transition is represented by the conjunction of the present control state value, the gating predicate, the next control state value, and the update function, as follows:[2]

$$(s = 1) \wedge (a_x \geq b_x + d_x) \wedge (s' = 0)$$
$$\wedge(i' = 0 \wedge a'_x = a_x \wedge a'_y = a_y \wedge b'_x = b_x \wedge b'_y = b_y \wedge d'_x = d_x \wedge d'_y = d_y)$$

The entire transition relation T is just the disjunction of the formulas for each of the 10 transitions of this EFSM.

3.2 Implicit State Enumeration

In order to compute the reachable states of an EFSM, we need an effective way to represent and manipulate sets of states and relations between states. An overly rich set of arithmetic operators for the gating predicates and update functions would make calculations too difficult, while an overly constrained set would make it impossible to treat many practical EFSMs. As a compromise, we restrict our attention to EFSMs whose gating predicates and update functions are definable in Presburger arithmetic. Presburger arithmetic is the first-order theory of natural numbers with comparison and addition, and specifically without multiplication. The advantage of this restriction is that decision procedures exist for Presburger arithmetic, which permit one to mechanically check whether any Presburger formula is satisfiable. If we include multiplication in our language we will have Peano arithmetic for which no such decision procedure is possible.

---

[1] N represents the set of natural numbers, and B the set $\{0, 1\}$.

[2] The input is a don't care on this transition. The data variables $a_x, a_y, b_x, b_y, d_x, d_y$ keep their previous values.

Given an EFSM, we want to calculate the set of reachable states $R \subseteq N^{k_s}$. We use the usual iterative approach:

$$R_0(x_s) = I(x_s)$$
$$R_{j+1}(x'_s) = R_j(x'_s) \vee \exists x_s x_i x_o(R_j(x_s) \wedge T(x_i, x_s, x'_s, x_o))$$

The sequence of R_j is monotonically increasing but may not converge, since we are dealing with infinite state systems. However, all states x_s for which any $R_j(x_s)$ is true are indeed reachable, so if any of these were not intended to be reachable, then a true design error has been exposed.

The foundation of our approach is the observation that, if I and T are Presburger formulas, then the various R_j are also Presburger formulas. When the datapath of an EFSM is composed of addition and comparison operations acting on natural numbers, T can indeed be expressed as a Presburger formula. In fact Presburger arithmetic is powerful enough to express transition relations that include other functions such as division by a constant or remainder from dividing by a constant.

3.3 Automata-Based Decision Procedures

The effectiveness of implicit state enumeration depends on the underlying decision procedure. Automata-based Presburger decision procedures rely on the fact, demonstrated by Büchi [4], that for every Presburger formula there is a finite automaton that recognizes the natural number tuples that satisfy the formula. This link between finite automata and Presburger arithmetic is what makes automata useful for verification of EFSMs. By using automata to represent Presburger formulas, we can perform implicit state enumeration for EFSMs whose gating predicates and update functions are definable in Presburger arithmetic.

To see how automata can represent Presburger formulas, first consider regular languages consisting of strings over the alphabet of bit vectors B^k. These are the languages recognized by finite automata that read bit vectors. An example of such a machine, for $k = 2$, is shown in Figure 2. This machine recognizes the language

$$\begin{pmatrix} 0 \\ 1 \end{pmatrix} + \begin{pmatrix} 1 \\ 1 \end{pmatrix}) \begin{pmatrix} 1 \\ 0 \end{pmatrix} \begin{pmatrix} 0 \\ 0 \end{pmatrix}^*$$

Given automata representing languages f and g, there are well known algorithms to build new automata to represent the sets $\overline{f}$, $f \vee g$, and $f \wedge g$ [21]. We can also existentially quantify particular components of the bit vectors, forming an automaton with an alphabet of smaller dimension. For example, existentially quantifying the first component of the above language yields a new language in only one dimension, the language 100*. As we will see, these operations are the basic building blocks of implicit state enumeration. Also crucial is the fact that the minimum state automaton for a language is unique, and hence is a canonical form.

Each transition of an automaton must specify for which inputs the automaton is to follow that transition. When the letters of the alphabet are bit vectors with large dimension, the transition labels can get very complex. Following [12], we use BDDs

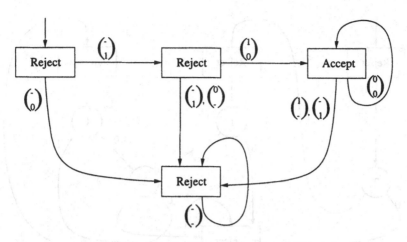

Fig. 2. An automaton. "-" means "0 or 1".

with multiple terminals as a practical implementation technique to efficiently represent the transitions of an automaton. Specifically, each state of an automaton points to a BDD that determines the next state as a function of the incoming bit vector. The terminal nodes of the BDD are the possible next states, and the BDDs for different states can share common subgraphs. Figure 3 shows the same automaton as in Figure 2, with its transitions represented by BDDs.

Now that we have an efficient method to represent automata that read bit vectors, we can discuss how to use automata to represent the set of natural number tuples satisfying a Presburger formula. Given a bit string $b = b_0 b_1 \ldots b_{n-1} \in B^*$, we can interpret it as a natural number with low order bits first:

$$x(b) = \sum_{i=0}^{n-1} b_i 2^i$$

This mapping is easily extended to multiple dimensions, $x : (B^k)^* \to N^k$. For example, the string of bit vectors

$$
\begin{array}{ll}
\text{string} \longrightarrow \\
x_1 \colon 0\ 0\ 1 & 0 \quad \text{bit} \\
x_2 \colon 1\ 1\ 1 & 0 \quad \text{vector} \\
x_3 \colon 1\ 1\ 0 & 1 \quad \downarrow
\end{array}
$$

represents the tuple $(x_1, x_2, x_3) = (4, 7, 11)$. Thus, an automaton operates over inputs that denote a bit slice of an arbitrary width datapath.

Using this encoding for natural numbers, Boudet and Comon have given a simple procedure for constructing automata from Presburger formulas [2]. The constructed automaton accepts just those natural number tuples that satisfy the corresponding formula. Figure 4 shows the automaton, with transitions represented by BDDs, that recognizes the tuples satisfying the Presburger formula $x_1 + x_2 = x_3$. To illustrate its operation, consider again the tuple $(4, 7, 11)$, represented by the string of bit vectors shown above.

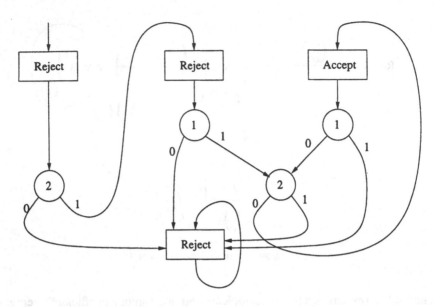

Fig. 3. Using BDDs for transitions of automata.

When reading this string, the automaton starts at the initial state. The first input has 0 for x_1, 1 for x_2, and 1 for x_3, which leads the automaton back to the initial state, ready to read the second bit vector in the string. After reading all the bit vectors, the automaton will end in the accepting state, reflecting the fact that $4 + 7 = 11$.

We use automata to represent Presburger formulas, so the iterative calculation of reachable states is performed by union, intersection, and existential quantification of automata. An efficient minimization procedure, essentially the same as that of Klarlund [17], generates canonical forms of the resulting automata. This keeps the representation as small as possible and allows testing for convergence to a fixed point.

3.4 Automata vs. BDDs

We use an automata-based engine to analyze infinite state systems. The real systems we wish to verify have finite, fixed width datapaths. The states of these systems can simply be mapped to strings of a finite fixed length. Once we have an automaton R representing the set of reachable states for the unbounded EFSM, to decide if a given real state is reachable we can just check if R accepts the string that represents the real state.

In order for this computation to be accurate, the transition relation T must accurately model the real system. The most common difficulty in building an accurate model is arithmetic overflow. Addition in an 8-bit system could be modeled in an EFSM by the formula

$$(x + y < 256 \land z = x + y) \lor (x + y \geq 256 \land z = x + y - 256)$$

The effect of this, however, is to limit the computation on the EFSM to systems of a single width, which in fact unrolls the automaton and results in computations isomorphic to BDDs.

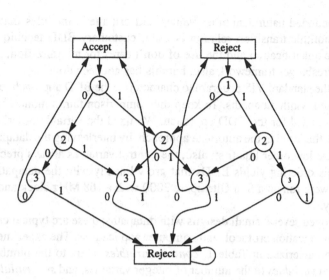

Fig. 4. The automaton representing $x_1 + x_2 = x_3$, with transitions represented by BDDs.

In systems where overflow can occur, one can instead introduce a new parameter variable to specify the bound:

$$(x + y < b \land z = x + y) \lor (x + y \geq b \land z = x + y - b)$$

With this form of the transition relation, one is again representing an infinite class of systems and the compact looping structure of the automata is preserved. When checking for the reachability of a state of a real fixed width system, one represents the state by a string that includes the actual value of the bound. In the experiments that follow, none of the circuits exhibit rollover, so the modeling as EFSMs was straightforward.

Note that for a fixed w-width datapath, the BDD representing a state set or transition relation is never more than w larger than the corresponding automaton. This is because "unrolling" an automaton w times yields an acyclic automaton, essentially a BDD, representing the language of the original automaton restricted to length w strings. This fact strictly bounds the improvement that automata can achieve relative to BDDs.

Automata thus provide a useful link between datapath structure and BDDs. Any relation definable in Presburger arithmetic can be represented by a BDD that grows linearly with the width of the datapath, by using a variable ordering that interleaves the bits of the various data words.

4 Experimental Results

The purpose of our experiments is to compare the BDD-based approach to state reachability of systems with fixed width datapaths to the automata-based approach to state reachability for the corresponding systems with unbounded width datapaths. We use the automata-based Shasta engine of [23]. We incorporated this into the VIS program [3], along with a front-end package that interprets a dialect of Verilog that includes wires carrying unbounded natural number values and arithmetic modules that operate on them. A monolithic transition relation is used; customary BDD techniques, such as early variable quantification and the use of don't cares for minimization, could be applied in the Presburger framework also, but this has not been done.

We used the standard VIS program to characterize the BDD approach to state reachability for fixed width datapaths. To keep the comparison fair, a monolithic transition relation is also used for the BDD approach. We fixed the variable ordering in VIS to correspond to that used in the automata approach, by interleaving the datapath variables and putting the low order bits first; also, the control variables always precede the data variables. This ordering yields BDDs that grow linearly with the datapath width. All experiments were run on a Sun Ultrasparc 3000 with a 168 MHz clock and 512 MB of main memory.

We developed several small designs with datapaths; these are typical circuits found in DSP, communication protocol, and computer applications. The experimental circuits are briefly characterized in Table 1; *control variables* refers to the number of binary variables, *data variables* to the number of integer variables, and *sequential depth* to the greatest lower bound on the path length from the initial state to any reachable state.

- "movavg" reads in a stream of numbers and keeps the sum of the last eight numbers read. It keeps these numbers in data registers, and on each cycle overwrites with the current input the register with the oldest number. A one-hot coded control word keeps track of which data register to update next. In each cycle the accumulator is decremented by the oldest number and incremented by the new number.
- "sdiv" is a serial divider. A numerator and denominator are read in, and then the denominator is repeatedly subtracted from the numerator. The sequential depth of this machine grows exponentially with the width of the datapath, and is unbounded for unbounded integers. The EFSM reachable state calculation thus does not reach a fixed point, nor does the BDD reachability calculation converge in a practical time even for moderate sized datapaths.
- "bound" is the EFSM shown in Figure 1. It reads the x, y coordinates of two points and a difference vector, and checks whether the two points are closer than that difference. The checks are performed one at time. The control states are one-hot encoded.
- "euclid" implements Euclid's greatest common divisor algorithm. Two numbers are read in and saved. These numbers are copied to two working registers. At each cycle the smaller number is subtracted from the larger number. This process halts when the two numbers are equal, at which point they will have the value of the GCD of the numbers initially read in. This machine has a sequential depth that grows exponentially with the width of the datapath.

Table 2 shows the BDD size and automaton size[3] for the reachable state set for a given step of reachability. For "movavg", data is given for step 8 (where the peak sizes are seen) and for "bound" step 5 (the fixed point and peak). For "sdiv", the sizes grow gradually—step 19 is shown as a representative example, and for "euclid" they grow

---

[3] For an automaton, the size is taken as the sum of the number of BDD nodes that represent the transitions, and the number of terminal nodes that represent the states of the automaton.

Table 1. Characteristics of experimental circuits.

example	control variables	data variables	sequential depth	
			fixed width w	unbounded width
movavg	9	9	15	15
sdiv	0	4	2^w	∞
bound	6	6	5	5
euclid	0	4	2^w	∞

Table 2. Sizes of automata and BDDs for reached state sets.

example	automata size	BDD size for bit width				ratio for 32 bits
		8	16	24	32	
movavg	134	796	1660	2524	3388	25.3
sdiv	3407	2002	11911	22423	32935	9.7
bound	2380	7013	17045	27077	37109	15.6
euclid	19196	36408	172090	307914	443738	23.1

quickly—step 6 is shown. For most of our examples, the BDD representation of the reached state set for 8-bit datapaths is already larger than the automata representation, and continues to grow roughly linearly with the width of the datapath. At 32 bits the BDDs are roughly 10 times larger or more; the final column shows the ratio of BDD sizes for the 32-bit case to the automaton sizes.

Table 3. Sizes of automata and BDDs for monolithic transition relation.

example	automata size	BDD size for bit width			
		8	16	24	32
movavg	38680	20809	43489	66169	88849
sdiv	180	1036	2252	3468	4684
bound	26096	3740	8236	12732	17228
euclid	290	1804	3980	6156	8332

The sizes of the monolithic transition relations (after the primary input variables have been quantified) for the different circuits are shown in Table 3. Here the automata-based representation does not show as consistent an advantage over BDDs as was shown for reachable state sets.

The above analysis is in terms of "representation size". When we compare run time and memory use, the picture is quite different. Table 4 shows compute time and memory for building the transition relation and performing reachability. In all cases the automata approach takes significantly more resources. Much of this inefficiency is undoubtedly due to the unoptimized prototype nature of our code. But there are more fundamental underlying causes as well. While the looping structure of automata allows more com-

Table 4. Computational costs for implicit state enumeration by automata and by BDDs for several datapath widths. (CPU sec. / Megabytes)

example	automata cost	BDD cost for bit width			
		8	16	24	32
movavg	2043/ 317	17/ 12	52/ 20	100/ 34	170/ 39
sdiv	229/ 16	2/ 2	11/ 9	25/ 13	31/ 18
bound	379/ 49	6/ 8	17/ 14	29/ 22	43/ 32
euclid	252/ 63	9/ 8	46/ 22	153/ 36	174/ 51

pact representation, it can make manipulation less efficient. With conventional BDDs, the acyclic structure means that operations near the root can take advantage of lower level canonicalization. With automata, an entire operation must be completed before canonicalization can be performed. This can result in larger intermediate results during calculations. In particular, existential quantification requires the determinization of an intermediate nondeterministic automaton; although the size of the final result is usually reasonable, the size of the determinized, but unminimized automaton, can be huge. Finally, the cost of canonicalization for automata is $O(n \log n)$, somewhat worse than the linear cost of canonicalization for BDDs.

In summary, when comparing representation size, we see that the automata approach is superior to the BDD approach, in most cases, once datapath widths exceed 16 bits. However, when comparing run time and memory, this crossover probably does not take place until 64 bits or beyond.

5   Conclusions and Future Work

We have shown how BDD-based implicit state enumeration of FSMs can be generalized to automata-based implicit state enumeration of infinite state EFSMs. Our experiments demonstrate that an automaton representation of an unbounded width datapath can be 10x more compact than the BDD representation of the corresponding 32-bit datapath. On the other hand, the run times tend to be much worse, for two reasons. The first is intrinsic: since automata have cycles, the graph algorithms have higher complexity. The second is that our implementation of the automata package is probably equivalent in sophistication to the first BDD packages of a decade ago. Just as significant improvements were achieved by more mature BDD packages, we believe that an order of magnitude performance improvement may be possible for our automata package. There are simplifications of data structures that should be straightforward. A very significant improvement would be provided by finding a way to perform partial state minimization of an automaton before the entire automaton is constructed. Developing such techniques is an important part of our future plans. Beyond improving the automata package, we would like to apply our techniques to more examples, and to add a temporal logic model checker.

Another interesting avenue of research is the analysis of iterated systems, which are systems with arrays of identical FSMs. An automata-based approach can also be used to perform implicit state enumeration of these systems [16]. We have performed an experiment on the Dining Philosophers with Encyclopedia example of Kurshan and McMillan [20]. We found that safety properties can be verified for an arbitrary n-philosopher system more efficiently (in time and total memory) than for a BDD-based approach for a fixed n greater than or equal to 12. This is contrary to the outcome of the EFSM experiments, and deserves further investigation.

Acknowledgments We thank Kurt Keutzer for suggesting to us the use of Presburger arithmetic for analyzing EFSMs.

References

[1] D. A. Basin and N. Klarlund. Hardware verification using monadic second-order logic. In P. Wolper, editor, *Proc. Computer Aided Verification*, volume 939 of *LNCS*, pages 31–41. Springer-Verlag, July 1995.

[2] A. Boudet and H. Comon. Diophantine equations, Presburger arithmetic and finite automata. In H. Kirchner, editor, *Trees and Algebra in Programming - CAAP*, volume 1059 of *LNCS*, pages 30–43. Springer-Verlag, 1996.

[3] R. K. Brayton, G. D. Hachtel, A. Sangiovanni-Vincentelli, F. Somenzi, A. Aziz, S.-T. Cheng, S. Edwards, S. Khatri, Y. Kukimoto, A. Pardo, S. Qadeer, R. K. Ranjan, S. Sarwary, T. R. Shiple, G. Swamy, and T. Villa. VIS: A system for verification and synthesis. In R. Alur and T. A. Henzinger, editors, *Proceedings of the Conference on Computer-Aided Verification*, volume 1102 of *LNCS*, pages 428–432, New Brunswick, NJ, July 1996. Springer-Verlag.

[4] J. R. Büchi. On a decision method in restricted second order arithmetic. In *Proc. Int. Congress Logic, Methodology, and Philosophy of Science*, pages 1–11, Berkeley, CA, 1960. Stanford University Press.

[5] T. Bultan, R. Gerber, and W. Pugh. Symbolic model checking of infinite state programs using Presburger arithmetic. In O. Grumberg, editor, *Proc. Computer Aided Verification*, volume 1254 of *LNCS*, pages 400–411, Haifa, June 1997. Springer-Verlag.

[6] K.-T. Cheng and A. Krishnakumar. Automatic functional test generation using the extended finite state machine model. In *Proc. 30th Design Automat. Conf.*, pages 86–91, June 1993.

[7] E. M. Clarke, O. Grumberg, and D. E. Long. Model checking and abstraction. In *Proc. Principles of Programming Language*, Jan. 1992.

[8] O. Coudert, C. Berthet, and J. C. Madre. Verification of synchronous sequential machines based on symbolic execution. In J. Sifakis, editor, *Proceedings of the Workshop on Automatic Verification Methods for Finite State Systems*, volume 407 of *LNCS*, pages 365–373. Springer-Verlag, June 1989.

[9] D. Cyrluk and P. Narendran. Ground temporal logic: A logic for hardware verification. In D. L. Dill, editor, *Proc. Computer Aided Verification*, volume 818 of *LNCS*, pages 247–259, Stanford, CA, June 1994. Springer-Verlag.

[10] P. Godefroid and D. E. Long. Symbolic protocol verification with queue BDDs. In *Proc. Logic in Computer Science*, pages 198–206, July 1996.

[11] A. Gupta. *Inductive Boolean Function Manipulation: A Hardware Verification Methodology for Automatic Induction*. PhD thesis, Carnegie Mellon University, 1994. Memorandum No. CMU-CS-94-208.

[12] J. G. Henriksen, J. Jensen, M. Jørgensen, N. Klarlund, R. Paige, T. Rauhe, and A. Sandholm. Mona: Monadic second-order logic in practice. In *Tools and Algorithms for the Construction and Analysis of Systems, First International Workshop, TACAS '95*, volume 1019 of *LNCS*, pages 89–110. Springer-Verlag, May 1995.

[13] R. Hojati and R. K. Brayton. Automatic datapath abstraction in hardware systems. In P. Wolper, editor, *Proc. Computer Aided Verification*, volume 939 of *LNCS*, pages 98–113. Springer-Verlag, July 1995.

[14] P. Kelb, T. Margaria, M. Mendler, and C. Gsottberger. MOSEL: A flexible toolset for monadic second-order logic. In *TACAS '97: Int'l Workshop on Tools and Algorithms for the Construction and Analysis of Systems*, volume 1217 of *LNCS*, pages 183–202. Springer-Verlag, Apr. 1997.

[15] W. Kelly, V. Maslov, W. Pugh, E. Rosser, T. Shpeisman, and D. Wonnacott. The Omega library (Version 1.1.0) interface guide. http://www.cs.umd.edu/ projects/omega, Nov. 1996.

[16] Y. Kesten, O. Maler, M. Marcus, A. Pnueli, and E. Shahar. Symbolic model checking with rich assertional languages. In O. Grumberg, editor, *Proc. Computer Aided Verification*, volume 1254 of *LNCS*, pages 424–435. Springer-Verlag, June 1997.

[17] N. Klarlund. An $n\log n$ algorithm for online BDD refinement. In O. Grumberg, editor, *Proc. Computer Aided Verification*, volume 1254 of *LNCS*, pages 107–118. Springer-Verlag, June 1997.

[18] A. Krishnakumar and K.-T. Cheng. On the computation of the set of reachable states of hybrid models. In *Proc. 31st Design Automat. Conf.*, pages 615–621, June 1994.

[19] R. P. Kurshan. Analysis of discrete event coordination. In J. W. de Bakker, W.-P. de Roever, and G. Rozenberg, editors, *Proceedings of the REX Workshiop on Stepwise Refinement of Distributed Systems, Models, Formalisms, Correctness*, volume 430 of *LNCS*, pages 414–453. Springer-Verlag, 1989.

[20] R. P. Kurshan and K. L. McMillan. A structural induction theorem for processes. In *Proc. Eighth Symp. Princ. of Distributed Computing*, pages 239–247, 1989.

[21] H. R. Lewis and C. H. Papadimitriou. *Elements of the Theory of Computation*. Prentice Hall, Englewood Cliffs, New Jersey, 1981.

[22] W. Pugh. A practical algorithm for exact array dependence analysis. *Communications of the ACM*, 35(8):102–114, Aug. 1992.

[23] T. R. Shiple, J. H. Kukula, and R. K. Ranjan. A comparison of Presburger engines for EFSM reachability. In A. Hu and M. Vardi, editors, *Proc. Computer Aided Verification*, volume 1427 of *LNCS*, pages 280–292, Vancouver, June 1998. Springer-Verlag.

[24] P. Wolper and B. Boigelot. An automata-theoretic approach to Presburger arithmetic constraints. In *Proc. of Static Analysis Symposium*, volume 983 of *LNCS*, pages 21–32. Springer-Verlag, Sept. 1995.

[25] Z. Zhou, X. Song, S. Tahar, E. Cerny, F. Corella, and M. Langevin. Formal verification of the island tunnel controller using multiway decision graphs. In M. Srivas and A. Camilleri, editors, *Proc. Formal Methods in Computer-Aided Design*, volume 1166 of *LNCS*, pages 233–247. Springer-Verlag, Nov. 1996.

Model Checking on Product Structures*

Klaus Schneider

University of Karlsruhe, Department of Computer Science,
Institute for Computer Design and Fault Tolerance (Prof. Dr.-Ing. D. Schmid),
P.O. Box 6980, 76128 Karlsruhe, Germany,
email: Klaus.Schneider@informatik.uni-karlsruhe.de,
http://goethe.ira.uka.de/~schneider

Abstract *We present an algorithm for checking* CTL *formulas in Kripke structures with side conditions, where the side conditions define new variables in terms of path formulas. Given any* CTL *formula where the defined variables may occur, the presented algorithm will determine the set of states where the* CTL* *formula holds that is obtained by replacing each new variable defined by a side condition by its definition.*

The basic idea of our algorithm is to translate each side condition to a Kripke structure that encodes precisely the definition of the new variable. After that, we compute the products of these structures with the given structure and use a generalization of the well-known CTL *model checking procedure. The presented model checking procedure can still be implemented as a symbolic model checking procedure (e.g. with BDDs).*

We moreover show how each CTL* *model checking problem can be translated efficiently to a* CTL *model checking problem with side conditions, and hence show that the method can be used to construct efficient* CTL* *and* LTL *model checking procedures. Moreover, it is shown that for* LTL *model checking, we can still use standard* CTL *model checking procedures instead of our generalized version.*

1 Introduction

Temporal logics as e.g. CTL [1], LTL [2], and as a superset of both CTL* [3] are convenient means for specifying concurrent systems. In particular, efficient verification tools for CTL have become standard [4,5,6] and are already used in industrial applications to quickly detect design errors. While LTL and CTL* have exponential verification procedures, CTL can be checked in linear time w.r.t. the length of the specification and the size of the system [7,8]. On the other hand, some properties can be specified in a more succinct and more readable manner in LTL. For example, take the formula $A \bigvee_{i=1}^{n} G a_i$, which specifies that there for each computation run of the system at least one properties a_i must hold forever. It can be shown that checking the negation of these formulas is an $\mathcal{NP}$-complete problem [8] and that the formula can also be expressed in CTL [17]. As CTL model checking is polynomial, and we believe that $\mathcal{P} \neq \mathcal{NP}$ holds, all CTL formulas that are equivalent to this formula must be more then polynomially longer than

---

* This work has been financed by DFG project Automated System Design, SFB358 and also by the priority program 'Design and Design Methodology of Embedded Systems'.

the considered LTL formula. Hence, the higher runtime complexity of LTL verification may be compensated by shorter specifications.

However, LTL suffers much more from the state-explosion problem than CTL does. To see this, note that all known decision procedures for LTL are based on translations to ω-automata [9,10,11]: Having computed an ω-automaton $\mathcal{A}_\Phi$ for the specification Φ with the acceptance condition φ and the automaton $\mathcal{A}_S$ for the system's behavior, the remaining problem is to check whether φ holds for the product automaton $\mathcal{A}_\Phi \times \mathcal{A}_S$. However, the computation of the product automaton $\mathcal{A}_\Phi \times \mathcal{A}_S$ usually yields in very large state spaces the number of states of the system is multiplied with the number of states of the ω-automaton. For this reason, sophisticated techniques as on-the-fly methods [12,13] and partial-order reductions [14,15,16] have been developed.

In this paper, we follow an alternative approach: In [17], a new temporal logic called LeftCTL* has been defined that can be translated to CTL to use standard CTL model checking for its verification. The advantage of LeftCTL* in comparison to CTL is that it allows to describe some properties much more readable[1] than it is possible in CTL. It can be proved (unpublished so far) that the translation can be performed in time $O(|\Phi| 2^{2|\Phi|})$ and that the resulting CTL formulas are in the worst case of size $O(|\Phi| 2^{2|\Phi|})$.

In this paper, we show how these results of [17] can be used to obtain efficient decision procedures for the logics LTL and CTL*. The basic idea of this paper is roughly the following: We extract from a given CTL* formula Φ over the variables V_Σ the largest LeftCTL* formula Φ' by replacing successively each subformula φ_i that violates the grammar rules of LeftCTL* by a new variable ℓ_i. It follows that φ_i contains at most the variables $V_\Sigma \cup \{\ell_1, \ldots, \ell_{i-1}\}$ and contains none of the path quantifiers E and A, and in particular that $\Phi = [\Phi']_{\ell_1 \ldots \ell_n}^{\varphi_1 \ldots \varphi_n}$ holds[2]. As we can translate the LeftCTL* formula Φ' to an equivalent CTL formula Φ'', the remaining problem is to check the CTL formula Φ'' in Kripke structures $\mathcal{K}$ with some side conditions $\ell_i = \varphi_i$ that must hold on every fair path of $\mathcal{K}$. We call this problem the *model checking problem for* CTL *under side conditions*. To solve this problem, we present in this paper a new algorithm that works as follows: first we construct[3] for each side condition $\ell_i = \varphi_i$ a structure $\mathcal{D}_{\ell_i}^{\varphi_i}$ that precisely encodes the information that ℓ_i behaves on each fair path as φ_i. Then we compute the product of these structures with the given structure $\mathcal{K}$ and check the formula Φ'' in that product structure. The problem is however that we are interested in the truth value of Φ in $\mathcal{K}$ rather than the truth value of Φ'' in the product structure. For this reason, our algorithm must translate the results back to $\mathcal{K}$. This is essentially done by a quotient construction, which is implicitly performed by our algorithm after the fixpoint iterations.

Therefore we can in general not use existing CTL model checking tools for our solution. However, the use of the standard CTL model checking procedure is correct for the verification of LTL properties. This is not really surprising, since translations from LTL to ω-automata allow also to use existing CTL model checking tools [9]. Also,

---

[1] For example, A $\bigvee_{i=1}^n$ Ga$_i$ is a LeftCTL* formula that has only large equivalent CTL formulas.

[2] This denotes replacement of ℓ_i by φ_i.

[3] This is nothing else than traditional translations from LTL to ω-automata. However, we can reduce the problem to definitions $\ell_i = \varphi_i$, where φ_i contains exactly one temporal operator.

if we consider the overall complexities, we obtain in both cases exponential decision procedures. The advantage of the presented method is however that only parts of the specification (namely the subformulas φ_i of the definitions) contribute to the blow up of the state space while the remaining part is retained in a CTL formula. This is important as usually the size of the Kripke structure limits the application of model checking procedures. Consequently, our approach for LTL model checking behaves much better with large Kripke structures.

The paper is organized as follows: In the next section, we introduce the syntax and semantics of temporal logics we consider in the paper. We also define product, quotients and bisimulation relation on Kripke structures. Then, we consider the model checking problem with side conditions and show its application in the succeeding section to CTL* and LTL model checking. The paper concludes with experimental results.

2 Temporal Logics

In general, formulas of a temporal logic consist of temporal operators, boolean operators and path quantifiers. The detailed syntax of the temporal logics CTL* [3], LeftCTL* [17], CTL [1], and LTL [2], over a fixed set of operators and a set of variables V_Σ is given by the grammar rules in figure 1.

$$
\begin{aligned}
\text{CTL}^* : \quad & S ::= V_\Sigma \mid \neg S \mid S \wedge S \mid S \vee S \mid \text{E}P \mid \text{A}P \\
& P ::= S \mid \neg P \mid P \wedge P \mid P \vee P \mid \text{X}P \mid \text{G}P \mid \text{F}P \\
& \quad \mid [P \text{ W } P] \mid [P \text{ U } P] \mid [P \text{ B } P] \\
& \quad \mid [P \text{ \underline{W} } P] \mid [P \text{ \underline{U} } P] \mid [P \text{ \underline{B} } P] \\[4pt]
\text{LeftCTL}^* : \quad & S ::= V_\Sigma \mid \neg S \mid S \wedge S \mid S \vee S \mid \text{E}P_E \mid \text{A}P_A \\
& P_E ::= S \mid \neg P_A \mid P_E \wedge P_E \mid P_E \vee P_E \mid \text{X}P_E \mid \text{G}S \mid \text{F}P_E \\
& \quad \mid [P_E \text{ W } S] \mid [S \text{ U } P_E] \mid [P_E \text{ B } S] \\
& \quad \mid [P_E \text{ \underline{W} } S] \mid [S \text{ \underline{U} } P_E] \mid [P_E \text{ \underline{B} } S] \\
& P_A ::= S \mid \neg P_E \mid P_A \wedge P_A \mid P_A \vee P_A \mid \text{X}P_A \mid \text{G}P_A \mid \text{F}S \\
& \quad \mid [P_A \text{ W } S] \mid [P_A \text{ U } S] \mid [S \text{ B } P_E] \\
& \quad \mid [P_A \text{ \underline{W} } S] \mid [P_A \text{ \underline{U} } S] \mid [S \text{ \underline{B} } P_E] \\[4pt]
\text{CTL} : \quad & S ::= V_\Sigma \mid \neg S \mid S \wedge S \mid S \vee S \mid \text{E}P \mid \text{A}P \\
& P ::= \text{X}S \mid \text{G}S \mid \text{F}S \\
& \quad \mid [S \text{ W } S] \mid [S \text{ U } S] \mid [S \text{ B } S] \\
& \quad \mid [S \text{ \underline{W} } S] \mid [S \text{ \underline{U} } S] \mid [S \text{ \underline{B} } S] \\[4pt]
\text{LTL} : \quad & S ::= \text{A}P \\
& P ::= V_\Sigma \mid \neg P \mid P \wedge P \mid P \vee P \mid \text{X}P \mid \text{G}P \mid \text{F}P \\
& \quad \mid [P \text{ W } P] \mid [P \text{ U } P] \mid [P \text{ B } P] \\
& \quad \mid [P \text{ \underline{W} } P] \mid [P \text{ \underline{U} } P] \mid [P \text{ \underline{B} } P]
\end{aligned}
$$

Figure 1. Some sublanguages of CTL*

The nonterminals S and P with or without indices in the grammar rules of figure 1 describe sets of *state* and *path formulas*, respectively. LeftCTL* is the subset of CTL* where one of the arguments of binary temporal operators are restricted to state formulas. In the grammar of LeftCTL*, we distinguish between path formulas that occur after the path quantifiers A (P_A) and E (P_E).

To simplify the presentation, the following equations are used to remove redundant temporal operators (0 is always false and 1 is always true): $GP_A = [0\ B\ (\neg P_A)]$, $FS = [1\ \underline{U}\ S]$, $[P_A\ W\ S] = [S\ B\ (\neg P_A \wedge S)]$, $[P_A\ \underline{W}\ S] = [S\ B\ (\neg P_A \wedge S)] \wedge [1\ \underline{U}\ S]$, $[P_A\ U\ S] = [P_A\ \underline{U}\ S] \vee [0\ B\ (\neg P_A)]$, and $[S\ \underline{B}\ P_E] = [S\ B\ P_E] \wedge [1\ \underline{U}\ S]$. *It is important to see that these equations do preserve membership in* LeftCTL*, *i.e. each* LeftCTL* *formula still belongs to* LeftCTL* *after rewriting it with these equations.* On the other hand, we will sometimes use the abbreviations $a \rightarrow b := \neg a \vee b$, $a = b := (a \rightarrow b) \wedge (b \rightarrow a)$, and $(a \Rightarrow b|c) := (a \rightarrow b) \wedge (\neg a \rightarrow c)$. Moreover, we denote the set of variables that occur in a formula φ as $VAR(\varphi)$ and the replacement of each occurrence of a variable x in a formula Φ by a formula φ as $[\Phi]_x^\varphi$. The semantics of temporal logic is given for Kripke structures:

Definition 1 (Kripke Structures). *A Kripke structure* $\mathcal{K} = (\mathcal{I}, \mathcal{S}, \mathcal{R}, \mathcal{L}, \mathcal{F})$ *for a set of variables* V_Σ *is given by a finite set of states* $\mathcal{S}$, *a set of initial states* $\mathcal{I} \subseteq \mathcal{S}$, *a transition relation* $\mathcal{R} \subseteq \mathcal{S} \times \mathcal{S}$, *a label function* $\mathcal{L} : \mathcal{S} \rightarrow 2^{V_\Sigma}$ *that maps each state to a set of variables, and a set of fairness constraints* $\mathcal{F} = \{F_1, \ldots, F_f\}$, *where* $F_i \subseteq \mathcal{S}$.

A path π is a function that maps natural numbers to states of a Kripke structure such that $(\pi^{(t)}, \pi^{(t+1)}) \in \mathcal{R}$ holds. A path π is said to be fair w.r.t. $\mathcal{F}$, iff for each $F_i \in \mathcal{F}$ there is a state $s_i \in F_i$ that occurs infinitely often on π. We denote the set of fair paths starting in s with $FPaths_\mathcal{K}(s)$. The k-th state of a path π is denoted by $\pi^{(k-1)}$ and the suffix where the first δ states are cut off is written as $\lambda x.\pi^{(x+\delta)}$.

Definition 2 (Semantics of CTL*). *Given a path* π *through a Kripke structure* $\mathcal{M}$, *the following rules define the semantics of* CTL* *path formulas:*

- $(\mathcal{M}, \pi) \models \varphi$ *iff* $(\mathcal{M}, \pi^{(0)}) \models \varphi$ *for each state formula* φ
- $(\mathcal{M}, \pi) \models \neg\varphi$ *iff not* $(\mathcal{M}, \pi) \models \varphi$
- $(\mathcal{M}, \pi) \models \varphi \wedge \psi$ *iff* $(\mathcal{M}, \pi) \models \varphi$ *and* $(\mathcal{M}, \pi) \models \psi$
- $(\mathcal{M}, \pi) \models \varphi \vee \psi$ *iff* $(\mathcal{M}, \pi) \models \varphi$ *or* $(\mathcal{M}, \pi) \models \psi$
- $(\mathcal{M}, \pi) \models X\varphi$ *iff* $(\mathcal{M}, \lambda x.\pi^{(x+1)}) \models \varphi$
- $(\mathcal{M}, \pi) \models [\varphi\ \underline{U}\ \psi]$ *iff there is a* $\delta \in \mathbb{N}$ *such that* $(\mathcal{M}, \lambda x.\pi^{(x+\delta)}) \models \psi$ *and* $(\mathcal{M}, \lambda x.\pi^{(x+t)}) \models \varphi \wedge \neg\psi$ *for all* $t < \delta$.
- $(\mathcal{M}, \pi) \models [\varphi\ B\ \psi]$ *iff if* $\delta \in \mathbb{N}$ *is the smallest number with* $(\mathcal{M}, \lambda x.\pi^{(x+\delta)}) \models \psi$, *then there is a* $\delta' < \delta$ *with* $(\mathcal{M}, \lambda x.\pi^{(x+\delta')}) \models \varphi$

For a given state s *of a structure* $\mathcal{M}$, *the semantics of a state formula is given by the following definitions:*

- $(\mathcal{M}, s) \models x$ *iff* $x \in \mathcal{L}(s)$
- $(\mathcal{M}, s) \models \neg\varphi$ *iff not* $(\mathcal{M}, s) \models \varphi$
- $(\mathcal{M}, s) \models \varphi \wedge \psi$ *iff* $(\mathcal{M}, s) \models \psi$ *and* $(\mathcal{M}, s) \models \psi$
- $(\mathcal{M}, s) \models \varphi \vee \psi$ *iff* $(\mathcal{M}, s) \models \psi$ *or* $(\mathcal{M}, s) \models \psi$
- $(\mathcal{M}, s) \models A\varphi$ *iff for each fair path* π *through* $\mathcal{M}$ *starting in* s, $(\mathcal{M}, \pi) \models \varphi$ *holds*
- $(\mathcal{M}, s) \models E\varphi$ *iff* $(\mathcal{M}, s) \models \neg A\neg\varphi$

In the following, we must reason about equivalent Kripke structures. Hence, it is necessary to clarify when two structures are considered to be equivalent or not. For this reason, we use the notion of bisimulation equivalence as used in [18] extended to fairness constraints [?,20]. In particular, it has been shown that two structures are bisimilar iff they satisfy the same set of CTL* formulas [19,20].

Definition 3 (Bisimulation Equivalence). *Given two structures* $\mathcal{K}_1 = (\mathcal{I}_1, \mathcal{S}_1, \mathcal{R}_1, \mathcal{L}_1, \mathcal{F}_1)$, $\mathcal{K}_2 = (\mathcal{I}_2, \mathcal{S}_2, \mathcal{R}_2, \mathcal{L}_2, \mathcal{F}_2)$ *over a signature* V_Σ. *A relation* $\sim$ *over* $\mathcal{S}_1 \times \mathcal{S}_2$ *is called a bisimulation relation iff the following holds:*

- **BISIM1:** $s_1 \sim s_2$ *implies* $\mathcal{L}_1(s_1) = \mathcal{L}_2(s_2)$
- **BISIM2a:** *for all* $s_1 \in \mathcal{S}_1$ *and* $s_2 \in \mathcal{S}_2$ *with* $s_1 \sim s_2$ *and any* $\pi_1 \in \mathsf{FPaths}_{\mathcal{K}_1}(s_1)$, *there is a* $\pi_2 \in \mathsf{FPaths}_{\mathcal{K}_2}(s_2)$ *such that* $\pi_1^{(t)} \sim \pi_2^{(t)}$ *holds for all* $t \in \mathbb{N}$
- **BISIM2b:** *for all* $s_1 \in \mathcal{S}_1$ *and* $s_2 \in \mathcal{S}_2$ *with* $s_1 \sim s_2$ *and any* $\pi_2 \in \mathsf{FPaths}_{\mathcal{K}_2}(s_2)$, *there is a* $\pi_1 \in \mathsf{FPaths}_{\mathcal{K}_1}(s_1)$ *such that* $\pi_1^{(t)} \sim \pi_2^{(t)}$ *holds for all* $t \in \mathbb{N}$
- **BISIM3a:** *for any* $s_1 \in \mathcal{I}_1$, *there is a* $s_2 \in \mathcal{I}_2$ *with* $s_1 \sim s_2$
- **BISIM3b:** *for any* $s_2 \in \mathcal{I}_2$, *there is a* $s_1 \in \mathcal{I}_1$ *with* $s_1 \sim s_2$.

Two paths π_1 *and* π_2 *through* $\mathcal{K}_1$ *and* $\mathcal{K}_2$, *respectively, are called bisimulation equivalent iff* $\forall t \in \mathbb{N}.\pi_1^{(t)} \sim \pi_2^{(t)}$ *holds. We also write* $\mathcal{K}_1 \sim \mathcal{K}_2$ *for structures, if* $\sim$ *is a bisimulation relation* $\sim$ *between* $\mathcal{K}_1$ *and* $\mathcal{K}_2$.

Moreover, $\sim$ is a simulation relation iff **BISIM1**, **BISIM2a**, and **BISIM3a** are satisfied. Usually, a system is not given as a homogeneous finite-state machine, rather it consists of several subsystems that run concurrently and interact with each other. For this reason, we need the product of Kripke structures to determine the overall behavior:

Definition 4 (Product Structures). *Given two Kripke structures* $\mathcal{K}_1 = (\mathcal{I}_1, \mathcal{S}_1, \mathcal{R}_1, \mathcal{L}_1, \mathcal{F}_1)$ *and* $\mathcal{K}_2 = (\mathcal{I}_2, \mathcal{S}_2, \mathcal{R}_2, \mathcal{L}_2, \mathcal{F}_2)$ *over the sets of variables* V_{Σ_1} *and* V_{Σ_2}, *respectively, where* $\mathcal{F}_1 = \{F_{1,1}, \dots, F_{1,n}\}$ *and* $\mathcal{F}_2 = \{F_{2,1}, \dots, F_{2,m}\}$. *The product model* $\mathcal{K}_1 \times \mathcal{K}_2 = (\mathcal{I}_\times, \mathcal{S}_\times, \mathcal{R}_\times, \mathcal{L}_\times, \mathcal{F}_\times)$ *is defined as follows:*

- $\mathcal{S}_\times := \{(s_1, s_2) \in \mathcal{S}_1 \times \mathcal{S}_2 \mid \mathcal{L}_1(s_1) \cap V_{\Sigma_2} = \mathcal{L}_2(s_2) \cap V_{\Sigma_1}\}$
- $\mathcal{I}_\times := \mathcal{S}_\times \cap (\mathcal{I}_1 \times \mathcal{I}_2)$
- $\mathcal{R}_\times := \{((s_1, s_2), (s_1', s_2')) \in \mathcal{S}_\times \times \mathcal{S}_\times \mid (s_1, s_1') \in \mathcal{R}_1 \wedge (s_2, s_2') \in \mathcal{R}_2\}$
- $\mathcal{L}_\times((s_1, s_2)) := \mathcal{L}_1(s_1) \cup \mathcal{L}_2(s_2)$
- $\mathcal{F}_\times := \left(\bigcup_{i=1}^{n} (F_{1,i} \times \mathcal{S}_2) \cap \mathcal{S}_\times \right) \cup \left(\bigcup_{i=1}^{m} (\mathcal{S}_1 \times F_{2,i}) \cap \mathcal{S}_\times \right)$

Note that $\mathcal{K}_1 \times \mathcal{K}_2$ *is a structure on* $V_{\Sigma_1} \cup V_{\Sigma_2}$.

Note that $(s_1, s_2) \in \mathcal{S}_1 \times \mathcal{S}_2$ iff the labels $\mathcal{L}_1(s_1)$ and $\mathcal{L}_2(s_2)$ do not differ on the common variables $V_{\Sigma_1} \cap V_{\Sigma_2}$. Hence, it may be the case that some states $s \in \mathcal{S}_1$ have no product states and it may also be the case that deadend states, i.e. states without successor states can arise. We can also compute quotients of structures, and in some cases this allows to recompute a factor $\mathcal{K}_i$ of a product structure $\mathcal{K}_1 \times \mathcal{K}_2$.

Definition 5 (Quotient Structures). *Let* $\mathcal{K} = (\mathcal{I}, \mathcal{S}, \mathcal{R}, \mathcal{L}, \mathcal{F})$ *be a Kripke structure over a set of variables* V_Σ *and let* $\sim$ *be an equivalence relation on* $\mathcal{S}$ *that preserves labels (i.e.* $s_1 \sim s_2$ *implies* $\mathcal{L}(s_1) = \mathcal{L}(s_2)$*). Then, the quotient structure* $\mathcal{K}_{/\sim} = (\widetilde{\mathcal{I}}, \widetilde{\mathcal{S}}, \widetilde{\mathcal{R}}, \widetilde{\mathcal{L}}, \widetilde{\mathcal{F}})$ *of* $\mathcal{K}$ *by* $\sim$ *is given as follows:*

- $\widetilde{\mathcal{I}} := \{\{s' \in \mathcal{S} \mid s \sim s'\} \mid s \in \mathcal{I}\}$
- $\widetilde{\mathcal{S}} := \{\{s' \in \mathcal{S} \mid s \sim s'\} \mid s \in \mathcal{S}\}$
- $(\widetilde{s}_1, \widetilde{s}_2) \in \widetilde{\mathcal{R}}$ *iff* $\exists (s_1', s_2') \in \mathcal{R}.s_1' \sim s_1 \wedge s_2' \sim s_2$
- $\widetilde{\mathcal{L}}(\widetilde{s}) := \mathcal{L}(s)$
- $\widetilde{\mathcal{F}} := \{\widetilde{F}_i \mid F_i \in \mathcal{F}\}$, *where* $\widetilde{F}_i := \{\{t \in \mathcal{S} \mid s \sim t\} \mid s \in F_i\}$

3   Model Checking on Product Structures

In this section, we establish a model checking procedure for CTL that respects side conditions: We consider a given structure $\mathcal{K} = (\mathcal{I}, \mathcal{S}, \mathcal{R}, \mathcal{L}, \mathcal{F})$ over the variables V_Σ, a finite number of side conditions $\ell_i = \varphi_i$ for $i \in \{1, \dots, n\}$ and a CTL formula over the variables $V_\Sigma \cup \{\ell_1, \dots, \ell_n\}$. We assume for each i that $\ell_i \notin V_\Sigma$ and that $\text{VAR}(\varphi_i) \subseteq V_\Sigma \cup \{\ell_1, \dots, \ell_{i-1}\}$. Additionally, it must hold that each occurrence of each ℓ_i inside Φ is in the scope of a path quantifier[4]. Our task is to compute all states s of $\mathcal{S}$ where $(\mathcal{K}, s) \models [\Phi]_{\ell_1 \dots \ell_n}^{\varphi_1 \dots \varphi_n}$ holds. Clearly, the latter is a CTL* model checking problem and it is shown in the next section that each CTL* model checking problem can be presented in this form. Therefore, our algorithm can be applied to construct a CTL* model checking procedure.

There is a similar algorithm of Emerson and Lei [21,22] that appears at a first glance similar to the one presented here. However, in their approach the formulas φ_i were state (LTL) formulas such that we can successively compute the sets of states in $\mathcal{K}$ where φ_i holds and relabel these states with ℓ_i. This can however not be done for our problem since our formulas φ_i are path formulas. In this case, it is in general not possible to label the structure in the desired manner[5], since there might be a state s_0 with two paths ξ_1 and ξ_2 starting in s_0 such that φ_i holds on ξ_1, but not on ξ_2. Then, there is no possibility to label s_0 such that $\ell_i = \varphi_i$ is respected (i.e. $\text{G}[\ell_i = \varphi_i]$ should hold along each fair path of the structure).

For example, consider the structure $\mathcal{K}$ given in figure 2 with the two paths $\xi_1 := (s_0, s_1, s_2, \dots)$ and $\xi_2 := (s_0, \dots)$ starting in state s_0. It can be easily seen that $(\mathcal{K}, \xi_1) \models \text{F}a$ holds, but on the other hand $(\mathcal{K}, \xi_2) \not\models \text{F}a$. Hence, the truth value of $\text{F}a$ depends on the path starting in s_0 and therefore there is no label for a new variable ℓ such that $(\mathcal{K}, s_0) \models \text{AG}[\ell = \text{F}a]$ would hold.

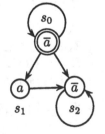

Figure 2: Example Kripke Structure

For this reason, we suggest the following procedure: *each state with paths that contradict in the truth value of a formula* φ_i *has to be split up into two substates, such*

---

[4] This is necessary to assure that $[\Phi]_{\ell_1 \dots \ell_n}^{\varphi_1 \dots \varphi_n}$ is a state formula.

[5] For path formulas we should label the structure such that in any state $(\mathcal{K}, s) \models \text{AG}[\ell_i = \varphi_i]$ holds

that one of them inherits the paths that satisfy φ_i and the other inherits the paths that falsify φ_i. We will see in the following, how the necessary computations to split the states can be described by computing certain product structures with a structure $\mathcal{D}_{\ell_i}^{\varphi_i}$ as given in definition 4.

The relationship between a side condition $\ell_i = \varphi_i$ and the structure $\mathcal{D}_{\ell_i}^{\varphi_i}$ we use for the product is that $\mathcal{D}_{\ell_i}^{\varphi_i}$ precisely encodes the information that ℓ_i should behave as φ_i on every fair path. This means that for each state of the structure $\mathcal{D}_{\ell_i}^{\varphi_i}$, we have $(\mathcal{D}_{\ell_i}^{\varphi_i}, s) \models \mathsf{AG}[\ell_i = \varphi_i]$. This has the consequence that all paths starting in any state of $\mathcal{D}_{\ell_i}^{\varphi_i}$ agree on the truth value of φ_i. This information is then carried over to a given structure $\mathcal{K}$ over the variables V_Σ by computing the product structure $\mathcal{K} \times \prod_{i=1}^n \mathcal{D}_{\ell_i}^{\varphi_i}$: It is well-known [23] that each factor structure simulates the product structure if the labels of the latter are restricted to the variables of the former. In particular, each $\mathcal{D}_{\ell_i}^{\varphi_i}$ simulates $\mathcal{K} \times \prod_{i=1}^n \mathcal{D}_{\ell_i}^{\varphi_i}$. This means that each path starting in any product state[6] (s, q) consists of factor paths starting in q_i that run through the factor structures $\mathcal{D}_{\ell_i}^{\varphi_i}$. It follows that also each fair path through $\mathcal{K} \times \prod_{i=1}^n \mathcal{D}_{\ell_i}^{\varphi_i}$ must satisfy $\ell_i = \varphi_i$ for each i. Therefore, all fair paths of a state of the product structure agree on φ.

Apart from the property that all fair paths starting in a state of the product structure should agree on φ_i, we also wish that the structure $\mathcal{K}$ is not damaged by the product computation. However, we do not want that $\mathcal{K} \times \prod_{i=1}^n \mathcal{D}_{\ell_i}^{\varphi_i}$ simulates $\mathcal{K}$ because then we would have not achieved anything interesting since the structures were bisimilar, and hence essentially the same. Nevertheless, no information of $\mathcal{K}$ should be lost, which means that we must be able to recompute at least a structure $\mathcal{K}'$ that is bisimilar to $\mathcal{K}$ from the product $\mathcal{K} \times \prod_{i=1}^n \mathcal{D}_{\ell_i}^{\varphi_i}$. For this reason, we give the following definition:

Definition 6 (φ-Defining Structures). *Let φ be a quantifier-free CTL^* path formula over the variables V_Σ and let $\ell \notin V_\Sigma$ be a variable that does not occur in φ. Then, each structure $\mathcal{D}_\ell^\varphi = (\mathcal{I}_\ell^\varphi, \mathcal{S}_\ell^\varphi, \mathcal{R}_\ell^\varphi, \mathcal{L}_\ell^\varphi, \mathcal{F}_\ell^\varphi)$ over the variables $\mathrm{VAR}(\varphi) \cup \{\ell\}$ is called a φ-defining structure iff the following properties hold:*

D1: $\forall L : \mathbb{N} \to 2^{\mathrm{VAR}(\varphi)}.\exists s \in \mathcal{S}_\ell^\varphi.\exists \pi \in \mathsf{FPaths}_{\mathcal{D}_\ell^\varphi}(s).\forall t \in \mathbb{N}.L^{(t)} = \mathcal{L}_\ell^\varphi(\pi^{(t)}) \setminus \{\ell\}$
D2: $\forall L \subseteq \mathrm{VAR}(\varphi).\exists s \in \mathcal{I}_\ell^\varphi.L = \mathcal{L}_\ell^\varphi(s) \setminus \{\ell\}$
D3: $\forall s \in \mathcal{S}_\ell^\varphi.(\mathcal{D}_\ell^\varphi, s) \models \mathsf{AG}[\ell = \varphi]$

Clearly, property **D3** guarantees that the structure $\mathcal{D}_\ell^\varphi$ encodes the information that ℓ behaves on each fair path of $\mathcal{D}_\ell^\varphi$ exactly as φ. The condition that $\mathsf{AG}[\ell = \varphi]$ has to hold in each state can also be expressed in different manners. In particular, it can be shown to be equivalent to $\mathsf{AG}(\ell \Rightarrow \mathsf{A}\varphi | \mathsf{A}\neg\varphi)$ which means that for each state, all paths leaving the state agree on the satisfaction of φ. If a state is labeled with ℓ, it means that each fair path leaving this state must satisfy φ; otherwise all fair path leaving this state falsify φ. *This property leads to the fact that when products are computed with defining structures $\mathcal{D}_\ell^\varphi$, states that have paths that do not agree on φ are split into two substates such that the one inherits all paths that satisfy φ while the other one inherits all paths that falsify φ.*

For example, consider the structures given in figure 3. It is easily seen that the structure $\mathcal{D}_\ell^{\mathsf{F}a}$ in the middle of figure 3 is a $\mathsf{F}a$-defining structure (hence we already

---

[6] We write (s, q) instead of $(s, q_1, \ldots, q_n)$.

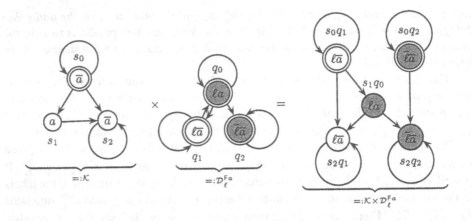

Figure3. Example for a product structure $\mathcal{K} \times \mathcal{D}_\ell^{\mathsf{Fa}}$

denoted it as $\mathcal{D}_\ell^{\mathsf{Fa}}$). The product of this structure with the one given in figure 2 is given on the right hand side of figure 3. We have already seen that the paths $\xi_1 := (s_0, s_1, s_2, \dots)$ and $\xi_2 := (s_0, \dots)$ starting in state s_0 of $\mathcal{K}$ do not agree on Fa. These paths do also occur as factors of paths in the product structure: the corresponding paths are $\varrho_1 := (s_0 q_1, s_1 q_0, s_2 q_2, \dots)$ and $\varrho_2 := (s_0 q_2, \dots)$. However, these paths start in different product states of s_0.

We now turn to the second important property that we mentioned before the definition of φ-defining structures. We must assure that no computation of $\mathcal{K}$ is lost during the product computation. This is guaranteed by **D1**: for each fair path π_1 of $\mathcal{K}$ there is a fair path π_2 through $\mathcal{D}_\ell^\varphi$ such that for each point of time t the labels of the states $\pi_1^{(t)}$ and $\pi_2^{(t)}$ are consistent, i.e. for any $t \in \mathbb{N}$ we have $\mathcal{L}(\pi_1^{(t)}) \cap \mathrm{VAR}(\varphi) = \mathcal{L}_\ell^\varphi(\pi^{(t)}) \setminus \{\ell\}$ (to see this instantiate L in **D1** by $\lambda t.\mathcal{L}(\pi_1^{(t)}) \cap \mathrm{VAR}(\varphi)$). Hence, the path $\lambda t.(\pi_1^{(t)}, \pi_2^{(t)})$ is a path through the product structure and no path π_1 of $\mathcal{K}$ is lost in the product computation. **D2** moreover guarantees that no state, and in particular no initial state gets lost in the product. This is almost redundant to **D1**, apart in the case when there is a state that has no fair paths at all. The main results concerning the relationship of a structure $\mathcal{K}$ and a number of φ_i-defining structures $\mathcal{K} \times \mathcal{D}_{\ell_i}^{\varphi_i}$ is formally stated in the next theorem.

Theorem 7 (Products with Defining Structures). *Given a Kripke structure $\mathcal{K} = (\mathcal{I}, \mathcal{S}, \mathcal{R}, \mathcal{L}, \mathcal{F})$ over some set of variables V_Σ, and for $i \in \{1, \dots, n\}$ the quantifier-free CTL* path formulas φ_i over $V_\Sigma \cup \{\ell_1, \dots, \ell_{i-1}\}$ where $\ell_i \notin V_\Sigma$. Let $\mathcal{D}_{\ell_i}^{\varphi_i}$ for $i \in \{1, \dots, n\}$ be a φ_i-defining structure and compute $\mathcal{D}_\ell^\varphi := \prod_{i=1}^n \mathcal{D}_{\ell_i}^{\varphi_i}$ and $\mathcal{K} \times \mathcal{D}_\ell^\varphi = (\mathcal{I}_\times, \mathcal{S}_\times, \mathcal{R}_\times, \mathcal{L}_\times, \mathcal{F}_\times)$. Then, the following holds:*

- $(\mathcal{K} \times \mathcal{D}_\ell^\varphi, (s, q)) \models \mathsf{AG}[\ell_i = \varphi_i]$ *for $i \in \{1, \dots, n\}$*
- *For any state $s \in \mathcal{S}$ and any path $\pi \in \mathrm{FPaths}_\mathcal{K}(s)$, there is a state $(s, q) \in \mathcal{S}_\times$ and a path $\xi \in \mathrm{FPaths}_{\mathcal{D}_\ell^\varphi}((s, q))$ such that for all $t \in \mathbb{N}$ there is a $\sigma^{(t)} \in \mathcal{S}_\ell^\varphi$ such that $\xi^{(t)} = (\pi^{(t)}, \sigma^{(t)})$.*

- For any state $s \in S$ there is a product state $(s, q) \in S_\times$ where each q_i is an initial state of $\mathcal{D}_{\ell_i}^{\varphi_i}$.
- Define the relation $\sim$ on $S_\times \times S_\times$ as $(p_1, q_1) \sim (p_2, q_2) :\Leftrightarrow p_1 = p_2$. Then $\sim$ is an equivalence relation that preserves labels up to the variables ℓ_i. Moreover, let $\mathcal{K}_\varphi$ be the structure $\mathcal{K} \times \mathcal{D}_\ell^\varphi$ restricted to V_Σ, i.e. $\mathcal{K}_\varphi = (\mathcal{I}_\times, S_\times, \mathcal{R}_\times, \mathcal{L}'_\times, \mathcal{F}_\times)$, where $\mathcal{L}'_\times((s_1, s_2)) := \mathcal{L}_\times((s_1, s_2)) \setminus \{\ell_1, \ldots, \ell_n\}$. Then, the quotient $\mathcal{K}_{\varphi/\sim}$ of $\mathcal{K}_\varphi$ by $\sim$ is bisimilar to $\mathcal{K}$.

Note that the first item follows since $\mathsf{AG}[\ell_i = \varphi_i]$ is a ACTL* formula and each ACTL* formula that holds in a structure holds also in each product with that structure. The third item means that no state of $\mathcal{K}$ is lost and the second item assures that no path of $\mathcal{K}$ is lost by the product computation. Note that this does not mean that $\mathcal{K}$ and $\mathcal{K} \times \mathcal{D}_\ell^\varphi$ were bisimilar, because for different paths $\pi_1, \pi_2 \in \text{FPaths}_\mathcal{K}(s)$ there might be different states $s_1', s_2' \in S_\ell^\varphi$ where the corresponding paths σ_1, σ_2 through $\mathcal{D}_\ell^\varphi$ start. However, we can combine all the product states of s by a quotient computation using the equivalence relation $\sim$ as given in the last item of the above theorem. Product states are equivalent w.r.t. $\sim$ iff their first component is the same. As the quotient structure $\mathcal{K}_{\varphi/\sim}$ is bisimilar [7] to $\mathcal{K}$ it clearly follows that the structure $\mathcal{K}$ has not been damaged too much by the product computation and this in turn means that we can hope to check formulas in $\mathcal{K} \times \mathcal{D}_\ell^\varphi$ instead of checking them in $\mathcal{K}$.

This means that we must combine the usual CTL model checking procedure with the quotient construction w.r.t $\sim$ to be able to check CTL formulas in $\mathcal{K} \times \prod_{i=1}^n \mathcal{D}_{\ell_i}^{\varphi_i}$ instead of checking them in $\mathcal{K}$. This combination with the quotient construction is described in the algorithm of figure 4. Before we prove its correctness, we consider how it works by an example computation. Consider the formula $[\mathsf{EF}a] \wedge [\mathsf{EG}\neg a]$ and the product structure in figure 3. The following facts can be easily seen:

$$(\mathcal{K}, s_0) \models [\mathsf{EF}a] \wedge [\mathsf{EG}\neg a]$$

$$(\mathcal{K} \times \mathcal{D}_\ell^{\mathsf{F}a}, s_0 q_1) \models \mathsf{EF}a \qquad (\mathcal{K} \times \mathcal{D}_\ell^{\mathsf{F}a}, s_0 q_2) \not\models \mathsf{EF}a$$

$$(\mathcal{K} \times \mathcal{D}_\ell^{\mathsf{F}a}, s_0 q_1) \not\models \mathsf{EG}\neg a \qquad (\mathcal{K} \times \mathcal{D}_\ell^{\mathsf{F}a}, s_0 q_2) \models \mathsf{EG}\neg a$$

Hence, although $(\mathcal{K}, s_0) \models [\mathsf{EF}a] \wedge [\mathsf{EG}\neg a]$ holds in s_0, the formula does not hold in any product state $s_0 q_i$ of s_0 in the product structure. This is due to the splitting of the state s_0. If the model checking is done on the product structure, then all substates of a state s_0 have to be considered for checking whether a formula beginning with a path quantifier holds in the state s_0 or not. This means we must carry over the information to the quotient under $\sim$.

---

[7] As $\mathcal{K}_{\varphi/\sim}$ and $\mathcal{K}$ are bisimilar, they satisfy the same CTL* formulas over V_Σ. We must emphasize here that *only the quotient $\mathcal{K}_{\varphi/\sim}$ is bisimilar to $\mathcal{K}$, while the product (even restricted to V_Σ) is in general not bisimilar to $\mathcal{K}$*. To see this, reconsider the example given in figure 3. The classes under the relation $\sim$ are clearly $\{(s_0 q_1), (s_0, q_2)\}$, $\{(s_1 q_0)\}$, and $\{(s_2 q_1), (s_2, q_2)\}$ and it is easily seen that the quotient of $\mathcal{K} \times \mathcal{D}_\ell^{\mathsf{F}a}$ by $\sim$ is even isomorphic to $\mathcal{K}$. However, $\mathcal{K}$ and $\mathcal{K} \times \mathcal{D}_\ell^{\mathsf{F}a}$ are not bisimilar, even if we restrict the latter to variables of $\mathcal{K}$. To see this, note that $(\mathcal{K} \times \mathcal{D}_\ell^{\mathsf{F}a}, s_i q_j) \models (\mathsf{AF}a) \vee (\mathsf{AG}\neg a)$ holds for any state $s_i q_j$ of the product structure $\mathcal{K} \times \mathcal{D}_\ell^{\mathsf{F}a}$. However, we have $(\mathcal{K}, s_0) \not\models (\mathsf{AF}a) \vee (\mathsf{AG}\neg a)$, and as bisimilar states satisfy the same CTL* formulas, it follows that there is no state in the product structure that is bisimilar to the initial state s_0. Hence, $\mathcal{K}$ and $\mathcal{K} \times \mathcal{D}_\ell^{\mathsf{F}a}$ are not bisimilar.

```
FUNCTION Quan∃(SQ) =                    FUNCTION Quan∀(SQ) =
  P := {};                                P := {};
  FOR (s, q) ∈ SQ DO                      FOR (s₁, q) ∈ SQ DO
    P := P ∪ {(s, p) | (s, p) ∈ S×};        F_{s₁} := {s ∈ S× | ∃p.s = (s₁, p)};
  END;                                      IF F_{s₁} ⊆ SQ THEN P := P ∪ F_{s₁};
  return P;                                 SQ := SQ \ F_{s₁};
                                          END;
                                          return P;
```

```
FUNCTION States×(Φ) = /* compute the set of states, where Φ holds */
/* GFP(f), and LFP(f) compute greatest and least fixpoint of f */
/* suc∃(M) := {s ∈ S | ∃t ∈ M.(s, t) ∈ R} */
/* suc∀(M) := {s ∈ S | ∀t ∈ M.(s, t) ∈ R} */
  CASE Φ of
    is_var(Φ):  return {(s, q) | Φ ∈ L×((s, q))};
    ¬φ       :  return S \ States×(φ);
    φ ∧ ψ    :  return States×(φ) ∩ States×(ψ);
    φ ∨ ψ    :  return States×(φ) ∪ States×(ψ);
    EXφ      :  M_φ = States×(φ); return Quan∃(suc∃(M_φ));
    E[φ B ψ] :  M_φ = States×(φ); M_ψ = States×(ψ);
                return Quan∃(GFP(λx.¬M_ψ ∧ [M_φ ∨ suc∃(x)]));
    E[φ U ψ] :  M_φ = States×(φ); M_ψ = States×(ψ);
                return Quan∃(LFP(λx.M_ψ ∨ [M_φ ∧ suc∃(x)]));
    AXφ      :  M_φ = States×(φ); return Quan∀(suc∀(M_φ));
    A[φ B ψ] :  M_φ = States×(φ); M_ψ = States×(ψ);
                return Quan∀(GFP(λx.¬M_ψ ∧ [M_φ ∨ suc∃(x)]));
    A[φ U ψ] :  M_φ = States×(φ); M_ψ = States×(ψ);
                return Quan∀(LFP(λx.M_ψ ∨ [M_φ ∧ suc∀(x)]));
  END;
```

Figure4. Checking CTL formulas in Product Models with Side Conditions

If the function $States_\times(\Phi)$ given in figure 4 is called with the formula $\Phi := [EFa] \wedge [EG \neg a]$ and the product model of figure 3, then it will first compute the set of states $\{s_0 q_1, s_1 q_0\}$ where EFa holds by usual fixpoint iterations. After that, if a state $s_i q_j$ is in the computed state set, then *all other product states of s_i are added* to the set by function $Quan_\exists()$. This is correct since if one state $s_i q_j$ is in the computed state set, we know that there must be a path through the product structure that satisfies Fa. Hence, there is also a path running through $\mathcal{K}$ that satisfies Fa and as $\mathcal{K}$ is bisimilar to the quotient, there must also be a fair path starting in the quotient state $\widetilde{(s, q)}$ that satisfies Fa. Hence, we must include not only $s_i q_j$, but all members of its class w.r.t. to $\sim$. This computation is performed by the function $Quan_\exists()$. An analogous transformation has to be performed for state formulas $A\varphi$ by function $Quan_\forall()$ in figure 4. Hence, the functions $\text{Quan}_\exists()$ and $\text{Quan}_\forall()$ switch to the quotient structure which is bisimilar to $\mathcal{K}$, and so these functions make our state sets consistent with $\mathcal{K}$. In case of $\{s_0 q_1, s_1 q_0\}$, the new set $\{s_0 q_1, s_0 q_2, s_1 q_0\}$ is obtained by the application of the function $\text{Quan}_\exists()$. This means that the formula EFa holds in the states $\{s_0, s_1\}$ of the original structure (as the classes of these states are included in the mentioned state set). Analogously, the set of states $\{s_0 q_2, s_2 q_2\}$ where $EG\neg a$ holds is computed and the application of $\text{Quan}_\exists()$ yields in $\{s_0 q_1, s_0 q_2, s_2 q_1, s_2 q_2\}$. Finally, the conjunction of the two formulas is computed by intersecting the two sets: the result is $\{s_0 q_1, s_0 q_2\}$, which means that $[EFa] \wedge [EG\neg a]$ holds exactly in state s_0 of the original structure.

The algorithm of figure 4 applies after each fixpoint iteration one of the functions $\text{Quan}_\exists()$ or $\text{Quan}_\forall()$ in order to make the found set of states consistent with the quotient structure. The correctness of the algorithm given in figure 4 is stated in the following theorem:

Theorem 8 (CTL Model Checking on Product Structures). *Given a structure* $\mathcal{K} = (\mathcal{I}, \mathcal{S}, \mathcal{R}, \mathcal{L}, \mathcal{F})$ *over the variables* V_Σ, *variables* $\ell_1, \ldots, \ell_n$ *that do not belong to* V_Σ, *and the quantifier-free* CTL* *path formulas* $\varphi_1, \ldots, \varphi_n$ *with* $\text{VAR}(\varphi_i) \subseteq V_\Sigma \cup \{\ell_1, \ldots, \ell_{i-1}\}$. *Let moreover* Φ *be a* CTL *formula over* $V_\Sigma \cup \{\ell_1, \ldots, \ell_n\}$ *such that each occurrence of each* ℓ_i *in* Φ *is inside the scope of a path quantifier. Then, it follows for any* φ_i-*defining structures* $\mathcal{D}_{\ell_i}^{\varphi_i} = (\mathcal{I}_{\ell_i}^{\varphi_i}, \mathcal{S}_{\ell_i}^{\varphi_i}, \mathcal{R}_{\ell_i}^{\varphi_i}, \mathcal{L}_{\ell_i}^{\varphi_i}, \mathcal{F}_{\ell_i}^{\varphi_i})$ *that*

$$\text{States}_\times(\Phi) = \left\{ (s, q) \in \mathcal{S} \times \prod_{i=1}^n \mathcal{S}_{\ell_i}^{\varphi_i} \,\middle|\, (\mathcal{K}, s) \models [\Phi]_{\ell_1 \ldots \ell_n}^{\varphi_1 \ldots \varphi_n} \right\}$$

The proof is done by a simultaneous induction on the structure of Φ where it is proved that the following is equivalent

1. $(\mathcal{K}, s) \models [\Phi]_{\ell_1 \ldots \ell_n}^{\varphi_1 \ldots \varphi_n}$
2. $\forall q \in \prod_{i=1}^n \mathcal{S}_{\ell_i}^{\varphi_i}.(s, q) \in \mathcal{S} \times \prod_{i=1}^n \mathcal{S}_{\ell_i}^{\varphi_i} \to (s, q) \in \text{States}_\times(\Phi)$
3. $(\mathcal{K}, s) \models [\Phi]_{\ell_1 \ldots \ell_n}^{\varphi_1 \ldots \varphi_n} \Leftrightarrow \exists q \in \prod_{i=1}^n \mathcal{S}_{\ell_i}^{\varphi_i}.(s, q) \in \text{States}_\times(\Phi)$

Note that the function $\text{States}_\times()$ behaves exactly as the usual model checking algorithm for CTL [24], except that after the fixpoint iterations one of the functions $\text{Quan}_\exists()$ and $\text{Quan}_\forall()$ are called to make the result consistent with the quotient structure $\mathcal{K}_{\varphi/\sim}$. So, what $\text{States}_\times()$ really does is to model check a CTL formula on $\mathcal{K}_{\varphi/\sim}$, but without really having $\mathcal{K}_{\varphi/\sim}$. Instead, $\text{States}_\times()$ works on $\mathcal{K} \times \prod_{i=1}^n \mathcal{D}_{\ell_i}^{\varphi_i}$, but nevertheless computes results for $\mathcal{K}_{\varphi/\sim}$ which also hold for $\mathcal{K}$.

Clearly, it is a disadvantage to use larger structures for model checking since this makes the state explosion already appearing in $\mathcal{K}$ even worse. However, in our case it has the advantage to be able to check an arbitrary CTL* formula Ψ, if Ψ is split into a CTL formula Φ and some side conditions $\ell_i = \varphi_i$ such that $\Psi = [\Phi]_{\ell_1 \ldots \ell_n}^{\varphi_1 \ldots \varphi_n}$. We will consider in the next section how such a transformation of a given CTL* formula is computed. However, we will not directly compute a CTL formula, instead we compute a LeftCTL* formula and translate this formula to an equivalent CTL formula. As LeftCTL* is syntactically much richer than CTL, it follows that less side conditions are generated and hence, that less products are to be computed. Hence, the blow-up problem is reduced to some extent by the extraction of LeftCTL*.

4 CTL* and LTL Model Checking by Extraction of LeftCTL*

In [17] it is shown that each LeftCTL* can be translated to an equivalent CTL formula such that the usual CTL model checking procedure can be applied also for LeftCTL* model checking. In this section, we consider how the results of [17] and the previous section can be used to construct model checking procedures for CTL* and LTL by the help of a LeftCTL* model checking procedure. Note first that there are LTL and CTL* formulas that are not equivalent to any CTL formula and hence these formulas are also not equivalent to any LeftCTL* formula (for example, take the LTL formula AFGa [3]). In contrast to CTL*, path formulas are not allowed to occur arbitrarily in LeftCTL* formulas. However, we can extract a LeftCTL* formula together with some side conditions from a given CTL* formula as follows: if an occurrence of a path formula φ_i violates the grammar rules of LeftCTL*, then each occurrence of φ_i is replaced by a new variable ℓ_i. Applied repeatedly, this leads finally to a LeftCTL* formula (see figure 5). Of course, it has to be ensured that the thereby introduced new variables ℓ_i behave in each case equivalent to φ_i. This means that we must respect the side conditions $\ell_i = \varphi_i$. The algorithm given in figure 5 performs the mentioned extraction together with a translation of the resulting LeftCTL* formula to CTL. For this reason, it makes use of functions E_conj and A_disj that translate LeftCTL* formulas of the form $E(\varphi \wedge \psi)$ and $A(\varphi \vee \psi)$, respectively. remove$_{E,A}$ removes all path quantifiers in a given formula.

If φ_i would be a state formula, then we could relabel the structure such that the equation AG$[\ell_i = \varphi_i]$ holds in each state: Simply compute the set of states where φ_i holds and label them with ℓ_i. This has already been suggested by the algorithm of Emerson and Lei [21,22] to reduce CTL* model checking problems to LTL model checking problems. However, the algorithms given in figure 5 only generate side conditions $\ell_i = \varphi_i$, where φ_i are path formulas. We have already shown by the structure given in figure 2 that in this case, we can in general not hope to relabel the structure in the desired manner. However, we can use the results of the previous section to compute a product with φ_i-defining structures and using the product model checking algorithm of figure 4. This yields in the following theorem:

Theorem 9 (CTL* Model Checking Using Product Structures). *Given a structure $\mathcal{K}$ for a signature V_Σ, an arbitrary CTL* formula Φ over V_Σ, the function extract_S() given in figure 5 computes as a side effect (we assume that at the beginning, $\mathcal{E} = \{\}$ holds) a set of side conditions $\mathcal{E} = \{\ell_1 = \varphi_1, \ldots, \ell_n = \varphi_n\}$ with $\ell_i \notin V_\Sigma$ and VAR(φ_i) $\subseteq V_\Sigma \cup \{\ell_1, \ldots, \ell_{i-1}\}$ and yields in a LeftCTL* formula Φ' over the signature $V_\Sigma \cup \{\ell_1, \ldots, \ell_n\}$, such that the following holds:*

1. *$\Phi \equiv [\Phi']^{\varphi_1 \ldots \varphi_n}_{\ell_1 \ldots \ell_n}$*
2. *All occurrences of each variable ℓ_i inside Φ' are in the scope of a path quantifier.*
3. *The formulae φ_i start with a temporal operator whose arguments are propositional formulas.*

Hence, the translation of Φ' to an equivalent CTL formula Φ'' and the application of the algorithm of figure 4 computes all states of $\mathcal{K}$ where the CTL formula Φ holds.*

```
VAL ε := {};

FUNCTION extract_S(Φ) = /* Φ ∈ P*_Σ */
CASE Φ of
  is_var(Φ) : return Φ;
  ¬φ        : return ¬extract_S(φ);
  φ ∧ ψ     : return extract_S(φ) ∧ extract_S(ψ);
  φ ∨ ψ     : return extract_S(φ) ∨ extract_S(ψ);
  Aφ        : return extract_P_A(φ);
  Eφ        : return extract_P_E(φ);
  default   : return addeq((remove_{E,A}(Φ));
```

```
FUNCTION addeq(Φ) return l* Φ ∈ P*_Σ */
IF prop(Φ) return Φ;
ℓ = new_var;
CASE Φ of
  Xφ      : ε := ε ∪ {ℓ = X(extract_S(φ))};
  Gφ      : ε := ε ∪ {ℓ = G(extract_S(φ))};
  Fφ      : ε := ε ∪ {ℓ = F(extract_S(φ))};
  [φ U ψ] : ε := ε ∪ {ℓ = [(extract_S(φ)) U (extract_S(ψ))]};
  [φ U ψ] : ε := ε ∪ {ℓ = [(extract_S(φ)) U (extract_S(ψ))]};
  [φ B ψ] : ε := ε ∪ {ℓ = [(extract_S(φ)) B (extract_S(ψ))]};
  [φ B ψ] : ε := ε ∪ {ℓ = [(extract_S(φ)) B (extract_S(ψ))]};
return ℓ;
```

```
FUNCTION extract_P_E(Φ) =
CASE Φ of
  is_var(Φ): return Φ;
  ¬φ       : return extract_P_A(φ);
  φ ∧ ψ    : return E_conj(Φ);
  φ ∨ ψ    : return extract_P_E(φ) ∨ extract_P_E(ψ);
  Xφ       : return EX extract_P_E(φ);
  Gφ       : return EG extract_S(φ);
  Fφ       : return EF extract_P_E(φ);
  [φ U ψ]  : return E[extract_S(φ) U extract_P_E(ψ)];
  [φ U ψ]  : return E[extract_S(φ) U extract_P_E(ψ)];
  [φ B ψ]  : return E[extract_P_E(φ) B extract_S(ψ)];
  [φ B ψ]  : return E[extract_P_E(φ) B extract_S(ψ)];
```

```
FUNCTION extract_P_A(Φ) =
CASE Φ of
  is_var(Φ): return Φ;
  ¬φ       : return ¬extract_P_E(φ);
  φ ∧ ψ    : return extract_P_A(φ) ∧ extract_P_A(ψ);
  φ ∨ ψ    : return A_disj(Φ);
  Xφ       : return AX extract_P_A(φ);
  Gφ       : return AG extract_P_A(φ);
  Fφ       : return AF extract_S(φ);
  [φ U ψ]  : return A[extract_P_A(φ) U extract_S(ψ)];
  [φ U ψ]  : return A[extract_P_A(φ) U extract_S(ψ)];
  [φ B ψ]  : return A[extract_S(φ) B extract_P_E(ψ)];
  [φ B ψ]  : return A[extract_S(φ) B extract_P_E(ψ)];
```

Figure5. Algorithm for extracting LeftCTL* with side conditions from CTL*

The proof of the theorem is not very hard and follows by an easy inspection of the algorithm given in figure 5 for all cases of the CTL* formula Φ. Hence, our new CTL* model checking procedure works as follows: first extract from the given CTL* formula Φ a LeftCTL* formula Φ' and some side conditions $\ell_1 = \varphi_1, \ldots, \ell_n = \varphi_n$ by the algorithm given in figure 5. Then compute the product of $\mathcal{K}$ with φ_i-defining structures and perform the model checking using the algorithm of figure 4. This yields in a set of product states (s, q) in which we ignore the second components q such that as a result the set of states of $\mathcal{K}$ is obtained where Φ holds.

It remains to compute for any side condition $\ell_i = \varphi_i$ a φ_i-defining structure. Theorem 9 tells us that it is sufficient to find φ_i-defining structures for formulas φ_i that start with a temporal operator that has propositional arguments. In case that the arguments are even variables, we can use the following lemma.

Lemma 10 (Defining Structures). *The structures* $\mathcal{D}_\ell^\varphi$ *as given in figure 6 for* $\varphi \in \{[a \underline{U} b], [a \, B \, b], Xa\}$ *are defining structures of the formula* φ. *All states are initial states and in the structures* $\mathcal{D}_\ell^{[a \underline{U} b]}$ *and* $\mathcal{D}_\ell^{[a \, B \, b]}$ *there is a single fairness constraint that consists of the gray shaded states.*

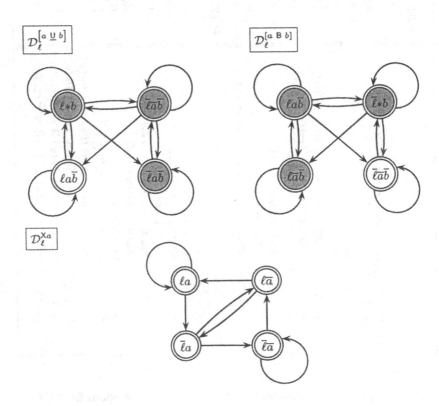

Figure6. Defining structures for $[a \underline{U} b]$, $[a \, B \, b]$, and Xa.

The lemma is easily verified by checking that the structures fulfill the requirements **D1**, **D2**, and **D3** listed in definition 6. For the more general case when the arguments a and b are propositional formulas, we propose the following: once again, we can abbreviate these propositional formulas by new variables ϑ_a and ϑ_b, compute the set of states where a and b hold and relabel these states with ϑ_a and ϑ_b, respectively.

For implementations with symbolic model checking using BDDs, it is very simple to compute products with φ_i-defining structures, even when the arguments are arbitrary propositional formulas. We just have to add the following formulas to the transition relation and to the fairness constraints for any propositional logic formulas a and b:

φ_i	Transition Relation	Fairness Constraint
Xa	$\ell = X\ell$	$--$
Ga	$\ell = a \wedge X\ell$	$GF(\ell \to a)$
Fa	$\ell = a \vee X\ell$	$GF(a \to \ell)$
$[a\ W\ b]$	$\ell = (b \Rightarrow a \vert X\ell)$	$GF(\ell \vee b)$
$[a\ \underline{W}\ b]$	$\ell = (b \Rightarrow a \vert X\ell)$	$GF(\ell \to b)$
$[a\ U\ b]$	$\ell = b \vee a \wedge X\ell$	$GF(\ell \vee \neg a \vee b)$
$[a\ \underline{U}\ b]$	$\ell = b \vee a \wedge X\ell$	$GF(\ell \to \neg a \vee b)$
$[a\ B\ b]$	$\ell = \neg b \wedge (a \vee X\ell)$	$GF(\ell \vee a \vee b)$
$[a\ \underline{B}\ b]$	$\ell = \neg b \wedge (a \vee X\ell)$	$GF(\ell \to a \vee b)$

Also, the functions $\mathsf{Quan}_\exists()$ and $\mathsf{Quan}_\forall()$ can be easily implemented using BDDs. They simply correspond to quantification over the boolean variables $\ell_1, \ldots, \ell_n$ that have been introduced for the side conditions. Hence, the product model checking approach as presented in the previous section can be easily implemented in a symbolic model checking procedure using BDDs.

If we are only interested in LTL model checking, it turns out that the product model checking procedure can be simplified in that we do not need the functions $\mathsf{Quan}_\exists()$ and $\mathsf{Quan}_\forall()$. This has the consequence that any existing CTL model checker can be used to check LTL formulas with our product structure approach.

Theorem 11 (LTL Model Checking Using Product Structures). *Given a structure $\mathcal{K}$ for a signature V_Σ, an arbitrary quantifier free formula Φ over V_Σ, the function call* extract_$S(A\Phi)$ *(cf. figure 5) computes as a side effect (we assume that at the beginning $\mathcal{E} = \{\}$ holds) a set of side conditions $\mathcal{E} = \{\ell_1 = \varphi_1, \ldots, \ell_n = \varphi_n\}$ with $\ell_i \notin V_\Sigma$ and $\mathrm{VAR}(\varphi_i) \subseteq V_\Sigma \cup \{\ell_1, \ldots, \ell_{i-1}\}$ and yields in a LeftCTL* formula Φ' over the signature $V_\Sigma \cup \{\ell_1, \ldots, \ell_n\}$ such that the following holds for arbitrary φ_i-defining structures $\mathcal{D}_{\ell_i}^{\varphi_i}$:*

$$(\mathcal{K}, s) \models A\Phi \quad iff \quad \forall q \in \prod_{i=1}^{n} \mathcal{D}_{\ell_i}^{\varphi_i}.(s, q) \in \mathcal{S} \times \prod_{i=1}^{n} \mathcal{D}_{\ell_i}^{\varphi_i} \to (\mathcal{K} \times \prod_{i=1}^{n} \mathcal{D}_{\ell_i}^{\varphi_i}, (s, q)) \models \Phi'$$

By the above theorem, we do not need the functions $\mathsf{Quan}_\exists()$ and $\mathsf{Quan}_\forall()$ in figure 4 when a LTL formula is to be checked (note that $(\mathcal{K} \times \prod_{i=1}^{n} \mathcal{D}_{\ell_i}^{\varphi_i}, (s, q)) \models \Phi'$ is a CTL model checking problem). Instead, a single call to $\mathsf{Quan}_\forall()$ after the CTL model checking is sufficient and this can even be replaced by a universal quantification over q

as given in the above theorem. Note that quantification over boolean variables is simple for BDDs. Hence, the above theorem gives us the key to use any existing CTL model checker for LTL model checking.

Translations from LTL to ω-automata allow us to do the same. However, the results using a translation to ω-automata are not as good as in the procedure presented here. This is simply for the reason, that the approach presented here retains as much as possible of the given LTL formula in the extracted LeftCTL* formula and generates side conditions only when necessary. Translations to ω-automata must however generate much more 'side conditions' because the 'CTL formulas they extract' are simple temporal formulas (e.g. GFφ in case of Büchi automata). Having this view, we can interpret the presented translations also as a translation of LTL formulas to ω-automata whose acceptance conditions are general CTL formulas.

5 Experimental Results

The presented translation procedure has been implemented in C++ and can be freely tested via a CGI interface over the WWW under the URL of the author. At the moment, only LTL can be checked by transformation into a CTL model checking problem for the well-known CTL model checker SMV [4] as a back end. In this section, some experimental results obtained by that implementation and the original implementation of [9] are given.

The example presented here shows that exponential savings in terms of reachable states that in turn leads to enormous savings in runtime and space requirements are possible. The example is essentially a counter circuit that counts events that occur at its boolean input a until a maximal number $e \in \mathbb{N}$ of events has been reached. In this case the only output o is set. The circuit has also a reset input r and additionally $\log(e) + 1$ internal state variables[8].

The specification Φ_e that is to be checked is defined as G$[r \to \varphi_e]$, where φ_e is defined recursively as follows: $\varphi_0 := \mathsf{X}o$ and $\varphi_{e+1} := \mathsf{X}[((r \lor \varphi_e)) \mathbin{\mathsf{W}} (r \lor a)]$. Clearly, as $\Phi_e \in P_A$, using the translation of [17] the specification can be translated in an equivalent CTL formula of the same size. A detailed analysis of the product automaton of the tableau and the circuit leads to the results in the table below:

	Reach. States	Poss. States	BDD nodes
Tableaux	$2^{2e+2}2^{3+\log(e)}$	$2^{2e+2}2^{3+\log(e)}$	$31e + 15 + 13\log(e)$
Closed Prefix Tableaux	$(2e+6)2^{3+\log(e)}$	$2^{3e+2}2^{3+\log(e)}$	$12e + 31 + 9\log(e)$
LeftCTL*	$2^{3+\log(e)}$	$2^{3+\log(e)}$	$8 + 9\log(e)$

The first row contains the data for usual tableaux, e.g. as given in [9] and the second one for an enhanced tableau procedure also developed by the author. It can be seen that the presented translation leads to only $O(e)$ reachable states in contrast to $O(e2^{2e})$ reachable states of the tableau method. Moreover, ordinary tableaux generate $O(e)$ fairness constraints, whereas none is required for the presented method. Consequently[9], the

---

[8] $\log(e)$ is the greatest natural number that is less than the real $\log(e)$.

[9] The worst case complexity of checking whether a CTL formula φ holds in a model with S states with a fairness constraint φ is $O(S^2 |\varphi| |\psi|)$.

worst case complexities are $O(e^2)$ for the presented method in comparison to $O(e^3 2^{2e})$ for tableaux. The actual runtimes and space requirements are given in figure 7 (SUN UltraSparc, Solaris 5.5.1, SMV 2.4.3).

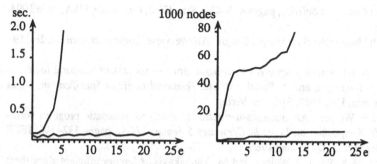

Figure7. Runtime and space requirements for the identification example

While the above example can still be seen as a pathological one (even if it appears together with its specification in practical protocols), one might suppose that this is not the average case. For this reason, a list of roughly 300 LTL theorems has been collected and the procedures have been applied to them. It turned out that the presented procedure outperforms other existing tools. The theorems are available also under the URL of the author and can be tested publicly.

References

1. E.M. Clarke and E.A. Emerson. Design and Synthesis of Synchronization Skeletons using Branching Time Temporal Logic. In D. Kozen, editor, *Workshop on Logics of Programs*, volume 131 of *Lecture Notes in Computer Science*, pages 52–71, Yorktown Heights, New York, May 1981. Springer-Verlag.
2. A. Pnueli. The temporal logic of programs. In *Symposium on Foundations of Computer Science*, volume 18, pages 46–57, New York, 1977. IEEE.
3. E.A. Emerson and J.Y. Halpern. "sometimes" and "not never" revisited: On branching versus linear time temporal logic. *Journal of the ACM*, 33(1):151–178, January 1986.
4. K.L. McMillan. *Symbolic Model Checking*. Kluwer Academic Publishers, Norwell Massachusetts, 1993.
5. A. Aziz, F. Balarin, S.-T. Cheng, R. Hojati, T. Kam, S.C. Krishnan, R.K. Ranjan, T.R. Shiple, V. Singhal, S. Tasiran, H.-Y. Wang, R.K. Brayton, and A.L. Sangiovanni-Vincentelli. HSIS: A BDD-Based Environment for Formal Verification. In *ACM/IEEE Design Automation Conference (DAC)*, San Diego, CA, June 1994. San Diego Convention Center.
6. J. Dingel and T. Filkorn. Model checking for infinite state systems using data abstraction, assumption-commitment style reasoning and theorem proving. In P. Wolper, editor, *Conference on Computer Aided Verification (CAV)*, volume 939 of *Lecture Notes in Computer Science*, pages 54–69, Liege, Belgium, July 1995. Springer Verlag.
7. E.M. Clarke, E.A. Emerson, and A.P. Sistla. Automatic Verification of Finite-State Concurrent Systems Using Temporal Logic Specifications. *ACM Transactions on Programming Languages and Systems*, 8(2):244–263, April 1986.

8. A.P. Sistla and E.M. Clarke. The complexity of propositional linear temporal logics. *Journal of Assoc. Comput. Mach.*, 32(3):733–749, July 1985.
9. E.M. Clarke, O. Grumberg, and K. Hamaguchi. Another look at LTL model checking. In David L. Dill, editor, *Conference on Computer Aided Verification (CAV)*, volume 818 of *Lecture Notes in Computer Science*, pages 415–427, Standford, California, USA, June 1994. Springer-Verlag.
10. P. Wolper. The tableau method for temporal logic: An overview. *Logique et Analyse*, 28:119–136, 1985.
11. P. Wolper. On the relation of programs and computations to models of temporal logic. In B. Banieqbal, H. Barringer, and A. Pnueli, editors, *Temporal Logic in Specification*, pages 75–123, Altrincham, UK, 1987. Springer-Verlag.
12. M.Y. Vardi and P. Wolper. An automata-theoretic approach to automatic program verification. In *IEEE Symposium on Logic in Computer Science (LICS)*, pages 332–344. IEEE Computer Society Press, D.C., June 1986.
13. C. Courcoubetis, M.Y. Vardi, P. Wolper, and M. Yannakakis. Memory efficient algorithms for the verification of temporal properties. *Formal Methods in System Design*, 1:275–288, 1992.
14. A. Valmari. A Stubborn Attack on the State Explosion Problem. In R. P. Kurshan and E.M. Clarkes, editors, *Workshop on Computer Aided Verification (CAV)*, June 1990.
15. P. Godefroid and P. Wolper. Using partial orders for the efficient verification of deadlock freedom and safety properties. *Formal Methods in System Design*, 2(2):149–164, April 1993.
16. D. Peled. Combining partial order reductions with on-the-fly model-checking. In David L. Dill, editor, *Conference on Computer Aided Verification (CAV)*, volume 818 of *Lecture Notes in Computer Science*, pages 377–390, Standford, California, USA, June 1994. Springer-Verlag.
17. K. Schneider. CTL and equivalent sublanguages of CTL*. In C. Delgado Kloos, editor, *IFIP Conference on Computer Hardware Description Languages and their Applications (CHDL)*, pages 40–59, Toledo,Spain, April 1997. IFIP, Chapman and Hall.
18. R. Milner. *Communication and Concurrency*. Prentice-Hall International, London, 1989.
19. M.C. Browne, E.M. Clarke, and O. Grumberg. Characterizing Finite Kripke Structures in Propositional Temporal Logic. *Theoretical Computer Science*, 59(1-2), July 1988.
20. A. Aziz, V. Singhal, F. Balarin, R.K. Brayton, and A.L. Sangiovanni-Vincentelli. Equivalences for fair kripke structures. In *International Colloquium on Automata, Languages and Programming (ICALP)*, Jerusalem, Israel, July 1994.
21. E.A. Emerson and C.-L. Lei. Modalities for model checking: Branching time strikes back. In *ACM Symposium on Principles of Programming Languages (POPL)*, pages 84–96, New York, January 1985. ACM.
22. E.A. Emerson and C.-L. Lei. Modalities for model checking: Branching time strikes back. *Science of Computer Programming*, 8:275–306, 1987.
23. E. Clarke, O. Grumberg, and D. Long. Model checking and abstraction. *ACM Transactions on Programming Languages and systems*, 16(5):1512–1542, September 1994.
24. E. Clarke, O. Grumberg, and D. Long. Verification Tools for Finite State Concurrent Systems. In J.W. de Bakker, W.-P. de Roever, and G. Rozenberg, editors, *A Decade of Concurrency-Reflections and Perspectives*, volume 803 of *Lecture Notes in Computer Science*, pages 124–175, Noordwijkerhout, Netherlands, June 1993. REX School/Symposium, Springer-Verlag.

BDDNOW: A Parallel BDD Package *

Kim Milvang-Jensen[1] and Alan J. Hu[2]

[1] Dept. of Computer Science, Univ. of Copenhagen (DIKU)
lordtime@diku.dk, http://www.diku.dk/students/lordtime/
[2] Dept. of Computer Science, Univ. of British Columbia
ajh@cs.ubc.ca, http://www.cs.ubc.ca/spider/ajh/

Abstract. BDDs (binary decision diagrams) are ubiquitous in formal verification tools, and the time and memory used by the BDD package is frequently the constraint that prevents application of formal verification. Accordingly, several researchers have investigated using parallel processing for BDDs. In this paper, we present a parallel BDD package with several novel features. The parallelization scheme strives for minimal communication overhead, so we are able to demonstrate speed-up even running on networked commodity PC workstations. Average memory utilization per node is comparable to that of efficient sequential packages. In addition, the package supports dynamic variable reordering, and simultaneous computation of multiple BDD operations. Finally, the package is designed for portability – providing a subset of the CUDD API for the application programmer, and running on the widely available PVM package.

1   Introduction

Binary decision diagrams (BDD) [1] have become the dominant data structure for representing propositional logic formulas in formal verification and CAD. In practice, the memory required by large BDDs is typically the limiting factor for formal verification tools. In some cases, especially with the advent of dynamic variable reordering [5], run time is also a limiting factor. Accordingly, considerable research has been invested in more efficient BDD implementations.

An obvious line of research is to use parallel processing to accelerate BDD operations, but BDDs are not easily parallelizable. Nevertheless, several groups have produced parallel BDD packages for various different architectures. For space reasons, we discuss here only the work of Stornetta and Brewer [8], which is the prior work most similar to ours in that we both target distributed memory machines. We refer the interested reader to their paper for references to other work.

Networked workstations are currently the most widely available parallel computing resource, so we designed our package for this platform. Especially important for BDDs is the use of the pooled RAM of the networked workstations to work on BDDs too large to fit into the memory of any one workstation. Stornetta and Brewer also mention networks of workstations as a promising platform for their package. However, their package is more suitable for a parallel distributed memory supercomputer, as their scheme of

---

* This work was performed while the first author was visiting the University of British Columbia. The second author was partially supported by an NSERC research grant.

randomly distributing nodes to processors (which gives excellent load balancing) results in very high communication overhead. Indeed, their reported results using a 32-node Meiko CS-2 showed speed-up over the sequential code only when the sequential code had exhausted its physical memory and was swapping to disk. In contrast, our approach trades off inferior load-balancing for lower communication. We are able to show speed-up over optimized sequential code with as few as four nodes, running on networked PCs. Both their package and ours can use the combined memory of networked workstations to avoid thrashing.

Our package has several other features. The package supports limited dynamic variable reordering — adjacent variables that are assigned to the same processor can be swapped reasonably efficiently, so we have implemented a parallel version of block-restricted sifting [3]. When not using reordering, the package is nearly as memory-efficient as CUDD [7], the optimized sequential package on which it is based. With reordering, some additional memory overhead is incurred, similar to the technique Ranjan *et al.* [4] used to add reordering to the breadth-first CAL package [6]. Another feature is the ability to perform several different BDD operations in parallel. Finally, the package is designed for portability: the API is a subset of the CUDD API, to simplify porting applications between our package and the popular CUDD package, and the parallel code uses the widely supported and freely available PVM package [2] for machine independence.

2 Design Principles

Our parallel BDD package was designed according to three main criteria: (1) Optimize for large BDDs, not small ones, (2) target networks of workstations as the platform, and (3) make the package portable.

The package runs on multiple machines using PVM, but the user program is just a sequential program, using the package exactly as it would use any other BDD package. The user can simply recompile the program with our package instead of the CUDD package, provided that only the supported subset of operations is used. For our experimental results, we used the same test program compiled with either our package or CUDD.

2.1 Workload Distribution

Manipulating BDDs typically consists of applying the following steps recursively:

1. Check for base cases.
2. Do a result-cache lookup to see if this operation has been done already.
3. Recurse on the left and right children.
4. Otherwise, find/create the node *ite(x,left_result,right_result)*.

This computation can be represented as a DAG, where the nodes represent operations and the edges represent recursive calls and cache hits. The base cases are the sinks. See Figure 1. This DAG can be traversed in any order, which allows doing many operations in parallel. The hard part lies in the large shared data structures (e.g., result cache and unique table). Since we want good memory utillization, data replication must be minimized.

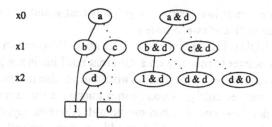

Fig. 1. A BDD and the resulting computation DAG for an operation on nodes *a* and *d*.

Because we are targeting the package at networks of workstations, we expect communication to be expensive, so we want to exploit any locality of reference we can find. As explained in [6], the part of the unique table that will be accessed by an operation can be determined by the topmost variable of the BDDs involved. The cache can also be divided into parts storing only results for where the topmost variable falls in a certain range. Therefore, each machine stores the nodes for some variable indices, and each machine does all operations where the topmost variable is one of the indices it has responsibility for. The advantage of this scheme is that all memory reference become local, and only the work moves between processors. To further limit communication, we have chosen to have each processor manage a contiguous range of indices. We experimented briefly with assigning indices in other ways, but did not achieve good results.

When a processor finds out that a recursive call has to be done on another machine, it sends the job, and goes on doing other work. This means that unlike in a depth-first traversal, the program needs to keep track of what work is in progress, and what state it is in. In other words, the computation DAG must be represented somehow. Fortunately, only the parts of the DAG where computation has not yet completed need to be represented, and it is possible to limit the size of that part by having a processor stop and wait for the jobs it has sent to other machines to complete, before going on to more work.

The main problem with this approach is the lack of efficient ways to balance the work. The only flexibility is the choice of what processors have responsibility for what indices. The package does load balancing whenever the size of the BDD doubles (starting at 100000 nodes). The indices are redistributed so that each processor gets approximately the same number of nodes, keeping neighbouring indices together. This is suboptimal: we have seen that the most work is not done where the most nodes are, but rather at indices slightly later in the order. A better strategy should take into account both where the work is done, as well as the amount of memory available at each processor. Our current choice does, however, give good distribution of memory use, and for very large BDDs, memory distribution is more important than work distribution.

2.2 Memory Usage

The CUDD package has fairly low memory usage, and we have largely maintained this in our package. To be able to determine the topmost variable in a call, we add the index to all pointers, just as is done in breadth-first BDD operations. This increases the size of each node slightly. Offsetting this increase, we eliminate the reference count. Keeping track of reference counts in a distributed graph is costly, and garbage collection should be done by using mark and sweep instead.

Normally a BDD is just a pointer to a BDD node. However, in a parallel implementation, having a pointer to memory on a different machine is not good design and may cause problems running on heterogeneous machines. It also makes moving indices from one machine to another costly, as you cannot guarantee that a node will get the same address on the new machine as it had on the old. Alternatively, one could modify the virtual memory system to create one linear address space over all the machines, but this too would be difficult on heterogeneous setups. To avoid these problems, we introduce a different addressing scheme. Within each index we create an array of arrays of nodes. By numbering the nodes for each index in an array, we get machine-independent addressing. The double indirection structure gives finer granularity over memory allocation. Each node is uniquly identified by the triple (index,array number,number within array). There is some waste of space, because node space is allocated per level, and an extra memory reference is required for each node access, but this reference will be to the small array of array addresses. The nice features of this scheme are that it is clean, moving indices is easy, and nodes require only 8 shorts to represent.

While the package does use memory of all the processors well, the size of the BDDs that can be handled is still limited by the fact that the index with the most nodes has to fit on a single machine.

2.3 Dynamic Variable Reordering

Dynamic variable reordering [5] is becoming increasingly more important, and a good BDD package must support it. The basic operation in reordering is the swapping of two adjecent variables. Though doing each swap parallelized would be interesting, we have concentrated our efforts in doing the individual swaps locally, but making it possible to do several independent swaps at the same time.

Some of the problems with doing reordering in our case are the same as when doing breadth-first BDD manipulation, as they spring from having the index on the ingoing pointer, rather then in the node itself, and from wanting to keep nodes with the same index located on the same pages in memory. Just as in [4], after trying different schemes for using forwarding pointers, we concluded that the best way is to change representation during reordering, and after reordering move the nodes so they are located where you want them. Unfortunately, moving nodes results in a memory overhead.

There are also challenges in reordering unique to the distributed storing of the BDD. Exactly what they are and how we have overcome them is beyond the scope of this paper. We will just note that they do result in some memory overhead, but to a large degree, we can use the same space as is needed for moving nodes, so the total memory overhead is not that much bigger.

A more interesting problem is the need of an algorithm that takes advantage of the fact that several independant swaps can be done in parallel. For example, block-restricted sifting [3], which was originally suggested as a faster sequential reordering algorithm, allows individual blocks to be reordered in parallel, so we have implemented this algorithm. The current implementation can swap variables only within a single machine, so more complicated reordering algorithms will have to iterate: (1) define blocks, (2) swap within blocks, and (3) cleanup. Currently, cleanup requires synchronization of all the machines, and the cleanup operation is itself fairly expensive, so depending on the algorithm used it may be nessecary to include support for redefining blocks for a certain part of the BDD without involving the rest.

2.4 Multiple Independent BDD Operations

The way the user program sends tasks to the processes running the BDD package is essentially no different from the way the processes send tasks to each other. Since internally many jobs are done at the same time, it is possible for the user process to start new BDD operations before the old ones have completed. This is currently implemented.

Ideally, the user would just send a sequence of operations to the BDD package, and the package would then figure out which ones did not depend on the results of the others, and have several operations executing at the same time. User calls to the BDD package would be non-blocking — the application program would continue to execute, doing useful work, and perhaps generating additional calls to the BDD package. The user program would not need to know about the parallel execution. The only change to the user program would be that BDD comparison would be done through a call to the BDD package that guaranteed that both operands were done before the pointers were compared. Intuitively, the BDD package would behave like a superscalar processor, dynamically extracting parallelism from sequential code.[1] Doing such superscalar execution is beyond the scope of our project, but integrating it with the current package, which does support multiple simultaneous BDD operations, should be simple.

3 Experimental Results

Experimental results were obtained from running the package on a cluster of 16 266MHz Pentium II computers, each with 128MB of RAM running FreeBSD, connected by a high speed network consisting of 32-bit LANai 4.1 boards and two M2M-dual 8 port SAN switches. We use a standard PVM implementation running on top of TCP/IP.

Due to some operating system constraints we were only able to get 64MB of memory on each machine, exept on one machine, so all runs with 2 or more machines only have 64MB per machine available. Even without the OS memory limit, there was only about 100MB of physical memory available for user processes.

3.1 Combinational Circuits

To give the customary benchmark example, we ran a simple combinational circuit comparison program on some ISCAS'85 circuits (comparing redundant and non-redundant

---

[1] Sanghavi *et al.* [6] present a similar idea under the monikers "pipelining" and "superscalarity" in the context of their breadth-first sequential BDD package. By introducing blocking comparisons, we propose dynamically extracting parallelism from an unmodified application program, whereas their method requires specially written code to use. By analogy, their approach resembles a statically scheduled VLIW processor more than a superscalar processor. Furthermore, in our parallel BDD package, we could actually take advantage of the parallelism, whereas they were simply trying to reduce page faults.

versions, when available) and compared the execution times with our parallel BDD package to those with the CUDD package[2], which is a highly-efficient sequential BDD package. Times are in seconds.

Circuit	Nodes	Operations	Sequential Exec. Time	Parallel Exec. Time w/ n Procs.				
				1	2	4	6	8
c1355	199859	583970	2.5	4.7	6.3	8.5		13.0
c1908	233471	752402	3.3	6.1	8.3	10.9		16.0
c880	1370138	1798896	10.5	14.9	22.2	23.5	22.4	
c3540	3448095	14286323	57.9	101.8	105.2	90.5	88.9	94.5

These results are not good, but building the BDD for combinational circuits performs many operations on small BDDs. Upon closer examination, we found that although the processors do about the same amount of work, they spend considerable time idling, which suggests that superscalar operation may help. Also, although the parallel package performs poorly on small circuits, it gets better as the BDDs involved increase in size. We did not find an ISCAS'85 circuit that fit in 100MB but not in 64MB, so we cannot give a fair comparison (i.e., against non-thrashing sequential code) requiring the parallel package to combine memory from several machines. Obviously, the parallel package vastly outperforms the sequential code once the sequential code runs out of physical memory.

3.2 Big BDDs

Recall that our primary design goal was handling large BDDs. To get something scalable, we build BDDs for the n functions $g_0, \ldots, g_{n-1}$, where $g_i = \bigwedge_{j=0}^{n-1} (a_i \vee b_{i+j \bmod n})$.

This gives an exponential size BDD, and we can set n as we need to get a appropriate size BDD. For $n = 20$, the BDD has 5.25M nodes and the CUDD package uses 93MB of memory. Times are in seconds.

Sequential Exec. Time	Parallel Exec. Time w/ n Procs.				
	1	2	4	6	8
70.5	108.8	77.9	68.1	61.0	57.6

We believe this is the first reported speed-up of a distributed-memory parallel BDD package over efficient sequential code running in main memory (not thrashing).

We also ran with $n = 21$. The BDD had 10.5M nodes, and thus requires around 200MB. Due to the granularity of the memory balancing, we were not able to fit this BDD on 4 machines, but we did build it using 6 machines, taking 151 seconds. The sequential package obviously cannot build this BDD on a single machine, demonstrating how the parallel package can harness the combined memory of several workstations.

Some of the speed-up is achieved because each processor in the parallel version allocates as much cache as a single processor in the sequential version, so the parallel version has more total cache available. Also, the parallel version running on one processor

---

[2] Results for CUDD reported are for version 2.1.2, which was the most current available and the code on which our BDDNOW package is based. By publication time, CUDD 2.2.0 had become available. Our techniques should be equally applicable to parallelizing the new version.

is clearly much slower than the sequential code. Some of this slow-down may be from necessary parallelization overhead, and some may be from insufficient code tuning in our preliminary implementation.

4 Future Work

The package described in this paper is still a preliminary proof-of-concept implementation. Some obvious next steps include ruggedizing the code to ready it for distribution, and implementing more of the CUDD API. More interesting future research involves load balancing and communication overhead. As we mentioned earlier, the package sacrifices some load balancing in order to reduce communication overhead. Unfortunately, in extreme cases, the workload can become so unbalanced that no useful parallel work is performed. Moreover, substantial time is still spent on communication. Further research is definitely warranted.

Implementing true superscalar execution as described in Section 2.4 would be a practical way to improve performance on certain applications.

Dynamic variable reordering remains a challenging problem. Although we were able to parallelize block-resticted sifting, sometimes there are not enough blocks to gain significant speed-up. Also, block-restricted sifting does not always find a good variable order. More research is needed into variable reordering algorithms that are parallelizable.

In sum, this BDD package is an efficient and portable BDD package that demonstrates speed-up over optimized sequential code and effectively uses the memory of networked workstations. It will be an excellent platform for further research.

References

1. R. E. Bryant. Graph-based algorithms for boolean function manipulation. *IEEE Transactions on Computers*, C-35(8):677–691, August 1986.
2. A. Geist, A. Beguelin, J. Dongarra, W. Jiang, R. Manchek, and V. Sunderam. *PVM: Parallel Virtual Machine — A Users' Guide and Tutorial for Networked Parallel Computing*. MIT Press, 1994. See also http://www.epm.ornl.gov/pvm/pvm_home.html.
3. C. Meinel and A. Slobodová. Speeding up variable reordering of OBDDs. In *International Conference on Computer Design*, pages 338–343. IEEE, October 1997.
4. R. K. Ranjan, W. Gosti, R. K. Brayton, and A. Sangiovanni-Vincentelli. Dynamic reordering in a breadth-first manipulation based BDD package: Challenges and solutions. In *International Conference on Computer Design*, pages 344–351. IEEE, October 1997.
5. R. Rudell. Dynamic variable ordering for ordered binary decision diagrams. In *International Conference on Computer-Aided Design*, pages 42–47. IEEE, 1993.
6. J. V. Sanghavi, R. K. Ranjan, R. K. Brayton, and A. Sangiovanni-Vincentelli. High performance BDD package by exploiting memory hierarchy. In *33th Design Automation Conference*, pages 635–640. ACM/IEEE, 1996.
7. F. Somenzi. CUDD: CU decision diagram package. Available from ftp://vlsi.colorado.edu/pub/.
8. T. Stornetta and F. Brewer. Implementation of an efficient parallel BDD package. In *33th Design Automation Conference*, pages 641–644. ACM/IEEE, 1996.

Model Checking VHDL with CV *

David Déharbe[1], Subash Shankar[2], and Edmund M. Clarke[2]

[1] Universidade Federal do Rio Grande do Norte,
Natal, Brazil
e-mail: david@dimap.ufrn.br
[2] Carnegie Mellon University,
Pittsburgh, USA
e-mail: {sshankar, emc}+@cs.cmu.edu

Abstract. This article describes a prototype implementation of a symbolic model checker for a subset of VHDL. The model checker applies a number of techniques to reduce the search space, thus allowing for efficient verification of real circuits. We have completed an initial release of the VHDL model checker and have used it to verify complex circuits, including the control logic of a commercial RISC microprocessor.

1 Introduction

Ensuring the correctness of computer circuits is an extremely important and difficult task. The most commonly used verification technique is simulation. However, in many cases simulation can miss important errors. A more powerful approach is the use of formal methods such as temporal logic model checking [3], which can guarantee correctness. However, most model checkers can verify only circuits that are written in their own language. This makes it difficult to apply the technique in practice, since very few designers use these languages. For model checking to be accepted by industry, we believe that it is essential to provide an interface between the verification tools and some widely used hardware description language. VHDL is an obvious choice for such a language: it is used as input for many CAD systems, it provides a variety of descriptive styles, and it is an IEEE standard [6].

Much research has been conducted to give a formal semantics to VHDL and apply formal verification techniques (see e.g. [7, 1]). While there exist formal VHDL semantics in a number of formalisms, it is difficult to use many of these in formal verification. Whereas many operational semantics approaches tend to lead to excessively large models, many axiomatic semantics either restrict the VHDL subset too much or are in logics difficult to reason in. There have been

* This research is sponsored by the Semiconductor Research Corporation (SRC) under Contract No. 97-DJ-294, the National Science Foundation (NSF) under Grant No. CCR-9505472, and the Defense Advanced Research Projects Agency (DARPA) under Contract No. DABT63-96-C-0071. Any opinions, findings and conclusions or recommendations expressed in this material are those of the authors and do not necessarily reflect the views of SRC, NSF, DARPA, or the United States Government.

some attempts to apply model checking to VHDL specifications, though it is still desirable to reduce the state space generated from the VHDL programs.

We have developed a VHDL verification system called CV. The verification system uses symbolic model checking. Our approach allows for a number of optimizations that result in dramatically smaller state spaces. We have incorporated several of these into CV, and our results show state space reductions of several orders of magnitude using these optimizations. The optimizations include cone of influence reduction to eliminate irrelevant portions of the circuit along with a new scheme for approximating the set of reachable states. This approximation is simple to compute, and results in additional large reductions in state space size. Finally, our models represent whole VHDL simulation cycles with single transitions, unlike several other VHDL model checkers, thus affording additional savings.

An initial prototype of CV has been implemented and used to verify several complex designs, including the control logic of a commercial RISC processor. In this demonstration, we describe this prototype and present results that demonstrate the applicability of the method to actual industrial designs.

2 Overview of the system

In the context of this work, a VHDL description can be considered to be an *implementation* that must be checked against a (partial) *specification* composed of a set of temporal logic formulas. Initially, the VHDL description is compiled into a state-transition graph represented internally by BDDs [2]. Model checking techniques are then used to determine if the specification holds in the circuit model.

We want the temporal logic specification and the VHDL implementation to be physically independent and reside in separate files. This independence property is quite important, since we might want to modify the specification (e.g. in incremental verification) without recompiling the VHDL implementation. Also, when using a structural description style, the same VHDL design can be instantiated several times as a component of a larger model. The system must be flexible enough to avoid recompiling the same description in such cases. Consequently, the compilation process is split into a front-end that checks that the description is legal VHDL and a back-end that generates the symbolic model.

The system is composed of:

1. a general-purpose VHDL analyzer, called cva, which takes as input a text file that contains VHDL descriptions and generates library units in the intermediate format .cv. This compiler is constructed in much the same way as commercial VHDL front-ends. A library of C functions has been developed to access the intermediate format files. Thus, the front-end can be used independently of the model checker. CV does not require a commercial compiler. In Section 3 we describe the current VHDL subset.

2. a VHDL model checker, called **cvc**, which takes as input a specification file, builds the model of the corresponding VHDL description, and applies the verification algorithms to this model.

   (a) The model elaborator implements the semantics of VHDL in terms of state-transition graphs as described in [5]. It reads the intermediate format files produced by the compiler **cva** and builds a symbolic, BDD-based model of the VHDL design. This step is crucial for the entire verification process since the complexity of the model checking algorithms heavily depends on the size of the model (i.e. the number of state variables, and the size of the BDDs). To this end, considerable effort has been devoted to generating a compact model.

   (b) The consistency checker takes as input a specification file composed of temporal logic properties. The specification language is defined in Section 4.

   (c) The model checker itself determines if the implementation satisfies the specification. When an error is detected, the model checker can produce a counterexample as a VHDL testbench. The testbench can be directly used with a conventional simulator to show an execution trace that violates the property. This information can then be used to debug the circuit. It is also possible to output the counterexample as a sequence of states of the VHDL architecture. Details about the model checker implementation are provided in Section 5.

The structure of **CV** is depicted in Figure 1. Solid lines correspond to commands given by the user whereas dashed lines represent data automatically read or written by **cva** and **cvc**.

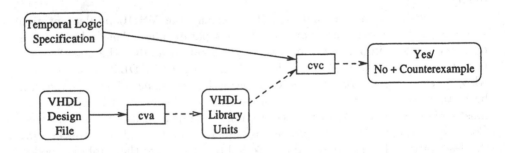

Fig. 1. CV Structure

3   Description of the VHDL Subset

VHDL is a very rich language. Some of its constructs generate infinite models and cannot be modeled by a finite-state tool such as **CV**, such as, for instance, combination of a increasingly delayed signal assignment nested in an infinite loop, or a non-halting recursive function with local variables. Also, many constructs can be expressed in terms of more basic constructs, while maintaining the same semantics. It is also desirable to concentrate on a synthesizable subset of VHDL, as most such subsets eschew some parts of VHDL (e.g. either delta delays or unit time delays). Our approach has been to identify and implement a core subset of the language and incrementally extend this subset based on these guidelines. The elements of the core subset are

Design entities: Entity declaration, architecture body, package declaration.
Concurrent statements: Block statement and process statement.
Sequential statements: Wait statement, signal and variable assignment statements, if statement.
Libraries: Library, library clause and use clause.
Declarations: Signal (input, output and local), variable, constant and type declarations.
Types: Enumerated types.

The following extensions have already been implemented:

Concurrent statements: Selected signal assignment statement, concurrent signal assignment statement, processes with sensitivity lists.
Sequential statements: Case statement, while statement, null statement.
Types: Integer, array, and record types.

4 Specification Language

The language used in CV for specifying the expected behavior of a VHDL design in a given environment is essentially the logic CTL with fairness constraints. The environment is described in terms of assumptions on the values taken by the input signals of the design. A valid simulation of the design is a simulation where all assumptions on the input signals are verified. The behavior of a VHDL design in an environment is then defined to be the set of all valid simulations with respect to this environment. Therefore, a specification is composed of a set of *assumptions* and *commitments* about a VHDL description. An assumption is a condition on the inputs of the design under verification. Commitments describe the expected behavior of the system provided all assumptions hold. It is the role of the model checker to verify that the VHDL description satisfies the commitments provided the assumptions hold.

For convenience purposes, the specification language also makes it possible to define abbreviations. An abbreviation is an identifier that denotes an expression and may be used to simplify assumptions and commitments. For instance:

```
abbreviation PUSHED is (PUSH and PUSH_RDY);
```

Two categories of assumptions are possible: invariant and fairness. An invariant is defined with the keyword **always** followed by a condition on the signals of in mode of the specified VHDL design. The effect of an invariant definition is to restrict the behavior to the set of simulations where the associated condition holds for every cycle. For instance, the following invariant assumption states that inputs PUSH and POP are never simultaneously true.

```
assume always not ((PUSH = '1' ) and (POP = '1'));
```

The effect of a fairness definition is to restrict the behavior to the set of simulations where the associated condition holds infinitely often. Fairness is often needed to prove properties about the progress of a system. Generally, fairness constraints increase verification time and memory consumption. For instance, the following fairness assumptions can be used to state that CLOCK infinitely often changes value:

```
assume recurring (CLOCK = '0');
assume recurring (CLOCK = '1');
```

Commitments are expressed in the temporal logic CTL (Computation Tree Logic). A CTL operator is composed of a path quantifier ($\mathbf{A}$, $\mathbf{E}$) followed by a linear temporal operator ($\mathbf{X}$, $\mathbf{G}$, $\mathbf{F}$, $\mathbf{U}$, $\mathbf{W}$). Since the (standard) semantics of a VHDL design is defined in terms of simulation, we use temporal logic to express properties of the possible simulations of a design. The path quantifier $\mathbf{A}$ ($\mathbf{E}$) selects all (some) simulations, and The linear temporal operator $\mathbf{X}$ ($\mathbf{G}$, $\mathbf{F}$, $\mathbf{U}$, $\mathbf{W}$) selects the next simulation cycle (all cycles, some cycle, until some cycle, unless some cycle) in a given simulation. A formula is true for a description if it holds in all initial states. For instance, the CTL formula $\mathbf{AG}f$ states that f must hold at all states reachable from the initial states in all possible simulations. f must always hold.

For example, suppose it is necessary to check that a bad situation never happens. An abbreviation can be used to denote this bad situation:

```
abbreviation BAD_SITUATION is <some condition> ;
```

The following commitment states that BAD_SITUATION never happens:

```
commit safe: ag not BAD_SITUATION;
```

Suppose it is necessary to show that whatever the current state of the system is, it can reach the restart state. Let RESTART be the condition that identifies the restart state:

```
commit RESTARTABLE: ag ef RESTART;
```

More complex properties, involving for instance, events on signals can also be expressed in the logic.

5   The Model Checker

Transitions in the BDD-based models built by CV correspond to whole simulation cycles in the VHDL program. This allows our models to have fewer variables and fewer bits to represent the program counter, compared to approaches that use multiple transitions for each simulation cycle.

The models built by CV use a boolean functional vector representation for the transitions [4]. This limits considerably the explosion of the size of the transition representation for large systems. The representation also makes it easy to eliminate parts of the model that are not relevant with respect to the specification. The model checker needs to consider only the transition functions of the variables that are in the cone of influence of the specification. This cone of influence is simply constructed from the true support set of the transition functions.

Computing the reachable states of the model proves useful in practice for enhancing the performance of symbolic model checking. We have devised a heuristic based on the observation that a conservative approximation, or *overestimation*, of the reachable states, can also be used to simplify the transition representation, and consequently simplify the computation of the reachable states. We have devised and implemented a heuristic to compute overestimations of the valid states much faster than the valid states themselves. Indeed, an overestimation can be simply and efficiently computed considering the valid states of the submachine corresponding to a subset of the vector of transition functions.

To the best of our knowledge, these optimizations are unique for a VHDL-based model checker. Our experience was that, for some examples, each of the optimizations reduces the computation time by an order of magnitude and decreases the amount of space used. It is possible to handle significantly larger examples when both optimizations are combined.

Therefore, given a specification, the model checker performs the following steps:

1. Parse the specification file.
2. Load the corresponding VHDL library unit, and build its BDD-based model using a boolean functional vector representation for the transitions.
3. Invoke dynamic variable reordering to reduce the size of the BDDs.
4. Compute overestimations of the reachable states and simplify transition representations.
5. Compute the reachable states of the model, and use the result to further reduce the BDDs representing the transitions.
6. Check the specification, using symbolic model checking techniques.
7. Report the results and generate counterexamples if necessary.

6   Conclusion

In its current version, our system is already able to handle some realistic designs. One of the most interesting examples is a model of the control logic of a commercial RISC processor. The VHDL description counts 25 different processes, and spans several hundred lines of code.

Building the BDD-based model of this description takes only about 5 seconds. Then dynamic variable reordering is invoked to reduce the size of the model. This step takes approximatively 75 seconds, and reduces the size of the transition function to only about 5,000 BDD nodes, despite the fact that the model has 157 variables. We then compute an approximation of the set of reachable states which further reduces this figure to 4,000 BDD nodes. The model has about 10^{28} reachable states out of 10^{47} possible states. By using the optimization that considers only the variables in the model that affect the specification, a simple property can be checked in just 30 seconds. Without this optimization, verification would take two orders of magnitude longer.

We are currently continuing the development of CV and expect further developments along two lines:

1. performance: automated abstraction, compositional reasoning, low-level implementation improvements.
2. usability: extension of the VHDL subset (timing aspects, structural design), design of a graphical user interface.

References

1. R.K. Brayton, E.M. Clarke, and P.A. Subrahmanyam, editors. *Formal Methods in System Design*, volume 7, (1/2). Kluwer, August 1995. Special Issue on VHDL Semantics – Guest Editor: D. BORRIONE.
2. R.E. Bryant. Graph-based algorithm for boolean function manipulation. *IEEE Transactions on Computers*, C(35):1035–1044, 1986.
3. E.M. Clarke, E.A. Emerson, and A.P. Sistla. Automatic verification of finite-state concurrent systems using temporal logic specifications. *ACM Transactions on Programming Languages and Systems*, 8(2):244–263, April 1986.
4. O. Coudert and J.-C. Madre. Symbolic computation of the valid states of a sequential machine : algorithms and discussion. In *International workshop on formal methods for correct VLSI design*, Miami, January 1991. ACM/IFIP WG10.2.
5. D. Déharbe and D. Borrione. Semantics of a verification-oriented subset of VHDL. In P.E. Camurati and H. Eveking, editors, *CHARME'95: Correct Hardware Design and Verification Methods*, volume 987 of *Lecture Notes in Computer Science*, Frankfurt, Germany, October 1995. Springer Verlag.
6. IEEE. *IEEE Standard VHDL Language Reference Manual*, 1987. Std 1076-1987.
7. C. Delgado Kloos and P. Breuer, editors. *Formal Semantics for VHDL*, volume 307 of *Series in Engineering and Computer Science*. Kluwer Academic Publishers, 1995.

Alexandria: A Tool for Hierarchical Verification

Annette Bunker, Trent N. Larson,
Michael D. Jones*, and Phillip J. Windley**
{bunker, larson, windley}@cs.byu.edu, mjones@cs.utah.edu

Brigham Young University, Provo, UT 84602-6576

Abstract. Alexandria is an implementation of the hierarchical verification methodology for the Higher-Order Logic (HOL) theorem prover. The main contribution of Alexandria is the reduction of effort required by the user to create and use hierarchical hardware proofs in HOL. We discuss the implementation and use of Alexandria with an example and outline our future work.

1 Introduction

Hierarchical decomposition of verification is an accepted practice in hardware verification [LA92] [GW92]. Hierarchical verification assists practitioners with a simpler division of proof efforts for collaborating researchers, as well as a means to reuse old proofs in new verifications.

Alexandria is a tool designed to support the hierarchical decomposition methodology. It is based on the Higher-Order Logic (HOL) theorem prover and uses abstract theories and predicate types to enforce the proof decomposition. Alexandria provides functions for creating parameterized hardware modules and proving correctness theorems about the modules.

Circuits in Alexandria are described in the BOLT hardware description language (HDL) and are then translated to an internal logical representation. BOLT was chosen as the tool's first implementation language for its simplicity. However, efforts are currently underway to make support for other HDLs, such as VHDL and Verilog, available.

---

* Current affiliation: PhD student, University of Utah
** This work was sponsored by the National Science Foundation under NSF grant MIP–9412581

When proving properties of circuits, Alexandria adds only the specifications of subcomponents to the proof as assumptions. This enforces the hierarchical nature of Alexandria proofs, by not allowing an information about the implementation enter the proof process.

Alexandria also includes an HOL tactic for automatically proving the correctness of purely combinational circuits given a specification. Predicate types in Alexandria ensure that subcomponents meet their specifications before they are used in a later design. Abstract theories in Alexandria ensure that verified devices can be instantiated with any correct submodule without affecting the overall correctness proof. However, these concepts are entirely hidden from the user by a few simple functions that allow users to create specifications and manage proof structure.

In the next section, we discuss the implementation and use of Alexandria in more detail. Section 3 contains examples from the verification of a parameterized ALU. In section 4, we outline our plans for future development of Alexandria.

2 Overview of Alexandria

Alexandria is a suite of tools that encourage hierarchical design and verification. Modular design and verification in Alexandria requires three steps:

1. The specification for each component and primitive is entered into Alexandria .
2. System components are designed hierarchically, with explicit dependencies on subcomponents or primitive devices.
3. Finally, each component implementation is entered into Alexandria , which generates correctness conditions to be proven.

The second step may be done on paper or with CAD tools. The first and third rely on the following two functions which make up the main interface to Alexandria :

 - Users enter a hardware device specification with `create_hardware_type`, using HOL terms. This information is used to create predicate types and abstract theories that manage the verification effort.

– Users enter a description of the implementation with `create_bolt_module` using the netlist HDL, BOLT. This function generates verification correctness conditions based on the intended specification for the device and the specifications of its subcomponents which are then proven by the user.

In addition to these simple interface functions, Alexandria also maintains dependency information which is used to create libraries of components and primitives, generate correctness conditions, and even verify simple devices automatically.

Most of the benefits of modular, hierarchical design still rely on the skill of the designers. However, once the designer is ready to formalize the design, Alexandria provides structure for entering this information into a theorem prover. Alexandria associates the specification with a predicate type in the type system, which is how it maintains the dependencies between hardware device types. Once the user has verified implementations using those types, the proofs can be reused anywhere the specified types are utilized. The proofs may also be shared among designers who are interested in the same device types. By using Alexandria to organize the proof effort, hardware proofs can be shared or modified, in whole or in part, with no new proof effort.

3   Using Alexandria to Verify the Birtwistle ALU

Alexandria has been used to verify the ALU presented in [BCG] in an effort involving three people. The ALU is an eight command, n-bit wide ALU developed as an example of verification. In addition, portions of the implementation have been modified and reverified for the present work. These modifications did not impact other parts of the correctness proof. In this section, we overview the verification effort and how made use of Alexandria.

We chose to verify the Birtwistle ALU because the correctness proof had already been done for it, providing for a strong understanding of the ALU with little effort on the parts of the verifiers, as well as a basis example for comparison purposes. There is ample documentation for the Birtwistle ALU. We also anticipated that it would provide a good example for hierarchical design and proof.

3.1 Initial Design and Verification

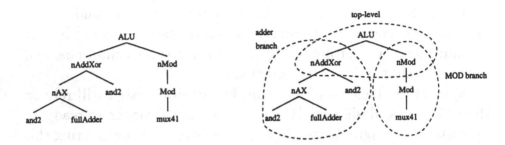

Fig. 1. ALU component hierarchy, and division of verification effort

Figure 1 shows the division of verification effort between the verifiers assigned to the ALU project. The leaves of the tree are primitive components which are specified but not implemented in HOL. All other components are implemented in terms of these primitive components.

Specification We first developed specifications for all components and subcomponents in the ALU. For example, the component nAddXor (see figure 1) is specified as follows:

```
create_hardware_type
  name = "nAddXor_spec",
  library = "birt_alu",
  input_vars = ["n", "E", "a", "b", "cin"],
  input_tys = [=='':num''==,=='':bool''==,=='':word''==,=='':word''==,=='':bool''==],
  output_vars = ["s", "c"],
  output_tys = [=='':word''==,=='':bool''==],
  spec = (--'E => nAdder_spec (n,a,b,cin) (s,c)
                | (nXOR_spec (n,a,b,s)) ∧ (c = F)'--);
```

create_hardware_type takes seven arguments: the name of the device; the name of the current project; input_vars and output_vars, which are names for the input and output lines respectively; input_tys and output_tys are the types of the inputs and outputs; and the specification for this type of device. In the example above, input n is a number, input E is a boolean value, a is an n-bit word, and so forth. The outputs and their types match in like manner. The specification for the device states that if the enable line (E) is true, then the ALU performs arithmetic operations, as defined in the nAdder branch of the tree, otherwise, it performs logical operations, as defined in the nXOR subtree.

create_hardware_type creates a type space, associated with a name (nAddXor_spec in our example). Implementations are then created to populate each existing type space. The type space creation is the only information that needs to be shared among collaborators during a proof project that uses Alexandria. These type spaces enforce the boundaries between abstraction levels in the proof.

Implementation The Birtwistle ALU example was already implemented, so we were able to reuse much of the work that had been done. However, we found that the earlier design was not cleanly split between levels in the abstraction hierarchy. Also, some of the implementation details of subcomponents were important in the implementation of their parents. These facts resulted in the redesign of some of the ALU components and their specifications in order to ensure that they fit the hierarchical verification paradigm.

The design of the nAddXor_1 in Alexandria looks like this:

```
create_bolt_module
  {library = "birt_alu",
   devs = [("nAX", nAX_spec), ("and2", and_2_spec)],
   spec = nAddXor_spec,
   bolt_desc = (~~'MODULE s c .nAddXor_1 m E a b cin;
               BEGIN
               s q .nAX m E a b cin;
               c .and2 E q;
               END;'~~)};
```

The call to create_bolt_module requires four parameters: the name of the current project, a listing of the devices used to build the ALU, the specification for this device, and BOLT description of the connections between the two subcomponent devices. Again, the association between implementation module and its specification must be made explicit by the user in the **devs** parameter.

create_bolt_module constructs an HOL term representing this device, then asserts that it has the given specification of nAddXor_spec. It is not obvious that this implementation conforms to that specification, so Alexandria generates correctness conditions that the user must prove.

Verification After creating the hardware types and describing our implementation of each, the verifiers were able to independently verify each device against its specification.

For example, after implementing the nAddXor as shown above (our nAddXor_1), Alexandria generates the following correctness condition:

```
(--'(let nAX = dest_nAX_ty nAX'
    and and2 = dest_and_2_ty and2'
    in ?q. nAX (n,E,a,b,cin) (s,q) ∧ and2 (E,q) c) ==>
        nAddXor_spec (n,E,a,b,cin) (s,c)'--)
===================================
(--'nAddXor_1 nAX' and2' =
    nAddXor_ty_of
    (fn (m,E,a,b,cin) (s,c) =>
      let nAX = dest_nAX_ty nAX'
      and and2 = dest_and_2_ty and2'
      in ?q. nAX (m,E,a,b,cin) (s,q) ∧ and2 (E,q) c)'--)
```

In English, this says that after assuming the nAX and and2 components meet their specifications, the nAddXor_1 implementation of the nAddXor meets the nAddXor_ty type restriction. The nAddXor_ty type is Alexandria's name for the nAddXor specification.

The proof followed a form very common in hardware proofs; namely, a rewrite with the definitions and followed by a case split on each of the inputs. We included one specialized tactic called ABS_HW_VERF_TAC which handles many of the trivial devices; it is based on work found in [ALW93].

The methodology enforced by Alexandria pointed out one problem early in our work: there was a flaw in our initial specification of nMod. Since we all used the same hardware specifications, and since the top level of the proof relies on the specification of each branch, the researcher working on the top level discovered the error in the specification. The error originated from a typographical error in an early online version of the specification. The error was corrected and disaster was averted, with only minor effort and inconvenience to a couple of the researchers.

Collaborative proof was supported in a practical way using Alexandria . Almost all of the discussions between the verifiers dealt with component specifications. Each collaborator had very little or no knowledge of other's design or proof internals. Alexandria is a simple tool that that supports modular, hierarchical design and proof.

3.2 Adjustments to the Implementation

After the ALU was verified, we substituted a different implementation of the nAddXor module in our design. Doing this is a proof written in the more common, ad hoc manner would have required us to verify the full ALU again from the ground up, possibly modifying the proof in many places.

With Alexandria, we designed the nAddXor to conform to the specification from our previous design. While one team member had to verify that the nAddXor met its specification, no other part of the design or proof needed to be modified to show that the new ALU still met its specification.

This experiment showed that Alexandria can be an effective tool for organizing proofs modularly. It facilitated collaboration among the members of the verification team, as well as allowed for changes in the lower-level designs without impacting the high-levels of the hierarchy.

4   Conclusions

While it is certainly possible to verify hardware components in a hierarchical manner without tools like Alexandria, such a strategy generally involves a great deal of organization and discipline among the verification team members. The goal of Alexandria was to provide the functionality essential for enforcing hierarchical verification. We have shown here how Alexandria can be used to specify and verify hardware designs, and how it has actually eased collaboration and proof reuse.

Future plans for Alexandria include:

- verification of several larger examples
- support for state-holding devices
- support for more hardware description languages
- support for iterated structures

References

[ALW93] Mark Aagard, Miriam Leeser, and Phillip J. Windley. Towards a superduper hardware tactic. In Jeffery J. Joyce and Carl Seger, editors, *Proceedings of the 1993 International Workshop on the HOL Theorem Prover and its Applications.*, August 1993.

[BCG] Graham Birtwistle, Shiu-Kai Chin, and Brian Graham. new_theory 'hol';; an introduction to hardware verification in higher order logic. Available at http://lal.cs.byu.edu/lal/holdoc/birtwistle/html/current/all/all.html.

[GW92] Jody Gambles and Phillip J. Windley. Integrating formal verification with CAD tool environments. In *Proceedings of the Fourth Annual IEEE/NASA Symposium on VLSI Design*, October 1992.

[LA92] Miriam Leeser and Mark Aagard. A methodology for reusable hardware proofs. In Luc Claeson and Michael J.C. Gordon, editors, *Proceedings of the 1992 International Workshop on the HOL Theorem Prover and its Applications*, November 1992.

PV: An Explicit Enumeration Model-Checker[*]

Ratan Nalumasu and Ganesh Gopalakrishnan

Department of Computer Science
University of Utah, Salt Lake City, UT 84112
{ratan,ganesh}@cs.utah.edu
WWW: http://www.cs.utah.edu/~{ratan,ganesh}

Abstract. PV (Protocol Verifier) is an explicit enumeration based model-checker for verifying finite state systems for next-time free LTL (LTL-X) properties. It implements a new partial order reduction algorithm, called Two-phase, that works in conjunction with selective caching to combat the state explosion problem faced by model-checkers.

1 Introduction

PV (Protocol Verifier) is an explicit enumeration based model-checker for distributed system protocols. It uses a subset of PROMELA [6], an *asynchronous* concurrent language ("interleaving semantics") with non-deterministic guarded command, as its modeling language. SPIN [8] is a popular model-checker used for verifying PROMELA models. It implements a partial order reduction algorithm, based on [7,10], to reduce the size of the state graph of the model *without affecting the truth value of the properties*. Partial order reductions reduce the size of the state graph by exploiting the fact that in realistic protocols, there are many transitions that "commute" with each other, and hence it is sufficient to explore those transitions in any *one order* to preserve the truth value of the temporal property under consideration. In essence, from every state, a partial order reduction algorithm selects a *subset* of transitions to explore, while a normal graph traversal such as depth first search (DFS) algorithm would explore all transitions. Partial order reduction algorithms play a very important role in mitigating state explosion, often reducing the computational and memory cost by an exponential factor.

On a number of examples, we noticed that SPIN does not bring sufficient reductions, and traced the reason to a step of the algorithm called *proviso*. We formulated a new algorithm, called Two-phase that does not

---

[*] Supported in part by ARPA Order #B990 Under SPAWAR Contract #N0039-95-C-0018 (Avalanche), DARPA under contract #DABT6396C0094 (UV), and NSF MIP MIP-9321836.

use proviso, and implemented it in a tool called PV. On many practical examples, we found that PV generates a much smaller graph than SPIN. Another novel aspect of the Two-phase algorithm is that it naturally supports selective caching, which reduces further memory requirements of the model-checker. As the amount of available memory is typically limits the size of the model a model-checker can handle, selective caching may offer an advantage over other partial order reduction algorithms. The details of the algorithm and the proof of correctness of the algorithm can be found in [9]. The tool can be obtained by contacting the authors.

2 PROMELA Description Language

A PROMELA description consists of a set of processes that run in an interleaving fashion, and communicate using global variables and communication channels. The interleaving semantics allow one to model the fact that, in most distributed systems, processes run at arbitrary speeds.

A process in PROMELA consists of a set of local variables and a set of transitions (transitions in PROMELA resemble statements in C). PROMELA also provides a special transition ''assert(expr)''. It is an error for expr to evaluate to 0 when the transition can be executed.

A PROMELA model also may contain a Büchii automaton specified as a never claim. The never claim represents "bad" paths in the description. Any linear temporal logic (LTL) formulae ϕ can be translated into a Büchii automaton A_ϕ. Hence, the model-checking problem of verifying $M \models \phi$ can be decided by expressing M as a PROMELA model, and using $A_{\neg\phi}$ as the never claim.

3 Partial Order Reductions in PV

The major difference between Two phase and other partial order reduction algorithms (including the SPIN algorithm) is the way the algorithms expand a given state. Other partial order reduction algorithms attempt to expand each state visited during the search using a subset of enabled transitions at that state and postponing the rest of the transitions. Algorithm must ensure that no transition is postponed forever, referred to as *ignoring problem*. To address the ignoring problem, the algorithms use a proviso (or a condition very similar to the provisos). The proviso ensures that when a state x is expanded using only a subset of transitions, then none of resulting states is on the stack maintained by the DFS algorithm that implements the graph exploration algorithm.

Two phase search strategy is completely different: when it encounters a new state x, it first expands the state using only *deterministic* transitions in its first phase resulting in a state y. (Informally, deterministic transitions are the transitions that can be taken at the state without effecting the truth property of the property being verified. A process has a deterministic transition when its ample-set [10] is of size 1.) Then in the second phase, y is expanded *completely*. The advantage of this search strategy is that it is not necessary to use a proviso. As the results in Section 6 show, this often results in a much smaller graph.

4 On-the-fly model-checking in PV

PV is an on-the-fly model-checker, i.e., it verifies if the model contains any violations as it constructs the state graph. PV finds three kinds of errors in the PROMELA model: deadlocks, assertion violations, and **never** claim violations.

Tarjan [11] presented a DFS based on-the-fly algorithm to compute SCCs *without storing any edge information*. Since space is at a premium for most verification problems, not having to store the edge information can be a major benefit of using this algorithm. This algorithm uses one word overhead per state visited and traverses the graph twice.

[4] presents a more efficient on-the-fly model-checking algorithm to find if a graph has at least one infinite path satisfying a Büchii acceptance condition. This algorithm uses 1-bit overhead per state and traverses the graph at most twice. PV uses this algorithm.

5 Selective Caching

When a state S_0 is expanded in the first phase resulting in states S_1, S_2 ... S_n, a straight-forward implementation would enter all these states into the state graph. However, as presented in [9], it is not necessary to save *all* states. Instead a state S_{i+1} can be entered into the state graph if and only if $S_{i+1} < S_i$ where $i \in \{1, \ldots, n-1\}$, where $<$ is any total order on the reachable states of the model. PV uses bit-wise comparison as $<$. This results in a selective caching algorithm which has been found to give a state-space reduction by a factor of 3, in practice.

6 Experimental Results

Two phase algorithm implemented in PV outperforms the proviso based algorithm implemented in SPIN on examples where there proviso is in-

Protocol	Spin	PV	
		all	Selective
B5	243/0.34	11/0.33	1/0.3
W5	63/0.33	243/0.39	243/0.3
SC3	17,741/4.6	2,687/1.6	733/1.4
SC4	749,094/127	102,345/41.0	47,405/21.9
Mig	113,628/14	22,805/2.6	9,185/1.7
Inv	961,089/37	60,736/5.2	27,600/3.0
Pftp	95,241/11.0	187,614/30	70,653/19
Snoopy	16,279/4.4	14,305/2.7	8,611/2.4
WA	4.8e+06/340	706,192/31	169,680/21
UPO	4.9e+06/210	733,546/32	176,618/21
ROWO	5.2e+06/330	868,665/44	222,636/32

Table 1. Number of states visited and the time taken in seconds by SPIN and PV on various protocols

voked often, as confirmed by the results in Table 1. This table shows results of running SPIN and PV (with and without selective caching enabled) on various protocols. The column "Spin" shows the number of states entered into the state graph and the time taken in seconds by the SPIN. The column "all" column in PV shows the number of states entered into the state graph and the time taken in seconds by PV without the selective caching. The column "Selective" in PV shows the number of states entered into the state graph and time taken in seconds by PV with the selective caching. All verification runs are conducted on an Ultra-SPARC-1 with 512MB of DRAM.

Contrived examples: B5 is a trivial system with that shows that the graph generated by PV may be exponentially smaller than the graph generated by SPIN. W5 is a contrived example that shows the converse is also true. W5 has no deterministic states; hence PV degenerates to a full search, while SPIN can find significant reductions. SC is a server/client protocol. This protocol consists of n servers and n clients. A client chooses a server and requests for a service. A service consists of a two round trip messages between server and client and some local computations. SPIN cannot complete the graph construction for $n = 4$, when the memory is limited to 64MB; when the memory limit is increased to 128MB it generates 750k states.

DSM protocols: *Mig* and *inv* are two cache coherency protocols used in the implementation of distributed shared memory (DSM) using a directory based scheme in Avalanche multiprocessor [2]. In a directory based DSM implementation, each cache line has a designated node that acts as its "home", i.e., the node that is responsible for maintaining the coherency of the line. When a node needs to access the line, if it does not have the required permissions, it contacts the home node to obtain the permissions. Both *mig* and *inv* have two cache lines and four processes; two processors act as home nodes for the cache lines and the other two processors access the cache lines. Both algorithms can complete the analysis of *Mig* within 64MB of memory, but on *inv*, SPIN requires 128MB of memory PV on the other hand finishes comfortably generating a modest 27,600 states (with selective caching) or 60,736 states (without selective caching) in 64MB.

Protocols in SPIN distribution: *Pftp* and *snoopy* protocols are provided as part of SPIN distribution. On *pftp*, SPIN generates fewer states than PV without state caching. The reason is that there is very little determinism in this protocol. Since PV depends on determinism to bring reductions, it generates a larger state space. However, with state caching, the number of states in the hash table goes down by a factor of 2.7. On *snoopy*, even though PV generates fewer states, the number of states generated by SPIN and PV (without selective caching) is too close to obtain any meaningful conclusion. The reason for this is two-fold. First, this protocol contains some determinism, which helps PV. However, there are a number of deadlocks in this protocol. Hence, the proviso is not invoked many times. Hence the number of states generated is very close.

Memory model verification examples: *WA, UPO*, and *ROWO* test the interaction of PA (Precision Architecture from Hewlett-Packard) memory ordering rules with the runway bus protocol [1,5]. Runway is a high-performance split-transaction bus designed to support cache coherency protocols required to implement a symmetric multiprocessor (SMP). These three protocols consist of two HP PA models connected to the runway bus, executing read and write instructions. These property of interest is whether the PA/runway system correctly implements memory consistency rules called write atomicity (WA), and uniprocessor ordering (UPO), and read-order, write-order (ROWO) [3]. On these protocols, the number of states saved by SPIN is approximately 25 times larger than the number of states saved by PV (with selective caching).

References

1. William R. Bryg, Kenneth K. Chan, and Nicholas S.Fiduccia. A high-performance, low-cost multiprocessor bus for workstations and midrange servers. *Hewlett-Packard Journal*, pages 18–24, February 1996.

2. John B. Carter, Chen-Chi Kuo, and Ravindra Kuramkote. A comparison of software and hardware synchronization mechanisms for distributed shared memory multiprocessors. Technical Report UUCS-96-011, University of Utah, Salt Lake City, UT, USA, September 1996.

3. W. W. Collier. *Reasoning About Parallel Architectures*. Prentice-Hall, Englewood Cliffs, NJ, 1992.

4. C. Courcoubetis, M. Vardi, P. Wolper, and M. Yannakakis. Memory efficient algorithms for the verification of temporal properties. In *Computer Aided Verification*, pages 233–242, June 1990.

5. G. Gopalakrishnan, R. Ghughal, R. Hosabettu, A. Mokkedem, and R. Nalumasu. Formal modeling and validation applied to a commercial coherent bus: A case study. In Hon F. Li and David K. Probst, editors, *CHARME*, Montreal, Canada, 1997.

6. Gerard Holzmann. *Design and Validation of Computer Protocols*. Prentice Hall, 1991.

7. Gerard Holzmann and Doron Peled. An improvement in formal verification. In *Proceedings of Formal Description Techniques*, Bern, Switzerland, October 1994.

8. Gerard J. Holzmann and Doron Peled. The state of SPIN. In Rajeev Alur and Thomas A. Henzinger, editors, *Computer Aided Verification*, volume 1102 of *Lecture Notes in Computer Science*, pages 385–389, New Brunswick, New Jersey, July 1996. Springer-Verlag. Tool demo.

9. Ratan Nalumasu and Ganesh Gopalakrishnan. A partial order reduction algorithm without the proviso. Technical Report UUCS-98-017, University of Utah, Salt Lake City, UT, USA, August 1998.

10. Doron Peled. Combining partial order reductions with on-the-fly model-checking. *Journal of Formal Methods in Systems Design*, 8 (1):39–64, 1996. also in Computer Aided Verification, 1994.

11. R. Tarjan. Depth-first search and linear graph algorithms. *SIAM Journal on Computing*, 1(2):146–160, June 1972.

Author Index

Springer
and the
environment

At Springer we firmly believe that an international science publisher has a special obligation to the environment, and our corporate policies consistently reflect this conviction.
We also expect our business partners – paper mills, printers, packaging manufacturers, etc. – to commit themselves to using materials and production processes that do not harm the environment. The paper in this book is made from low- or no-chlorine pulp and is acid free, in conformance with international standards for paper permanency.

Springer

Lecture Notes in Computer Science

For information about Vols. 1–1436

please contact your bookseller or Springer-Verlag